Multicriteria Decision Making and Differential Games

MATHEMATICAL CONCEPTS AND METHODS IN SCIENCE AND ENGINEERING

Series Editor: Angelo Miele
Mechanical Engineering and Mathematical Sciences
Rice University, Houston, Texas

1976:

INTRODUCTION TO VECTORS AND TENSORS
Volume 1: Linear and Multilinear Algebra
Ray M. Bowen and C.-C. Wang

INTRODUCTION TO VECTORS AND TENSORS
Volume 2: Vector and Tensor Analysis
Ray M. Bowen and C.-C. Wang

MULTICRITERIA DECISION MAKING AND DIFFERENTIAL GAMES
Edited by George Leitmann

Multicriteria Decision Making and Differential Games

Edited by

George Leitmann
University of California, Berkeley

Plenum Press · New York and London

Library of Congress Cataloging in Publication Data

Main entry under title:

Multicriteria decision making and differential games.

(Mathematical concepts and methods in science and engineering)
Includes index.
1. Differential games. 2. Decision-making. I. Leitmann, George.
QA272.M84 519.5'4 76-22775
ISBN 0-306-30920-3

227 West 17th Street, New York, N.Y. 10011

Printed in the United States of America

Acknowledgments

Many of the contributions in this volume have appeared in essentially similar form in the *Journal of Optimization Theory and Applications*. Chapters marked with asterisks have undergone fairly extensive revision.

Chapter	Published in JOTA
I	Vol. 14, No. 3, pp. 319-377
II	Vol. 14, No. 5, pp. 573-584
III	Vol. 18, No. 1, pp. 3-13
IV	Vol. 13, No. 3, pp. 362-378
VI	Vol. 18, No. 1, 119-140
VII	Vol. 13, No. 3, pp. 290-302
VIII	Vol. 18, No. 1, pp. 93-102
IX	Vol. 11, No. 5, pp. 533-555
X	Vol. 11, No. 6, pp. 613-626
XI*	Vol. 10, No. 3, pp. 160-177
XII	Vol. 9, No. 6, pp. 399-425
XIII	Vol. 9, No. 5, pp. 344-358
XIV	Vol. 14, No. 6, pp. 613-631
XV	Vol. 14, No. 4, pp. 419-424
XVII	Vol. 14, No. 5, pp. 557-571
XVIII	Vol. 13, No. 3, pp. 343-361
XIX	Vol. 18, No. 1, pp. 153-163
XXI	Vol. 13, No. 3, pp. 334-342
XXII	Vol. 13, No. 3, pp. 275-289
XXIII	Vol. 13, No. 3, pp. 319-333
XXIV*	Vol. 18, No. 1, pp. 103-118
XXV	Vol. 18, No. 1, pp. 65-71
XXVI*	Vol. 9, No. 5, pp. 324-343
XXVII	Vol. 18, No. 1, pp. 15-29

Contributors

K. Bergstresser, Department of Mathematics, Washington State University, Pullman, Washington 99163, USA

J. V. Breakwell, Department of Aeronautics and Astronautics, Stanford University, Stanford, California 94305, USA

A. Charnes, Department of General Business, University of Texas, Austin, Texas 78712, USA

S. Clemhout, Department of Consumer Economics and Public Policy, Cornell University, Ithaca, New York 14850, USA

J. B. Cruz, Jr., Coordinated Science Laboratory and Department of Electrical Engineering, University of Illinois, Urbana, Illinois 61801, USA

M. C. Delfour, Centre de Recherches Mathématiques, Université de Montréal, Montréal, Canada

P. Hagedorn, Fachbereich Mechanik, Technische Hochschule Darmstadt, Darmstadt, Germany

W. H. Hartman, Bell Laboratories, Holmdel, New Jersey 07733, USA

A. Haurie, École des Hautes Études Commerciales, Montréal, Canada

Y. C. Ho, Division of Engineering and Applied Physics, Harvard University, Cambridge, Massachusetts 02138, USA

P. R. Kleindorfer, International Institute of Management, Wissenschaftszentrum Berlin, Berlin 33, Germany

G. Leitmann, Department of Mechanical Engineering, University of California, Berkeley, California 94720, USA

J. G. Lin, Department of Electrical Engineering, Columbia University, New York, New York 10027, USA

P. T. Liu, Department of Mathematics, University of Rhode Island, Kingston, Rhode Island 02881, USA

A. W. Merz, Aerophysics Research Corporation, Mountain View, California 94040, USA

K. Mori, Department of Electrical Engineering, Waseda University, Tokyo, Japan

G. Moriarty, Department of Electrical Engineering, Illinois Institute of Technology, Chicago, Illinois 60616, USA

U. R. Prasad, Department of Aeronautical Engineering, Indian Institute of Science, Bangalore, India

I. G. Sarma, School of Automation, Indian Institute of Science, Bangalore, India

W. Schmitendorf, Department of Mechanical Engineering, Northwestern University, Evanston, Illinois 60201, USA

M. R. Sertel, International Institute of Management, Wissenschaftszentrum Berlin, Berlin 33, Germany

E. Shimemura, Department of Electrical Engineering, Waseda University, Tokyo, Japan

M. Simaan, Coordinated Science Laboratory and Department of Electrical Engineering, University of Illinois, Urbana, Illinois 61801, USA

A. Sprzeuzkouski, Laboratoire d'Automatique Théorique, Université Paris 7, Paris, France

W. Stadler, Department of Mechanical Engineering, University of California, Berkeley, California 94720, USA

F. K. Sun, Division of Engineering and Applied Physics, Harvard University, Cambridge, Massachusetts 02138, USA

H. Y. Wan, Jr., Department of Economics, Cornell University, Ithaca, New York 14850, USA

D. J. Wilson, Department of Mathematics, University of Melbourne, Parkville, Victoria, Australia

P. L. Yu, Department of General Business, University of Texas, Austin, Texas 78712, USA

Preface

This volume is a collection of contributions to the subject of multicriteria decision making and differential games, all of which are based wholly or in part on papers that have appeared in the Journal of Optimization Theory and Applications. The authors take this opportunity to revise, update, or enlarge upon their earlier publications.

The theory of multicriteria decision making and differential games is concerned with situations in which a single decision maker is faced with a multiplicity of usually incompatible criteria, performance indices or payoffs, or in which a number of decision makers, or players, must take into account criteria each of which depends on the decisions of all the decision makers.

The first six chapters are devoted to situations involving a single decision maker, or a number of decision makers in complete collaboration and thus being in effect a single decision maker. Chapters I–IV treat various topics in the theory of domination structures and nondominated decisions. Chapter V presents a discussion of efficient, or Pareto-optimal, decisions. The approach to multicriteria decision making via preference relations is explored in Chapter VI.

When there is more than one decision maker, cooperation, as well as noncooperation, is possible. Chapters VII and VIII deal with the topic of coalitions in a dynamic setting, while Chapters IX and X address the situation of two unequal decision makers, a leader and a follower.

When there are many decision makers who do not cooperate in a dynamic decision process, we are in the realm of noncooperative differential games. Chapters XI–XV deal with various aspects of both two-person, zero-sum and many-person, nonzero-sum differential games.

Whenever a decision is to be taken there arises the vital question of the information on which the decision can be based. Chapters XVI–XX are devoted to two-person, zero-sum games under uncertainty. Chapter XXI is a contribution to the theory of team decision making under uncertainty. Various topics in noncooperative decision making in an uncertain environment are explored in Chapters XXII–XXIV. An example of cooperative decision making under uncertainty is presented in Chapter XXV.

Appropriately, this collection concludes with two contributions to a topic that gave much impetus to the development of differential game theory, that of pursuit and evasion.

Contents

Multicriteria Decision Making and Differential Games

I

Cone Convexity, Cone Extreme Points, and Nondominated Solutions in Decision Problems with Multiobjectives[1]

P. L. Yu

Abstract. Although there is no universally accepted solution concept for decision problems with multiple noncommensurable objectives, one would agree that a *good* solution must not be dominated by the other feasible alternatives. Here, we propose a structure of domination over the objective space and explore the geometry of the set of all nondominated solutions. Two methods for locating the set of all nondominated solutions through ordinary mathematical programming are introduced. In order to achieve our main results, we have introduced the new concepts of cone convexity and cone extreme point, and we have explored their main properties. Some relevant results on polar cones and polyhedral cones are also derived. Throughout the paper, we also pay attention to an important special case of nondominated solutions, that is, Pareto-optimal solutions. The geometry of the set of all Pareto solutions and methods for locating it are also studied. At the end, we provide an example to show how we can locate the set of all nondominated solutions through a derived decomposition theorem.

[1] The author would like to thank Professors J. Keilson and M. Zeleny for their helpful discussion and comments. Thanks also go to an anonymous reviewer for his helpful comments concerning the author's previous working paper (Ref. 1). He is especially obliged to Professors M. Freimer and A. Marshall for their careful reading of the first draft and valuable remarks. The author is also very grateful to Professor G. Leitmann and Dr. W. Stadler for their helpful comments.

1. Introduction

In daily decision problems, one is quite likely to face problems with multiple noncommensurable objectives. For this type of decision problems, a logically sound and universally accepted solution concept is not yet existent (see Refs. 2–4). Nevertheless, one would agree that a *good* decision should not be dominated by the other alternatives, in the sense that there should not be other feasible alternatives which yield a greater satisfaction for the decision maker.

To be more precise, let $f(x) = (f_1(x), f_2(x),..., f_l(x))$ be l noncommensurable objectives defined over $X \subset R^n$. Let $Y = f[X] = \{f(x) | x \in X\}$. X and Y will be called the *decision space* and *objective space*, respectively. Now, given a point $y \in Y$, one could associate it with a set of domination factors, denoted by $D(y)$, such that, if $y^1 \neq y$ and $y^1 \in y + D(y)$, then y^1 is dominated by y. As an example, let $y = (y_1, y_2) = (f_1(x), f_2(x))$, with y_1 and y_2 respectively representing the index of the income and the prestige which are resulting from a decision x. Assume that the decision maker wants to increase his income and his prestige as much as possible. Then, we could define

$$D(y) = \{(d_1, d_2) | d_1, d_2 \leqq 0\}.$$

Obviously, if $y^1 \neq y$ and $y^1 \in y + D(y)$, then y^1 is dominated by y, in the sense that y^1 can yield a worse satisfaction than that of y. *A nondominated solution* of our decision problems is one that is not dominated by any other feasible choice. It is seen that a *good* decision must be a nondominated solution; and to screen out *the set of all nondominated solutions*, if it is nonempty, essentially is the first step toward a good decision.

In this paper, we shall discuss the domination structure, $\{D(y)\}$, and derive methods to locate the set of all nondominated solutions in Section 5. We show that a nondominated solution must be a cone extreme point and derive a sequential approximation technique to locate the set of all nondominated solutions by a sequence of sets of cone extreme points. Because of its significance, we devote Section 4 to discussion of the concept of cone extreme points in the objective space. Two fundamental techniques for locating the set of all cone extreme points through ordinary mathematical programming are derived. The results are then applied in Section 5 to find the set of all cone extreme points in the decision space. The methods that we introduce are based on the concepts of polar cones and cone convexity of Y. Because of this, we devote Section 2 and Section 3 to the discussion of polar cones and cone convexity respectively. Only the relevant results in polar cones are derived in

Section 2. Quite a few of them, we think, have not been explored before. Cone convexity, a new concept, can be regarded as a generalization of the concept of convex set. We study its general properties and derive some relevant results for application in the subsequent sections. Sufficient conditions for $f[X]$ to be cone convex are derived through the concepts of polar cones and polyhedral cones.

Throughout the paper we have also paid attention to an important special case of the nondominated solutions, that is, *Pareto-optimal solutions*. Formally, a point $x^0 \in X$ is Pareto optimal (or efficient) if there is no other point $x^1 \in X$ such that[2]

$$f(x^1) \geqq f(x^0), \qquad f(x^1) \neq f(x^0).$$

From this definition we see that $x^0 \in X$ is a Pareto-optimal solution iff it is a nondominated solution, with $D(y)$ the nonpositive cone in R^l for all $y \in Y$. For convenience, we will denote the set of all Pareto-optimal solutions by P^0.

The conditions for a point in X to be Pareto optimal have been expounded in the literature (see Refs. 5–9).[3] However, the geometry of the entire set P^0 and the methods to find P^0 have not yet been well developed. As a special case, the geometry of P^0 in the objective space and methods to locate P^0 are studied in both Sections 4 and 5.

Before we go further, for convenience let us introduce the following notation. Let $x = (x_1, x_2, \ldots, x_n)$ and $y = (y_1, y_2, \ldots, y_n)$. Then,

(i) $x > y$ iff $x_j > y_j$ for all $j = 1, \ldots, n$;

(ii) $x \geqslant y$ iff $x_j \geqq y_j$ for all $j = 1, \ldots, n$ and $x \neq y$;

(iii) $x \geqq y$ iff $x_j \geqq y_j$ for all $j = 1, \ldots, n$.

Usually, we shall denote a set by a capital character and use superscripts to indicate the index of a set of vectors. Given a set S, its closure, interior, and relative interior[4] will be denoted by $\bar{S}$, Int S, and S^I respectively. Given two sets S and T in R^n, their addition is defined by

$$S + T = \{s + t \mid s \in S, t \in T\}.$$

---

[2] The meaning of the inequality is given by (iii) below.

[3] In statistical decision content, a Pareto solution is also known as an admissible solution (or strategy). See Refs. 10–11.

[4] With respect to the relative topology induced in

$$M(S) = \left\{ y = \sum_i \lambda_i x_i \mid x_i \in S, \sum_i \lambda_i = 1 \right\},$$

the manifold generated by S.

Also, $c \cdot x$ will denote the inner product of c and x as well as the matrix multiplication of c and x with proper orders. The context should make clear which is intended. Let I be an index set. Then,

$$\cup \{S(i) | \, i \in I\} \qquad \text{or} \qquad \cap \{S(i) | \, i \in I\}$$

will denote the union or the intersection of all $S(i)$, $i \in I$.

2. Cones and Their Polar Cones

In this section, we will discuss and derive some relevant results regarding cones and their polar cones for later development. Besides the general concept of polyhedral cones and polar cones, our main interest in this section is to derive Theorem 2.1, a necessary and sufficient condition for a cone to have a nonempty interior in its polar cone, and a closed expression for that interior. Theorem 2.1 plays an important role in Section 4. For the purpose of the subsequent sections, we cannot assume all cones to be closed. We do not assume the cones to be convex either, unless otherwise specified, because in some applications, such as tangent cones, cones may not be convex. The derivation in this section has been kept at an elementary level. The same results could also be derived through a more sophisticated method.[5]

A set S is convex if x_1, $x_2 \in S$ implies that

$$\lambda x_1 + (1 - \lambda) x_2 \in S$$

for all $0 \leqslant \lambda \leqslant 1$. A nonempty set Λ is a cone if $x \in \Lambda$ and $\eta \geqq 0$ imply that $\eta x \in \Lambda$. Λ is a convex cone if Λ is convex and is a cone. It is seen that Λ is a convex cone iff x_1, $x_2 \in \Lambda$ and $(\eta_1, \eta_2) \geqq (0, 0)$ imply that $\eta_1 x_1 + \eta_2 x_2 \in \Lambda$. Note that a cone always contains 0. For an arbitrary set S in R^l, we define its *polar cone* by

$$S^* = \{y \in R^l \mid y \cdot x \leqq 0 \quad \text{for all} \quad x \in S\}.$$

Clearly, S^* is a closed convex cone and $S_1 \subset S_2$ implies that $S_1^* \supset S_2^*$. It is also clear that $S^* = (\bar{S})^*$. The following lemma was proved in Ref. 12, p. 95.

Lemma 2.1. Let Λ be a convex cone. Then,

(i) $(\Lambda^*)^* = \bar{\Lambda}$;
(ii) Λ is closed iff $(\Lambda^*)^* = \Lambda$.

[5] See Footnote 7.

A cone which is a closed polyhedron is called a *polyhedral cone.* It can be shown (see Ref. 12, Chapter 2) that Λ is a polyhedral cone iff there exists a finite number of vectors $\{v^1, v^2,\ldots, v^m\}$ such that

$$\Lambda = \left\{\sum_{i=1}^{m} a_i v^i \mid a_i \in R^1,\ a_i \geqq 0\right\}.$$

The set $V = \{v^i \mid i = 1,\ldots, m\}$ is called a *generator* of the cone Λ.

Lemma 2.2. (i) Let

$$\Lambda = \left\{\sum_{i=1}^{m} a_i v^i \mid a_i \in R^1{}_t\ a_i \geqq 0\right\}$$

be a polyhedral cone. Then,

$$\Lambda^* = \{u \mid u \cdot v^i \leqq 0 \qquad \text{for} \quad i = 1,\ldots, m\};$$

so Λ^* is also a polyhedral cone.

(ii) Let Λ be a convex cone. Then, Λ^* is a polyhedral cone iff $\bar{\Lambda}$ is a polyhedral cone.

Proof. (i) Suppose that $u \in \Lambda^*$. Then, $u \cdot v^i \leqq 0$ for all $i = 1,\ldots,m$. Thus,

$$\Lambda^* \subset \{u \mid u \cdot v^i \leqq 0 \qquad \text{for} \quad i = 1,\ldots, m\}.$$

On the other hand, if $u \cdot v^i \leqq 0$ for $i = 1,\ldots, m$, then

$$u \cdot \sum_{i=1}^{m} a_i v^i = \sum_{i=1}^{m} a_i (u \cdot v^i) \leqq 0$$

if all $a_i \geqq 0$. Thus, $u \in \Lambda^*$ and

$$\Lambda^* \supset \{u \mid u \cdot v^i \leqq 0,\ i = 1,\ldots, m\}.$$

The sufficiency of (ii) comes from (i) and $\Lambda^* = (\bar{\Lambda})^*$. For the necessity of (ii), observe that $(\Lambda^*)^* = \bar{\Lambda}$, because of Lemma 2.1, and that Λ is a convex cone. In view of (i), we see that $\bar{\Lambda} = (\Lambda^*)^*$ is a polyhedral cone.

Definition 2.1.[6] A cone Λ in R^l, not necessarily convex, is said to be *acute* if there is an open half-space

$$H = \{x \in R^l \mid a \cdot x > 0, a \neq 0\}$$

such that

$$\bar{\Lambda} \subset H \cup \{0\}. \tag{1}$$

For simplicity and without confusion, whenever we say that an acute cone Λ is contained by an open half-space H, it should be understood that we mean in the sense (1).

Lemma 2.3. Let Λ be an acute polyhedral cone in R^l with dim $\Lambda = s$. Then, for each containing open half-space H of Λ, there exists s independent vectors

$$U = \{u^i \mid i = 1,..., s\}$$

such that the polyhedral cone Λ_1 generated by U has the following property:

$$\Lambda \subset \Lambda_1 \subset H \cup \{0\}.$$

Proof. Let

$$H = \{x \mid c \cdot x > 0\}, \qquad \text{and} \qquad V = \{v^j \mid j = 1,..., r\}$$

be a generator of Λ. Let

$$B = \{b^k \mid k = 1,..., l\}$$

be a basis of R^l such that (*i*) b^1 is the projection of c into L, which is the linear space spanned by V, (ii) $\{b^k \mid 1 \leqq k \leqq s\}$ are independent vectors in L, and (iii) $\{b^k \mid s \leqq k \leqq l\}$ are independent vectors in the orthogonal space of L. Then, with respect to B, each v^j could be written as

$$v^j = (x_1^{\ j}, x_2^{\ j},..., x_s^{\ j}, 0,..., 0), \quad \text{with} \quad x_1^{\ j} > 0 \quad \text{and} \quad c = (1, 0,..., 0, c_{s+1},..., c_l).$$

Since we are dealing with cones, without loss of generality, we could assume $x_1^{\ j} = 1$.

---

[6] A pointed cone (see Ref. 12) may not be acute. Nevertheless, one could show that a closed convex cone is acute iff it is pointed. This observation yields another equivalent definition for acute cones: a cone is acute iff the closure of the convex hull of the cone is pointed. Some results of pointed, closed, convex cones can then be used to derive (i) of Theorem 2.1. We shall not stop to do so.

Let us consider $U(M) = \{u^i \mid 1 \leqq i \leqq s\}$, with

$$\begin{aligned} u^1 &= (1, M, 0, \ldots, 0, 0, \ldots, 0), \\ u^2 &= (1, 0, M, \ldots, 0, 0, \ldots, 0), \\ &\cdots\cdots\cdots\cdots\cdots\cdots \\ u^{s-1} &= (1, 0, 0, \ldots, M, 0, \ldots, 0), \\ u^s &= (1, -M, -M, \ldots, -M, 0, \ldots, 0). \end{aligned}$$

Note that the jth component u_j^i of the ith vector, $1 \leqq i \leqq s - 1$, is given by

$$u_j^i = \begin{cases} 1 & \text{if } j = 1, \\ M & \text{if } j = i + 1, \\ 0 & \text{otherwise.} \end{cases}$$

Obviously, $U(M)$ is a set of s independent vectors. Observe that $U(M) \subset H$, because

$$c \cdot u^j = 1 > 0, j = 1, \ldots, s.$$

Thus, the cone generated by $U(M)$ is also contained by $H \cup \{0\}$. It suffices now to show that

$$v^j = \sum_{i=1}^{s} \lambda_i^j u^i, \ \lambda_i^j \in R, \ \lambda_i^j \geqq 0 \qquad \text{for } \ j = 1, 2, \ldots, r, \tag{2}$$

when M is large enough.

Now, given

$$v^j = (1, x_2^j, \ldots, x_s^j, 0, \ldots, 0)$$

and $U(M)$, by treating λ_i^j as variable, (2) could be written as

$$\sum_{i=1}^{s} \lambda_i^j = 1, \tag{3}$$

$$\lambda_k^j M - \lambda_s^j M = x_{k+1}^j, \qquad k = 1, 2, \ldots, s - 1. \tag{4}$$

Note that (4) implies that

$$\lambda_k^j = \lambda_s^j + x_{k+1}^j / M, \qquad k = 1, 2, \ldots, s - 1. \tag{5}$$

By (3) and (5), we get

$$s\lambda_s^j + \sum_{k=2}^{s} x_k^j / M = 1 \qquad \text{or} \quad \lambda_s^j = 1/s - \sum_{k=2}^{s} x_k^j / sM. \tag{6}$$

By (5) and (6), it is clear that, when M is large enough, we could have a nonnegative solution λ_i^j for (2). Since $V = \{v^j\}$ is a finite set, by selecting a large value for M, (2) could be satisfied for all $j = 1, 2, \ldots, r$.

Lemma 2.4. Let

$$U = \{u^i \mid 1 \leqq i \leqq r\}$$

be a set of r independent vectors in R^l and

$$\Lambda = \left\{ \sum_{i=1}^{r} \lambda_i u^i \mid -\infty < \lambda_i < \infty, i = 1, \ldots, s; 0 \leqq \lambda_i, i = s+1, \ldots, r \right\}.$$

Then, Λ^* contains an $l - r$ dimensional subspace and $\dim \Lambda^* = l - s$.

Proof. Select

$$\{u^k \mid r+1 \leqq k \leqq l\}$$

so that

$$U_1 = \{u^i \mid 1 \leqq i \leqq l\}$$

is a set of l independent vectors in R^l. Let U_1 also denote the square matrix with u^i in its ith column. Since U_1 is nonsingular, $U_1^T U_1$ is positive definite and nonsingular. Let A be the inverse of $U_1^T U_1$ and set $W = U_1 A$. Then,

$$U_1^T W = U_1^T U_1 A = [U_1^T U_1] [U_1^T U_1]^{-1} = I.$$

Thus, the columns of W and columns of U_1 satisfy

$$u^i \cdot w^j = \begin{cases} 1 & \text{if } i = j, \\ 0 & \text{otherwise.} \end{cases}$$

Since A is nonsingular, W is also a basis in R^l. Thus, every $x \in R^l$ can be written as

$$x = \sum_{i=1}^{l} \alpha_i w^i.$$

If $x \in \Lambda^*$ and $v \in \Lambda$, then

$$x \cdot v = \left(\sum_{i=1}^{l} \alpha_i w^i \right) \left(\sum_{i=1}^{r} \lambda_i u^i \right) = \sum_{i=1}^{r} \alpha_i \lambda_i \leqq 0.$$

Since $-\infty < \lambda_i < \infty$, $i = 1, \ldots, s$, and $\lambda_i \geqq 0$, $i = s+1, \ldots, r$, the inequality can be held for all $v \in \Lambda$ iff $\alpha_i = 0$ for $i = 1, \ldots, s$, $\alpha_i \leqq 0$ for

$i = s + 1,\ldots, r$, and $-\infty < \alpha_i < \infty$ for $i = r + 1,\ldots, l$. This shows that Λ^* contains an $(l - r)$-dimensional subspace and $\dim \Lambda^* = l - s$.

Corollary 2.1. Let Λ be a polyhedral cone in R^l. Then,

(i) if L is the maximal subspace contained by Λ and $\dim L = s$, then $\dim \Lambda^* = l - s$;

(ii) if Λ is acute, then $\dim \Lambda^* = l$ and $\operatorname{Int} \Lambda^* \neq \varnothing$.

Proof. (ii) Note that, by Lemma 2.3, there are r, $r \leqslant l$, independent vectors which generate a polyhedral cone $\Lambda_1 \supset \Lambda$. In view of Lemma 2.4, $\dim \Lambda_1{}^* = l$. Since $\Lambda_1{}^* \subset \Lambda^*$, this shows that

$$\dim \Lambda^* = l, \qquad \operatorname{Int} \Lambda^* \neq \varnothing .$$

(i) By the decomposition theorem (see Ref. 12, p. 60), we can write $\Lambda = L \oplus \Lambda'$, where Λ' is an acute polyhedral cone contained by $L^\perp$, where $L^\perp$ is the orthogonal space of L. Note that, if $\dim \Lambda = r$, then $\dim \Lambda' = r - s$. In view of Lemma 2.3, we can find $r - s$ independent vectors which generate Λ_0 such that $\Lambda' \subset \Lambda_0 \subset L^\perp$. On the other hand, we could select $r - s$ independent vectors from the generators of Λ' and let Λ_2 be the cone generated by those vectors. Note that

$$\Lambda_0 \supset \Lambda' \supset \Lambda_2 , \qquad L \oplus \Lambda_0 \supset L \oplus \Lambda' \supset \Lambda' \supset L \oplus \Lambda_2 .$$

Observe that both $L \oplus \Lambda_0$ and $L \oplus \Lambda_2$ can be written in the form stated in Lemma 2.4, and

$$\dim(L \oplus \Lambda_0)^* = \dim(L \oplus \Lambda_2)^* = l - s.$$

Since

$$(L \oplus \Lambda_2)^* \supset (L \oplus \Lambda')^* = \Lambda^* \supset (L \oplus \Lambda_0)^*,$$

we see that $\dim \Lambda^* = l - s$.

Theorem 2.1. Let Λ be a cone (not necessarily convex) in R^l. Then,

(i) $\operatorname{Int} \Lambda^* \neq \varnothing$ iff Λ is acute;

(ii) when Λ is acute, $\operatorname{Int} \Lambda^* = \{y \mid y \cdot x < 0 \text{ for all nonzero } x \in \bar{\Lambda}\}$.

Proof. (i) Suppose that Int $\Lambda^* \neq \varnothing$. Let $y^0 \in \operatorname{Int} \Lambda^*$ and

$$H = \{x \mid y^0 \cdot x < 0\}.$$

We show that $\bar{\Lambda} \subset H \cup \{0\}$ or $y^0 \cdot x < 0$ for all $x \in \bar{\Lambda}$ and $x \neq 0$. Suppose that there is $x_0 \in \bar{\Lambda}$, $x_0 \neq 0$, such that

$$y^0 \cdot x_0 \geqq 0. \tag{7}$$

Since $\Lambda^* = (\bar{\Lambda})^*$ and $y^0 \in \operatorname{Int} \Lambda^*$, there is N_{y^0}, a neighborhood of y^0, such that, for all $y \in N_{y^0}$,

$$y \cdot x_0 \leqq 0. \tag{8}$$

If we treat $y \cdot x_0$ as a linear function in y, (7) and (8) show that the linear function assumes its maximum point at y^0, an interior point of N_{y^0}. This is impossible, unless $y \cdot x_0$ is constant over N_{y^0}. Obviously, the latter is also impossible, because $x_0 \neq 0$. This proves the necessity.

In order to show the sufficiency, let $\bar{\Lambda}$ be contained by

$$H = \{x \mid a \cdot x > 0\},$$

that is,

$$\bar{\Lambda} \subset H \cup \{0\}.$$

Let

$$H_1 = \{x \mid a \cdot x = 1\}.$$

We first show that $\bar{\Lambda} \cap H_1$ is bounded.

Suppose that $\bar{\Lambda} \cap H_1$ is not bounded. We could select a sequence $\{x_n\}$ from $\bar{\Lambda} \cap H_1$ such that $n \leqq |x_n|$. Let $y_n = x_n/|x_n|$. We see that

$$a \cdot y_n = a \cdot x_n/|x_n| = 1/|x_n|.$$

Since $\{y_n\}$ is a sequence on the compact unit sphere, we can find a subsequence $\{y_n{}^1\}$ which converges to a point y_0 in the unit sphere. Note that $a \cdot y_0 = 0$. On the other hand, since each $y_n{}^1 \in \bar{\Lambda}$ and $\bar{\Lambda}$ is closed, $y_0 \in \bar{\Lambda}$ or $a \cdot y_0 > 0$. This leads to a contradiction.

Since $\bar{\Lambda} \cap H_1$ is closed and bounded, we could select a polygon $P \supset \bar{\Lambda} \cap H_1$ on H_1. The cone which is generated by the extreme points of P is an acute polyhedral cone. Recall that $S_1 \subset S_2$ implies that $S_1{}^* \supset S_2{}^*$. In view of Corollary 2.1, we see that

$$\operatorname{Int}(\bar{\Lambda})^* = \operatorname{Int} \Lambda^* \neq \varnothing.$$

(ii) The fact that

$$\text{Int } \Lambda^* \subset \{y \mid y \cdot x < 0, \text{ for all nonzero } x \in \bar{\Lambda}\}$$

can be proved by a contradiction similar to that for the necessity of (i). In order to show that

$$\text{Int } \Lambda^* \supset \{y \mid y \cdot x < 0 \text{ for all nonzero } x \in \bar{\Lambda}\},$$

let y^0 be such that $y^0 \cdot x < 0$ for all nonzero $x \in \bar{\Lambda}$. It suffices to show that there is a neighborhood N_{y^0} of y_0 such that $y \cdot x \leqq 0$ for all $x \in \bar{\Lambda}$ and $y \in N_{y^0}$. Assume the contrary. Let

$$\Lambda_1 = \{x \mid x \in \bar{\Lambda}, \mid x \mid = 1\}.$$

Since a cone is uniquely determined by its unit vectors, we could find two sequences: $\{y_n\} \to y^0$ and $\{x_n\} \subset \Lambda_1$ such that $y_n \cdot x_n > 0$. Since Λ_1 is compact, there is a convergent subsequence $\{x_{n_k}\} \to x_0 \in \Lambda_1$. We see that

$$\lim_{n_k \to \infty} y_{n_k} x_{n_k} = y^0 \cdot x_0 \geqq 0.$$

This contradicts the statement that $y^0 \cdot x < 0$ for all nonzero $x \in \bar{\Lambda}$.

3. Cone Convexity

The concept of convex set has played an important role in optimization problems, such as in mathematical programming, optimal control, and game theory. In many cases, convexity on the entire set is not necessary. Some partial convexity such as *directional convexity* and the convexity to be introduced may be enough for particular problems (as an example, see Ref. 13).

Note that a set S is directionally convex in the direction of $u \neq 0$, if x_1, $x_2 \in S$ and $0 \leqq \lambda \leqq 1$ implies that there is $\mu \geqq 0$ such that

$$\lambda x_1 + (1 - \lambda)x_2 + \mu u \in S.$$

Thus, if S is convex, then it is directionally convex in any arbitrary direction.

Definition 3.1. Given a set S and a convex cone Λ in E^l, S is said to be *Λ-convex* iff $S + \Lambda$ is a convex set.

Remark 3.1. (i) A set is convex iff it is $\{0\}$-convex; and (ii) a set is directionally convex in $u \neq 0$, iff it is U-convex, where

$$U = \{\lambda u \mid \lambda \leqq 0\}$$

is the negative half line of u.

Proof. Assertion (i) is obvious. For the necessity of (ii), let $x_1, x_2 \in S + U$ and $0 \leqq \lambda \leqq 1$. We want to show that

$$\lambda x_1 + (1 - \lambda) x_2 \in S + U$$

whenever S is directionally convex in u.

By definition, there are $y_1, y_2 \in S$ and $\mu_1, \mu_2 \geqq 0$ such that

$$x_1 = y_1 - \mu_1 u, \qquad x_2 = y_2 - \mu_2 u.$$

Let

$$w = \lambda y_1 + (1 - \lambda) y_2 .$$

Then, there are $\mu_0 \geqq 0$ and $w_0 \in S$ such that $w = w_0 - \mu_0 u$. Now,

$$\begin{aligned}
\lambda x_1 + (1 - \lambda) x_2 &= \lambda(y_1 - \mu_1 u) + (1 - \lambda)(y_2 - \mu_2 u) \\
&= \lambda y_1 + (1 - \lambda) y_2 - [\lambda \mu_1 + (1 - \lambda)\mu_2] u \\
&= w - [\lambda \mu_1 + (1 - \lambda)\mu_2] u \\
&= w_0 - [\mu_0 + \lambda \mu_1 + (1 - \lambda)\mu_2] u.
\end{aligned}$$

Since

$$\mu_0 + \lambda \mu_1 + (1 - \lambda)\mu_2 \geqq 0,$$

one has

$$\lambda x_1 + (1 - \lambda) x_2 \in S + U.$$

For the sufficiency, let $x_1, x_2 \in S$ and $0 \leqq \lambda \leqq 1$. Since $S + U$ is convex,

$$\lambda x_1 + (1 - \lambda) x_2 \in S + U.$$

Thus, there are $y \in S$ and $\mu \geqq 0$ such that

$$\lambda x_1 + (1 - \lambda) x_2 = y - \mu u.$$

That is,

$$y = \lambda x_1 + (1 - \lambda) x_2 + \mu u.$$

So S is directionally convex in u.

The above remark as well as the following lemma warrant that Λ-convexity is a generalization of ordinary convexity.

Lemma 3.1. Let Λ_1 and Λ_2 be two convex cones in R^l such that $\Lambda_1 \subset \Lambda_2$. Then, if set S in R^l is Λ_1-convex, it is also Λ_2-convex.

Proof. Since S is Λ_1-convex, $S + \Lambda_1$ is a convex set. Since $\Lambda_1 \subset \Lambda_2$,

$$S + \Lambda_2 = (S + \Lambda_1) + \Lambda_2 .$$

Observe that $S + \Lambda_1$ and Λ_2 are convex, and so is their sum. Thus, $S + \Lambda_2$ is convex. That is, S is Λ_2-convex.

Because our work is primarily related to convex cones, unless otherwise stated, all cones from now on will mean convex cones.

Remark 3.2. Immediately from Lemma 3.1, we see that, if S is Λ-convex, then it is also $\bar{\Lambda}$-convex. The converse is not generally true. As an example, in R^2 let

$$S_1 = \{(x, y) | \; x + y = 1, \; x, y \geqq 0\}, \qquad S = (0, 0) \cup S_1$$

and

$$\Lambda = \{(x, y) | \; x, y > 0\} \cup \{(0, 0)\}.$$

It is seen that S is $\bar{\Lambda}$-convex. However, because the line connecting (0, 0) and (1, 0) is not contained in $S + \Lambda$, S is not Λ-convex. By defining

$$\Lambda = \{(x, y) | \; x, y \geqq 0\},$$

the same example shows that, although S is Λ-convex, it is not $(\{0\} \cup \Lambda^I)$-convex. Note that $\bar{\Lambda}$ and Λ^I are the closure and the relative interior of Λ, respectively.

Remark 3.3. The closure and the relative interior of a convex set are also convex sets (see Chapter 3 of Ref. 11). This nice property is not preserved in cone convexity. More precisely, that S is Λ-convex does not imply that $\bar{S}$ or S^I is also Λ-convex. As an example, in R^2 let

$$S_1 = \{(x, y) \mid (x - 1)^2 + (y - 3)^2 < 1\},$$
$$S_2 = \{(x, y) \mid (x - 3)^2 + (y - 1)^2 < 1\},$$
$$S = (0, 0) \cup S_1 \cup S_2 .$$

Thus,

$$\bar{S} = \{(0, 0)\} \cup \bar{S}_1 \cup \bar{S}_2 , \qquad S^I = S_1 \cup S_2 .$$

Now, let

$$\Lambda = \{(x, y) | \ x, y > 0\} \cup \{(0, 0)\}.$$

Then, S is Λ-convex. However, because the line connecting (0, 0) and (0, 3) is not contained in $\bar{S} + \Lambda$, $\bar{S}$ is not Λ-convex. It can also be easily checked that S^I is not Λ-convex.

Let $f = (f_1, f_2, ..., f_l)$ be a vector function defined over $X \subset R^n$. Usually, we do not know the shape of $f[X]$, the image of X through f. This fact makes it difficult to have a geometric analysis over $f[X]$.

Theorem 3.1.[7] $f[X]$ is Λ-convex iff, for every $x_1, x_2 \in X$ and $\mu \in R$, $0 \leqq \mu \leqq 1$,

$$\mu f(x_1) + (1 - \mu) f(x_2) \in f[X] + \Lambda.$$

Proof. The necessity is obvious. In order to see the sufficiency, let

$$y_1 = f(x_1) + \lambda_1, \qquad y_2 = f(x_2) + \lambda_2,$$

where $x_1, x_2 \in X$ and $\lambda_1, \lambda_2 \in \Lambda$. Then,

$$\begin{aligned} \mu y_1 + (1 - \mu) y_2 &= \mu f(x_1) + (1 - \mu) f(x_2) \\ &\quad + (\mu\lambda_1 + (1 - \mu)\lambda_2) \in f[X] + \Lambda + \Lambda = f[X] + \Lambda. \end{aligned}$$

Thus, $f[X] + \Lambda$ is a convex set, and $f[X]$ is Λ-convex.

Note that the theorem does not require X to be convex.

Corollary 3.1. (i) $f[X]$ is Λ-convex iff, for any $x_1, x_2 \in X$ and $\mu \in R$, $0 \leqq \mu \leqq 1$, there is $x_3 \in X$ and $\lambda \in \Lambda$ such that

$$\mu f(x_1) + (1 - \mu) f(x_2) = f(x_3) + \lambda.$$

(ii) Let

$$\Lambda^{\leqq} = \{\lambda \mid \lambda \in R^l, \lambda \leqq 0\}.$$

Then, $f[X]$ is $\Lambda^{\leqq}$-convex iff, for every $x_1, x_2 \in X$ and $\mu \in R$, $0 \leqq \mu \leqq 1$, there is $x_3 \in X$ such that

$$\mu f(x_1) + (1 - \mu) f(x_2) \leqq f(x_3).$$

[7] Through a similar proof, we have the following result: S is Λ-convex iff, for every $s_1, s_2 \in S$ and $\mu \in R$, $0 \leqq \mu \leqq 1$,

$$\mu s_1 + (1 - \mu) s_2 \in S + \Lambda.$$

In particular, if there is $x_0 \in X$ such that $f(x_0) \geqq f(x)$ for all $x \in X$, then $f[X]$ is $\Lambda^{\leqq}$-convex.

(iii) Let

$$\Lambda^{\geqq} = \{\lambda \mid \lambda \in R^l, \lambda \geqq 0\}.$$

Then, $f[X]$ is $\Lambda^{\geqq}$-convex iff, for every x_1, $x_2 \in X$ and $\mu \in R$, $0 \leqq \mu \leqq 1$, there is $x_3 \in X$ such that

$$\mu f(x_1) + (1 - \mu) f(x_2) \geqq f(x_3).$$

In particular, if there is $x_0 \in X$ such that $f(x_0) \leqq f(x)$ for all $x \in X$, then $f[X]$ is $\Lambda^{\geqq}$-convex.

Proof. By Theorem 3.1, it is obvious.

Theorem 3.2. Let $X \subset R^n$ be a convex set and $\Lambda \subset R^l$ be a convex cone. Suppose that $f(x)$ is defined over X. Then, (i) if $\lambda \cdot f(x)$ is concave for each $\lambda \in \Lambda$, then $f[X]$ is Λ^*-convex; (ii) if $\lambda \cdot f(x)$ is convex for each $\lambda \in \Lambda$, then $f[X]$ is $-\Lambda^*$-convex.

Proof. (i) Let x_1, $x_2 \in X$ and $\mu \in R$, $0 \leqq \mu \leqq 1$. For any $\lambda \in \Lambda$, by assumption, we have

$$\mu(\lambda \cdot f(x_1)) + (1 - \mu)(\lambda \cdot f(x_2)) \leqq \lambda \cdot f(\mu x_1 + (1 - \mu)).$$

Thus,

$$\lambda \cdot [\mu f(x_1) + (1 - \mu) f(x_2) - f(\mu x_1 + (1 - \mu) x_2)] \leqq 0$$

for all $\lambda \in \Lambda$. That is,

$$\mu f(x_1) + (1 - \mu) f(x_2) - f(\mu x_1 + (1 - \mu) x_2) \in \Lambda^*,$$

or

$$\mu f(x_1) + (1 - \mu) f(x_2) \in f(\mu x_1 + (1 - \mu) x_2) + \Lambda^* \subset f[X] + \Lambda^*.$$

By invoking Theorem 3.1, our assertion is clear.

(ii) By the assumptions, $-\lambda \cdot f(x)$ is concave for each $\lambda \in \Lambda$. Thus, $\lambda \cdot f(x)$ is concave for each $\lambda \in -\Lambda$. The assertion (ii) follows from (i), because $(-\Lambda)^* = -\Lambda^*$.

In view of Lemma 2.1, we have the following corollary.

Corollary 3.2. Let $X \subset R^n$ be a convex set and $\Lambda \subset R^l$ be a closed convex cone. Suppose that $f(x)$ is defined over X. Then,

(i) if $\lambda \cdot f(x)$ is concave for each $\lambda \in \Lambda^*$, then $f[X]$ is Λ-convex;
(ii) if $\lambda \cdot f(x)$ is convex for each $\lambda \in -\Lambda^*$ then $f[X]$ is Λ-convex.

Corollary 3.3. Let $X \subset R^n$ be a convex set and $\Lambda \subset R^l$ be a polyhedral cone with a generator $\{h^j \mid 1 \leqq j \leqq r\}$. Suppose that $f(x)$ is defined over X. Then,

(i) if each $h^j \cdot f(x)$, $1 \leqq j \leqq r$, is concave, then $f[X]$ is Λ^*-convex;
(ii) if each $h^j \cdot f(x)$, $1 \leqq j \leqq r$, is convex, then $f[X]$ is $-\Lambda^*$-convex.

Proof. For each $\lambda \in \Lambda$, we can write

$$\lambda = \sum_{j=1}^{r} \mu_j h^j, \qquad \mu_j \geqq 0.$$

Thus,

$$\lambda \cdot f(x) = \sum_{j=1}^{r} \mu_j h^j \cdot f(x).$$

Since $\mu_j \geqq 0$, $\lambda \cdot f(x)$ is concave (or convex) whenever each $h^j \cdot f(x)$ is concave (or convex). The assertions follow immediately from Theorem 3.2.

Similarly, from Corollary 3.2, we have the following corollary.

Corollary 3.4. Let $X \subset R^n$ be a convex set and $\Lambda \subset R^l$ be a polyhedral cone. Let

$$\{H^j \mid 1 \leqq j \leqq r\}$$

be a generator for Λ^*. Suppose that $f(x)$ is defined over X. Then,

(i) if each $H^j \cdot f(x)$, $1 \leqq j \leqq r$, is concave, then $f[X]$ is Λ-convex;
(ii) if each $-H^j \cdot f(x)$, $1 \leqq j \leqq r$, is convex, then $f[X]$ is Λ-convex.

Recall that

$$\Lambda^{\geqq} = \{\lambda \mid \lambda \in R^l, \lambda \geqq 0\}, \qquad \Lambda^{\leqq} = \{\lambda \mid \lambda \in R^l, \lambda \leqq 0\}.$$

We see that

$$(\Lambda^{\geqq})^* = \Lambda^{\leqq}.$$

Let e^j, $1 \leqq j \leqq l$, be the vector in R^l with the jth component equal one and all the other components zeros. We get immediately that

$$\{e^j \mid 1 \leqq j \leqq l\}, \qquad \{-e^j \mid 1 \leqq j \leqq l\}$$

are generators for $\Lambda^{\geqq}$ and $\Lambda^{\leqq}$, respectively. Now, suppose that $f(x)$ is concave or convex[8] over a convex set X. Then, $\lambda \cdot f(x)$ is concave or convex over X for all $\lambda \in \Lambda^{\geqq}$. Note that

$$e^j \cdot f(x) = f_j(x).$$

By invoking Corollary 3.4, we have the following corollary.

Corollary 3.5. Suppose X is a convex set. Then,

(i) $f[X]$ is $\Lambda^{\leqq}$-convex when f is concave over X;

(ii) $f[X]$ is $\Lambda^{\geqq}$-convex when f is convex over X.

Note that Corollary 3.5 can also be easily derived from Corollary 3.1.

In view of Lemma 3.1 and Corollary 3.5, we have the following corollary.

Corollary 3.6. Suppose that $f(x)$ is defined over a convex set X. Then, $f[X]$ is Λ-convex whenever f is concave over X and $\Lambda \supset \Lambda^{\leqq}$ or whenever f is convex over X and $\Lambda \supset \Lambda^{\geqq}$.

Remark 3.4. Suppose that Λ is an arbitrary convex cone in R^l. In order to check the Λ-convexity of $f[X]$, we could select a polyhedral cone $\Lambda_1 \subset \Lambda$. If $f[X]$ is Λ_1-convex (by Corollary 3.4, say), then, by Lemma 3.1, we know that $f[X]$ is also Λ-convex.

4. Cone Extreme Points

4.1. General Concept. As will be seen in Section 5, if we have a constant set of domination factors, a nondominated solution can be regarded as a cone extreme point. If the above assumption does not hold, we can locate the set of all nondominated solutions through a sequence of sets of cone extreme points. In this section, we will study the geometric structure of cone extreme points in objective space Y (if desired, Y may be regarded as an arbitrary set in R^l). Two methods for locating the set of all cone extreme points on Y through ordinary mathematical programs are explored. The results will then be applied in the next section.

[8] That is, each f_j is concave or convex.

Definition 4.1.[9] Let Y and Λ denote, respectively, a set and a cone in R^l. A point y^0 is a *Λ-extreme point* of Y if $y^0 \in Y$ and there is no y^1 in Y, $y^1 \neq y^0$, such that $y^0 \in y^1 + \Lambda$. We will denote the set of all Λ-extreme points of Y by $\text{Ext}[Y \mid \Lambda]$.

Note that if Y contains a single point, then $Y = \text{Ext}[Y \mid \Lambda]$. This is not an interesting case. From now on, we shall assume that Y contains at least two points.

Remark 4.1. Immediately from the definition we see that, except when $\Lambda = \{0\}$,

$$y \in \text{Ext}[Y \mid \Lambda]$$

implies that $y \in \partial Y$ (the boundary of Y). Thus, if Y is an open set and $\Lambda \neq \{0\}$, then

$$\text{Ext}[Y \mid \Lambda] = \varnothing.$$

Sufficient conditions for $\text{Ext}[Y \mid \Lambda]$ to be nonempty will be given in Corollary 4.6. Recall that, unless otherwise specified, we are dealing with convex cones.

Lemma 4.1. The following results hold:

(i) $\text{Ext}[Y \mid \Lambda] = \begin{cases} Y & \text{if } \Lambda = \{0\}, \\ \phi & \text{if } \Lambda = R^l; \end{cases}$

(ii) $\text{Ext}[Y \mid \Lambda_2] \subset \text{Ext}[Y \mid \Lambda_1]$ if $\Lambda_1 \subset \Lambda_2$;

(iii) $\text{Ext}[Y + \Lambda \mid \Lambda] \subset \text{Ext}[Y \mid \Lambda]$;

(iv) $\text{Ext}[Y \mid \Lambda] \subset \text{Ext}[Y + \Lambda \mid \Lambda]$ if Λ contains no nontrivial subspace.

Proof. (i) The statement is clearly true.

(ii) Suppose that

$$y \in \text{Ext}[Y \mid \Lambda_2].$$

if

$$y \notin \text{Ext}[Y \mid \Lambda_1],$$

[9] Given Y and Λ, we could define $y_1 \succ y_2$ iff $y_2 \in y_1 + \Lambda$. Suppose this Λ contains no nontrivial subspace. It is seen that $\succ$ is a partial ordering over Y. It is also seen that y^0 is a Λ-extreme point of Y iff y^0 is a maximal element of Y with respect to $\succ$. Although the concept of cone extreme points is important in decision problems and in mathematical programming, it has been explored rarely.

then there is $y^1 \in Y$, $y^1 \neq y$, such that

$$y \in y^1 + \Lambda_1 \subset y^1 + \Lambda_2 .$$

Thus,

$$y \notin \operatorname{Ext}[Y \mid \Lambda_2].$$

This leads to a contradiction.

(iii) Suppose that

$$y \in \operatorname{Ext}[Y + \Lambda \mid \Lambda].$$

It suffices to show that $y \in Y$. Suppose that $y \notin Y$. Then, there are $y^1 \in Y$ and $h \neq 0$, $h \in \Lambda$ such that $y = y^1 + h$. Since $0 \in \Lambda$, $Y \subset Y + \Lambda$. We see that

$$y \notin \operatorname{Ext}[Y + \Lambda \mid \Lambda].$$

This leads to a contradiction.

(iv) Let

$$y \in \operatorname{Ext}[Y \mid \Lambda]$$

Since $0 \in \Lambda$, $y \in Y + \Lambda$. Suppose that

$$y \notin \operatorname{Ext}[Y + \Lambda \mid \Lambda].$$

There are $y^1 \in Y + \Lambda$ and $h^1 \neq 0$, $h^1 \in \Lambda$ such that $y = y^1 + h^1$. Since $y^1 \in Y + \Lambda$, we could write $y^1 = y^0 + h^0$, $y^0 \in Y$ and $h^0 \in \Lambda$. Thus, $y = y^0 + (h^0 + h^1)$. Since Λ contains no subspace $h^0 + h^1 \neq 0$. It is seen that

$$y \notin \operatorname{Ext}[Y \mid \Lambda].$$

This leads to a contradiction.

Remark 4.2. Immediately from (ii), we see that

$$\operatorname{Ext}[Y \mid \bar{\Lambda}] \subset \operatorname{Ext}[Y \mid \Lambda].$$

The converse is not generally true. As an example, let

$$Y = \{(x, y) \mid 0 \leqq x \leqq 1, 0 \leqq y \leqq 1\}, \qquad \Lambda = \{(x, y) \mid x < 0\} \cup \{(0, 0)\}.$$

It is seen that

$$\operatorname{Ext}[Y \mid \Lambda] = \{(x, y) \mid x = 1, 0 \leqq y \leqq 1\},$$

but

$$\operatorname{Ext}[Y \mid \bar{\Lambda}] = \varnothing .$$

Remark 4.3. The assumption in (iv) of Lemma 4.1 cannot be relaxed. In order to see that Λ cannot contain any subspace, let us consider the following example. In R^2, let

$$Y = \{(x, y) \mid x = y, 0 \leqq x, y \leqq 1\}, \qquad \Lambda = \{(x, y) \mid x \leqq 0\}.$$

We see that

$$\operatorname{Ext}[Y \mid \Lambda] = \{(1, 1)\}, \qquad \operatorname{Ext}[Y + \Lambda \mid \Lambda] = \varnothing .$$

Observe that, if L is the maximum linear subspace contained by Λ, then, by the decomposition theorem, we can write $\Lambda = L \oplus \Lambda^{\perp}$, where $\Lambda^{\perp} = \Lambda \cap L^{\perp}$ and $L^{\perp}$ is the orthogonal space of L. Note that, if $L = \{0\}$, then $\Lambda = \Lambda^{\perp}$. Also, note that $\Lambda^{\perp}$ is uniquely determined for Λ.

Theorem 4.1. A necessary and sufficient condition for $y_0 \in \operatorname{Ext}[Y \mid \Lambda]$ is that

(i) $Y \cap (y_0 + L) = \{y_0\}$,
(ii) $y_0^{\perp} \in \operatorname{Ext}[Y^{\perp} \mid \Lambda^{\perp}]$,

where L is the maximum linear space contained by Λ, and $y_0^{\perp}$, $Y^{\perp}$, $\Lambda^{\perp}$ are the projections of y_0, Y, and Λ into $L^{\perp}$, respectively.

Proof. *Necessity.* We shall show that, if (i) or (ii) does not hold, then $y_0 \notin \operatorname{Ext}[Y \mid \Lambda]$.

Suppose that (i) does not hold; clearly, $y_0 \notin \operatorname{Ext}[Y \mid \Lambda]$.

Now, suppose that (ii) does not hold. Then, there are $y^{\perp} \in Y^{\perp}$ and $h^{\perp} \neq 0$, $h^{\perp} \in \Lambda^{\perp}$ such that

$$y_0^{\perp} = y^{\perp} + h^{\perp}. \tag{9}$$

Let $y \in Y$ be a point such that its projection point into $L^{\perp}$ is $y^{\perp}$. Thus, $y = y^{\perp} + y^{L}$, with $y^{L} \in L$. Set $y_0 = y_0^{\perp} + y_0^{L}$, with $y_0^{L} \in L$. By (9), we have

$$y_0 = y^{\perp} + h^{\perp} + y_0^{L} = (y - y^{L}) + h^{\perp} + y_0^{L} = y + (y_0^{L} - y^{L}) + h^{\perp}.$$

Note that, because $h^{\perp} \neq 0$ and $y_0^{L} - y^{L} \in L$, we have $y_0^{L} - y^{L} + h^{\perp} \neq 0$ Thus, $y_0 \neq y$. Also,

$$y_0^{L} - y^{L} + h^{\perp} \in L + h^{\perp} \subset L + \Lambda^{\perp} = \Lambda.$$

We see that $y_0 \in y + \Lambda$ and $y_0 \neq y$. Thus,

$$y_0 \notin \text{Ext}[Y \mid \Lambda].$$

Sufficiency. Suppose that

$$y_0 \notin \text{Ext}[Y \mid \Lambda].$$

We shall see that (i) or (ii) cannot hold. By the assumption, there are $y \in Y$ and $h \neq 0$, $h \in \Lambda$ such that

$$y_0 = y + h. \tag{10}$$

Write

$$y_0 = y_0^{\perp} + y_0^{L}, \qquad y = y^{\perp} + y^{L}, \qquad h = h^{\perp} + h^{L},$$

where

$$y_0^{L}, y^{L}, h^{L} \in L, \qquad y_0^{\perp}, y^{\perp}, h^{\perp} \in L^{\perp}.$$

From (10), we get

$$y_0^{\perp} = y^{\perp} + h^{\perp}. \tag{11}$$

Observe that, $h \neq 0$, $h^{\perp}$ and h^{L} cannot be both zero. Let us consider two possible cases.

Case 1: $h^{\perp} \neq 0$. Then, (ii) cannot hold, because of (11).

Case 2: $h^{\perp} = 0$. Then, $h^{L} = h \neq 0$ and, thus $y_0 = y + h^{L}$ or $y = y_0 - h^{L}$. Thus, (i) cannot hold, because $y \in y_0 + L$.

Corollary 4.1. Suppose that $\Lambda^{\perp} \neq \{0\}$. Then,

$$y_0 \in \text{Ext}[Y \mid \Lambda]$$

implies that y_0 is a boundary point of $Y + \Lambda$.

Proof. By (ii) of Theorem 4.1, we have

$$y_0^{\perp} \in \text{Ext}[Y^{\perp} \mid \Lambda^{\perp}].$$

Note that $\Lambda^{\perp}$ contains no subspace. Thus, by (iv) of Lemma 4.1,

$$y_0^{\perp} \in \text{Ext}[Y^{\perp} + \Lambda^{\perp} \mid \Lambda^{\perp}].$$

By Remark 4.1, we see that y_0 must be a boundary point (with respect to the topology induced by $L^{\perp}$) of $Y^{\perp} + \Lambda^{\perp}$, that implies that y_0 is a boundary point of $Y + \Lambda$.

Remark 4.4. The assumption that $\Lambda^{\perp} \neq \{0\}$ cannot be eliminated. As an example, in R^2 let

$$Y = \{(x, y) | \, y = 1\}, \qquad \Lambda = \{(x, y) | \, x = 0\}.$$

Note that $\Lambda^{\perp} = \{0\}$ and $Y = \mathrm{Ext}[Y \mid \Lambda]$. But $R^2 = Y + \Lambda$, each point of Y is an interior point of $Y + \Lambda$.

Corollary 4.2. Suppose that $\Lambda^{\perp} \neq \{0\}$ and $Y + \Lambda = R^l$. Then, $\mathrm{Ext}[Y \mid \Lambda] = \varnothing$.

Proof. It comes directly from Corollary 4.1.

Remark 4.5. In the example stated in the previous remark, $Y + \Lambda = R^2$. But

$$\Lambda^{\perp} = \{0\}, \qquad \mathrm{Ext}[Y \mid \Lambda] \neq \varnothing.$$

The present Corollary is useful in checking whether

$$\mathrm{Ext}[Y \mid \Lambda] = \varnothing.$$

As an example, in R^2 let

$$Y = \{(x, y) | \, x = y\}, \qquad \Lambda = \{(x, y) | \, y \geqq 0\}.$$

Note that $\Lambda^{\perp} \neq \{0\}$ and $Y + \Lambda = R^2$. Thus,

$$\mathrm{Ext}[Y \mid \Lambda] = \varnothing.$$

4.2. First Method to Locate the Set of All Cone Extreme Points. Given a cone, by Lemma 2.2, we know that Λ^* is a polyhedral cone iff $\bar{\Lambda}$ is a polyhedral cone. Given $y^0 \in Y$, define

$$Y_j(y^0) = \{y \mid y \in Y, H^k \cdot y \geqq H^k \cdot y^0, k = 1,..., q, k \neq j\}. \tag{12}$$

Lemma 4.2. Let Λ be such that Λ^* is a polyhedral cone with a generator $\{H^k \mid k = 1,..., q\}$. Then,

(i) if $y^0 \in \mathrm{Ext}[Y \mid \bar{\Lambda}]$, then, for each j, $1 \leqq j \leqq q$, $H^j \cdot y^0 > H^j \cdot y$ for all $y \in Y_j(y^0)$, $y \neq y^0$;

(ii) if, for some j, $1 \leqq j \leqq q$, $H^j \cdot y^0 > H^j \cdot y$ for all $y \in Y_j(y^0)$, and $y \neq y^0$, then $y^0 \in \mathrm{Ext}[Y \mid \Lambda]$.

Proof. (i) Suppose that there is $y^1 \subset Y_j(y^0)$, $y^1 \neq y^0$, such that

$$H^j \cdot y^0 \leqq H^j \cdot y^1.$$

Then,

$$H^k \cdot (y^0 - y^1) \leqq 0, \qquad k = 1,\ldots, q.$$

Thus, $y^0 - y^1 \in (\Lambda^*)^* = \bar{\Lambda}$ or $y^0 \in y^1 + \bar{\Lambda}$. Since $y^1 \neq y^0$, $y^0 \notin \text{Ext}[Y \mid \bar{\Lambda}]$, resulting in a contradiction.

(ii) Suppose that $y^0 \notin \text{Ext}[Y \mid \Lambda]$. Then, there is $y^1 \in Y$, $y^1 \neq y^0$ such that $y^0 \in y^1 + \Lambda$. Write $y^0 = y^1 + h$, $h \in \Lambda$, $h \neq 0$. Since $\{H^k\} \subset \Lambda^*$,

$$H^k \cdot h \leqq 0 \qquad \text{for all} \quad k = 1,\ldots, q.$$

Thus,

$$H^k \cdot y^0 = H^k \cdot (y^1 + h) = H^k \cdot y^1 + H^k \cdot h \leqq H^k \cdot y^1$$

for all $k = 1,\ldots, q$, resulting in a contradiction.

Immediately from the lemma, we have the following theorem.

Theorem 4.2. Let Λ be such that $\bar{\Lambda}$ is a polyhedral cone. Let $\{H^k \mid 1 \leqq k \leqq q\}$ be a generator of Λ^*. Suppose that

$$\text{Ext}[Y \mid \Lambda] = \text{Ext}[Y \mid \bar{\Lambda}].$$

Then, $y^0 \in \text{Ext}[Y \mid \Lambda]$ iff, for any arbitrary j, $1 \leqq j \leqq q$,

$$H^j y^0 > H^j y$$

for all $y \in Y_j(y^0)$ and $y \neq y^0$.

Corollary 4.3. Suppose that Λ is a polyhedral cone. Let $\{H^k \mid 1 \leqq k \leqq q\}$ be a generator of Λ^*. Then,

$$y^0 \in \text{Ext}[Y \mid \Lambda]$$

iff, for any arbitrary j, $1 \leqq j \leqq q$,

$$H^j y^0 > H^j y$$

for all $y \in Y_j(y^0)$ and $y \neq y^0$.

Remark 4.6. The assumption in Theorem 4.2 cannot be eliminated. As an example, let

$$Y_1 = \left\{(x, y) \middle| \begin{matrix} 0 \leqq x \leqq 1 \\ 0 \leqq y \leqq 2 \end{matrix}\right\},$$

$$Y_2 = \left\{(x, y) \middle| \begin{matrix} 0 \leqq x \leqq 2 \\ 0 \leqq y \leqq 1 \end{matrix}\right\},$$

and

$$Y = Y_1 \cup Y_2 .$$

Suppose that

$$\Lambda = \{(x, y) | \ (x, y) < (0, 0)\} \cup \{(0, 0)\}.$$

Then, Λ^* could be generated by the vectors (1, 0) and (0, 1). Note that

$$\text{Ext}[Y \mid \Lambda] \supset \{(x, y) | \ x = 1, 1 \leqq y \leqq 2\} \cup \{(x, y) | \ y = 1, 1 \leqq x \leqq 2\}.$$

The point

$$(1, 1) \in \text{Ext}[Y \mid \Lambda]$$

could never satisfy the condition of the theorem. However,

$$\text{Ext}[Y \mid \bar{\Lambda}] = \{(1, 2), (2, 1)\}$$

does satisfy the theorem.

By Theorem 4.2, we can transfer the problem of locating $\text{Ext}[Y \mid \Lambda]$ into ordinary mathematical programs. Indeed, if we set

$$M^0(j) = \{y^0 \mid H^j y^0 > H^j y \qquad \text{for all} \quad y \in Y_j(y^0), y \neq y^0\}, \tag{13}$$

we get the following corollary.

Corollary 4.4. If

$$\text{Ext}[Y \mid \Lambda] = \text{Ext}[Y \mid \bar{\Lambda}],$$

or Λ is a polyhedral cone so that $M^0(j)$ can be defined as above, then

$$\text{Ext}[Y \mid \Lambda] = M^0(j), \qquad j = 1,..., q.$$

In order to compute[10] $M^0(j)$, let $\underline{r}_k$ and $\bar{r}_k$ be a lower and an upper bound of $\{H^k \cdot y \mid y \in Y\}$. Note that $\underline{r}_k$ and $\bar{r}_k$ may be $-\infty$ and $+\infty$. Let $r(j)$ be the vector in R^{q-1} representing $\{r_k \in R \mid k \neq j,$

[10] It is difficult to construct $M^0(j)$ directly.

$1 \leqq k \leqq q\}$, and $\underline{r}(j)$ (or $\bar{r}(j)$) representing those $\underline{r}_k$ (or $\bar{r}_k$), $k \neq j$. Given $\underline{r}(j) \leqq r(j) \leqq \bar{r}(j)$, define

$$Y(r(j)) = \{y \in Y \mid H^k \cdot y \geqq r_k, k \neq j, k = 1,..., q\}, \tag{14}$$

and let $y^0(r(j))$ be the unique solution, if any, of the problem

$$\max H^j \cdot y, y \in Y(r(j)).$$

Set

$$Y^0(j) = \{y^0(r(j)) \mid \underline{r}(j) \leqq r(j) \leqq \bar{r}(j)\}. \tag{15}$$

Lemma 4.3. $Y^0(j) = M^0(j)$.

Proof. The fact that $M^0(j) \subset Y^0(j)$ is obvious from the definitions (13) and (15), by setting $r_k = H^k. y_0$, if necessary.

In order to see that $M^0(j) \supset Y^0(j)$, let $y^0 \in Y^0(j)$. Then, for some $r(j)$, $\underline{r}(j) \leqq r(j) \leqq \bar{r}(j)$, $y^0 \in Y(r(j))$ and $H^j \cdot y^0 > H^j y$ for all $y \neq y^0$ and $y \in Y(r(j))$. Since $y^0 \in Y(r(j))$,

$$H^k \cdot y^0 \geqq r_k, \qquad k \neq j.$$

By (12) and (14), $Y_j(y^0) \subset Y(r(j))$. Note that $y^0 \in Y_j(y^0)$. Since y^0 uniquely maximizes $H^j. y$ over $Y(r(j))$, it must also uniquely maximize $H^j. y$ over $Y_j(y^0)$. This shows that $y^0 \in M^0(j)$.

As a consequence of Corollary 4.4 and Lemma 4.3, we have the following theorem.

Theorem 4.3. Let Λ be such that $\bar{\Lambda}$ is a polyhedral cone and $\{H^k \mid 1 \leqq k \leqq q\}$ is a generator of Λ^*. If

$$\text{Ext}[Y \mid \Lambda] = \text{Ext}[Y \mid \bar{\Lambda}],$$

then

$$\text{Ext}[Y \mid \Lambda] = M^0(j) = Y^0(j), j = 1,..., q,$$

where $M^0(j)$ and $Y^0(j)$ are defined by (13) and (15), respectively.

Remark 4.7. Theorem 4.3 essentially converts the problem of finding $\text{Ext}[Y \mid \Lambda]$ into ordinary mathematical programming problems. Observe that we do not make any assumption regarding Y. The only assumption needed is that $\bar{\Lambda}$ is a polyhedral cone and

$$\text{Ext}[Y \mid \Lambda] = \text{Ext}[Y \mid \bar{\Lambda}].$$

Of course, when Λ is a polyhedral cone, the assumption is satisfied.

Remark 4.8. Suppose that $\bar{\Lambda}$ is not a polyhedral cone. In view of Lemma 2.2, Λ^* cannot be a polyhedral cone. In this case, we could construct two polyhedral cones Λ_1 and Λ_2 so that $\Lambda_1 \subset \Lambda \subset \Lambda_2$. Our theorem can then be applied to locate $\mathrm{Ext}[Y \mid \Lambda_1]$ and $\mathrm{Ext}[Y \mid \Lambda_2]$. By invoking (ii) of Lemma 4.1, we get

$$\mathrm{Ext}[Y \mid \Lambda_2] \subset \mathrm{Ext}[Y \mid \Lambda] \subset \mathrm{Ext}[Y \mid \Lambda_1].$$

This observation yields the following corollary.

Corollary 4.5. Let $\{H_1^k \mid k = 1,..., q_1\}$ and $\{H_2^k \mid k = 1,..., q_2\}$ be the generators of Λ_1^* and Λ_2^*, respectively. For any arbitrary i, j such that $1 \leqq i \leqq q_1$ and $1 \leqq j \leqq q_2$, define

$$M_1^0(i) = \{y^0 \mid H_1^i y^0 > H_1^i y \quad \text{for all} \quad y \in Y_{i1}(y^0), y \neq y^0\},$$

where

$$Y_{i1}(y^0) = \{y \mid y \in Y, H_1^k y \geqq H_1^k y^0, k = 1,..., q_1, k \neq i\},$$

and

$$M_2^0(j) = \{y^0 \mid H_2^j y^0 > H_2^j y \quad \text{for all} \quad y \in Y_{j2}(y^0), y \neq y^0\},$$

where

$$Y_{j2}(y^0) = \{y \mid y \in Y, H_2^k y \geqq H_2^k y^0, k = 1,..., q_2, k \neq j\}.$$

Then,

$$M_1^0(i) \supset \mathrm{Ext}[Y \mid \Lambda] \supset M_2^0(j).$$

Remark 4.9. As we did for Lemma 4.3, there is no difficulty in converting $M_1^0(\mathrm{i})$ and $M_2^0(j)$ into ordinary mathematical programs. We shall not stop to do so.

4.3. Second Method to Locate the Set of All Cone Extreme Points. Let Y be a set in R^l and $y_0 \in \bar{Y}$. Suppose that there is a nonzero vector λ in R^l such that $\lambda \cdot y_0 = \alpha$ and $\lambda \cdot y \leqq \alpha$ for all $y \in Y$. Then, y_0 is known as a *supporting point* of Y, and $\lambda \cdot y$, *a supporting functional* of Y at y. Since we are going to emphasize the vector λ, we shall call λ a supporting functional as well. Thus, we see that λ is a supporting functional of Y at y_0 iff

$$\lambda \cdot y_0 = \sup\{\lambda \cdot y \mid y \in Y\}.$$

Lemma 4.4. Let Y be a set and Λ a cone in R^l. Suppose that (i) $y_0 \in Y$ is a supporting point of Y with a supporting functional

$\lambda \in \operatorname{Int} \Lambda^*$; or (ii) $y_0 \in Y$ is the unique supporting point of Y with a supporting functional $\lambda \in \Lambda^*$. Then, $y_0 \in \operatorname{Ext}[Y \mid \Lambda]$.

Proof. Suppose that $y_0 \notin \operatorname{Ext}[Y \mid \Lambda]$. Then, there is $y_1 \in Y$ such that

$$y_0 = y_1 + h, \qquad h \neq 0, h \in \Lambda.$$

When Assumption (i) prevails, since $\lambda \in \operatorname{Int} \Lambda^*$, Λ must be acute and $\lambda \cdot h < 0$ (by Theorem 2.1). Thus,

$$\lambda y_0 = \lambda y_1 + \lambda \cdot h < \lambda y_1,$$

and λ cannot be a supporting function of Y at y_0, resulting in a contradiction. Now, suppose that Assumption (ii) prevails. Since $\lambda \cdot h \leqq 0$, we get

$$\lambda y_0 \leqq \lambda y_1 .$$

Thus, y_0 cannot be the unique supporting point of Y with respect to λ. This leads to a contradiction again.

Now, given $\lambda \in \Lambda^*$, let us define

$$Y^0(\lambda) = \{y_0 \in Y \mid \lambda y_0 = \sup \lambda y, y \in Y\}.$$

Thus, $Y^0(\lambda)$ denotes the set of supporting points of Y in Y, so that λ is a supporting functional at each point of $Y^0(\lambda)$. Note that $Y^0(\lambda)$ may be empty even if $\lambda \in \operatorname{Int} \Lambda^*$.

Corollary 4.6. The following results hold:

(i) $\bigcup \{Y^0(\lambda) \mid \lambda \in \operatorname{Int} \Lambda^*\} \subset \operatorname{Ext}[Y \mid \Lambda]$;

(ii) $\operatorname{Ext}[Y \mid \Lambda] \neq \varnothing$ if there is $y_0 \in Y$ so that y_0 is the unique supporting point of Y with respect to a supporting functional $\lambda \in \Lambda^*$;

(iii) $\operatorname{Ext}[Y \mid \Lambda] \neq \varnothing$ if Y is compact and Λ is acute.

Proof. (i) and (ii) come immediately from Lemma 4.4.

(iii) Since Λ is acute, $\operatorname{Int} \Lambda^* \neq \varnothing$ (by Theorem 2.1). By compactness of Y, for each $\lambda \in \operatorname{Int} \Lambda^*$, we could find at least a supporting point in Y. In view of (i), Assertion (iii) is clear.

Remark 4.10. The Assumptions in (iii) of the corollary may not be eliminated as shown by the following two examples.

Example 4.1. In R^2, let

$$Y = \{(x, y) \mid x, y \geqq 0\}, \qquad \Lambda = \{(x, y) \mid x \leqq 0, y \leqq 0\}.$$

Note that both Y and Λ are closed and convex. Λ is acute, but Y is not compact. We see that $\mathrm{Ext}[Y \mid \Lambda] = \varnothing$.

Example 4.2. In R^2, let

$$Y = \{(x, y) \mid 0 \leqq x \leqq 1, 0 \leqq y \leqq 1\}, \qquad \Lambda = \{(x, y) \mid x \leqq 0\}.$$

Note that Y is compact, but Λ is not acute. We see that $\mathrm{Ext}[Y \mid \Lambda] = \varnothing$.

Lemma 4.5. Let Y be a set and Λ with $\Lambda^{\perp} \neq \{0\}$, a cone in R^l. Suppose that Y is Λ-convex. Then, every point of $\mathrm{Ext}[Y \mid \Lambda]$ is a supporting point of Y with a supporting functional in Λ^*.

Proof. Let $y_0 \in \mathrm{Ext}[Y \mid \Lambda]$. From Corollary 4.1, we see that it is necessary that y_0 is on the boundary of $Y + \Lambda$. By assumption, $Y + \Lambda$ is convex. So is $\overline{Y + \Lambda}$. Thus, there is a supporting functional λ of $\overline{Y + \Lambda}$ at y_0 (for the existence, see Chapter 3, p. 103, of Ref. 12). Since $\lambda y_0 \geqq \lambda y$ for all $y \in \overline{Y + \Lambda}$, we see that $\lambda \in \Lambda^*$. Otherwise, let $h \in \Lambda$ such that $\lambda \cdot h > 0$. Then, $\lambda \cdot (y_0 + h) > \lambda y_0$ would lead to a contradiction. Since $Y \subset Y + \Lambda \subset \overline{Y + \Lambda}$, we see that λ is also a supporting functional of Y.

In view of Lemmas 4.4 and 4.5 we have the following corollary.

Corollary 4.7.[11] Suppose that Y is Λ-convex and $\Lambda^{\perp} \neq \{0\}$. Then,

(i) $\bigcup \{Y^0(\lambda) \mid \lambda \in \mathrm{Int}\, \Lambda^*\} \subset \mathrm{Ext}[Y \mid \Lambda] \subset \bigcup \{Y^0(\lambda) \mid \lambda \in \Lambda^*, \lambda \neq 0\}$;

(ii) if for all boundary points of Λ^*, except 0, $Y^0(\lambda)$ is either empty or contains only a single point, then

$$\mathrm{Ext}[Y \mid \Lambda] = \bigcup \{Y^0(\lambda) \mid \lambda \in \Lambda^*, \lambda \neq 0\}.$$

Remark 4.11. When Y is not Λ-convex, (i) is not generally true. Consider the following example. In R^2, let

$$\begin{aligned} Y_1 &= \{(x, y) \mid 2x + y \leqq 5, \quad 0 \leqq x \leqq 2, \quad 0 \leqq y \leqq 3\}, \\ Y_2 &= \{(x, y) \mid x + 2y \leqq 5, \quad 0 \leqq x \leqq 3, \quad 0 \leqq y \leqq 2\}, \\ Y &= Y_1 \cup Y_2, \\ \Lambda &= \{(x, y) \mid x, y \leqq 0\}. \end{aligned}$$

[11] (i) of this corollary is a generalization of the decision theory theorem that every admissible strategy is Bayes, and that every strategy which is Bayes against positive weights is admissible.

We see that

$$\operatorname{Ext}[Y \mid \Lambda] = \{(x, y) |\ 2x + y = 5,\ 1 \leqq x \leqq 5/3\},$$
$$\cup \{(x, y) |\ x + 2y = 5,\ 5/3 \leqq x \leqq 3\}.$$

But (see Fig. 1)

$$\bigcup \{Y(\lambda) |\ \lambda \in \Lambda^*,\ \lambda \neq 0\} = \{(1, 3), (3, 1)\}.$$

Remark 4.12. The condition that $\Lambda^\perp \neq \{0\}$ in Corollary 4.7 cannot be eliminated. As an example, in R^2 let (see Fig. 2)

$$Y = \{(x, y) |\ x + y \leqq 0,\ x - y \leqq 0\} \cup \{(x, y) |\ x + y \geqq 0,\ x - y \geqq 0\},$$
$$\Lambda = \{(x, y) |\ x = 0\}.$$

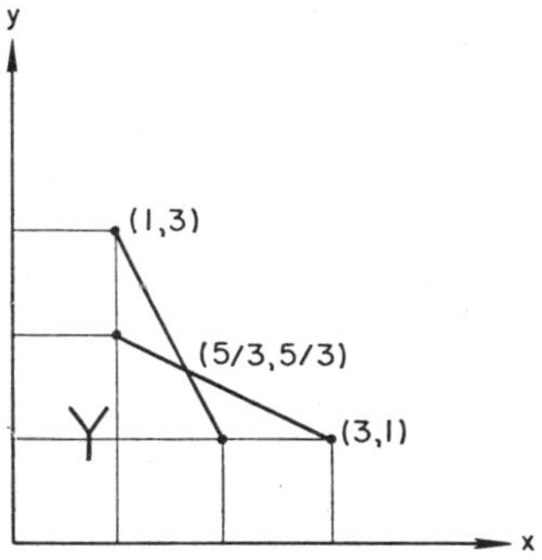

Fig. 1

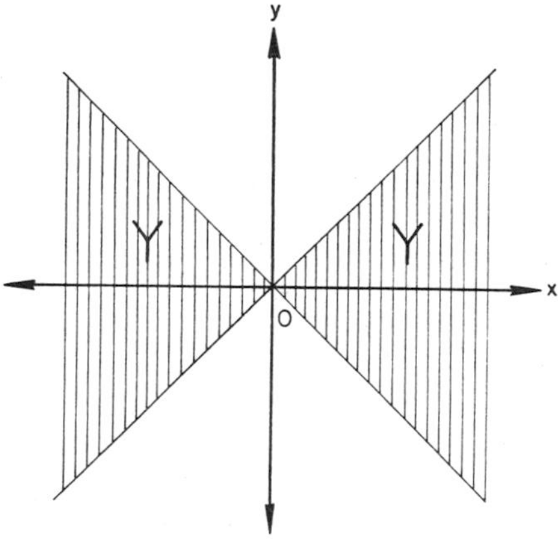

Fig. 2

Note that Y is represented by the shaded area, and

$$\mathrm{Ext}[Y \mid \Lambda] = \{(0, 0)\}.$$

Also,

$$\Lambda^* = \{(x, y) \mid y = 0\}, \qquad Y^0(\lambda) = \varnothing$$

for each $\lambda \in \Lambda^*$. We see that (i) of Corollary 4.7 cannot hold. This is because $\Lambda^\perp = \{0\}$.

Let Y and $\mathscr{D} = \{D_k \mid k = 1,..., m\}$ be sets in R^l. We say that $\mathscr{D}$ is a *partition* of Y if

$$\bigcup_{k=1}^{m} D_k = Y \qquad \text{and} \qquad D_i \cap D_j = \varnothing \qquad \text{if} \quad i \neq j.$$

Definition 4.2. A set Y is *piecewise Λ-convex* if there is a finite partition $\mathscr{D} = \{D_k \mid k = 1,..., m\}$ of Y such that each D_k is Λ-convex.

Corollary 4.8. The following results hold:

(i) let $\mathscr{D} = \{D_k \mid k = 1, 2,..., m\}$ be a partition of Y. Then,

$$\bigcup \{Y(\lambda) \mid \lambda \in \mathrm{Int}\, \Lambda^*\} \subset \mathrm{Ext}[Y \mid \Lambda] \subset \bigcup_{k=1}^{m} \mathrm{Ext}[D_k \mid \Lambda].$$

(ii) if $\Lambda^\perp \neq \{0\}$ and Y is piecewise Λ-convex with respect to $\mathscr{D}$, then

$$\bigcup_{k=1}^{m} \mathrm{Ext}[D_k \mid \Lambda] \subset \bigcup_{k=1}^{m} [\bigcup \{Y_k^0(\lambda) \mid \lambda \in \Lambda^*, \lambda \neq 0\}],$$

where

$$Y_k^0(\lambda) = \{y_0 \in D_k \mid \lambda y_0 = \sup \lambda y, y \in D_k\}.$$

Proof. In view of Corollaries 4.6 and 4.7, it suffices to show that

$$\mathrm{Ext}[Y \mid \Lambda] \subset \bigcup_{k=1}^{m} \mathrm{Ext}[D_k \mid \Lambda].$$

Let $y^1 \in \mathrm{Ext}[Y \mid \Lambda]$. Then, $y^1 \in D_k$ for some k. It is seen that $y^1 \in \mathrm{Ext}[D_k \mid \Lambda]$.

Remark 4.13. In locating $\mathrm{Ext}[Y \mid \Lambda]$, Corollary 4.8 could be used to resolve some difficulties of non Λ-convexity. Consider the example in Remark 4.11. We could set

$$D_1 = \{(x, y)|\ 2x + y \leqq 5,\quad 0 \leqq x \leqq 5/3,\quad 0 \leqq y < 3\},$$

$$D_2 = \{(x, y)|\ x + 2y \leqq 5,\quad 5/3 \leqq x \leqq 3,\quad 0 \leqq y\}.$$

It is seen that $\{D_1, D_2\}$ is a partition of Y, and each D_k, $k = 1, 2$, is Λ-convex. We see that

$$\text{Ext}[D_1 \mid \Lambda] = \{(x, y)|\ 2x + y = 5,\quad 1 \leqq x \leqq 5/3\},$$

$$\text{Ext}[D_2 \mid \Lambda] = \{(x, y)|\ x + 2y = 5,\quad 5/3 \leqq x \leqq 3\},$$

and thus

$$\text{Ext}[Y \mid \Lambda] = \bigcup_{k=1,2} \text{Ext}[D_k \mid \Lambda].$$

5. Nondominated Solutions

5.1. Structure of Domination and Nondominated Solutions in Objective Space. Recall, from Section 1, that we have a decision space X and an objective space $Y = f[X]$. In order to avoid confusion, we will use x and y to represent the elements of X and Y, respectively, unless otherwise stated.

Given two outcomes y_1 and y_2 in Y, $y_1 \neq y_2$, if y_2 is preferred to y_1 we will denote it by $y_2 \succ y_1$.

Definition 5.1. A nonzero vector $d \in R^l$ is a *domination factor* for $y \in Y$ iff $y_1 = y + \lambda d$ and $\lambda > 0$ implies that $y \succ y_1$. The set of all domination factors for y, together with the zero vector in R^l, will be denoted by $D(y)$. The family $\{D(y)|\ y \in Y\}$ will be called the *structure of domination* of our decision problem. For simplicity, the structure will be denoted by $D(\cdot)$.

Note that $y + D(y)$ may not be contained by Y.

Assumption 5.1. For each $y \in Y$, $D(y)$ is a convex set. Thus, by Definition 5.1, it is also a convex cone.

Remark 5.1. From Definition 5.1, given a domination factor for y, then any fraction of it or any positive multiple of it is also a domination factor. Assumption 5.1 states that $D(y)$ is a convex cone. Thus, given $d_1, d_2 \in D(y)$ and $\lambda_1, \lambda_2 \geqq 0$, then

$$\lambda_1 d_1 + \lambda_2 d_2 \in D(y).$$

Essentially it says that the sum of two domination factors for y is also a domination factor for y. Definition 5.1 and Assumption 5.1, thus, coincide with our intuition that, given a *bad factor*, then any part of it, or multiple of it, is also bad and that the sum of two bad factors must be bad too. In view of this intuitive concept, we see that our assumption is not too restrictive.

Example 5.1. When $l = 2$, let $y = (y_1, y_2)$. Suppose that the decision maker has a differentiable utility function $u(y)$ such that $\nabla u(y) \cdot (y' - y) \leqq 0$ implies that $u(y') < u(y)$. Note that this implies that $u(y)$ is a pseudo-concave function. As an example, $u(y) = y_1 y_2$ for $y_1, y_2 > 0$. Let the isovalued curves $\{I_p\}$ of $u(y)$ be as depicted in Fig. 3. We see that, for each $y \in Y$, we could assign

$$D(y) = \{d \mid \nabla u(y) \cdot d \leqq 0\}.$$

Observe that, although in Fig. 3 $y^0 > y'$, $y' - y^0 \notin D(y^0)$. Also, observe that $D(y)$ is not a constant cone for all $y \in Y$. Note that, usually, the utility function is unknown. This example is used to explain how the structure of domination could be obtained and to show a relation between a utility function and its induced domination structure. Note that, given a utility function defined over Y, we can always construct its structure of domination; the converse is not generally true.

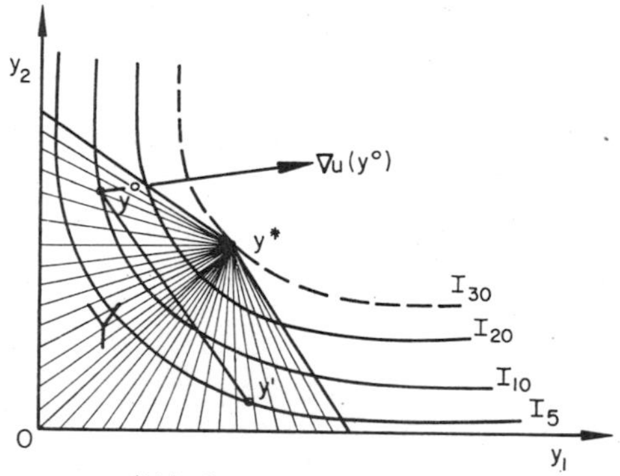

(Y is the shaded region.)

Fig. 3

Definition 5.2. Given Y and $D(\cdot)$, a point $y_0 \in Y$ is a *nondominated solution* with respect to $D(\cdot)$ if there is no $y_1 \in Y$, $y_1 \neq y_0$, such that $y_0 \in y_1 + D(y_1)$. Likewise, given X and $f(x)$, a point $x_0 \in X$ is a *nondominated solution* if there is no x_1 in X so that

$$f(x_0) \neq f(x_1) \qquad \text{and} \qquad f(x_0) = f(x_1) + D(f(x_1)).$$

The set of all nondominated solutions in the decision and objective spaces will be denoted by $N_X(D(\cdot))$ and $N_Y(D(\cdot))$, respectively.

Note that, in decision-making problems with multiobjectives, obviously none of the dominated solutions

$$\{y \mid y \in Y, y \notin N_Y(D(\cdot))\}$$

are desirable. Since all good decisions must be contained in $N_X(D(\cdot))$ or $N_Y(D(\cdot))$, to find $N_X(D(\cdot))$ and $N_Y(D(\cdot))$, if they are nonempty, essentially is the first step toward good decision.

Example 5.2. In Fig. 3 (Example 5.1), clearly y^* is the unique nondominated solution, which is also the optimal point with respect to the given utility function.

Example 5.3. Suppose that there is a constant vector $\boldsymbol{a}$ such that, if $y_1, y_2 \in Y$ and $a \cdot y_1 > a \cdot y_2$, then $y_1 \succ y_2$. Note that this does not imply that we have a linear utility function over Y. We could set

$$D(y) = \{0\} \cup \{d \mid a \cdot d < 0\} = D$$

for all $y \in Y$. Note that D is a cone and $N_Y(D(\cdot)) = \mathrm{Ext}[Y \mid D]$. If $N_Y(D(\cdot)) \neq \varnothing$, it would not be difficult to find it, at least theoretically (because it reduces to a mathematical programming problem).

Example 5.4. Suppose that

$$D(y) = \{d \mid d \leqq 0, d \in R^l\} = \Lambda^{\leqq}$$

for all $y \in Y$. By Sections 1 and 4, we see immediately that

$$N_Y(D(\cdot)) = P^0 = \mathrm{Ext}[Y \mid \Lambda^{\leqq}] = \mathrm{Ext}[f[X] \mid \Lambda^{\leqq}].$$

Recall P^0 is the set of all Pareto-optimal solutions.

From the above examples, it is not difficult to see that the concepts of one-objective optimization and of Pareto optimality are some special cases of our domination structure.

Given $y \in Y$, define

$$Y_y = [y + D(y)] \cap Y.$$

Thus, Y_y is the subset of Y which contains y and those points which are dominated by y.

Lemma 5.1. If each $D(y)$ is convex and contains no nontrivial subspace, then

(i) given $y_0 \in Y$, then $\{y_0\} = \text{Ext}[Y_{y_0} \mid D(y_0)]$;

(ii) given $N_0 \subset N_Y(D(\cdot))$ and $Y_{N_0} = \bigcup \{Y_y \mid y \in N_0\}$, then

$$N_0 \subset \text{Ext}[Y_{N_0} \mid \Lambda],$$

where

$$\Lambda = \bigcap \{D(y) \mid y \in N_0\};$$

(iii) if $Y = \bigcup \{Y_y \mid y \in N_Y(D(\cdot))\}$, then

$$N_Y(D(\cdot)) \subset \text{Ext}[Y \mid \Lambda],$$

where

$$\Lambda = \bigcap \{D(y) \mid y \in N_Y(D(\cdot))\};$$

(iv) $N_Y(D(\cdot)) \subset \text{Ext}[Y \mid \Lambda^0]$, where $\Lambda^0 = \bigcap \{D(y) \mid y \in Y\}$.

The no subspace assumption is not needed for (iv).

Proof. (i) Since every point of Y_{y_0} other than y_0 cannot be a point of $\text{Ext}[Y_{y_0} \mid D(y_0)]$, it suffices to show that

$$y_0 \in \text{Ext}[Y_{y_0} \mid D(y_0)].$$

Suppose that

$$y_0 \notin \text{Ext}[Y_{y_0} \mid D(y_0)].$$

Then, there is

$$y^1 = y_0 + d^1, \qquad d^1 \neq 0,\ d^1 \in D(y_0),$$

such that

$$y_0 = y^1 + d_0, \qquad d_0 \neq 0,\ d_0 \in D(y_0).$$

Thus,

$$y_0 = y_0 + d^1 + d_0.$$

Since $D(y_0)$ contains no subspace, $d^1 + d_0 \neq 0$, resulting in a contradiction.

(ii) Let $y_0 \in N_0 \subset Y_{N_0}$. Suppose that

$$y_0 \notin \mathrm{Ext}[Y_{N_0} \mid \Lambda].$$

Then, there is $y_1 \in Y_{N_0}$, $y_1 \neq y_0$, such that $y_0 \in y_1 + \Lambda$. Consider two possible cases.

Case 1: $y_1 \in N_0$. Then, since

$$y_0 \in y_1 + \Lambda \subset y_1 + D(y_1),$$

we have

$$y_0 \notin N_Y(D(\cdot)),$$

resulting in a contradiction, because $N_0 \subset N_Y(D(\cdot))$.

Case 2: $y_1 \notin N_0$. Then, there is $y_2 \in N_0$ such that $y_1 \in y_2 + D(y_2)$. Since $y_0 \in y_1 + \Lambda$,

$$y_0 \in y_2 + D(y_2) + \Lambda = y_2 + D(y_2),$$

because $D(y_2)$ is a convex cone and $\Lambda \subset D(y_2)$. Thus, $y_0 \notin N_Y(D(\cdot))$, again resulting in a contradiction. Note that $y_0 \neq y_2$ because $D(y_2)$ contains no subspace.

(iii) This part of the proof is a special case of (ii). We shall not repeat it.

(iv) Suppose that $y_0 \notin \mathrm{Ext}[Y \mid \Lambda^0]$. Then, for some $y_1 \neq y_0$,

$$y_0 = y_1 + \Lambda \subset y_1 + D(y_1).$$

Thus, $y_0 \notin N_Y(D(\cdot))$.

Remark 5.2. The assumption in (iii) of Lemma 5.1 cannot be eliminated. As an example, let $Y = Y_1 \cup Y_2$, with

$$Y_1 = \{(y_1, y_2) \mid y_1, y_2 \leqq 0\}, \qquad Y_2 = \{y_1, y_2) \mid y_1 = 1, y_2 < 1\},$$

and

$$D((y_1, y_2)) = \begin{cases} \{(d_1, d_2) \mid d_1, d_2 \leqq 0\} & \text{if } (y_1, y_2) \in Y_1, \\ \{(d_1, d_2) \mid d_1 = 0, d_2 \leqq 0\} & \text{if } (y_1, y_2) \in Y_2. \end{cases}$$

We see that

$$N_Y(D(\cdot)) = \{(0, 0)\},$$

$$\Lambda = \bigcap \{D(y) \mid y \in N_Y(D(\cdot))\} = \{(d_1, d_2) \mid d_1, d_2 \leqq 0\},$$

$$\mathrm{Ext}[Y \mid \Lambda] = \varnothing.$$

Thus, $\mathrm{Ext}[Y \mid \Lambda]$ does not contain $N_Y(D(\cdot))$

Now, suppose that

$$\Lambda^0 = \bigcap \{D(y) \mid y \in Y\} \neq \{0\}.$$

For $n = 0, 1, 2, \ldots$, we could construct two sequences $\{Y^n\}$ and $\{\Lambda^n\}$ as follows:

$$Y^{n+1} = \mathrm{Ext}[Y^n \mid \Lambda^n],$$

where

$$\Lambda^n = \bigcap \{D(y) \mid y \in Y^n\} \quad \text{and} \quad Y^0 = Y.$$

Since $Y^{n+1} \subset Y^n$, $\{Y^n\}$ has a limit

$$\tilde{Y} = \bigcap \{Y^n \mid 0 \leqq n < \infty\}.$$

Lemma 5.2. The following results hold:

(i) For each n, $Y^n \supset N_Y(D(\cdot))$;

(ii) $\tilde{Y} \supset N_Y(D(\cdot))$.

Proof. (i) We prove (i) by induction. In view of (iv) of Lemma 5.1, it suffices to show that, if $Y^n \supset N_Y(D(\cdot))$, then $Y^{n+1} \supset N_Y(D(\cdot))$. Suppose that

$$y^1 \in N_Y(D(\cdot)) \cap Y^n,$$

but

$$y^1 \notin Y^{n+1} = \mathrm{Ext}[Y^n \mid \Lambda^n].$$

There is $y^0 \in Y^n \subset Y$ such that $y^1 \neq y^0$ and

$$y^1 \in y^0 + \Lambda^n \subset y^0 + D(y^0).$$

But this shows that $y^1 \notin N_Y(D(\cdot))$, resulting in a contradiction.

(ii) Since $Y^n \supset Y^{n+1}$,

$$\tilde{Y} = \bigcap \{Y^n \mid 0 \leqq n < \infty\}.$$

The conclusion follows immediately from (i).

Remark 5.3. In the above procedure, because $\Lambda^{n+1} \supset \Lambda^n$, by Lemma 4.1

$$\mathrm{Ext}[Y \mid \Lambda^{n+1}] \subset \mathrm{Ext}[Y \mid \Lambda^n].$$

By induction on n, one has no difficulty in showing that

$$\mathrm{Ext}[Y \mid \Lambda^n] \subset \mathrm{Ext}[Y^n \mid \Lambda^n]$$

for $n = 1, 2, \ldots$. However, it is not generally true that

$$\mathrm{Ext}[Y \mid \Lambda^n] = \mathrm{Ext}[Y^n \mid \Lambda^n].$$

Consider the example given in Remark 5.2. We see that

$$\Lambda^0 = \{(d_1, d_2) \mid d_1 = 0, d_2 \leqq 0\},$$

$$Y^1 = \mathrm{Ext}[Y \mid \Lambda^0] = \{(y_1, y_2) \mid y_1 \leqq 0, y_2 = 0\}.$$

Thus,

$$\Lambda^1 = \{(d_1, d_2) \mid d_1, d_2 \leqq 0\}, \qquad \mathrm{Ext}[Y^1 \mid \Lambda^1] = \{(0, 0)\} = N_Y(D(\cdot)).$$

However,

$$\mathrm{Ext}[Y \mid \Lambda^1] = \varnothing, \qquad \mathrm{Ext}[Y \mid \Lambda^1] \neq \mathrm{Ext}[Y^1 \mid \Lambda^1].$$

Remark 5.4. The above procedure yields a sequential approximation method toward $N_Y(D(\cdot))$. Although $N_Y(D(\cdot)) \subset \tilde{Y}$, it is not generally true that $N_Y(D(\cdot)) = \tilde{Y}$ even if some strong conditions are satisfied. As an example, in R^2 let

$$Y = \{(y_1, y_2) \mid -1 \leqq y_1 \leqq 1, -1 \leqq y_2 \leqq 1\},$$

$$D((y_1, y_2)) = \begin{cases} \{(d_1, d_2) \mid d_1, d_2 \leqq 0\} & \text{if } y_1 \leqq 0, \\ \{(d_1, d_2) \mid d_1 \leqq 0, d_2 \leqq [y_1/(1 - y_1)]d_1\} & \text{if } 0 < y_1 < 1, \\ \{(d_1, d_2) \mid d_1 = 0, d_2 \leqq 0\} & \text{if } y_1 = 1. \end{cases}$$

See Fig. 4 for $D((y_1, y_2))$, $0 < y_1 < 1$.

We see that

$$\Lambda^0 = \bigcap \{D((y_1, y_2)) \mid (y_1, y_2) \in Y\} = \{(d_1, d_2) \mid d_1 = 0, d_2 \leqq 0\}$$

and

$$Y^1 = \mathrm{Ext}[Y \mid \Lambda^0] = \{(y_1, y_2) \mid -1 \leqq y_1 \leqq 1, y_2 = 1\}.$$

note that

$$\Lambda^1 = \bigcap \{D((y_1, y_2)) \mid (y_1, y_2) \in Y^1\} = \Lambda^0$$

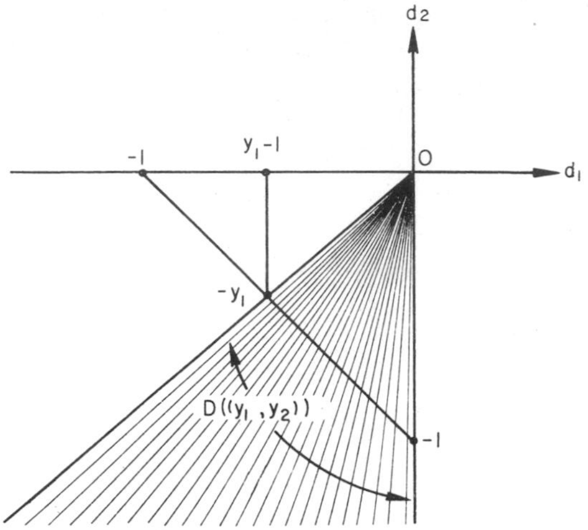

Fig. 4

and

$$Y^2 = \mathrm{Ext}[Y^1 \mid \Lambda^1] = Y^1.$$

It is seen that

$$\tilde{Y} = Y^1 = \{(y_1, y_2) \mid -1 \leqq y_1 \leqq 1, y_2 = 1\}.$$

Since

$$N_Y(D(\cdot)) = \{(y_1, y_2) \mid 0 \leqq y_1 \leqq 1, y_2 = 1\},$$

we have

$$N_Y(D(\cdot)) \subset \tilde{Y} \quad \text{but} \quad \tilde{Y} \not\subset N_Y(D(\cdot)).$$

Note that $D(y_1, y_2)$ is continuous over Y in the sense that, given a sequence

$$\{X_n = (y_{1n}, y_{2n})\} \to X_0 \in Y,$$

then

$$\lim\sup D(X_n) = D(X_0) = \lim\inf D(X_n).$$

The example shows that even if Y is a square and $D((y_1, y_2))$ is continuous, the limit set of our deduction procedure is not necessarily equal to the set of all nondominated solutions. Thus, when we get $\tilde{Y}$, it is still necessary to check whether $\tilde{Y} \subset N_Y(D(\cdot))$. The definition, as well as the results in Lemma 5.1, would be useful for the checking.

5.2. Nondominated Solutions in Decision Space: First Locating Method. In our decision problems, finding $N_X(D(\cdot))$ in the decision space is at least as important as finding $N_Y(D(\cdot))$ in the objective space. As shown in the previous section, we can use sequentially the sets of cone extreme points to approximate the set of all nondominated solutions. In this section, we shall discuss a first method to locate the cone extreme points in the decision space. We will also discuss some conditions for a nondominated solution in the decision space to be satisfied. This section is closely related with Section 4.2. In fact, our main results are based on those of Section 4.2.

Given a convex cone Λ in R^l, we define the set of all Λ-extreme points in the decision space by

$$X^0(\Lambda) = \{x \in X \mid f(x) \in \mathrm{Ext}[Y \mid \Lambda]\}.$$

We see that

$$\mathrm{Ext}[Y \mid \Lambda] = f[X^0(\Lambda)] = \{f(x) \mid x \in X^0(\Lambda)\}.$$

In order to make our work more precise, let us define the decision space by[12]

$$X = \{x \in R^n \mid g_i(x) \leqq 0,\ i = 1, 2,\dots, m\},$$

or

$$X = \{x \in R^n \mid g(x) \leqq 0\}, \qquad \text{with} \quad g = (g_1, g_2, \dots, g_m).$$

Let Λ be such that $\bar{\Lambda}$ is a polyhedral cone. By Lemma 2.2, we know that Λ^* is also a polyhedral cone. Let

$$\{H^k \mid k = 1,\dots, q\}$$

be a generator of Λ^*. Let $\underline{r}_k$ and $\bar{r}_k$ be a lower and an upper bound for

$$\{H^k \cdot f(x) \mid x \in X\},\ k = 1,\dots, q.$$

Corresponding to $Y(r(j))$, $y^0(r(j))$, and $Y^0(j)$ in Section 4.2, given $r(j)$, $\underline{r}(j) \leqq r(j) \leqq \bar{r}(j)$, we define

$$X(r(j)) = \{x \in X \mid H^k \cdot f(x) \geqq r_k,\ k \neq j,\ 1 \leqq k \leqq q\},$$

$$X^0(r(j)) = \left\{ x^0 \in X(r(j)) \;\middle|\; \begin{array}{l} H^j \cdot f(x^0) \geqq H^j \cdot f(x) \quad \text{for all} \quad x \in X(r(j)), \\ \text{equality holds only when } f(x) = f(x^0) \end{array} \right\}$$

$$X^0(j) = \bigcup \{X^0(r(j)) \mid \underline{r}(j) \leqq r(j) \leqq \bar{r}(j)\}.$$

---

[12] It is not difficult to extend the results to the case when X is defined by equality as well as inequality constraints (see Ref. 14). We shall not stop to do so.

Note that

$$Y(r(j)) = f[X(r(j))] = \{f(x) \mid x \in X(r(j))\},$$

$$\{y^0(r(j))\} = f[X^0(r(j))] \qquad \text{and} \qquad Y^0(j) = f[X^0(j)].$$

Thus, corresponding to Theorem 4.3 we have the following theorem.

Theorem 5.1. Let Λ be such that $\bar{\Lambda}$ is a polyhedral cone and $\{H^k \mid 1 \leqq k \leqq q\}$ is a generator of Λ^*. If

$$\operatorname{Ext}[Y \mid \Lambda] = \operatorname{Ext}[Y \mid \bar{\Lambda}],$$

then

$$X^0(\Lambda) = X^0(j), \qquad j = 1, 2, \ldots, q.$$

In view of Lemma 5.1 (iv), we have the following corollary.

Corollary 5.1. Let

$$\Lambda = \bigcap \{D(y) \mid y \in Y\}.$$

Suppose that $\bar{\Lambda}$ is a polyhedral cone, with $\{H^k \mid 1 \leqq k \leqq q\}$ being a generator of Λ^*, and

$$\operatorname{Ext}[Y \mid \Lambda] = \operatorname{Ext}[Y \mid \bar{\Lambda}].$$

Then, $x^0 \in N_X(D(\cdot))$ only if

$$x^0 \in X^0(\Lambda) = X^0(j), \qquad j = 1, 2, \ldots, q.$$

Remark 5.5. Observe that, if $x^0 \in X^0\ (r(j))$, then x^0 maximizes $H^j \cdot f(x)$ over the set $X(r(j))$. Thus, we have reduced the problem of locating $X^0(\Lambda)$, as well as $N_X(D(\cdot))$, to a family of mathematical programs. In order to find the entire set of $X^0(j)$, the parameter $r(j)$ of our mathematical programs must vary from $\underline{r}(j)$ to $\bar{r}(j)$ [treat $\underline{r}_k$ and $\bar{r}_k$ as the g.l.b. and l.u.b. of $H^k \cdot f(x)$ over X, if necessary]. This suggests that, given a polyhedral cone Λ^*, in order to save computation, it is a good idea to select a set containing a minimal number of vectors as its generator.

Observe that $X(r(j))$ could also be written as

$$X(r(j)) = \left\{ x \in R^n \,\middle|\, \begin{array}{l} g_i(x) \leqq 0,\ i = 1, \ldots, m \\ r_k - H^k \cdot f(x) \leqq 0,\ k \neq j,\ k = 1, \ldots, q \end{array} \right\}.$$

Let

$$L = \mu_j H^j \cdot f(x) - \sum_{k \neq j} \mu_k(r_k - H^k \cdot f(x)) - \sum_{i=1}^{m} \lambda_i g_i(x),$$

or

$$L = \sum_{k=1}^{q} \mu_k H^k \cdot f(x) - \sum_{i=1}^{m} \lambda_i g_i(x) - \sum_{k \neq j} \mu_k r_k , \tag{16}$$

where μ_k, λ_i, $1 \leqq k \leqq q$ and $1 \leqq i \leqq m$, are real numbers.

By Corollary 5.1 and Fritz John's theorem (see Ref. 14), we have the following theorem.

Theorem 5.2. In addition to the assumptions made in Corollary 5.1, suppose that $g(x)$ and $f(x)$ are differentiable over X [and so are over $X(r(j))$, $\underline{r}(j) \leqq r(j) \leqq \bar{r}(j)$]. Then, a necessary condition for $x^0 \in N_X(D(\cdot)) \subset X^0(\Lambda)$ is that there exists $r(j) \in R^{q-1}$, $\underline{r}(j) \leqq r(j) \leqq \bar{r}(j)$ $\mu_k \geqq 0$, $k = 1,\ldots, q$, and $\lambda_i \geqq 0$, $i = 1,\ldots, m$, not all μ_k and λ_i are zeros, such that, with respect to (16), the following conditions are satisfied:[13]

$$\nabla_x L(x^0) = 0, \tag{17}$$

$$\mu_k(r_k - H^k \cdot f(x^0)) = 0, \tag{18}$$

$$r_k - H^k \cdot f(x^0) \leqq 0, \qquad k = 1,\ldots, q, k \neq j, \tag{19}$$

$$\lambda_i \cdot g_i(x^0) = 0, \tag{20}$$

$$g_i(x^0) \leqq 0, \qquad i = 1,\ldots, m. \tag{21}$$

Recall that in R^l,

$$(\Lambda^{\leqq})^* = \Lambda^{\geqq},$$

and that $\{e^k \mid k = 1,\ldots, l\}$ is a generator for $\Lambda^{\geqq}$ (see the derivation for Corollary 3.5). Note that

$$e^k \cdot f(x) = f_k(x).$$

Thus, if

$$\Lambda = \Lambda^{\leqq},$$

we could write (16) as

$$\bar{L} = \sum_{k=1}^{q} \mu_k f_k(x) - \sum_{i=1}^{m} \lambda_i g(x) - \sum_{k \neq j} \mu_k r_k . \tag{22}$$

[13] $\nabla_x L(x^0)$ is the gradient of L w.r.t. x at x^0.

The conditions of (17)–(21) are reduced to [14]

$$\nabla_x \bar{L}(x^0) = 0, \tag{23}$$

$$\mu_k(r_k - f_k(x^0)) = 0, \tag{24}$$

$$r_k - f_k(x^0) \leqq 0, \qquad k = 1,..., l, \; k \neq j, \tag{25}$$

$$\lambda_i \cdot g_i(x^0) = 0, \qquad i = 1,..., m, \tag{26}$$

$$g_i(x^0) \leqq 0. \tag{27}$$

Corollary 5.2. Suppose that

$$\bigcap \{D(y) \mid y \in Y\} \supset \Lambda^{\leqq}$$

and that $f(x)$ and $g(x)$ are differentiable over X. Then, a necessary condition for

$$x^0 \in N_X(D(\cdot)) \subset X^0(\Lambda^{\leqq})$$

is that there are $r(j) \in R^{l-1}$, $\underline{r}(j) \leqq r(j) \leqq \bar{r}(j)$ [the components of $\underline{r}(j)$ and $\bar{r}(j)$ are the lower and upper bounds of $f_k(x)$ over X], $\mu_k \geqq 0$, $k = 1,..., l$, and $\lambda_i \geqq 0$, $i = 1,..., m$, not all μ_k and λ_i zeros, such that the conditions of (23)–(27) are satisfied.

Remark 5.6. In view of Example 5.4, Corollary 5.2 gives a necessary condition for a point in the decision space to be Pareto optimal.

Now, suppose that

$$\Lambda = \bigcap \{D(y) \mid y \in Y\}$$

is such that $\bar{\Lambda}$ is not a polyhedral cone. Then, by Lemma 2.2, Λ^* is not a polyhedral cone. In this case, as we did for Corollary 4.5, we could construct two polyhedral cones Λ_1 and Λ_2 so that $\Lambda_1 \subset \Lambda \subset \Lambda_2$.

Since

$$\text{Ext}[Y \mid \Lambda_2] \subset \text{Ext}[Y \mid \Lambda] \subset \text{Ext}[Y \mid \Lambda_1],$$

we have

$$X^0(\Lambda_2) \subset X^0(\Lambda) \subset X^0(\Lambda_1).$$

This observation yields the following corollary.

Corollary 5.3. Suppose that

$$\bigcap \{D(y) \mid y \in Y\} = \Lambda.$$

---

[14] $\bar{L}$ is given by (22).

Let Λ_1 be the polyhedral cone contained by Λ. Then, $x^0 \in N_X(D(\cdot))$ only if $x^0 \in X^0(\Lambda_1)$. In particular, let $\{H^k \mid 1 \leq k \leq q\}$ be a generator of $\Lambda_1{}^*$. Suppose that $f(x)$ and $g(x)$ are differentiable over X. Then,

$$x^0 \in N_X(D(\cdot)) \subset X^0(\Lambda_1)$$

only if the necessary conditions stated in Theorem 5.2 are satisfied.

5.3. Nondominated Solutions in Decision Space: Second Locating Method. In this section, we shall apply the results of cone convexity and introduce a second method for locating the cone extreme points in the decision space. Conditions for a point to be nondominated in the decision space will also be explored. The main results of this section are primarily based on those of Section 4.3.

Let $\Lambda = \bigcap_{y \in Y} D(y)$. Suppose that $\Lambda^\perp \neq \{0\}$ and Y is Λ-convex. Then, by Corollary 4.7, we have

$$\bigcup \{Y^0(\lambda) \mid \lambda \in \operatorname{Int} \Lambda^*\} \subset \operatorname{Ext}[Y \mid \Lambda] \subset \bigcup \{Y^0(\lambda) \mid \lambda \in \Lambda^*, \lambda \neq 0\},$$

where

$$Y^0(\lambda) = \{y_0 \in Y \mid \lambda y_0 = \sup \lambda y, y \in Y\} = \{y_0 \in Y \mid \lambda y_0 = \max \lambda y, y \in Y\}$$

Define

$$X_0(\lambda) = \{x_0 \in X \mid \lambda \cdot f(x_0) = \max \lambda \cdot f(x), x \in X\}, \tag{28}$$

$$X_A(\Lambda) = \bigcup \{X_0(\lambda) \mid \lambda \in \Lambda^*, \lambda \neq 0\}, \tag{29}$$

$$X_A{}^I(\Lambda) = \bigcup \{X_0(\lambda) \mid \lambda \in \operatorname{Int} \Lambda^*\}. \tag{30}$$

We see that

$$f[X_0(\lambda)] = Y^0(\lambda),$$

$$f[X_A(\Lambda)] = \bigcup \{Y^0(\lambda) \mid \lambda \in \Lambda^*, \lambda \neq 0\},$$

$$f[X_A{}^I(\Lambda)] = \bigcup \{Y^0(\lambda) \mid \lambda \in \operatorname{Int} \Lambda^*\}.$$

Since[15]

$$f[X^0(\Lambda)] = \operatorname{Ext}[Y \mid \Lambda],$$

by Corollary 4.7, we have the following theorem.

Theorem 5.3. If $\Lambda^\perp \neq \{0\}$ and Y is Λ-convex, then

$$X_A{}^I(\Lambda) \subset X^0(\Lambda) \subset X_A(\Lambda).$$

---

[15] Recall (Section 5.2) that $X^0(\Lambda)$ is the set of all Λ-extreme points in the decision space.

In view of this theorem, it is appropriate to call $X_A^I(\Lambda)$ and $X_A(\Lambda)$ the *inner* and *outer approximate sets* of $X^0(\Lambda)$. Since finding $X_0(\lambda)$, $\lambda \neq 0$, is a mathematical programming problem, Theorem 5.3 transfers the problem of finding cone extreme points into ordinary mathematical programs.

Recall that

$$X = \{x \mid g(x) \leqq 0\},$$

where $g(x) = (g_1(x),..., g_m(x))$. Suppose that $f(x)$ and $g(x)$ are differentiable over X. By Fritz John's-theorem (Ref. 14), $x_0 \in X_0(\lambda)$ only if there are $\mu_0 \in R$ and $\mu \in R^m$ such that

$$\mu_0\lambda \cdot \nabla f(x_0) - \mu \cdot \nabla g(x_0) = 0, \tag{31}$$

$$\mu \cdot g(x_0) = 0, \tag{32}$$

$$g(x_0) \leqq 0, \tag{33}$$

$$\mu_0 \geqq 0, \qquad \mu \geqq 0, \qquad (\mu_0, \mu) \neq (0, 0). \tag{34}$$

Let

$$I(x_0) = \{i \mid g_i(x_0) = 0\},$$

and let $\mu_{I(x_0)}$ and $g_{I(x_0)}(x_0)$ be the vectors derived from μ and $g(x_0)$ by deleting all components of μ and $g(x_0)$ which are not in $I(x_0)$. Define

$$F(x_0, \Lambda) = \{\lambda \cdot \nabla f(x_0) \mid \lambda \in \Lambda^*, \lambda \neq 0\},$$

$$F^I(x_0, \Lambda) = \{\lambda \cdot \nabla f(x_0) \mid \lambda \in \operatorname{Int} \Lambda^*\},$$

$$G(x_0) = \{\mu_{I(x_0)} \cdot \nabla g_{I(x_0)}(x_0) \mid \mu_{I(x_0)} \geqq 0\}.$$

It is understood that $G(x_0) = \{0\}$ if $I(x_0) = \phi$.

Note that $\mu \cdot g(x_0) = 0$ implies that $\mu_i = 0$ if $g_i(x_0) < 0$. By (31)–(34), we need to consider two possible cases.

Case 1: $\mu_0 = 0$. Then, $\mu \neq 0$, $\mu_{I(x_0)} \geqslant 0$. Thus, $x_0 \in X_A(\Lambda)$ or $x_0 \in X_A^I(\Lambda)$ only if

$$0 \in \{\mu_{I(x_0)} \cdot \nabla g_{I(x_0)}(x_0) \mid \mu_{I(x_0)} \geqslant 0\}. \tag{35}$$

Case 2: $\mu_0 > 0$. Without loss of generality we can set $\mu_0 = 1$ [divide (31) by μ_0, if necessary]. Then, $x_0 \in X_A(\Lambda)$ only if

$$F(x_0, \Lambda) \cap G(x_0) \neq \varnothing; \tag{36}$$

and $x_0 \in X_A{}^I(\Lambda)$ only if

$$F^I(x_0, \Lambda) \cap G(x_0) \neq \varnothing. \tag{37}$$

We formally summarize the results into the following theorem.

Theorem 5.4. Suppose that Y is Λ-convex,

$$\Lambda = \bigcap_{y \in Y} D(y),\ \Lambda^\perp \neq \{0\},$$

and $f(x)$ and $g(x)$ are differentiable over X. Then

(i) $x_0 \in X_A(\Lambda)$ only if $x_0 \in X$ and (35) or (36) are satisfied;
(ii) $x_0 \in X_A{}^I(\Lambda)$ only if $x_0 \in X$ and (35) or (37) are satisfied;
(iii) $x_0 \in N_X(D(\cdot))$ only if $x_0 \in X$ and (35) or (36) are satisfied.

Proof. (i) and (ii) are obvious from the previous discussion. (iii) comes from the fact that

$$N_X(D(\cdot)) \subset X^0(\Lambda) \subset X_A(\Lambda).$$

Theorem 5.5. (A) Suppose that $\Lambda^\perp \neq \{0\}$ and that over X, (a) $g(x)$ is quasiconvex and differentiable, (b) $\lambda \cdot f(x)$ is concave and differentiable for all $\lambda \in \Lambda^*$, and (c) the Kuhn–Tucker constraint qualification is satisfied. Then,

(i) $x_0 \in X_A(\Lambda)$ iff $x_0 \in X$ and $F(x_0, \Lambda) \cap G(x_0) \neq \varnothing$;
(ii) $x_0 \in X_A{}^I(\Lambda)$ iff $x_0 \in X$ and $F^I(x_0, \Lambda) \cap G(x_0) \neq \varnothing$.

(B) Suppose also that Λ is closed or $f[X]$ is Λ-convex. Then,

(iii) $X_A{}^I(\Lambda) \subset X^0(\Lambda) \subset X_A(\Lambda)$.

Proof. Assertions (i) and (ii) come from the Kuhn–Tucker theorem: under Assumption (A), a necessary and sufficient condition for $x_0 \in X_0(\lambda)$ is that there is $\mu \in R^m$, $\mu \geqq 0$ so that (31)–(34) is satisfied with $\mu_0 = 1$ (see Ref. 14).

By Hypothesis (A) and Assumption (B) that Λ is closed, we can conclude that X is a convex set and that, by Corollary 3.2, $f[X]$ is Λ-convex. From Theorem 5.3, we get Assertion (iii).

Remark 5.7. Suppose that $D(y) = \Lambda$ for all $y \in Y$. Then,

$$N_X(D(\cdot)) = X^0(\Lambda).$$

If Y is Λ-convex and $\Lambda^{\perp} \neq \{0\}$, then $X_A{}^I(\Lambda)$ and $X_A(\Lambda)$ are also the *inner* and *outer approximate sets* of $N_X(D(\cdot))$. Because $F(x, \Lambda)$, $F^I(x, \Lambda)$, and $G(x)$ are uniquely defined for each point of X, Theorem 5.5 can be used to locate $N_X(D(\cdot))$. As an example, see Section 5.4.

By invoking Corollary 3.6 and Theorem 5.5, we have the following corollary.

Corollary 5.4. Suppose that (A) $f(x)$ is concave and differentiable over X, (B) $g(x)$ is quasiconvex and differentiable over X, (C) the Kuhn–Tucker constraint qualification is satisfied throughout X, and (D) $\Lambda \supset \Lambda^{\leqq}$ and $\Lambda^{\perp} \neq \{0\}$. Then,

(i) $X_A{}^I(\Lambda) \subset X^0(\Lambda) \subset X_A(\Lambda)$;

(ii) $x_0 \in X_A{}^I(\Lambda)$ iff $x_0 \in X$ and $F^I(x_0, \Lambda) \cap G(x_0) \neq \varnothing$;

(iii) $x_0 \in X_A(\Lambda)$ iff $x_0 \in X$ and $F(x_0, \Lambda) \cap G(x_0) \neq \varnothing$.

Remark 5.8. The set of all Pareto-optimal solutions in the decision space is equal to $X^0(\Lambda^{\leqq})$. When Assumptions (A)–(C) of Corollary 5.4 are satisfied, we have

$$X_A{}^I(\Lambda^{\leqq}) \subset X^0(\Lambda^{\leqq}) \subset X_A(\Lambda^{\leqq}).$$

Thus, we can approximate $X^0(\Lambda^{\leqq})$ by applying Corollary 5.4. Note that

$$F(x_0, \Lambda^{\leqq}) = \{\lambda \cdot \nabla f(x_0) \mid \lambda \in R^l, \lambda \geqslant 0\}, \tag{38}$$

$$F^I(x_0, \Lambda^{\leqq}) = \{\lambda \cdot \nabla f(x_0) \mid \lambda \in R^l, \lambda > 0\}. \tag{39}$$

If we could successfully locate $X_A{}^I(\Lambda^{\leqq})$ and $X_A(\Lambda^{\leqq})$, then, by the definition and Corollary 4.7, we would have no difficulty in specifying the entire set of all Pareto-optimal solutions. As an example, see Section 5.4.

Remark 5.9. Suppose that Λ^* is a polyhedral cone with a generator $\{H^k \mid k = 1,..., q\}$. Let H be the matrix of order $q \times l$, with H^k in its kth row. Then (see Remark 5.10),

$$\Lambda^* = \{\lambda \cdot H \mid \lambda \in R^q, \lambda \geqq 0\},$$

$$\text{Int}\,\Lambda^* = \{\lambda \cdot H \mid \lambda \in R^q, \lambda > 0\} \quad \text{if} \quad \text{Int}\,\Lambda^* \neq \varnothing,$$

and

$$F(x_0, \Lambda) = \{\lambda \cdot H \cdot \nabla f(x_0) \mid \lambda \in R^q, \lambda \geqslant 0\}, \tag{40}$$

$$F^I(x_0, \Lambda) = \{\lambda \cdot H \cdot \nabla f(x_0) \mid \lambda \in R^q, \lambda > 0\}. \tag{41}$$

Remark 5.10. The relation

$$\text{Int}\,\Lambda^* = \{\lambda \cdot H \mid \lambda \in R^q, \lambda > 0\}$$

is a convenient expression. The fact that

$$\text{Int}\,\Lambda^* \supset \{\lambda \cdot H \mid \lambda \in R^q, \lambda > 0\}$$

is clear. In order to see that

$$\text{Int}\,\Lambda^* \subset \{\lambda \cdot H \mid \lambda \cdot R^q, \lambda > 0\},$$

let $d \in \text{Int}\,\Lambda^*$. Then, there is a closed neighborhood N of d such that $N \subset \Lambda^*$. Connect d with each H^k by a straight line and extend the line to a point $d^k \in \partial N$, so that we can write

$$d = (1 - \alpha_k)H^k + \alpha_k d^k \qquad 0 < \alpha_k < 1. \tag{i}$$

Note that

$$d^k = \sum_{j=1}^{q} \beta_{kj} \cdot H^j, \qquad \beta_{kj} \geqq 0.$$

Summation over k on both sides of (i) yields

$$qd = \sum_{k=1}^{q} [(1 - \alpha_k)H^k + \alpha_k d^k] = \sum_{k=1}^{q} \left[(1 - \alpha_k)H^k + \sum_{j=1}^{q} \alpha_k \beta_{kj} \cdot H^j\right]$$

$$= \sum_{k=1}^{q} \left[(1 - \alpha_k) + \sum_{j=1}^{q} \alpha_j \beta_{jk}\right] H^k. \tag{ii}$$

Since

$$(1 - \alpha_k) + \sum_{j=1}^{q} \alpha_j \beta_{jk} > 0,$$

our assertion is clear by (ii). Now, suppose that $\text{Int}\,\Lambda^* = \varnothing$. The same argument holds for the relative interior $(\Lambda^*)^I$ of Λ^*. That is,

$$(\Lambda^*)^I = \{\lambda \cdot H \mid \lambda \in R^q, \lambda > 0\}.$$

We now observe that, since H is a constant matrix, by treating $H \cdot \nabla f(x_0)$ as a factor, (38)–(39) and (40)–(41) have the same structure. By invoking Lemma 2.2 and Theorem 5.5, we see that, if Λ is such that $\bar{\Lambda}$ is a polyhedral cone and if Assumption (A) of Theorem 5.5 is satisfied, then finding $X^0(\Lambda)$ is exactly equivalent to finding the set of all Pareto-optimal solutions with objectives $H \cdot f(x)$ and $\Lambda^{\leqq}$ in R^q. In order to find the entire set of $X^0(\Lambda)$, the parameter λ of our mathematical program must vary over $\Lambda^{\geqq}$. In order to save computation, it is a good idea to select a set containing a minimal number of vectors for the generator of Λ^*.

Although Theorem 5.5. and Corollary 5.4 are the main results for locating $X_A(\Lambda)$ and $X_A{}^I(\Lambda)$, to speed up the procedure of locating them we still need some decomposition theorems on $X_A(\Lambda)$ and $X_A{}^I(\Lambda)$. Let $M = \{1, 2,..., m\}$ and $\mathcal{M} = \{I \mid I \subset M\}$. For $I \in \mathcal{M}$, define

$$X_I = \{x \mid g_I(x) \leqq 0\},$$

where $g_I(x)$ is derived from $g(x)$ by deleting all components of $g(x)$ except those in I. Let $X_I{}^0(\Lambda)$ be the set of all Λ-extreme points in X_I, and let $X_{IA}^I(\Lambda)$ and $X_{IA}(\Lambda)$ be the inner and outer approximate sets for $X_I{}^0(\Lambda)$, respectively, as defined by (28)–(30).

Theorem 5.6. The following results hold:

(i) $X_A(\Lambda) = \bigcup_{I \in \mathcal{M}} (X_{IA}(\Lambda) \cap X)$;

(ii) $X_A{}^I(\Lambda) = \bigcup_{I \in \mathcal{M}} (X_{IA}^I(\Lambda) \cap X)$.

Proof. (i) Because $X_M = X$ and $X_A(\Lambda) \subset X$, the fact that

$$X_A(\Lambda) \subset \bigcup_{I \in \mathcal{M}} (X_{IA}(\Lambda) \cap X)$$

is obvious. Now, suppose that $x_0 \in X_{IA}(\Lambda) \cap X$, for some $I \in \mathcal{M}$. Then, since $x_0 \in X_{IA}(\Lambda)$, x_0 maximizes $\lambda \cdot f(x)$ for some $\lambda \in \Lambda^*$ over $X_I \supset X$. Thus, $\lambda \cdot f(x_0) \geqq \lambda \cdot f(x)$ for all $x \in X$. Since $x_0 \in X$, we see that $x_0 \in X_A(\Lambda)$. Thus,

$$X_A(\Lambda) \supset \bigcup_{I \in \mathcal{M}} (X_{IA}(\Lambda) \cap X).$$

(ii) The proof is similar and will be omitted.

Let us denote the number of elements in $I \in \mathcal{M}$ by $[I]$, and let

$$\mathcal{I}^k = \{I \in \mathcal{M} \mid [I] = k\}, \; k = 0, 1,..., m.$$

For each $I \in \mathcal{M}$, define

$$\tilde{X}_I = \{x \mid g_I(x) = 0\}.$$

We have the following useful decomposition theorem.

Theorem 5.7. If Assumption (A) of Theorem 5.5 is satisfied over R^n (instead of X), then the following results hold:

$$\text{(i)}\quad X_A(\Lambda) = (X_{\phi A}(\Lambda) \cap X) \cup \left[\bigcup_{I \in \mathscr{I}^1} [X_{IA}(\Lambda) \cap \tilde{X}_I \cap X]\right]$$

$$\cup \left[\bigcup_{I \in \mathscr{I}^2} [X_{IA}(\Lambda) \cap \tilde{X}_I \cap X]\right] \cdots$$

$$\cup \left[\bigcup_{I \in \mathscr{I}^m} [X_{IA}(\Lambda) \cap \tilde{X}_I \cap X]\right]$$

$$= (X_{\phi A}(\Lambda) \cap X) \cup \left[\bigcup_{k=1}^{m} \bigcup_{I \in \mathscr{I}^k} [X_{IA}(\Lambda) \cap \tilde{X}_I \cap X]\right];$$

$$\text{(ii)}\quad X_A^I(\Lambda) = (X_{\phi A}^I(\Lambda) \cap X) \cup \left[\bigcup_{k=1}^{m} \bigcup_{I \in \mathscr{I}^k} [X_{IA}^I(\Lambda) \cap \tilde{X}_I \cap X]\right].$$

If Assumption (B) of Theorem 5.5 is also satisfied, then

$$\text{(iii)}\quad X_A^I(\Lambda) \subset X^0(\Lambda) \subset X_A(\Lambda).$$

Proof. (i) It is clear by Theorem 5.6 that the sets on the right-hand side are contained by $X_A(\Lambda)$. In order to see that $X_A(\Lambda)$ is contained by the union of the sets in the right-hand side, let $x_0 \in X_A(\Lambda)$. Recall that

$$I(x_0) = \{i \mid g_i(x_0) = 0\}.$$

By (i) of Theorem 5.5 [which depends only on Assumption (A)], we know that $x_0 \in X_A(\Lambda)$ iff $x_0 \in X$ and

$$F(x_0, \Lambda) \cap G(x_0) \neq \varnothing,$$

where

$$G(x_0) = \{\mu_{I(x_0)} \cdot \nabla g_{I(x_0)}(x_0) \mid \mu_{I(x_0)} \geqq 0\}.$$

Note that $x_0 \in X \subset X_{I(x_0)}$. By applying Theorem 5.5 to the set $X_{I(x_0)}$, because $F(x_0, \Lambda)$ does not depend on $I(x_0)$, we see that $x_0 \in X_{I(x_0)A}(\Lambda)$. Note that, $0 \leqq [I(x_0)] \leqq m$. When $[I(x_0)] = 0$, then $x_0 \in X_{\phi A}(\Lambda) \cap X$. Otherwise, because

$$x_0 \in \tilde{X}_{I(x_0)} \cap X,$$

we have

$$x_0 \in X_{I(x_0)A}(\Lambda) \cap \tilde{X}_{I(x_0)} \cap X.$$

Note that, in this case, $I(x_0) \in \mathscr{I}^k$ for some k, $1 \leq k \leq m$. Thus, we see that, in either case, x_0 is contained by the union of the sets on the right-hand side.

(ii) The proof is similar to that of (i) and will be omitted.

(iii) The proof follows directly from (iii) of Theorem 5.5.

Remark 5.11. The set $X_{IA}(\Lambda) \cap \tilde{X}_I \cap X$ could be located by first finding those points $X_{IA}(\Lambda)$ on $\tilde{X}_I$, that is, $X_{IA}(\Lambda) \cap \tilde{X}_I$, and then discarding those points of $X_{IA}(\Lambda) \cap \tilde{X}_I$ which violate the constraints $g_i(x) \leq 0$, $i \in I$. Thus, Theorem 5.7 could be used systematically to locate X_{IA}, X_{IA}^I, and $X^0(\Lambda)$. In addition, it could be used to produce, for all $x_0 \in X^0(\Lambda)$,

$$\Lambda^*(x_0) = \{\lambda \mid \lambda \in \Lambda^*, \lambda \neq 0, x_0 \in X_0(\lambda)\}.$$

Thus, if $\lambda \in \Lambda^*(x_0)$, then x_0 maximizes $\lambda \cdot f(x)$ over X. Of course, $\Lambda^*(x_0)$ is extremely useful in the final decision making. In Section 5.4, we provide an example to show the procedure for generating $X_A(\Lambda)$, $X_A{}^I(\Lambda)$, $X^0(\Lambda)$, and $\Lambda^*(x_0)$. When the objectives and the constraints are all linear, we could first search for the set of all nondominated extreme points and then use Theorem 5.7 to generate the entire set of all Pareto-optimal solutions. For a detailed derivation along this line, see Ref. 16. One should observe that, when the number of contraints gets larger and larger, the computation method according to Theorem 5.7 may demand a tremendous amount of time.

Example 5.5. Consider the following objective functions:

$$f_1(x) = x_1 + x_2 + x_4, \qquad f_2(x) = 2x_1 + x_3 - 2x_4,$$

and constraints:

$$g_1(x) = x_1{}^2 + x_2{}^2 + x_3{}^2 \leqq 1, \ g_2(x) = x_1 + x_4 \leqq 1, \ g_3(x) = x_2 + x_3 - x_4 \leqq 1.$$

We want to find $X^0(\Lambda^{\leqq})$, the set of all Pareto solutions.[16] Note that Assumptions (A) and (B) of Theorem 5.5 are satisfied throughout R^4. Since

$$\nabla f_1(x) = (1, 1, 0, 1), \qquad \nabla f_2(x) = (2, 0, 1, -2),$$

[16] Recall from Remark 5.9 that, if Λ^* is a polyhedral cone, then to find $X^0(\Lambda)$ is equivalent to finding $X^0(\Lambda^{\leqq})$ with $\Lambda^{\leqq}$ in R^q.

we have

$$F^I(X, \Lambda^{\leqq}) = F^{>} = \{\lambda_1(1, 1, 0, 1) + \lambda_2(2, 0, 1, -2) \mid \lambda_1 > 0, \lambda_2 > 0\}$$
$$= \{(\lambda_1 + 2\lambda_2, \lambda_1, \lambda_2, \lambda_1 - 2\lambda_2) \mid \lambda_1, \lambda_2 > 0\}, \tag{42}$$
$$F(X, \Lambda^{\leqq}) = F^{\geqslant} = \{(\lambda_1 + 2\lambda_2, \lambda_1, \lambda_2, \lambda_1 - 2\lambda_2) \mid (\lambda_1, \lambda_2) \geqslant (0, 0)\}. \tag{43}$$

Since

$$\nabla g_1(x) = (2x_1, 2x_2, 2x_3, 0), \quad \nabla g_2(x) = (1, 0, 0, 1), \quad \nabla g_3(x) = (0, 1, 1, -1),$$

we have

$$G_{\{1\}}(x) = \{\mu(x_1, x_2, x_3, 0) \mid \mu \geqq 0\}, \tag{44}$$
$$G_{\{2\}}(x) = \{\mu(1, 0, 0, 1) \mid \mu \geqq 0\}, \tag{45}$$
$$G_{\{3\}}(x) = \{\mu(0, 1, 1, -1) \mid \mu \geqq 0\}, \tag{46}$$
$$G_{\{1,2\}}(x) = \{\mu_1 x_1 + \mu_2, \mu_1 x_2, \mu_1 x_3, \mu_2) \mid \mu_1, \mu_2 \geqq 0\}, \tag{47}$$
$$G_{\{1,3\}}(x) = \{(\mu_1 x_1, \mu_1 x_2 + \mu_3, \mu_1 x_3 + \mu_3, -\mu_3) \mid \mu_1, \mu_3 \geqq 0\}, \tag{48}$$
$$G_{\{2,3\}}(x) = \{(\mu_2, \mu_3, \mu_3, \mu_2 - \mu_3) \mid \mu_2, \mu_3 \geqq 0\}, \tag{49}$$
$$G_{\{1,2,3\}}(x) = \{(\mu_1 x_1 + \mu_2, \mu_1 x_2 + \mu_3, \mu_1 x_3 + \mu_3, \mu_2 - \mu_3) \mid \mu_1, \mu_2, \mu_3 \geqq 0\} \tag{50}$$

We are now ready to find $X_A[\Lambda^{\leqq}]$ and $X_A{}^I[\Lambda^{\leqq}]$. Without confusion, we shall use X_{IA} and X_{IA}^I to represent $X_{IA}^I[\Lambda^{\leqq}]$ and $X_{IA}[\Lambda^{\leqq}]$, respectively.

Step (1). To find $X_{\phi A} \cap X$ and $X_{\phi A}^I \cap X$. In view of (42) and (43), we know that $0 \notin F^{>}$ and $0 \notin F^{\geqslant}$. Thus,

$$X_{\phi A} = \varnothing, \qquad X_{\phi A}^I = \varnothing, \qquad \text{and} \qquad X_{\phi A} \cap X = X_{\phi A}^I \cap X = \varnothing.$$

Step (2). For all I such that $[I] = 1$, there are three cases.

Case 1: $I = \{1\}$. We first find the set

$$X_1^{>} = \{x \mid F^{>} \cap G_I(x) \neq \varnothing\}.$$

In view of (42) and (44) for each $x \in X_I^{>}$, the following equalities must be satisfied for some $\lambda_1, \lambda_2 > 0$ and $\mu \geqq 0$:

$$\mu x_1 = \lambda_1 + 2\lambda_2 \Rightarrow x_1 = (\lambda_1 + 2\lambda_2)/\mu, \ \mu > 0,$$
$$\mu x_2 = \lambda_1 \Rightarrow x_2 = \lambda_1/\mu, \ \mu > 0,$$
$$\mu x_3 = \lambda_2 \Rightarrow x_3 = \lambda_2/\mu, \ \mu > 0,$$
$$0 = \lambda_1 - 2\lambda_2 \Rightarrow \lambda_1 = 2\lambda_2 .$$

Thus,

$$X_I^{>} = \{(4\lambda/\mu, 2\lambda/\mu, \lambda/\mu, x_4) \mid \lambda, \mu > 0\} = \{(4\bar{\lambda}, 2\bar{\lambda}, \bar{\lambda}, x_4) \mid \bar{\lambda} > 0\},$$

and

$$X_{IA}^I \cap X \cap \tilde{X}_I = X_I^{>} \cap X \cap \tilde{X}_I = \left\{ (4\bar{\lambda}, 2\bar{\lambda}, \bar{\lambda}, x_4) \;\middle|\; \begin{matrix} \bar{\lambda} > 0 \\ 16\bar{\lambda}^2 + 4\bar{\lambda}^2 + \bar{\lambda}^2 = 1 \\ 4\bar{\lambda} + x_4 \leqq 1 \\ 3\bar{\lambda} - x_4 \leqq 1 \end{matrix} \right\}$$

$$= \{(4/\sqrt{21}, 2/\sqrt{21}, 1/\sqrt{21}, x_4) \mid 3/\sqrt{21} - 1 \leqq x_4 \leqq 1 - 4/\sqrt{21}\}.$$

Similarly, one finds that

$$X_I^{\geqslant} = \{x \mid F^{\geqslant} \cap G_I(x) \neq \varnothing\} = X_I^{>} = \{(4\lambda, 2\lambda, \lambda, x_4) \mid \lambda > 0\}$$

and

$$X_{IA} \cap X \cap \tilde{X}_I$$
$$= \{(4/\sqrt{21}, 2/\sqrt{21}, 1/\sqrt{21}, x_4) \mid 3/\sqrt{21} - 1 \leqq x_4 \leqq 1 - 4/\sqrt{21}\}.$$

Note that, for all $X_{IA}^I \cap X \cap \tilde{X}_I$ and $X_{IA} \cap X \cap \tilde{X}_I$,

$$\Lambda^*(x_0) = \{(\lambda_1, \lambda_2) \mid \lambda_1 = 2\lambda_2, (\lambda_1, \lambda_2) > (0, 0)\}.$$

Case 2: $I = \{2\}$. Then, in view of (42) and (45) for each $x \in X_I^{>}$, the following equalities hold for some $\lambda_1, \lambda_2 > 0$ and $\mu \geqq 0$:

$$\mu = \lambda_1 + 2\lambda_2, \qquad 0 = \lambda_1, \qquad 0 = \lambda_2, \qquad \mu = \lambda_1 - 2\lambda_2.$$

It is seen that $X_I^{>} = \varnothing$. Similarly, $X_I^{\geqslant} = \varnothing$. Thus,

$$X_{IA}^I \cap X \cap \tilde{X}_I = X_{IA} \cap X \cap \tilde{X}_I = \varnothing.$$

Case 3: $I = \{3\}$. Then, in view of (42) and (46) for each $x \in X_I^{>}$, the following equalities hold for some $\lambda_1, \lambda_2 > 0$ and $\mu \geqq 0$:

$$0 = \lambda_1 + 2\lambda_2, \qquad \mu = \lambda_1, \qquad \mu = \lambda_2, \qquad -\mu = \lambda_1 - 2\lambda_2.$$

Obviously, $X_I^{>} = \varnothing$. Similarly, $X_I^{\geqslant} = \varnothing$. Thus,

$$X_{IA}^I \cap X \cap \tilde{X}_I = X_{IA} \cap X \cap \tilde{X}_I = \varnothing.$$

Step (3). For all I such that $[I] = 2$, there are three cases.

Case 1: $I = \{1, 2\}$. In view of (42) and (47) for each $x \in X_I^>$, the following equalities must hold for some $\lambda_1, \lambda_2 > 0$ and $\mu_1, \mu_2 \geqq 0$:

$$\begin{aligned} \mu_1 x_1 + \mu_2 = \lambda_1 + 2\lambda_2 &\Rightarrow x_1 = (\lambda_1 + 2\lambda_2 - \mu_2)/\mu_1 = 4\lambda_2/\mu_1\,, \\ \mu_1 x_2 = \lambda_1 &\Rightarrow x_2 = \lambda_1/\mu_1\,, \\ \mu_1 x_3 = \lambda_2 &\Rightarrow x_3 = \lambda_2/\mu_1\,, \\ \mu_2 = \lambda_1 - 2\lambda_2\,,\ \mu_2 \geqq 0 &\Rightarrow \lambda_1 \geqq 2\lambda_2\,. \end{aligned}$$

Note that μ_1 cannot be zero. Thus, by setting $\bar{\lambda}_1 = \lambda_1/\mu_1$ and $\bar{\lambda}_2 = \lambda_2/\mu_1$, we see that

$$X_I^> = \left\{(4\bar{\lambda}_2, \bar{\lambda}_1, \bar{\lambda}_2, x_4) \,\middle|\, \begin{matrix} \bar{\lambda}_1, \bar{\lambda}_2 > 0 \\ \bar{\lambda}_1 \geqq 2\bar{\lambda}_2 \end{matrix}\right\},$$

and

$$\begin{aligned} X_{IA}^I \cap X \cap \tilde{X}_I &= X_I^> \cap X \cap \tilde{X}_I \\ &= \left\{(4\lambda_2, \lambda_1, \lambda_2, x_4) \,\middle|\, \begin{matrix} 16\lambda_2^2 + \lambda_1^2 + \lambda_2^2 = 1 \\ 4\lambda_2 + x_4 = 1 \\ \lambda_1 + \lambda_2 - x_4 \leqq 1 \\ \lambda_1 \geqq 2\lambda_2 \\ \lambda_1, \lambda_2 > 0 \end{matrix}\right\} \\ &= \left\{(4\lambda_2, \lambda_1, \lambda_2, 1 - 4\lambda_2) \,\middle|\, \begin{matrix} 17\lambda_2^2 + \lambda_1^2 = 1 \\ \lambda_1 + 5\lambda_2 \leqq 2 \\ \lambda_1 \geqq 2\lambda_2 \\ \lambda_1, \lambda_2 > 0 \end{matrix}\right\} \\ &= \left\{(4\lambda_2, \lambda_1, \lambda_2, 1 - 4\lambda_2) \,\middle|\, \begin{matrix} 17\lambda_2^2 + \lambda_1^2 = 1 \\ 2\lambda_2 \leqq \lambda_1 \leqq 2 - 5\lambda_2 \\ \lambda_1, \lambda_2 > 0 \end{matrix}\right\}. \end{aligned}$$

Similarly,

$$X_I^{\geqslant} = \left\{(4\lambda_2, \lambda_1, \lambda_2, x_4) \,\middle|\, \begin{matrix} \lambda_1 \geqq 2\lambda_2 \\ (\lambda_1, \lambda_2) \geqslant (0, 0) \end{matrix}\right\},$$

and

$$\begin{aligned} X_{IA} \cap X \cap \tilde{X}_I &= X_I^{\geqslant} \cap X \cap \tilde{X}_I \\ &= \left\{(4\lambda_2, \lambda_1, \lambda_2, 1 - 4\lambda_2) \,\middle|\, \begin{matrix} 17\lambda_2^2 + \lambda_1^2 = 1 \\ 2\lambda_2 \leqq \lambda_1 \leqq 2 - 5\lambda_2 \\ (\lambda_1, \lambda_2) \geqslant (0, 0) \end{matrix}\right\}. \end{aligned}$$

Case 2: $I = \{1, 3\}$. In view of (42) and (48) for each $x \in X_I^>$, the following equalities must hold for some $\lambda_1, \lambda_2 > 0$ and $\mu_1, \mu_3 \geqq 0$:

$$\begin{aligned}
\mu_1 x_1 &= \lambda_1 + 2\lambda_2 \Rightarrow x_1 = \lambda_1/\mu_1 + 2\lambda_2/\mu_1 ,\\
\mu_1 x_2 + \mu_3 &= \lambda_1 \Rightarrow x_2 = \lambda_1/\mu_1 - \mu_3/\mu_1 = 2\lambda_1/\mu_1 - 2\lambda_2/\mu_1 ,\\
\mu_1 x_3 + \mu_3 &= \lambda_2 \Rightarrow x_3 = \lambda_1/\mu_1 - \lambda_2/\mu_1 ,\\
-\mu_3 &= \lambda_1 - 2\lambda_2 \leqq 0 \Rightarrow \lambda_1 \leqq 2\lambda_2 .
\end{aligned}$$

Note that $\mu_1 \neq 0$. By putting $\bar{\lambda}_1 = \lambda_1/\mu_1$ and $\bar{\lambda}_2 = \lambda_2/\mu_1$, we have

$$X_I^{>} = \left\{(\bar{\lambda}_1 + 2\bar{\lambda}_2 , 2\bar{\lambda}_1 - 2\bar{\lambda}_2 , \bar{\lambda}_1 - \bar{\lambda}_2 , x_4) \,\middle|\, \begin{array}{l} \bar{\lambda}_1 , \bar{\lambda}_2 > 0 \\ \bar{\lambda}_1 \leqq 2\bar{\lambda}_2 \end{array} \right\}.$$

Thus,

$$X_{IA}^{I} \cap X \cap \tilde{X}_I = X_I^{>} \cap X \cap \tilde{X}_I$$

$$= \left\{(\bar{\lambda}_1 + 2\bar{\lambda}_2 , 2\bar{\lambda}_1 - 2\bar{\lambda}_2 , \bar{\lambda}_1 - \bar{\lambda}_2 , x_4) \,\middle|\, \begin{array}{l} (\bar{\lambda}_1 + 2\bar{\lambda}_2)^2 + (2\bar{\lambda}_1 - 2\bar{\lambda}_2)^2 + (\bar{\lambda}_1 - \lambda_2)^2 = 1 \\ \bar{\lambda}_1 + 2\bar{\lambda}_2 + x_4 \leqslant 1 \\ 2\bar{\lambda}_1 - 2\bar{\lambda}_2 + \bar{\lambda}_1 - \bar{\lambda}_2 - x_4 = 1 \\ \bar{\lambda}_1 \leqq 2\bar{\lambda}_2 , (\bar{\lambda}_1 , \bar{\lambda}_2) > (0, 0) \end{array} \right\}$$

$$= \left\{(\lambda_1 + 2\lambda_2 , 2\lambda_1 - 2\lambda_2 , \lambda_1 - \lambda_2 , 3\lambda_1 - 3\lambda_2 - 1) \,\middle|\, \begin{array}{l} 6\lambda_1^2 - 6\lambda_1\lambda_2 + 9\lambda_2^2 = 1 \\ 4\lambda_1 - \lambda_2 \leqq 2 \\ \lambda_1 \leqq 2\lambda_2 \\ (\lambda_1 , \lambda_2) > (0, 0) \end{array} \right\}.$$

Similarly,

$$X_I^{\geqslant} = \left\{(\lambda_1 + 2\lambda_2 , 2\lambda_1 - 2\lambda_2 , \lambda_1 - \lambda_2 , x_4) \,\middle|\, \begin{array}{l} (\lambda_1 , \lambda_2) \geqslant 0 \\ \lambda_1 \leqq 2\lambda_2 \end{array} \right\},$$

and

$$X_{IA} \cap X \cap \tilde{X}_I = X_I^{\geqslant} \cap X \cap \tilde{X}_I$$

$$= \left\{(\lambda_1 + 2\lambda_2 , 2\lambda_1 - 2\lambda_2 , \lambda_1 - \lambda_2 , 3\lambda_1 - 3\lambda_2 - 1) \,\middle|\, \begin{array}{l} 6\lambda_1^2 - 6\lambda_1\lambda_2 + 9\lambda_2^2 = 1 \\ 4\lambda_1 - \lambda_2 \leqq 2 \\ \lambda_1 \leqq 2\lambda_2 \\ (\lambda_1 , \lambda_2) \geqslant (0, 0) \end{array} \right\}.$$

Case 3: $I = \{2, 3\}$. In view of (42) and (49) for each $x \in X_I^{>}$, the following equalities must hold for some $\lambda_1, \lambda_2 > 0$ and $\mu_2, \mu_3 \geqq 0$:

$$\mu_2 = \lambda_1 + 2\lambda_2 , \qquad \mu_3 = \lambda_1 , \qquad \mu_3 = \lambda_2 , \qquad \mu_2 - \mu_3 = \lambda_1 - 2\lambda_2 .$$

It is seen that there are no $(\lambda_1, \lambda_2) > (0, 0)$, $(\mu_2, \mu_3) \geqq (0, 0)$ which would satisfy the above system of equations. Thus, $X_I^{>} = \varnothing$ and

$$X_{IA}^I \cap X \cap \tilde{X}_I = X_I^{>} \cap X \cap \tilde{X}_I = \varnothing .$$

Similarly, $X_I^{\geqslant} = \varnothing$ and

$$X_{IA} \cap X \cap \tilde{X}_I = \varnothing .$$

Step (4). For $I = \{1, 2, 3\} = M$. In view of (42) and (50) for each $x \in X_M$, the following equalities must hold for some $(\lambda_1, \lambda_2) > (0, 0)$ and $(\mu_1, \mu_2, \mu_3) \geqq (0, 0, 0)$:

$$\begin{aligned}
\mu_1 x_1 + \mu_2 = \lambda_1 + 2\lambda_2 &\Rightarrow x_1 = \lambda_1/\mu_1 + 2\lambda_2/\mu_1 - \mu_2/\mu_1 = 4\lambda_2/\mu_1 - \mu_3/\mu_1 ,\\
\mu_1 x_2 + \mu_3 = \lambda_1 &\Rightarrow x_2 = \lambda_1/\mu_1 - \mu_3/\mu_1 ,\\
\mu_1 x_3 + \mu_3 = \lambda_2 &\Rightarrow x_3 = \lambda_2/\mu_1 - \mu_3/\mu_1 ,\\
\mu_2 - \mu_3 = \lambda_1 - 2\lambda_2 &\Rightarrow \mu_2 = \lambda_1 - 2\lambda_2 + \mu_3 \geqq 0.
\end{aligned}$$

Note that μ_1 cannot be zero. Thus, by putting $\bar{\lambda}_1 = \lambda_1/\mu_1$, $\bar{\lambda}_2 = \lambda_2/\mu_1$, $\bar{\mu} = \mu_3/\mu_1$, we have

$$X_M^{>} = \left\{ (4\bar{\lambda}_2 - \bar{\mu}, \bar{\lambda}_1 - \bar{\mu}, \bar{\lambda}_2 - \bar{\mu}, x_4) \;\middle|\; \begin{array}{l} (\bar{\lambda}_1, \bar{\lambda}_2) > (0, 0) \\ \bar{\mu} \geqq 0 \\ \bar{\lambda}_1 - 2\bar{\lambda}_2 + \bar{\mu} \geqq 0 \end{array} \right\},$$

and

$$X_{MA}^I \cap X \cap \tilde{X}_M = X_M^{>} \cap X \cap \tilde{X}_M$$

$$= \left\{ (4\lambda_2 - \mu, \lambda_1 - \mu, \lambda_2 - \mu, x_4) \;\middle|\; \begin{array}{l} (4\lambda_2 - \mu)^2 + (\lambda_1 - \mu)^2 + (\lambda_2 - \mu)^2 = 1 \\ 4\lambda_2 - \mu + x_4 = 1 \\ \lambda_1 - \mu + \lambda_2 - \mu - x_4 = 1 \\ \lambda_1 - 2\lambda_2 + \mu \geqq 0 \\ \lambda_1, \lambda_2 > 0, \mu \geqq 0 \end{array} \right\}$$

$$= \left\{ (4\lambda_2 - \mu, \lambda_1 - \mu, \lambda_2 - \mu, 1 - 4\lambda_2 + \mu) \;\middle|\; \begin{array}{l} (4\lambda_2 - \mu)^2 + (\lambda_1 - \mu)^2 + (\lambda_2 - \mu)^2 = 1 \\ \lambda_1 + 5\lambda_2 - 3\mu = 2 \\ \lambda_1 - 2\lambda_2 + \mu \geqq 0 \\ \lambda_1, \lambda_2 > 0, \mu \geqq 0 \end{array} \right\}.$$

Similarly,

$$X_M^{\geqslant} = \left\{ (4\bar{\lambda}_2 - \bar{\mu}, \bar{\lambda}_1 - \bar{\mu}, \bar{\lambda}_2 - \bar{\mu}, x_4) \;\middle|\; \begin{array}{l} (\bar{\lambda}_1, \bar{\lambda}_2) \geqslant (0, 0) \\ \bar{\mu} \geqq 0 \\ \bar{\lambda}_1 - 2\bar{\lambda}_2 + \bar{\mu} \geqq 0 \end{array} \right\},$$

and

$$X_{MA} \cap X \cap \tilde{X}_M$$

$$= \left\{ (4\lambda_2 - \mu, \lambda_1 - \mu, \lambda_2 - \mu, 1 - 4\lambda_2 + \mu) \;\middle|\; \begin{array}{l} (4\lambda_2 - \mu)^2 + (\lambda_1 - \mu)^2 + (\lambda_2 - \mu)^2 = 1 \\ \lambda_1 + 5\lambda_2 - 3\mu = 2 \\ \lambda_1 - 2\lambda_2 + \mu \geqq 0 \\ (\lambda_1, \lambda_2) \geqslant (0, 0), \mu \geqq 0 \end{array} \right\}.$$

Step (5). To find $X_A{}^I$, X_A, and $X^0(\Lambda^{\leqslant})$. Set $I_1 = \{1\}$, $I_2 = \{1, 2\}$, $I_3 = \{1, 3\}$, $I_4 = \{1, 2, 3\}$. In view of Theorem 5.7 and Steps (1)–(4), we get

$$X_A{}^I = \bigcup_{K=1}^{4} (X^I_{I_K A} \cap X \cap \tilde{X}_{I_K})$$

$$= \{(4/\sqrt{21}, 2/\sqrt{21}, 1/\sqrt{21}, x_4) |\; 3/\sqrt{21} - 1 \leqq x_4 \leqq 1 - 4/\sqrt{21}\}$$

$$\cup \left\{ (4\lambda_2, \lambda_1, \lambda_2, 1 - 4\lambda_2) \;\middle|\; \begin{array}{l} \lambda_1{}^2 + 17\lambda_2{}^2 = 1 \\ 2\lambda_2 \leqq \lambda_1 \leqq 2 - 5\lambda_2 \\ (\lambda_1, \lambda_2) > (0, 0) \end{array} \right\}$$

$$\cup \left\{ (\lambda_1 + 2\lambda_2, 2\lambda_1 - 2\lambda_2, \lambda_1 - \lambda_2, 3\lambda_1 - 3\lambda_2 - 1) \;\middle|\; \begin{array}{l} 6\lambda_1{}^2 - 6\lambda_1\lambda_2 + 9\lambda_2{}^2 = 1 \\ 4\lambda_1 - \lambda_2 \leqq 2 \\ \lambda_1 \leqq 2\lambda_2 \\ (\lambda_1, \lambda_2) > (0, 0) \end{array} \right\}$$

$$\cup \left\{ (4\lambda_2 - \mu, \lambda_1 - \mu, \lambda_2 - \mu, 1 - 4\lambda_2 + \mu) \;\middle|\; \begin{array}{l} (4\lambda_2 - \mu)^2 + (\lambda_1 - \mu)^2 + (\lambda_2 - \mu)^2 = 1 \\ \lambda_1 + 5\lambda_2 - 3\mu = 2 \\ \lambda_1 - 2\lambda_2 + \mu \geqq 0 \\ \lambda_1, \lambda_2 > 0, \mu \geqq 0 \end{array} \right\}$$

and

$$X_A = \bigcup_{K=1}^{4} (X_{I_K A} \cap X \cap \tilde{X}_{I_K})$$

$$= \{(4/\sqrt{21}, 2/\sqrt{21}, 1/\sqrt{21}, x_4) |\; 3/\sqrt{21} - 1 \leqq x_4 \leqq 1 - 4/\sqrt{21}\}$$

$$\cup \left\{ (4\lambda_2, \lambda_1, \lambda_2, 1 - 4\lambda_2) \;\middle|\; \begin{array}{l} \lambda_1{}^2 + 17\lambda_2{}^2 = 1 \\ 2\lambda_2 \leqq \lambda_1 \leqq 2 - 5\lambda_2 \\ (\lambda_1, \lambda_2) \geqslant (0, 0) \end{array} \right\}$$

$$\cup \left\{ (\lambda_1 + 2\lambda_2, 2\lambda_1 - 2\lambda_2, \lambda_1 - \lambda_2, 3\lambda_1 - 3\lambda_2 - 1) \;\middle|\; \begin{array}{l} 6\lambda_1{}^2 - 6\lambda_1\lambda_2 + 9\lambda_2{}^2 = 1 \\ 4\lambda_1 - \lambda_2 \leqq 2; \lambda_1 \leqq 2\lambda_2 \\ (\lambda_1, \lambda_2) \geqslant (0, 0) \end{array} \right\}$$

$$\cup \left\{ (4\lambda_2 - \mu, \lambda_1 - \mu, \lambda_2 - \mu, 1 - 4\lambda_2 + \mu) \,\middle|\, \begin{array}{l} (4\lambda_2 - \mu)^2 + (\lambda_1 - \mu)^2 + (\lambda_2 - \mu)^2 = 1 \\ \lambda_1 + 5\lambda_2 - 3\mu = 2 \\ \lambda_1 - 2\lambda_2 + \mu \geqq 0 \\ (\lambda_1, \lambda_2) \geqslant (0, 0), \mu \geqq 0 \end{array} \right\}.$$

Observe that

(i) $X^I_{I_1A} \cap X \cap \tilde{X}_{I_1} = X_{I_1A} \cap X \cap \tilde{X}_{I_1}$,

(ii) $X_{I_2A} \cap X \cap \tilde{X}_{I_2} \setminus X^I_{I_2A} \cap X \cap \tilde{X}_{I_2}$
$= \{(0, 1, 0, 1)\}$ by $(\lambda_1, \lambda_2) = (1, 0)$,

(iii) $X_{I_3A} \cap X \cap \tilde{X}_{I_3} \setminus X^I_{I_3A} \cap X \cap \tilde{X}_{I_3}$
$= \{(2/3, -2/3, -1/3, -2)\}$, by $(\lambda_1, \lambda_2) = (0, 1/3)$,

(iv) $X_{I_4A} \cap X \cap \tilde{X}_{I_4} = X^I_{I_4A} \cap X \cap \tilde{X}_{I_4}$,

because neither λ_1 nor λ_2 can be zero. Thus,

$$X_A \setminus X_A{}^I \subset \{(0, 1, 0, 1), (2/3, -2/3, -1/3, -2)\}.$$

Since (0, 1, 0, 1) and (2/3, −2/3, −1/3, −2) are the unique points in $X_0(\lambda_1, \lambda_2 = 0)$ and $X_0(\lambda_1 = 0, \lambda_2)$ respectively, by Corollary 4.7, we know that

$$\{(0, 1, 0, 1), (2/3, -2/3, -1/3, -2)\} \subset X^0(\Lambda^{\leqq}).$$

Thus, we get $X_A{}^I \subset X^0(\Lambda^{\leqq}) = X_A$. Observe that (0, 1, 0, 1) is the maximal point of $f_1(x)$ over X, while (2/3, −2/3, −1/3, −2) is the maximal point of $f_2(x)$ over X. Note that, for each $x^0 \in X(\Lambda^{\leqq})$, we have no difficulty in identifying its $\Lambda^*(x^0)$ (see Remark 5.11), because throughout our process we have either expressed $\Lambda^*(x^0)$ explicitly or expressed x^0 directly or indirectly in terms of λ_1 and λ_2.

6. Conclusions

We have proposed a structure of domination to resolve the difficulties of multiobjective decision problems. The concept of and methods for locating nondominated solutions have been explored. In order to achieve our goal, we have also introduced and studied the concepts of cone convexity and cone extreme points. These concepts are useful not only in our current problems, but also in optimization theory. The application of these concepts to mathematical programming is under study now. Throughout this paper, we have limited ourselves to one-stage decision problems. The extension of our results to dynamic multiobjective decision problems can be found in Ref. 17.

References

1. YU, P. L., *The Set of All Nondominated Solutions in Decision Problems with Multiobjectives*, University of Rochester, Systems Analysis Program, Working Paper Series, No. F-71-32, 1971.
2. LUCE, R. D., and RAIFFA, H., *Games and Decision*, John Wiley and Sons, Inc., New York, New York, 1967.
3. MACCRIMMON, K. R., *Decision Making Among Multiple-Attribute Alternatives: A Survey and Consolidated Approach*, The RAND Corporation, Memorandum No. RM-4823-ARPA, 1968.
4. RAIFFA, H., *Preferences for Multi-Attributed Alternatives*, The RAND Corporation, Memorendum No. RM-5868-DOT/RC, 1969.
5. DACUNHA, N. O., and POLAK, E., *Constrained Minimization Under Vector-Valued Criteria in Finite Dimensional Space*, Journal of Mathematical Analysis and Applications, Vol. 19, pp. 103–124, 1967.
6. GEOFFRION, A. M., *Strictly Concave Parametric Programming, Parts I and II*, Management Science, Vol. 13, pp. 244–253 and pp. 359–370, 1967.
7. GEOFFRION, A. M., *Solving Bicriterion Mathematical Programs*, Operations Research, Vol. 15, pp. 39–54, 1967.
8. GEOFFRION, A. M., *Proper Efficiency and The Theory of Vector Maximization*, Journal of Mathematical Analysis and Applications, Vol. 22, pp. 618–630, 1968.
9. VINCENT, T. L., and LEITMANN, G., *Control-Space Properties of Cooperative Games*, Journal of Optimization Theory and Applications, Vol. 6, pp. 91–113, 1970.
10. BLACKWELL, D., and GIRSHICK, M. A., *Theory of Games and Statistical Decisions*, John Wiley and Sons, New York, New York, 1954.
11. FERGUSON, T. S., *Mathematical Statistics, A Decision Theoretic Approach*, Academic Press, New York, New York, 1967.
12. STOER, J., and WITZGALL, C., *Convexity and Optimization in Finite Dimensions I*, Springer-Verlag, New York, New York, 1970.
13. YU, P. L., *A Class of Solutions for Group Decision Problems*, Management Science, Vol. 19, pp. 936–946, 1974.
14. MANGASARIAN, O. L., *Nonlinear Programming*, McGraw-Hill Book Company, New York, New York, 1969.

15. LEITMANN, G., ROCKLIN, S., and VINCENT, T. L., *A Note on Control Space Properties of Cooperative Games*, Journal of Optimization Theory and Applications, Vol. 9, pp. 379–390, 1972.
16. YU, P. L., and ZELENY, M., *The Set of All Nondominated Solutions in the Linear Case and a Multicriteria Simplex Method*, Journal of Mathematical Analysis and Applications, Vol. 49, No. 2, 1975.
17. YU, P. L., and LEITMANN, G., *Nondominated Decisions and Cone Convexity in Dynamic Multicriteria Decision Problems*, Journal of Optimization Theory and Applications, Vol. 14, No. 5, 1974.
18. YU, P. L., *Introduction to Domination Structures in Multicriteria Decision Problems*, Multiple Criteria Decision Making, Edited by J. L. Cochrane and M. Zeleny, University of South Carolina Press, Columbia, South Carolina, 1973.
19. YU, P. L., *Nondominated Investment Policies in Stock Markets Including an Empirical Study*, University of Rochester, Systems Analysis Program, F-7222, 1973.
20. YU, P. L., *Domination Structures and Nondominated Solutions*, Proceedings of the International Seminar on Multicriteria Decision Making, Sponsored by UNESCO at CISM, Udine, Italy, June 1974.
21. YU, P. L., and LEITMANN, G., *Compromise Solutions, Domination Structures, and Salukvadze's Solution*, Journal of Optimization Theory and Applications, Vol. 13, No. 3, 1974.
22. BERGSTRESSER, K., CHARNES, A., and YU, P. L., *Generalization of Domination Structures and Nondominated Solutions in Multicriteria Decision Making*, Journal of Optimization Theory and Applications, Vol. 18, No. 1, 1976.
23. BERGSTRESSER, K., and YU, P. L., *Domination Structures and Multicriteria Problems in N-Person Games*, University of Texas, Austin, Center for Cybernetic Studies, CCS234, 1975. (To appear in the *International Journal of Theory and Decision*)

II

Nondominated Decisions and Cone Convexity in Dynamic Multicriteria Decision Problems[1]

P. L. YU AND G. LEITMANN

Abstract. The concepts of domination structures and nondominated decisions are extended to dynamic decision problems. Previously derived results are utilized in conjunction with optimal control theory to deduce necessary as well as sufficient conditions for nondominated controls. Cone convexity for dynamic problems is discussed, and sufficient conditions are given for a class of linear systems.

1. Introduction

A common and basic problem in decision making is the absence, in general, of a single objective. This may occur when a single individual is responsible for making a decision, but more often it happens when two or more decision makers are involved. A number of schemes have been put forward to deal with this dilemma. The concepts of domination structures and of nondominated solutions were introduced in Ref. 1. Subsequent investigations of these concepts were reported in Refs. 2–6; most of these were limited to static cases. In the present paper, we extend the results to dynamic cases, and we state necessary conditions and sufficient conditions for a nondominated control. Specifically, in Section 2 we define a general decision problem with multicriteria and give some results which are valid for both static and dynamic cases.

[1] This paper is based on research supported in part by the Office of Naval Research.

In Section 3, we consider a particular dynamic case, namely, a fixed endpoint control problem. We convert the problem of determining nondominated controls into an optimal control problem with isoperimetric constraints and then give necessary conditions and sufficient conditions for a nondominated control. We also state a set of such necessary conditions and sufficient conditions in the presence of a certain cone convexity condition. Finally, in Section 4 we give conditions which assure this cone convexity.

2. Definitions and Preliminary Results

A multicriteria decision problem is defined by a given set of admissible decisions Δ and given mappings $I_j(\cdot)\colon \Delta \to R^1$, $j = 1, 2,..., m$, yielding for $\delta \in \Delta$ a performance vector[2]

$$I(\delta) = (I_1(\delta), I_2(\delta),..., I_m(\delta))^T.$$

In a dynamic case, such as a control problem, a decision δ is a control function (either open or closed-loop).

The discussion and results of this section are based on Ref. 1.

Definition 2.1. Let Λ be a convex cone in R^m. A decision $\delta^0 \in \Delta$ is *nondominated* in Δ with respect to Λ iff there is no $\delta \neq \delta^0$, $\delta \in \Delta$, such that

$$I(\delta^0) \in I(\delta) + \Lambda, \qquad I(\delta) \neq I(\delta^0).$$

Definition 2.2. Given a set S and a convex cone Λ in R^m, $y^0 \in S$ is a Λ-*extreme* point in S iff there is no $y \in S$, $y \neq y^0$, such that $y^0 \in y + \Lambda$. The set of all Λ-extreme points in S is denoted by $\mathrm{Ext}[S \mid \Lambda]$.

The *criteria space* for a given decision problem is

$$Y = \{I(\delta) | \ \delta \in \Delta\}.$$

Remark 2.1. It is an immediate consequence of Definitions 2.1 and 2.2 that δ is a nondominated decision in Δ with respect to Λ iff

$$I(\delta) \in \mathrm{Ext}[Y \mid \Lambda].$$

---

[2] Throughout this paper, all vectors are taken to be column vectors, and transpose is denoted by superscript T.

A cone is polyhedral if it is also a polyhedron. The *polar cone* of a cone Λ in R^m is defined by

$$\Lambda^* = \{\lambda \in R^m \mid \lambda^T d \leqq 0 \text{ for all } d \in \Lambda\}.$$

The polar cone of a polyhedral cone is also a polyhedral. Thus, it possesses a finite generator; that is, if Λ^* is polyhedral, then there is a set of vectors $\{H^1, H^2, ..., H^q\}$ such that

$$\Lambda^* = \left\{ \sum_{j=1}^{q} \alpha_j H^j \mid \alpha_j \geqq 0 \right\}.$$

In view of Corollary 4.3 of Ref. 1 and Remark 2.1, we have the following theorem.

Theorem 2.1. Suppose that Λ is a polyhedral cone in R^m and its polar cone Λ^* is generated by $\{H^1, H^2, ..., H^q\}$. Then, $I(\delta^0) \in \text{Ext}[Y \mid \Lambda]$[3] iff there exist $j \in \{1, 2, ..., q\}$ and $q - 1$ real numbers $\{r_k \mid k \neq j, k = 1, 2, ..., q\}$ such that $I(\delta^0)$ is the unique maximum of $H^{jT} I(\delta)$ for all

$$I(\delta) \in \{I(\delta) \in Y \mid H^{kT} I(\delta) \geqq r_k, \quad k \neq j, k = 1, 2, ..., q\}.$$

Remark 2.2. Note that Theorem 2.1 requires the uniqueness of the maximum of $H^{jT} I(\delta)$ with respect to $I(\delta)$; however, δ^0 need not be unique.

Let[4]

$$\Lambda^{\leqq} = \{d \in R^m \mid d \leqq 0\}.$$

Then,

$$(\Lambda^{\leqq})^* = \Lambda^{\geqq} = \{d \in R^m \mid d \geqq 0\}.$$

Thus, the set $\{e^j \mid j = 1, 2, ..., m\}$, where e^j is the jth column of the $m \times m$ identity matrix, is a finite generator for $\Lambda^{\geqq}$. Note that $e^{jT} I(\delta) = I_j(\delta)$. This observation and Theorem 2.1 yield the following theorem.[5]

---

[3] That is, δ^0 is a nondominated decision in Δ with respect to Λ.

[4] As in Ref. 1, for x and $y \in R^n$ we use

$$x > y \quad \text{iff} \quad x_j > y_j \quad \forall j \in \{1, 2, ..., n\},$$
$$x \geqslant y \quad \text{iff} \quad x_j \geqq y_j \quad \forall j \in \{1, 2, ..., n\} \quad \text{and} \quad x \neq y,$$
$$x \geqq y \quad \text{iff} \quad x_j \geqq y_j \quad \forall j \in \{1, 2, ..., n\}.$$

[5] In this case, δ^0 is referred to as *efficient* or *Pareto optimal.*

Theorem 2.2. $I(\delta^0) \in \text{Ext}[Y \mid \Lambda^{\leqq}]$ iff there exist $j \in \{1, 2,..., m\}$ and $m - 1$ real numbers $\{r_k \mid k \neq j,\ k = 1, 2,..., m\}$ such that $I_j(\delta^0)$ is the unique maximum of $I_j(\delta)$ for all

$$I(\delta) \in \{I(\delta) \in Y \mid I_k(\delta) \geqq r_k, \qquad k \neq j, j = 1, 2,..., m\}.$$

Remark 2.3. Suppose that Λ is not a polyhedral cone. Since $\text{Ext}[Y \mid \Lambda_1] \subset \text{Ext}[Y \mid \Lambda_2]$ if $\Lambda_1 \supset \Lambda_2$, one can use a polyhedral cone contained in Λ to derive necessary conditions for nondominated decisions. Thus, the assumption that Λ is polyhedral can be relaxed in deriving necessary conditions.

Lemma 4.4 of Ref. 1 leads to the following theorem.

Theorem 2.3. Let

$$Y^0(\lambda) = \{I(\delta^0) \in Y \mid \lambda^T I(\delta^0) = \max \lambda^T I(\delta), \qquad I(\delta) \in Y, \lambda \in R^m\}.$$

Then,

(i) $\bigcup\{Y^0(\lambda) \mid \lambda \in \text{Int}\, \Lambda^*\} \subset \text{Ext}[Y \mid \Lambda]$,

(ii) if $Y^0(\lambda)$, $\lambda \in \Lambda^*$, $\lambda \neq 0$, contains only one point, then $Y^0(\lambda) \subset \text{Ext}[Y \mid \Lambda]$.

Definition 2.3.[6] Given Y and Λ in R^m, Y is said to be Λ-convex iff $Y + \Lambda$ is a convex set.

If L is the maximum subspace contained in Λ, one can write

$$\Lambda = L + \Lambda^{\perp} \qquad \text{with} \qquad \Lambda^{\perp} = \Lambda \cap L^{\perp},$$

where $L^{\perp}$ is the orthogonal complement of L in R^m.

Now, the following theorem follows from Lemma 4.5 of Ref. 1.

Theorem 2.4. Let $Y^0(\lambda)$ be defined as in Theorem 2.3. Suppose that Y is Λ-convex and $\Lambda^{\perp} \neq \{0\}$. Then,

$$\text{Ext}[Y \mid \Lambda] \subset \{Y^0(\lambda) \mid \lambda \in \Lambda^*, \lambda \neq 0\}.$$

Remark 2.4. The preceding theorems are statements of necessary conditions and sufficient conditions for an admissible decision to be nondominated with respect to a constant domination cone. These conditions can be generalized (see Ref. 1). Application to static cases

---

[6] For a detailed discussion of cone convexity, see Ref. 1.

may be found in Ref. 1 and Ref. 5; under certain circumstances, it is computationally feasible to locate the set of all nondominated decisions. Here, we shall discuss the application of the preceding results to one class of optimal control problems.

3. Conditions for Nondominated Controls

Consider the differential system

$$\dot{x}(t) = f(x(t), t, u(t)), \qquad t \in [a, b], \tag{1}$$

where $x(\cdot)$: $[a, b] \to \mathscr{E} \subset R^n$ is absolutely continuous, where $\mathscr{E}$ is an open set, $u(\cdot)$: $[a, b] \to \mathscr{S}^p$ is piecewise continuous, where $\mathscr{S}^p$ is a p-dimensional sphere of radius $\rho =$ const, and $f(\cdot)$: $\mathscr{R} \to R^n$ is C^2, where $\mathscr{R}$ is a region of $R^n \times R^1 \times R^p$.

A *control* $u(\cdot)$ is admissible provided it generates a solution $x(\cdot)$ satisfying (1), with $x(t) \in \mathscr{E}$ for all $t \in [a, b]$, and

$$x(a) = x^a, \qquad x(b) = x^b, \tag{2}$$

where x^a, x^b are given. Thus, $\delta = u(\cdot)$ and Δ is the set of admissible controls.

Suppose now that the performance vector $I(u(\cdot))$ has components

$$I_i(u(\cdot)) = \int_a^b L_i(x(t), t, u(t))\, dt, \qquad i = 1, 2,\ldots, m, \tag{3}$$

where the $L_i(\cdot)$ are C^2 functions on $\mathscr{R}$. We shall write

$$L(x(t), t, u(t)) = (L_1(x(t), t, u(t)),\ldots, L_m(x(t), t, u(t)))^T.$$

Remark 3.1. If the performance index is a scalar ($m = 1$), the problem is a fixed endpoint optimal control problem. As we shall see, the problem with $m \geqq 2$ can be treated as a fixed endpoint optimal control problem with isoperimetric constraints. The discussion here is based on Chapter 6 of Ref. 7.[7] Problems with other boundary conditions and other constraints can be treated in similar fashion (see Section 3.10 of Ref. 8).

As a consequence of Theorem 2.1, we have the following theorem.

[7] Note, however, that the notation differs somewhat from that used in Ref. 7.

Theorem 3.1. Suppose that Λ is a polyhedral cone with Λ^* generated by $\{H^1, H^2, \ldots, H^q\}$. Then, (i) in order for $u^0(\cdot) \in \Delta$ to be a nondominated control in Δ with respect to Λ, it is necessary that there exist $j \in \{1, 2, \ldots, q\}$ and $q - 1$ real numbers $\{r_k \mid k \neq j, k = 1, 2, \ldots, q\}$ such that $u^0(\cdot)$ results in the maximum of

$$H^{jT}I(u(\cdot)) = \int_a^b H^{jT}L(x(t), t, u(t))\, dt \tag{4}$$

for problem (1)–(2) subject to the isoperimetric conditions

$$H^{kT}I(u(\cdot)) = \int_a^b H^{kT}L(x(t), t, u(t))\, dt \geqq r_k, \qquad k \neq j, k = 1, 2, \ldots, q; \tag{5}$$

and (ii) if $u^0(\cdot)$ is the unique maximizing control for the problem stated in (i), then $u^0(\cdot)$ is nondominated in Δ with respect to Λ.

The following theorem follows from (i) of Theorem 3.1 above and from Theorem 2.1 of Chapter 6 of Ref. 7. Let $x^0(\cdot)$ denote the solution of (1) generated by $u^0(\cdot)$. Then, we have the following theorem.

Theorem 3.2. Suppose that Λ is specified as in Theorem 3.1 and $u^0(\cdot)$ is nondominated in Δ with respect to Λ. Then, there exist $j \in \{1, 2, \ldots, q\}$, $q - 1$ real numbers $\{r_k \mid k \neq j, k = 1, 2, \ldots, q\}$, a constant multiplier $\lambda = (\lambda_1, \lambda_2, \ldots, \lambda_q)^T$, a continuous function $p(\cdot)\colon [a, b] \to R^n$ with $p(t) \neq 0$ on $[a, b]$, and a function

$$\mathscr{H}(\cdot) : R^n \times R^1 \times R^p \times R^n \times R^q \to R^1,$$

where

$$\mathscr{H}(x, t, u, p, \lambda) = p^T f(x, t, u) + \lambda^T HL(x, t, u)$$

and H is the matrix whose jth row is H^{jT}, such that

(i) $\lambda_k \geqq 0$ for all $k \in \{1, 2, \ldots, q\}$ and $\lambda_k = 0$ if $k \neq j$ and $H^{kT}I(u^0) > r_k$;

(ii) $p(\cdot)$ is a solution of

$$\dot{p}_i(t) = -\partial\mathscr{H}(x^0(t), t, u^0(t), p(t), \lambda)/\partial x_i, \qquad i = 1, 2, \ldots, n;$$

(iii) $\mathscr{H}(x^0(t), t, u^0(t), p(t), \lambda) \geqq \mathscr{H}(x^0(t), t, u, p(t), \lambda)$ for all u satisfying $|u| \leqq \rho$;

(iv) the function $M(\cdot)\colon [a, b] \to R^1$, where

$$M(t) = \mathscr{H}(x^0(t), t, u^0(t), p(t), \lambda),$$

is continuous on $[a, b]$ and satisfies

$$\dot{M}(t) = \partial \mathscr{H}(x^0(t), t, u^0(t), p(t), \lambda)/\partial t \qquad \text{on every continuity interval of } u^0(\cdot).$$

Remark 3.2. Theorem 3.2 is a *necessary* condition for the problem in (i) of Theorem 3.1. Given an admissible control $u^0(\cdot)$, one can use Theorem 3.2 to test whether or not $u^0(\cdot)$ can be nondominated. Note that, by setting

$$r_k = H^{kT} I(u^0(\cdot)), \qquad k \neq j,\ k = 1, 2,..., q,$$

the statement in (i) that $\lambda_k = 0$ if $k \neq j$ and $H^{kT} I(u^0(\cdot)) < r_k$ becomes superfluous. In locating candidates for nondominated control, the set $\{r_k\}$ is predetermined; for each choice of $\{r_k\}$, the theorem yields a candidate.

The following theorem follows from Theorem 2.2 and is a special case of Theorem 3.1.

Theorem 3.3. (i) In order for $u^0(\cdot)$ to be efficient (Pareto optimal)[8] in Δ, it is necessary that there exist $j \in \{1, 2,..., m\}$ and $m - 1$ real numbers $\{r_k \mid k \neq j,\ k = 1, 2,..., m\}$ such that $u^0(\cdot)$ results in the maximum of

$$I_j(u(\cdot)) = \int_a^b L_j(x(t), t, u(t))\, dt$$

for problem (1)–(2), subject to isoperimetric conditions

$$I_k(u(\cdot)) = \int_a^b L_k(x(t), t, u(t)\, dt \geqq r_k\,, \qquad k \neq j,\ k = 1, 2,..., m;$$

and (ii) if $u^0(\cdot)$ is the unique maximizing control for the problem stated in (i), then $u^0(\cdot)$ is efficient (Pareto optimal) in Δ.

The following theorem is then analogous to Theorem 3.2.

Theorem 3.4.[9] Suppose that $u^0(\cdot)$ is efficient (Pareto optimal) in Δ. Then, there exist $j \in \{1, 2,..., m\}$, $m - 1$ real numbers $\{r_k \mid k \neq j,$ $k = 1, 2,..., m\}$, a constant multipliers $\lambda = (\lambda_1, \lambda_2, ..., \lambda_m)^T$, a continuous

[8] That is, $u^0(\cdot)$ is nondominated in Δ with respect to $\Lambda^{\leqq}$.

[9] This theorem corresponds to Theorem 2 of Ref. 9.

function $p(\cdot)\colon [a, b] \to R^n$ with $p(t) \neq 0$ on $[a, b]$, and a function $\mathscr{H}(\cdot)\colon R^n \times R^1 \times R^p \times R^n \times R^m \to R^1$ where

$$\mathscr{H}(x, t, u, p, \lambda) = p^T f(x, t, u) + \lambda^T L(x, t, u),$$

such that (i)–(iv) of Theorem 3.2, apply, with q replaced by m.

Now, let us turn to Theorems 2.3 and 2.4. As a consequence of Theorem 2.3, we have the following *sufficient* conditions.

Theorem 3.5. The control $u^0(\cdot) \in \Delta$ is nondominated in Δ with respect to Λ if one of the following conditions is satisfied: (i) there is a $\lambda \in \operatorname{Int} \Lambda^*$ such that $u^0(\cdot)$ yields the maximum of $\lambda^T I(u(\cdot))$ for all $u(\cdot) \in \Delta$; or (ii) there is a $\lambda \in \Lambda^*$, $\lambda \neq 0$, such that $u^0(\cdot)$ is the unique control yielding the maximum of $\lambda^T I(u(\cdot))$ for all $u(\cdot) \in \Delta$.

The *necessary* conditions embodied in the following theorem are a consequence of Theorem 2.4.

Theorem 3.6. Suppose that Λ is such that $\Lambda^\perp \neq \{0\}$ and Y is Λ-convex. If $u^0(\cdot) \in \Delta$ is nondominated in Δ with respect to Λ, then there is a $\lambda \in \Lambda^*$, $\lambda \neq 0$, such that $u^0(\cdot)$ maximizes $\lambda^T I(u(\cdot))$ for all $u(\cdot) \in \Delta$.

Remark 3.3. The problems involving the maximization of $\lambda^T I(u(\cdot))$ in Theorems 3.5 and 3.6 are optimal control problems. Usual necessary conditions (see, for example, Refs. 7–8) and sufficient conditions (see, for example, Refs. 10–11) for optimal control problems are therefore applicable.

To illustrate the implications of cone convexity, we shall now state a theorem embodying necessary conditions. If Λ is as specified in Theorem 3.1, then it follows from Remark 5.9 of Ref. 1 that

$$\Lambda^* = \{\lambda^T H \mid \lambda \in R^q, \lambda \geqq 0\},$$

where H is as specified in Theorem 3.2. This observation, together with Theorem 3.6, leads to the following theorem.

Theorem 3.7. Suppose that Λ is as specified in Theorem 3.1 and that $u^0(\cdot) \in \Delta$ is nondominated in Δ with respect to Λ. Then, there exist a nonnegative vector $\lambda = (\lambda_1, \lambda_2, ..., \lambda_q)^T$, a continuous function $p(\cdot)\colon [a, b] \to R^n$ with $p(t) \neq 0$ on $[a, b]$, and a function

$$\mathscr{H}(\cdot)\colon R^n \times R^1 \times R^p \times R^n \times R^q \to R^1,$$

where

$$\mathscr{H}(x, t, u, p, \lambda) = p^T f(x, t, u) + \lambda^T H L(x, t, u)$$

and H is the matrix whose jth row is H^{jT}, such that (ii)–(iv) of Theorem 3.2 apply.

Remark 3.4. Note that Theorems 3.2 and 3.7 are very similar. However, as a consequence of the assumption of cone convexity, Theorem 3.7 is stronger. The condition requiring existence of $j \in \{1, 2,..., q\}$ and of $\{r_k \mid k \neq j, k = 1, 2,..., q\}$ is dropped in Theorem 3.7, as is part of condition (i).

4. Sufficient Conditions for Cone Convexity

Cone convexity plays an important role in decision-making theory (see, for example, Refs. 1–4, 6, and 12–14, as well as Section 3). This concept was introduced in Ref. 1, and some necessary conditions as well as some sufficient conditions for cone convexity can be found there. Here, we extend some of the sufficient conditions to the dynamic case treated earlier in this paper.

Following the proof of Corollary 3.4 of Ref. 1, one can exhibit the following theorem.

Theorem 4.1. Let Δ be a convex set, and let $I(\cdot): \Delta \to R^m$ be given. Suppose that $\Lambda \subset R^m$ is a polyhedral cone and that its polar cone Λ^* is generated by $\{H^j \mid j = 1, 2,..., q\}$. If each $H^{jT}I(u(\cdot)), j = 1, 2,..., q$, defines a concave function on Δ, then $Y = \{I(u(\cdot)) \mid u(\cdot) \in \Delta\}$ is Λ-convex.

Remark 4.1. If $\Lambda_0 \subset \Lambda_1$ and Y is Λ_0-convex, then Y is Λ_1-convex. Hence, in verifying that Y is Λ-convex by means of Theorem 4.1, one can use a polyhedral cone contained in Λ.

Now, let us consider the following class of control problems. The differential system is

$$\dot{x}(t) = A(t)\,x(t) + B(t)\,u(t) + v(t), \tag{6}$$

where $A(t)$ is an $n \times n$ matrix, $B(t)$ is an $n \times p$ matrix, $u(t)$ is a p-vector, $v(t)$ is an n-vector, and the components of $A(\cdot)$, $B(\cdot)$, and $v(\cdot)$ are integrable functions on a given interval $[a, b]$.

Let $\Omega \subset R^p$ be convex and compact, and let $C \subset R^n$ be convex.

A control $u(\cdot)$: $[a, b] \to \Omega$ is *preadmissible* iff $u(\cdot)$ is measurable and $u(t) \in \Omega$ for all $t \in [a, b]$. A control $u(\cdot)$: $[a, b] \to \Omega$ is *admissible* iff it is preadmissible and the corresponding solution $x(\cdot)$ of (6) satisfied $x(a) = x^a$, x^a given, and $x(b) \in C$. Let Δ denote the set of admissible controls.

Finally, let there be given a criterion vector function $I(\cdot)$, whose components are defined by

$$I_i(u) = \int_a^b L_i(x(t), t, u(t))\, dt, \qquad i = 1, 2, \ldots, m, \tag{7}$$

where the $L_i(\cdot)$ are continuous on $R^n \times [a, b] \times \Omega$.

Given a preadmissible control $u(\cdot)$, let $\phi(\cdot, u(\cdot))\colon [a, b] \to R^n$ be an absolutely continuous function satisfying (6); that is,

$$\phi(t, u(\cdot)) = x^a + \int_a^t [A(s)\, \phi(s, u(s)) + B(s)\, u(s) + v(s)]\, ds$$

for all $t \in [a, b]$.

Let $\Phi(t)$ denote the fundamental matrix of the homogeneous system

$$\dot{x}(t) = A(t)\, x(t).$$

Then,

$$\phi(t, u(\cdot)) = \Phi(t)\, x^a + \Phi(t) \int_a^t \Phi^{-1}(s)[B(s)\, u(s) + v(s)]\, ds. \tag{8}$$

Then, from (8), we have the following lemma.

Lemma 4.1. Let $u^1(\cdot)$ and $u^2(\cdot)$ be preadmissible controls, and let $\mu \in [0, 1]$. Then,

$$\phi(t, \mu u^1(\cdot) + (1 - \mu)\, u^2(\cdot)) = \mu\phi(t, u^1(\cdot)) + (1 - \mu)\, \phi(t, u^2(\cdot))$$

for all $t \in [a, b]$.

The following lemma is a consequence of Lemma 4.1.

Lemma 4.2. (i) The set of admissible controls Δ is convex; and (ii) the set $X(t) = \{\phi(t, u(\cdot)) \mid u(\cdot) \in \Delta\}$ is convex for all $t \in [a, b]$.

Proof. (i) follows from Lemma 4.1 and the convexity of set C; and (ii) is a result of Lemma 4.1 and (i).

Theorem 4.2. Let Δ and $I(\cdot)$ be as specified above and Ω as in Theorem 4.1, and let $L = (L_1, L_2, \ldots, L_m)^T$. Suppose that, for all $t \in [a, b]$, each $H^{jT}L(x, t, u)$, $j = 1, 2, \ldots, q$, defines a concave function on $X(t) \times \Omega$. Then, $Y = \{I(u(\cdot)) \mid u(\cdot) \in \Delta\}$ is Λ-convex.

Proof. In view of Theorem 4.1 and Lemma 4.2, it suffices to show that each $H^{jT}I(u(\cdot))$ defines a concave function on Δ. Given $u^1(\cdot)$, $u^2(\cdot) \in \Delta$, $\mu \in [0, 1]$, we wish to show that

$$H^{jT}I(\mu u^1(\cdot) + (1 - \mu)\, u^2(\cdot)) \geqq \mu H^{jT}I(u^1(\cdot)) + (1 - \mu)\, H^{jT}\, I(u^2(\cdot)).$$

Note that

$$\begin{aligned} &H^{jT}I(\mu u^1(\cdot) + (1 - \mu)\, u^2(\cdot)) \\ &\quad = \int_a^b H^{jT}L(\phi(t, \mu u^1(\cdot) + (1 - \mu)\, u^2(\cdot)), t, \mu u^1(t) + (1 - \mu)\, u^2(t))\, dt \\ &\quad = \int_a^b H^{jT}L(\mu\phi(t, u^1(\cdot)) + (1 - \mu)\, \phi(t, u^2(\cdot)), t, \mu u^1(\cdot) + (1 - \mu)\, u^2(\cdot))\, dt \\ &\quad \geqq \int_a^b [\mu H^{jT}L(\phi(t, u^1(\cdot)), t, u^1(\cdot)) + (1 - \mu)\, H^{jT}L(\phi(t, u^2(\cdot)), t, u^2(\cdot))]\, dt \\ &\quad = \mu H^{jT}I(u^1(\cdot)) + (1 - \mu)\, H^{jT}I(u^2(\cdot)), \end{aligned}$$

where we have utilized Lemma 4.1 and the concavity of $H^{jT}L(x, t, u)$. This concludes the proof.

Corollary 4.1. Let Δ, $I(u(\cdot))$, and Y be as in Theorem 4.2. Suppose that, for all $t \in [a, b]$, each $L_i(x, t, u)$, $i = 1, 2,..., m$, defines a concave function on $X(t) \times \Omega$. Then, Y is Λ-convex for every $\Lambda \supset \Lambda^{\leqq}$.

Proof. The corollary follows at once from Theorem 4.2, Remark 4.1, and the observation that the rows of the $m \times m$ identity matrix form a generator for $\Lambda^{\geqq} = (\Lambda^{\leqq})^*$.

References

1. YU, P. L., *Cone Convexity, Cone Extreme Points and Nondominated Solutions in Decision Problems with Multiobjectives*, Journal of Optimization Theory and Applications, Vol. 14, No. 3, 1974.
2. YU, P. L., *Introduction to Domination Structures in Multicriteria Decision Problems*, Selected Proceedings of the Seminar on Multicriteria Decision Making, Edited by J. L. Cochrane and M. Zeleny, University of South Carolina Press, Columbia, South Carolina, 1973.
3. FREIMER, M., and YU, P. L., *An Approach Toward Decision Problems with Multiobjectives*, University of Rochester, CSS-72-03, 1972.
4. YU, P. L., *Nondominated Investment Policies in Stock Markets (Including an Empirical Study)*, University of Rochester, Systems Analysis Program, F-7222, 1973.

5. YU, P. L., and ZELENY, M., *The Set of all Nondominated Solutions in the Linear Cases and a Multicriteria Simplex Method*, Journal of Mathematical Analysis and Applications, Vol. 49, No. 2, 1975.
6. YU, P. L., and LEITMANN, G., *Compromise Solutions, Domination Structures and Salukvadze's Solution*, Journal of Optimization Theory and Applications, Vol. 13, No. 3, 1974.
7. HESTENES, M., *Calculus of Variations and Optimal Control Theory*, John Wiley and Sons, New York, New York, 1966.
8. LEITMANN, G., *An Introduction to Optimal Control*, McGraw-Hill Book Company, New York, New York, 1966.
9. DA CUNHA, N. O., and POLAK, E., *Constrained Minimization under Vector-Valued Criteria in Linear Topological Spaces*, Mathematical Theory of Control, Edited by A. V. Balakrishnan and L. W. Neustadt, Academic Press, New York, New York, 1967.
10. STALFORD, H., *Sufficient Conditions for Optimal Control with State and Control Constraints*, Journal of Optimization Theory and Applications, Vol. 7, No. 2, 1971.
11. LEITMANN, G., and SCHMITENDORF, W., *Some Sufficient Conditions for Pareto-Optimal Control*, Journal of Dynamical Systems, Measurement and Control, Vol. 95, No. 3, 1973.
12. FREIMER, M., and YU, P. L., *Some New Results on Compromise Solutions for Group Decision Problems*, Vol. 22, No. 6, 1976.
13. YU, P. L., *A Class of Solutions for Group Decision Problems*, Management Science, Vol. 19, No. 8, 1973.
14. FREIMER, M., and YU, P. L., *The Application of Compromise Solutions to Reporting Games*, Game Theory as a Theory of Conflict Resolutions, Edited by A. Rapoport, D. Reidel Publishing Company, Dordrecht, Holland, 1974.

Additional Bibliography

15. SCHMITENDORF, W., and LEITMANN, G., *A Simple Derivation of the Necessary Conditions for Pareto Optimality*, IEEE Transactions on Automatic Control, Vol. AC-19, No. 5, 1974.

III

Generalization of Domination Structures and Nondominated Solutions in Multicriteria Decision Making

K. BERGSTRESSER, A. CHARNES, AND P. L. YU

Abstract. The concepts of domination structures and nondominated solutions are important in tackling multicriteria decision problems. We relax Yu's requirement that the domination structure at each point of the criteria space be a convex cone (Ref. 1) and give results concerning the set of nondominated solutions for the case where the domination structure at each point is a convex set. A practical necessity for such a generalization is discussed. We also present conditions under which a locally nondominated solution is also a globally nondominated solution.

1. Introduction

Decision problems with multiple noncommensurable objectives can often be abstracted in the following way: the set of possible decisions, denoted by $X(\subseteq R^n)$ is called the *decision space*; and a vector function $f: X \rightarrow Y \subseteq R^l$ indicates the value of each decision $x \in X$ on l different criteria scales. $Y = f(X)$ is called the *criteria space* or *objective space*.

Based on guidelines implicitly or explicitly established by the decision maker, to each point $y \in Y$ we may associate a set $D(y)$, called the set of *domination factors* at y, such that, if $y' \neq y$ and $y' \in y + D(y)$, then y' is dominated by y. In Ref. 3, Yu works primarily with the case where $D(y)$ is a convex cone for each $y \in Y$.

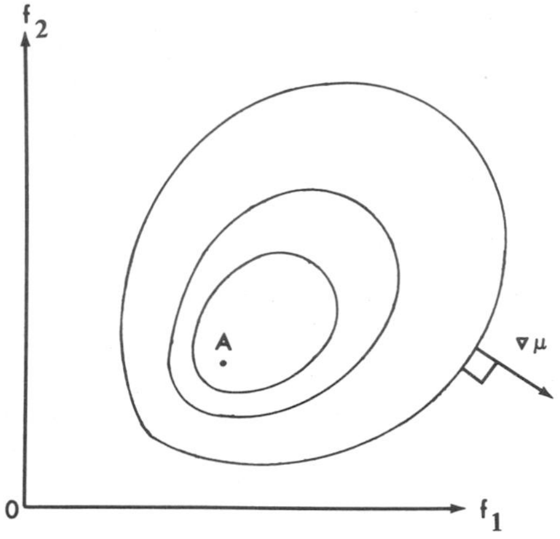

Fig. 1. Isovalued curves of the function $\mu(y)$.

In this paper, we treat the case where $D(y)$ is a convex set for all $y \in Y$. This extension will allow the concept of domination structures to be applied to a large class of decision-making problems.

The need for such an extension comes from the fact that there exists a large class of problems where, if we insist that $D(y)$ be a convex cone, then $D(y)$ must be $\{0\}$. As a consequence, each feasible choice will be nondominated, and our analysis will become useless. In order to see this point, assume that the decision maker has a convex utility[1] function $\mu(y)$ with isovalued curves described in Fig. 1.

Observe that, if we insist that each $D(y)$ be a convex cone, then $D(y) = \{0\}$ for each y. However, if we allow $D(y)$ to be a convex set which is not necessarily a cone, then for each $y \in Y$,

$$D(y) = \{d \mid \mu(d + y) < \mu(y)\},$$

a well-specified convex set. This domination structure captures almost all of the available information concerning the decision maker's preferences. In practice, if we have a *utility function* which is not unimodal or quasiconcave, and if we insist that $D(y)$ be a convex cone, then too much information concerning the decision maker's preferences will be lost.

[1] Such a situation arises, for instance, in the search for a house location near a nuclear testing area. The decision maker wants to stay as far away from the testing area as possible. If we use f_1 and f_2 as coordinates, we will have the indifference curves similar to those in Fig. 1 with the *center* point A as the testing point.

The generator scheduling problem for a utility company described in Ref. 2, pp. 696–699, is an example of this phenomenon. In order to avoid too much diversion from the results that we want to describe, we shall not elaborate the details of such an example.

2. Nondominated Points of Y

Given a decision space $X \subseteq R^n$, an associated criteria space $Y \subseteq R^l$, a function $f\colon X \to Y$ giving the value of each possible decision x on each criteria scale, and a domination structure $D(y)$, the first step toward a *good* decision is to identify the nondominated points of Y.

Definition 2.1. $y^0 \in Y$ is *globally nondominated* if there does not exist $y' \in Y$ such that $y' \neq y^0$ and $y' + d' = y^0$ for some $d' \in D(y')$. We denote the set of globally nondominated points of Y by $gN(Y, D(y))$.

Definition 2.2. $y^0 \in Y$ is *locally nondominated* if there exists a neighborhood N of y^0 such that there is no $y' \in N \cap Y$ with $y' \neq y^0$ and $y' + d' = y^0$ for some $d' \in D(y')$. We denote the set of locally nondominated points of Y by $lN(Y, D(y))$.

Remark 2.1. Clearly, $gN(Y, D(y)) \subseteq lN(Y, D(y))$.

One might naturally ask, when does $gN(Y, D(y)) = lN(Y, D(y))$? The next lemma partially answers this question.

Lemma 2.1. If $D(y)$ is a constant convex set D which contains 0 for each $y \in Y$ and if Y is convex, then $gN(Y, D(y)) = lN(Y, D(y))$.

Proof. Suppose that $y^0 \in lN(Y, D(y))$, but $y^0 \notin gN(Y, D(y))$. Therefore, there exists $y' \in Y$, $y' \neq y^0$, such that $y^0 = y' + d'$ for some $d' \in D$. Since Y is convex, $(y^0, y') \subseteq Y$.[2] Note that

$$\{y^0 - \alpha d' \mid 0 < \alpha < 1\} = (y^0, y').$$

Let N be any neighborhood of y^0. Therefore, there exists α', $0 < \alpha' < 1$, such that $y^0 - \alpha' d' \in N$. Put $y'' = y^0 - \alpha' d'$. Then, $y^0 = y'' + \alpha' d'$; and, since $\alpha' d' \in D$ (assuming $0 \in D$, $d' \in D$ implies $\alpha' d' \in D$ for all α', $0 < \alpha' < 1$), $y^0 \notin lN(Y, D(y))$. This contradicts $y^0 \in lN(Y, D(y))$.

[2] As in Stoer and Witzgall (Ref. 3, p. 32), $(y^0, y') = \{w = \alpha y^0 + (1 - \alpha) y' \mid 0 < \alpha < 1\}$ is the *open segment spanned by* y^0 *and* y'.

Remark 2.2. If y^0 is an isolated point of Y, i.e., there is a neighborhood N of y^0 in R^l such that $N \cap Y = \{y^0\}$, then $y^0 \in lN(Y, D(y))$. Let I be the set of isolated points of Y. In the light of Remarks 2.1 and 2.2, one might ask: does $lN(Y, D(y)) = gN(Y, D(y)) \cup I$? The following example demonstrates that the answer to this question is, in general, no. This example also illustrates the fact that, in general,

$$lN(Y, D(y)) \neq gN(Y, D(y)).$$

Example 2.1. Let

$$Y = \{(x, y) \mid x^2 + y^2 = 9 \quad \text{and} \quad x \geqq 0\}.$$

Let

$$D = \{(x, y) \mid x = 0,\ y \leqq 0\}.$$

Then, $(0, -3) \in lN(Y, D(y))$, but $(0, -3) \notin gN(Y, D(y))$, where $D(y) = D$ for all $y \in Y$. Also, $(0, -3)$ is not an isolated point of Y. We have

$$lN(Y, D(y)) = Y \qquad \text{and} \qquad gN(Y, D(y)) = \{(x, y) \in Y \mid y \geqq 0\}.$$

Remark 2.3. By Int D and Ext D we mean the set of interior points of D and the set of exterior points of D, respectively, both taken with respect to the usual topology on R^l. If $0 \in \operatorname{Int} D(y)$ for all $y \in Y$, then

$$lN(Y, D(y)) = gN(Y, D(y)) = \emptyset,$$

provided that Y is connected and contains at least two points. Further, if $0 \in \operatorname{Ext} D(y)$ for all $y \in Y$, then

$$lN(Y, D(y)) = Y.$$

In the light of these facts, for each $y \in Y$, we shall assume, from now on, that $0 \in \partial D(y)$, the boundary of $D(y)$.

Remark 2.4. Assuming that $0 \in \partial D(y)$, for each $y \in Y$, let $D'(y) = D(y) \backslash \{0\}$ and let $D^0(y) = D(y) \cup \{0\}$. Then,

$$gN(Y, D(y)) = gN(Y, D'(y)) = gN(Y, D^0(y))$$

and

$$lN(Y, D(y)) = lN(Y, D'(y)) = lN(Y, D^0(y)).$$

Therefore, we will assume that $0 \in \partial D(y)$; and, since it is then immaterial whether $0 \in D(y)$ or $0 \notin D(y)$, we will also assume that $0 \in D(y)$.

To simplify matters, we will assume for the remainder of this section that, for each $y \in Y$, $D(y) = D$, a fixed convex set satisfying $0 \in D$ and $0 \in \partial D$, and that Y contains at least two points. In this circumstance, we immediately obtain the following generalization of Lemma 4.1 in Yu (Ref. 1).

Lemma 2.2. The following results hold:

(i) $gN(Y, D) = \begin{cases} Y, & \text{if } D = \{0\}, \\ \emptyset, & \text{if } D = R^l; \end{cases}$

(ii) $lN(Y, D) = \begin{cases} Y, & \text{if } D = \{0\}, \\ I, & \text{the set of isolated points of } Y, \text{ if } D = R^l; \end{cases}$

(iii) $gN(Y, D_2) \subseteq gN(Y, D_1)$ if $D_1 \subseteq D_2$;

(iv) $lN(y, D_2) \subseteq lN(Y, D_1)$ if $D_1 \subseteq D_2$;

(v) $gN(Y + D, D) \subseteq gN(Y, D)$;

(vi) $gN(Y, D) = gN(Y + D, D)$ if there are no $d, d' \in D$ such that $d + d' = 0$;

(vii) $lN(Y, D) = lN(Y + D, D)$ if there are no $d, d' \in D$ such that $d + d' - 0$.

The following lemma shows that, unless $D = \{0\}$, a globally nondominated point must lie on the boundary of Y.

Lemma 2.3. If $D \neq \{0\}$, then $y^0 \in gN(Y, D)$ implies that $y^0 \in \partial(Y)$.

Proof. Suppose that $y^0 \in \text{Int } Y$. Therefore, there is a neighborhood of y^0, $N \subset Y$. $D \neq \{0\}$ implies that there is $d \in D$, $d \neq 0$. Therefore, the open segment $(0, d) \subseteq D$. Therefore, there is α, $0 < \alpha < 1$, such that $y^0 - \alpha d \in N$. Put $y' = y^0 - \alpha d$. Therefore, $y^0 = y' + \alpha d$ and $y^0 \notin gN(Y, D)$.

Definition 2.3. For an arbitrary set $S \subseteq R^n$, we define the *polar cone* of S by

$$S^* = \{y \in R^n \mid y \cdot s \leqq 0 \quad \text{for all } s \in S\}.$$

Lemma 2.4. The following result holds:

$$(y^0 - D^{**}) \cap Y = \{y^0\} \quad \text{implies that } y^0 \in gN(Y, D).$$

Proof. Suppose that $y^0 \notin gN(Y, D)$. Then, there exist $y' \in Y$, $y' \neq y^0$, and $d' \in D$ such that $y^0 = y' + d'$. Since $d' \in D \subseteq D^{**}$, we have $y' \in y^0 - D^{**}$ and $y' \in Y$. Therefore,

$$\{y^0, y'\} \subseteq (y - D^{**}) \cap Y,$$

contradicting the hypothesis.

The following lemma can be easily proved.

Lemma 2.5. $(y^0 - D^{**}) \cap Y = \{y^0\}$ iff there is no $y \in Y \backslash \{y^0\}$ such that $y^0 \in y + D^{**}$.

3. Globally Nondominated Solutions in Convex Polyhedral Domination Sets

Under the further assumption that D is a convex polyhedron, we give some necessary and sufficient conditions for $y^0 \in gN(Y, D)$. Following Stoer and Witzgall (Ref. 2, p. 36), we assume that a convex polyhedron is closed and can, therefore, be represented in the following way[3]:

$$D = \{d \mid A \cdot d \leqq c\} = \{d \in R^l \mid A_i \cdot d \leqq c_i, i = 1, \dots, q\}.$$

Theorem 3.1. Let $D = \{d \mid A \cdot d \leqq c\}$ be a convex polyhedron. Let

$$Y_j(y^0) = \{y \in Y \mid A_k \cdot y \geqq A_k \cdot y^0 - c_k, k = 1, \dots, q \quad \text{and} \quad k \neq j\}.$$

Then:

(i) $y^0 \in gN(Y, D)$ iff, for each $j \in \{1, \dots, q\}$, $A_j \cdot y^0 > A_j \cdot y + c_j$ for all $y \in Y_j(y^0)$, $y \neq y^0$;

(ii) $y^0 \in gN(Y, D)$ iff there is $j \in \{1, \dots, q\}$ such that $A_j \cdot y^0 > A_j \cdot y + c_j$ for all $y \in Y_j(y^0)$, $y \neq y^0$.

Proof. (i) For necessity, suppose that there is $j \in \{1, \dots, q\}$ and $y \in Y_j(y^0)$, $y \neq y^0$, such that $A_j \cdot y^0 \leqq A_j \cdot y + c_j$. Since $y \in Y_j(y^0)$, $A_k \cdot y^0 \leqq A_k \cdot y + c_k$ for $k \in \{1, \dots, q\}$, $k \neq j$. Therefore, $A_k \cdot (y^0 - y) \leqq c_k$ for all $k \in \{1, \dots, q\}$, and this implies that $y^0 - y \in D$. Therefore, $y^0 \in y + D$, implying that $y^0 \notin gN(Y, D)$. This is a contradiction.

---

[3] More generally, we could consider a *general convex polyhedron*, which we define as a set E such that $\bar{E}$ is a closed convex polyhedron. We can then represent E as follows: $E = \{d \in R^l \mid A \cdot d \leqq c \text{ and } B \cdot d < c\}$. Recall, we would assume, as above, that $0 \in \partial E$ and $0 \in E$.

Sufficiency follows immediately from sufficiency in (ii).

(ii) For sufficiency, suppose that $y^0 \notin gN(Y, D)$. Then, there is $y' \in Y$, $y' \neq y^0$, and $d' \in D$ such that $y^0 = d' + y'$. Therefore, for each $j \in \{1,..., q\}$, $y' \in Y_j(y^0)$, and $A_j \cdot y^0 \leqq A_j \cdot y' + c_j$. Again, this is a contradiction.

Necessity follows immediately from necessity in (i).

Remark 3.1. This theorem is a generalization of Lemma 4.2 and Theorem 4.2 in Yu (Ref. 1), since, if D is a (closed) convex cone Λ, a minimal generating set for Λ^*, $\{H^1,..., H^q\}$, actually forms a set, $\{A_1 ,..., A_q\}$, of contraints for D. Part (i) is a generalization of Corollary 4.3 in Yu (Ref. 1).

Theorem 3.2. Let $D = \{d \mid A \cdot d \leqq c\}$ be a convex polyhedron. Let

$$H_j(y^0) = \{y \in Y \mid A_j \cdot y \geqq A_j \cdot y^0 - c_j\}.$$

Then:

(i) if $y^0 \in gN(Y, D)$, then, for each $j \in \{1,..., q\}$ and for all $y \in H_j(y^0)$, $y \neq y^0$, there is $k(y) \in \{1,..., q\}\backslash\{j\}$ such that $A_{k(y)} \cdot y^0 > A_{k(y)} \cdot y + c_{k(y)}$;

(ii) if there exists $j \in \{1,..., q\}$ such that, for all $y \in H_j(y^0)$, $y \neq y^0$, $y \notin \bigcap_{k \neq j} H_k(y^0)$, then $y^0 \in gN(Y, D)$.

Proof. (i) Suppose that there exist $j \in \{1,..., q\}$ and $y \in H_j(y^0)$, $y \neq y^0$, such that, for all $k \in \{1,..., q\}\backslash\{j\}$, $A_k \cdot y^0 \leqq A_k \cdot y + c_k$. Since $y \in H_j(y^0)$, $A_j \cdot y^0 \leqq A_j \cdot y + c_j$. Therefore, $y^0 - y \in D$, implying that $y^0 \notin gN(y, D)$. This is a contradiction.

(ii) Suppose that $y^0 \notin gN(Y, D)$. Then, there exist $y' \in Y$, $y' \neq y^0$, and $d' \in D$ such that $y^0 = d' + y'$. Therefore, for each $j \in \{1,..., q\}$, $y' \in H_j(y^0)$ and, consequently, $y' \in \bigcap_{k \neq j} H_k(y^0)$. This is a contradiction.

Let

$$M^0(j) = \{y^0 \in Y \mid A_j \cdot y^0 > A_j \cdot y + c_j, \quad \text{for all } y \in Y_j(y^0), y \neq y^0\}.$$

Corollary 3.1. Let $D = \{d \mid A \cdot d \leqq c\}$ be a convex polyhedron. Then, $gN(Y, D) = M^0(j)$ for all $j \in \{1,..., q\}$.

Proof. Suppose that $y^0 \in gN(Y, D)$. By Theorem 3.1, $y^0 \in gN(Y,D)$ implies that, for $j \in \{1,..., q\}$, $A_j \cdot y^0 > A_j \cdot y + c_j$ for all $y \in Y_j(y^0)$, $y \neq y^0$. Therefore, $y^0 \in M^0(j)$.

Suppose that $y^0 \in M^0(j)$ but $y^0 \notin gN(Y, D)$. Then, there is $y \in Y$, $y \neq y^0$, and $d \in D$ such that $y + d = y^0$, i.e., $y^0 - y \in D$. Therefore, $y \in Y_j(y^0)$ for all j. Also, for all $j \in \{1,..., q\}$, $y^0 \notin M^0(j)$ since $A_j \cdot y^0 \leqq A_j \cdot y + c_j$. This contradicts $y^0 \in M^0(j)$ and, hence, $y^0 \in gN(Y, D)$.

Corollary 3.2. If $c_j > 0$ and $y^0 \in M^0(j)$, then y^0 must be an isolated point with respect to $Y_j(y^0)$ [i.e., there exists a neighborhood N of y^0 such that $N \cap Y_j(y^0) = \{y^0\}$].

Proof. Assume that y^0 is not an isolated point with respect to $Y_j(y^0)$. Define $f_A : R^l \to R$ by $f_A(y) = A_j \cdot y$ for all $y \in R^l$. Note that f_A is continuous. Therefore, letting $\epsilon = c_j/2$, there exists $\delta > 0$ such that, if $|y - y^0| < \delta$, then $|f_A(y) - f_A(y^0)| < \epsilon$. Let

$$N(\delta) = \{y \in R^l \mid |y - y^0| < \delta\}.$$

Therefore, there exists $y \neq y^0$ such that

$$y \in N(\delta) \cap Y_j(y^0)$$

and such that

$$|f_A(y) - f_A(y^0)| < \epsilon.$$

If $f_A(y) \geqq f_A(y^0)$, we have a contradiction, since $y^0 \in M^0(j)$ implies that

$$A_j \cdot y^0 > A_j \cdot y + c_j \geqq A_j \cdot y.$$

If $f_A(y^0) > f_A(y)$, we conclude that

$$f_A(y^0) - f_A(y) < \epsilon = c_j/2 < c_j,$$

a contradiction.

Remark 3.2. We observe that, if $D = \{x \mid A \cdot x \leqq c\}$, then $0 \in \partial D$ iff $c_j \geqq 0$ for all $j \in \{1,..., q\}$ and, for at least one $j \in \{1,..., q\}$, $c_j = 0$. Therefore, we will henceforth assume that $c_j \geqq 0$ for $j \in \{1,..., q\}$, consistent with our assumption that $0 \in \partial D$.

Corollary 3.3. If $y^0 \in gN(Y, D)$ and $c_j > 0$ and $Y_j(y^0)$ is connected, then $Y_j(y^0) = \{y^0\}$.

Proof. It follows immediately from Corollary 3.2.

To actually compute $M^0(j)$, we generalize Yu's work (Ref. 1) and proceed as follows. Let $\bar{r}_k$ be an upper bound (possibly $+\infty$) for the set

$\{A_k \cdot y \mid y \in Y\}$, and let $\underline{r}_k$ be a lower bound (possibly $-\infty$) for the set $\{A_k \cdot y \mid y \in Y\}$. Let $r(j) \in R^{q-1}$ be such that

$$r(j)_k = \begin{cases} r_k & \text{for} \quad k < j, \\ r_{k+1} & \text{for} \quad k \geqq j. \end{cases}$$

Let $\bar{r}(j) \in R^{q-1}$ be such that

$$\bar{r}(j)_k = \begin{cases} \bar{r}_k & \text{for} \quad k > j, \\ \bar{r}_{k+1} & \text{for} \quad k \geqq j. \end{cases}$$

Similarly, $\underline{r}(j) \in R^{q-1}$ is defined. For $\underline{r}(j) \leqq r(j) \leqq \bar{r}(j)$, define

$$Y(r(j)) = \{y \in Y \mid A_k \cdot y + c_k \geqq r_k\,,\, k \neq j, k = 1,\ldots, q\},$$

where

$$r_k = \begin{cases} r(j)_k & \text{if} \quad k < j, \\ r(j)_{k-1} & \text{if} \quad k > j. \end{cases}$$

Let $y^0(r(j))$ be the unique solution of

$$\max_{y \in Y(r(j))} A_j \cdot y,$$

if it exists, and if

(i) $A_j \cdot y^0(r(j)) > A_j \cdot y + c_j$ for all $y \in Y(r(j))$, $y \neq y^0(r(j))$,

(ii) $r_k = A_k \cdot y^0(r(j))$ for $k \neq j$ such that $c_k > 0$.

Put

$$Y^0(j) = \{y^0(r(j)) \mid \underline{r}(j) \leqq r(j) \leqq \bar{r}(j)\}.$$

Theorem 3.3. Assuming that $c_j \geqq 0$ for $j \in \{1,\ldots, q\}$, $gN(Y, D) = Y^0(j) = M^0(j)$ for $j \in \{1,\ldots, q\}$.

Proof. Using Corollary 3.3, we need only show that $M^0(j) = Y^0(j)$. Suppose that $y^0 \in M^0(j)$. Put $r_k = A_k \cdot y^0$. Then,

$$Y(r(j)) = \{y \in Y \mid A_k \cdot y + c_k \geqq r_k\,,\, k \neq j, k = 1,\ldots, q\} = Y_j(y^0).$$

Since $y^0 \in M^0(j)$,

$$A_j \cdot y^0 > A_j \cdot y + c_j$$

for all $y \in Y(r(j))$, $y \neq y^0$. Therefore, $y^0(r(j)) = y^0$ and $y^0 \in Y^0(j)$. Hence, $M^0(j) \subseteq Y^0(j)$.

Suppose that $y^0(r(j)) \in Y^0(j)$; i.e., for some $r(j)$, $\underline{r}(j) \leqq r(j) \leqq \bar{r}(j)$, $y^0(r(j))$ solves

$$\max_{y \in Y(r(j))} A_j \cdot y$$

and the following hold:

(i) $A_j \cdot y^0(r(j)) > A_j \cdot y + c_j$ for all $y \in Y(r(j))$, $y \neq y^0(r(j))$,

(ii) $r_k = A_k \cdot y^0(r(j))$ for $k \neq j$ such that $c_k > 0$.

We first show that $Y_j(y^0(r(j)) \subseteq Y(r(j))$. Note that $y^0(r(j)) \in Y(r(j))$ implies that, for $k \neq j$,

$$A_k \cdot y^0(r(j)) + c_k \geqq r_k ,$$

that is,

$$A_k \cdot y^0(r(j)) \geqq r_k - c_k ,$$

and

$$\text{if} \quad c_k = 0, \qquad A_k \cdot y^0(r(j)) \geqq r_k ,$$

$$\text{if} \quad c_k > 0, \text{ by (ii) } A_k \cdot y^0(r(j)) = r_k .$$

Suppose that $y' \in Y_j(y^0(r(j)))$. Then,

$$A_k \cdot y' + c_k \geqq A_k \cdot y^0(r(j)) \geqq r_k .$$

Thus, $y' \in Y(r(j))$. Since $y^0(r(j)) \in Y^0(j)$ and $Y_j(y^0(r(j))) \subseteq Y(r(j))$, it is seen that $y^0(r(j)) \in M^0(j)$.

Example 3.1. This example illustrates the necessity of condition (ii) in the definition of $y^0(r(j))$.

Let

$$D = \{y \in R \mid y \leqq 0, -y \leqq 1\}.$$

Thus,

$$A = \begin{bmatrix} A_1 \\ A_2 \end{bmatrix} = \begin{bmatrix} 1 \\ -1 \end{bmatrix} \quad \text{and} \quad C = \begin{bmatrix} C_1 \\ C_2 \end{bmatrix} = \begin{bmatrix} 0 \\ 1 \end{bmatrix}.$$

Let $Y = [0, 1] \cup \{3/2\}$. Therefore, $gN(Y, D) = \{3/2\}$. Let $j = 1$, $k = 2$, $r_2 = -1/4$. Then,

$$Y(r(1)) = \{y \in Y \mid -y \geqq (-1/4) - 1\} = \{y \in Y \mid y \leqslant 5/4\} = [0, 1].$$

Note that $y^0 = 1$ solves

$$\max_{y \in Y(r(1))} A_1 \cdot y$$

and $1 > y$ for all $y \in Y(r(1))$, $y \neq 1$. The difficulty arises since $A_2 \cdot 1 \neq r_2$; without condition (ii), we would have that

$$Y^0(1) = \{3/2, 1\} \neq gN(Y, D).$$

Remark 3.3. If D is an arbitrary convex set such that $\bar{D}$ is not a convex polyhedron, we could construct two convex polyhedra D_1 and D_1' such that $D_1 \subseteq D \subseteq D_1'$. By using Theorem 3.1 or 3.2, we can locate $gN(Y, D_1)$ and $gN(Y, D_1')$. Then, by Lemma 2.2(iii), we have that

$$gN(Y, D_1') \subseteq gN(Y, D) \subseteq gN(Y, D_1).$$

By this method, we can approximate $gN(Y, D)$, since every convex set D can be approximated from above and from below by a sequence of convex polyhedra

$$D_1 \subseteq D_2 \subseteq \cdots \subseteq D \subseteq \cdots \subseteq D_2' \subseteq D_1'.$$

References

1. Yu, P. L., *Cone Convexity, Cone Extreme Points, and Nondominated Solutions in Decision Problems with Multi-objectives*, Journal of Optimization Theory and Applications, Vol. 14, No. 3, 1974.
2. Charnes, A., and Cooper, W. W., *Management Models and Industrial Applications of Linear Programming*, Vol. 2, John Wiley and Sons, New York, New York, 1961.
3. Stoer, J., and Witzgall, C., *Convexity and Optimization in Finite Dimensions, I*, Springer-Verlag, Berlin, Germany, 1970.

IV

Compromise Solutions, Domination Structures, and Salukvadze's Solution

P. L. Yu and G. Leitmann

Abstract. We outline the concepts of compromise solutions and domination structures in such a way that the underlying assumptions and their implications concerning the solution concept suggested by Salukvadze may be clearer. An example is solved to illustrate our discussion.

1. Introduction

In daily decision making, we deal quite often with problems involving not only a single criterion. Rather, we may have multiple objective decision problems or decision problems involving more than one decision maker. For these types of decision problems, there is not as yet a universally accepted solution concept, even though there exist quite a few (see Refs. 1–38). Although we may use a recently derived concept of domination structures (Refs. 25, 27) to study the assumptions which underly each solution concept, there is no apparent reason to believe that a certain solution concept is superior to others. Although we, as researchers, consultants, or decision makers, can use the concept of domination structures to suggest a limitation of decisions to the set of nondominated solutions (which may contain more than one alternative), the final decision must depend on the judgment and/or the relative strength in negotiation or bargaining of the decision makers.

In the search for solution concepts, one often starts with static decision problems (one-stage decision problems) rather than with the much more complicated cases of dynamic decision problems. As a consequence, there is a natural gap among different researchers. Those dealing with static problems may be unaware of the implications of solution concepts in dynamic cases. On the other hand, those interested in dynamic problems may be unaware of current developments in solution concepts.

Recently, there appeared two interesting papers (Refs. 32–33) by Salukvadze, in which he considers dynamic decision problems with multiple objective functionals. He proposes a solution concept and works out the computational details. However, the discussion of the solution concept is not complete. The underlying assumptions as well as their implications are not entirely clear. The solution concept proposed by Salukvadze has been discussed independently in Ref. 24 and 34; in fact, it is a special case of the compromise solutions in Ref. 24.

In this note, we outline some underlying assumptions and their implications concerning the solution concept suggested by Salukvadze in order that its applicability may be clearer. In particular, in Section 2 we define the decision space and criteria space for both static and dynamic decision problems. We then focus on the concept of compromise solutions and list their known properties. Since Salukvadze's solution is a special case of compromise solutions, all properties described are applicable to his solution. In Section 3, we introduce the concept of domination structure and show how strong the assumption is that underlies compromise solutions (and hence Salukvadze's solution). Finally, in Section 4 we treat the example of Ref. 32 to illustrate the points made in Sections 2 and 3 so that the merits as well as the weaknesses of Salukvadze's solution may be more apparent.

2. Compromise Solutions

Suppose that we have to make a choice from a set of alternatives $X \subset R^n$ and that we can associate each alternative $x \in X$ with a set of criteria $(f_1(x),\dots, f_l(x))$. Let $f(x) = (f_1(x),\dots, f_l(x))$. We shall call X the *decision space*, while $Y = \{f(x) \mid x \in X\}$ is the *criteria space*. An element $y \in Y$ is often called the outcome of a decision.

Remark 2.1. Observe that the above definitions of decision and criteria spaces can be extended to dynamic cases; thus, the solution concepts that we describe here in fact are applicable to such cases. For instance, we can denote the set of all admissible strategies or controls μ by X (see Refs. 2, 4, 35 for admissible strategies or controls). Then, if for each μ we have $(f_1(\mu),..., f_l(\mu))$ as the performance indices, the criteria space can be specified by

$$Y = \{(f_1(\mu),..., f_l(\mu)) \mid \text{all admissible strategies (or controls) } \mu\}.$$

On the other hand, since we can convert integral payoffs into terminal payoffs by adding extra variables (see Refs. 2, 35), we may regard X as the set of all attainable terminal states in the enlarged space. Then, Y can be defined accordingly on X. Although conceptually we have no difficulty in defining X and Y in dynamic cases, to actually visualize them is a very difficult task. We shall describe the solution concepts in static cases and leave it to the reader to extend them to dynamic cases. However, in Section 4, we shall supply an example of a dynamic case.

Let

$$y_j^* = \sup\{f_j(x) \mid x \in X\}.$$

Then, $y^* = (y_1^*,..., y_l^*)$ is called the *utopia point* or *ideal point* of our problem with the interpretation that, whenever y^* is feasible, $f^{-1}(y^*)$ simultaneously maximizes each criterion. For simplicity, we shall assume that Y is compact. With some slight modification, the compactness assumption can be relaxed.

Given $y \in Y$, we define a class of *regret functions* by

$$R_p(y) = \| y^* - y \|_p = \Big[\sum_j (y_j^* - y_j)^p\Big]^{1/p}, \qquad p \geqslant 1. \tag{1}$$

Definition 2.1.[1] y^p and $f^{-1}(y^p)$ is the compromise solution with parameter $p \geqslant 1$ iff y^p minimizes $R_p(y)$ over Y.

For convenience, we shall call $R_p(y)$ the group regret of y with respect to p, while $y_j^* - y_j$ is the jth individual regret.

Remark 2.2. The solution concept proposed by Salukvadze is a compromise solution with $p = 2$. The corresponding group regret is associated with the Euclidean norm. The solution resulting from usual goal programming (Ref. 10) or simple majority rule can be regarded as a compromise solution with $p = 1$ (Ref. 28). Compromise solutions with $p = \infty$ correspond to a minimax criterion, because $y^{p=\infty}$ solves

$$\min_y \max_j \{y_j^* - y_j \mid j = 1 ..., l\}.$$

[1] Of course, $f^{-1}(\cdot)$ may be set-valued.

Remark 2.3. In group decision problems, we may use $y_j = f_j(x)$ to denote the jth decision maker's utility. It is reasonable to assume that compromise solutions cannot be acceptable to each decision maker unless each one has nonnegative utility. Thus, in Ref. 24, instead of X and Y, the decision and utility space for compromise solutions are defined respectively by

$$X_0 = \{x \in X \mid f(x) \geqq 0\} \qquad \text{and} \qquad Y_0 = \{f(x) \mid x \in X_0\}.$$

The utopia point y^* is then the point which has its jth component y_j^* maximizing $f_j(x)$ over X_0. With this modification, compromise solutions enjoy the property of *individual rationality* (i.e., nobody in the group has negative utility). In this note, we do not make such a modification because the individual rationality is an unnecessary nicety for one decision maker with multiple criteria. Observe that, except for individual rationality, all properties of compromise solutions (listed below) hold no matter whether or not we introduce the modification. Since compromise solutions can be applied to group decision problems or multicriteria decision problems with one decision maker, $y_j = f_j(x)$ in this note is called the jth objective, criterion, or utility, interchangeably. Also, since minimizing $f_j(x)$ over X is equivalent to maximizing $-f_j(x)$ over X, if $f_j(x)$ is the level of disutility we may interpret $-f_j(x)$ as the level of utility, and *vice versa*. Thus, without loss of generality, we can focus on the properties of compromise solutions defined in Definition 2.1. The extension to other cases is obvious.

Definition 2.2. Let

$$\Lambda^{\leqq} = \{d \in R^l \mid d \leqq 0\}.$$

We say that Y is $\Lambda^{\leqq}$-convex iff $Y + \Lambda^{\leqq}$ is convex.

For more discussion on cone convexity, see Ref. 25.

Compromise solutions enjoy the following properties (See Ref. 24).

(i) *Feasibility*. For each $p \geqq 1$, under the assumption that Y is compact, there is always a compromise solution. Observe that some solution concepts, such as *stable set* or *core*, may have no feasible solution.

(ii) *Least Group Regret*. Since y^p is the closest point over Y to the utopia point, the group regret is minimized in the sense of distance.

(iii) *No Dictatorship*. That is, the group decision is not completely determined by any one criterion $f_j(x)$. In contrast to lexicographical maximization[2] in which some f_j may not be considered in the final decision, each f_j is considered in the compromise solution.

(iv) *Pareto Optimality*. For $1 \leqq p < \infty$, each compromise solution is Pareto optimal. That is, there is no other $y \in Y$ such that $y \geqq y^p$ and $y \neq y^p$. This property comes directly from Definition 2.1.

(v) *Uniqueness*. Suppose that Y is $\Lambda^{\leqq}$-convex. Then, each y^p, $1 < p < \infty$, is unique.

(vi) *Symmetry or Principle of Equity*. If Y is convex and closed with respect to cyclical rotation, then for each p, $1 < p \leqslant \infty$, the $y_j^p, j = 1, \ldots, l$, are identical. For $p = 1$, there is at least one compromise solution y^1 such that the $y_j^1, j = 1, \ldots, l$, are identical. Thus, the principle of equity is implicit in the concept of compromise solutions.

(vii) *Independence of Irrelevant Alternatives*. Suppose that $X \subset X'$ and

$$\max_X f_j(x) = \max_{X'} f_j(x), \qquad j = 1, \ldots, l.$$

If a compromise solution with respect to X' happens to be x^0 in X, then x^0 is also a compromise solution with respect to X. Thus, the irrelevant alternatives may be discarded from the consideration for compromise solution. For instance, for $1 \leqq p < \infty$, we may discard those y of Y which are not Pareto optimal without affecting the final compromise solution.

If we treat p as a parameter of y^p, then $\{y^p \mid p \geqq 1\}$ enjoys the following properties.

(viii) *Continuity*. Suppose that Y is $\Lambda^{\leqq}$-convex. Then, as a function of p, y^p is continuous over $1 < p < \infty$. If y^1 (or y^∞) is unique, then y^p is continuous at $p = 1$ (or $p = \infty$) (see Ref. 36).

(ix) *Monotonicity and Bounds*. If $l = 2$ and Y is $\Lambda^{\leqq}$-convex, under a very mild condition it can be shown that $\{y_j^p\}$, $j = 1, 2$, is bounded by $[y_j^1, y_j^\infty]$; furthermore, it is a monotone function of p (see Ref. 24; for some generalization of this result for $l > 2$, see Ref. 36).

(x) *Monotonicity of the Group Utilities and the Individual Regrets*. Under the same assumptions as in (ix), it can be shown that both $\sum_j y_j^p$ and $\max_j\{y_j^* - y_j^p\}$ are decreasing functions of p. Observe that, if

[2] Let $X^0 = X$. For $k = 1, \ldots, l$, define

$$X^k = \{x^0 \in X^{k-1} \mid f_k(x^0) \geqq f_k(x),\ x \in X^{k-1}\}.$$

In lexicographical maximization, the final decision is on the set X^l. Observe that, if X^k contains only one point, the final decision is uniquely determined; $f_{k+1}, \ldots, f_l$ need not be considered.

$y_j = f_j(x)$ is the utility function for the jth decision maker, then $\sum_j y_j^p$ is the sum of the utilities and $\max_j\{y_j^* - y_j^p\}$ is the maximum individual regret. We may consider $\sum_j y_j^p$ and $\max_j\{y_j^* - y_j^p\}$ as group utility and individual regret, respectively, resulting from the compromise solution with parameter p. Our result says that, as p increases, the group utility decreases; however, individual regret reduces too. In this sense, the parameter p has a meaning for balancing group utility and individual regret. As a consequence, we see that simple majority rule (see Remark 2.2) is the rule which maximizes the group utility and most neglects the individual regret in the entire domain of compromise solutions (see Ref. 28).

The concept of compromise solutions can be generalized in several directions. For instance, we may replace $R_p(y)$ by

$$R_p(\alpha, y) = \Big[\sum_j \alpha_j^p (y_j^* - y_j)^p\Big]^{1/p}.$$

Then, most of the above properties, with a suitable modification, remain the same (see Ref. 24).

Although compromise solutions have merits, they are by no means perfect. Some associated assumptions are discussed in the next section. In addition, compromise solutions implicitly impose an intercomparison among the criteria or utilities through the group regret function (negative side of social welfare function). This imposition is not acceptable on some occasions. One should also observe that compromise solutions are not independent of a positive linear transformation of the $f_j(x)$. In fact, changing the scale of $f_j(x)$ has the same effect as changing the weight α in $R_p(\alpha, y)$. In applying this solution concept, one should be aware of this defect.

3. Domination Structures

In this section, we shall introduce briefly the concept of domination structures so that we can understand how strong an assumption has been imposed in compromise solutions. For more detailed discussion, we refer to Ref. 25; a simple summary can be found in Refs. 16, 27. An attempt to apply the concept to solve some practical problems can be found in Ref. 37.

Given two outcomes, y^1 and y^2, in the criteria space Y, we can write $y^2 = y^1 + d$, with $d = y^2 - y^1$. If y^1 is prefered to y^2, written $y^1 > y^2$, we can think of this preference as occurring because of d.

Now, suppose that the nonzero d has the additional property that, if $y = y^1 + \lambda d$ and $\lambda > 0$, then $y^1 \succ y$. Then, d will be called a *domination factor* for y^1. Note that, by definition, given a domination factor for y^1, any positive multiple of it is also a domination factor. It follows that, given $y^1 \succ y^2$, it is not necessarily true that $d = y^2 - y^1$ is a domination factor for y^1.

Let $D(y)$ be the set of all domination factors for y together with the zero vector of R^l. The family $\{D(y) | y \in Y\}$ is the *domination structure* of our decision problem. For simplicity, the structure will be denoted by $D(\cdot)$.

One important class of domination structures is $D(y) = \Lambda$, Λ a convex cone, for all $y \in Y$. In this case, we shall call Λ the *domination cone*.

Definition 2.3. Given Y, $D(\cdot)$, and two points y^1, y^2 of Y, by *y^1 is dominated by y^2* we mean

$$y^1 \in y^2 + D(y^2) = \{y^2 + d \mid d \in D(y^2)\}.$$

A point $y^0 \in Y$ is a *nondominated solution* iff there is no $y^1 \in Y$, $y^1 \neq y^0$, such that $y^0 \in y^1 + D(y^1)$. That is, y^0 is nondominated iff it is not dominated by any other outcomes. Likewise, in the decision space X, a point $x^0 \in X$ is a nondominated solution iff there is no x^1 in X such that

$$f(x^0) \neq f(x^1) \qquad \text{and} \qquad f(x^0) \in f(x^1) + D(f(x^1)).$$

The set of all nondominated solutions in the decision and criteria space will be denoted by $N_X(D(\cdot))$ and $N_Y(D(\cdot))$, respectively. Because of its geometric significance, a nondominated solution with respect to a domination cone Λ is also called a Λ-extreme point. The set of all Λ-extreme points is denoted by Ext $[Y \mid \Lambda]$.

Example 2.1. Let

$$\Lambda^{\leqq} = \{d \in R^l \mid d \leqq 0\}.$$

We see that y is Pareto optimal iff y is a $\Lambda^{\leqq}$-extreme point. That is, in the concept of Pareto optimality, one uses a constant domination cone $\Lambda^{\leqq}$. Observe that $\Lambda^{\leqq}$ is only $1/2^l$ of the entire space. When $l = 6$, for instance, $\Lambda^{\leqq}$ is only $1/64$ of R^6.

Example 2.2. In the additive weight method, one first finds a suitable weight $\lambda = (\lambda_1, \ldots, \lambda_l)$ and then maximizes

$$\lambda \cdot f(x) = \sum_j \lambda_j f_j(x)$$

over X or $\lambda \cdot y = \sum \lambda_j y_i$ over Y. Given λ, the solution concept implicitly uses the domination cone

$$\{d \in R^l \mid \lambda \cdot d < 0\}.$$

The concept of the additive weight method is closely related to that of ***trade-off*** in economic analysis. In fact, the latter can be a way of obtaining the weight λ. In order to illustrate this, let us limit ourselves to $l = 2$. The trade-off ratio of $f_2(x)$ with respect to $f_1(x)$ is defined by how many units of $f_2(x)$ we want to sacrifice in order to increase a unit of $f_1(x)$. Thus, the ratio gives the value of $f_2(x)$ in terms of $f_1(x)$. Although the ratio may be nonlinear in reality, in practice one often interprets it as a constant ratio. For the time being, assume that the ratio is constant and given by λ_2/λ_1, with $\lambda_1 > 0$. Clearly, our decision problem becomes one of maximizing $f_1(x) + (\lambda_2/\lambda_1)\, f_2(x)$ over X, which is equivalent to maximizing $\lambda_1 y_1 + \lambda_2 y_2$ over Y. The latter is essentially an additive weight method. To correctly predetermine the ratio λ_2/λ_1 is a very difficult task. In practice, one may first use his experience or judgement to set up its bounds. Once the bounds are set, we have partial information on preference. For instance, suppose that $1 < \lambda_2/\lambda_1 < 3$ is given. Then, we obtain valuable information, because it implies that, by using

$$\Lambda = \{(d_1, d_2) \mid d_1 + d_2 \leqq 0,\ d_1 + 3d_3 \leqq 0\}$$

as our domination cone, the optimal solution is contained in Ext $[Y \mid \Lambda]$ (see Fig. 1 and Ref. 27).

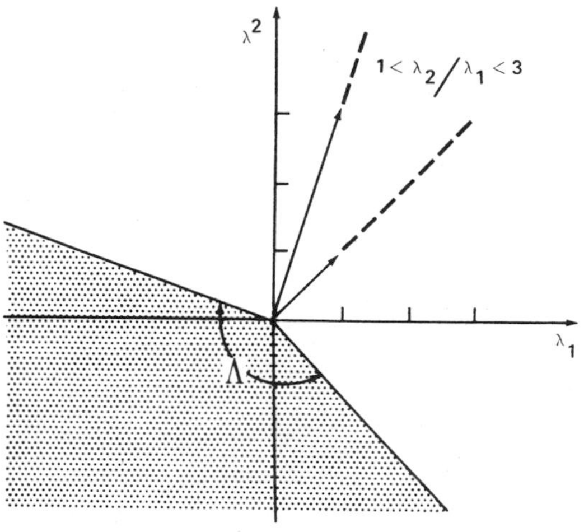

Fig. 1

Example 3.1. For compromise solutions with $1 \leqq p < \infty$, we can define the related domination structures by

$$D(y) = \{d \in R^l \mid \nabla R_p(y) \cdot d > 0\}$$

for $y \in Y$, where $\nabla R_p(y)$ is the gradient of $R_p(y)$, which is given by (1). This is because, in compromise solutions, the smaller $R_p(y)$ is, the more it is preferred. Since $R_p(y)$ is a differentiable convex function, $D(y)$ forms a domination cone for y. When $p = \infty$, we can again specify its domination structure; we shall not do so here in order not to distract from the main ideas. Observe that each $D(y)$ contains a half space, no matter what the parameter p is.

Remark 3.1. The domination cone induced by Pareto optimality (that is, $\Lambda^{\leqq}$) (see Example 3.1) is sometimes not large enough to encompass all information possessed by the decision makers. For instance, in Example 3.2, Λ is larger than $\Lambda^{\leqq}$; $\Lambda - \Lambda^{\leqq}$ is the valuable information missed by Pareto optimality. On the other hand, the domination structure induced by compromise solutions (Example 3.3) is such that each $D(y)$ contains a half space. To make this possible, too strong an assumption may have been imposed. For instance, in Example 3.2, although Λ is much larger than $\Lambda^{\leqq}$, it is much less than a half space. The difference between Λ and the half space of $D(y)$ induced by compromise solutions (when they are comparable) depends on how strong the assumptions are or how much information we have on the preference of the outcomes in Y. From the computational point of view, one has no difficulty in computing a compromise solution because the problem reduces to one of mathematical programming or to a control problem. However, because of the strong assumptions imposed or the information required, the compromise solutions may not be acceptable. A researcher may be able to find the compromise solutions, but these solutions may not be actually desirable for the decision maker (see the next section for further discussion). In daily decision problems, we are usually faced with problems of partial information. Both Pareto and compromise solutions cannot suitably explain the decision situation. The former does not adequately employ the partial information, while the latter may impose too strong an assumption, which is not conformable with the partial information. We believe that one should try to make suitable shoes (mathematical models) for the feet (the decision problems), rather than to cut off the feet in order to wear a given pair of shoes (one can use domination structures to attack some partial information decision problems, see Refs. 16, 27, 37).

4. Example

In order to illustrate the discussion of the previous section, consider a problem treated by Salukvadze (Ref. 32):

(i) *Playing Space*: $\{(z_1, z_2) \mid |z_1| \leqq 3\}$;

(ii) *Dynamic System*: $\dot{z}_1 = z_2, \quad \dot{z}_2 = u, \quad t \in [0, T]$;

(iii) *Control Set*: $u \in [-1, 1]$;

(iv) *Initial Condition*: $z_1(0) = 1, \quad z_2(0) = 0$;

(v) *Terminal Condition*: $z_1(T) = 0$, T not specified;

(vi) *Criteria*: (A) minimize $f_1(u(\cdot)) = T$,
(B) maximize $f_2(u(\cdot)) = z_2(T)$.

As mentioned in Remark 2.1, the decision space X and criteria space Y may be very difficult to visualize. However, in this example, letting X be the set of all admissible controls, the set of all Pareto-optimal solutions can be found. Note that (A) requires minimization; to permit use of the results in the earlier sections, we can consider maximizing $-f_1(u(\cdot))$. Rather than doing so, we shall consider the problem as is; the simple modifications required in the results are obvious.

It is readily established that the set of terminal points reachable from (1, 0) along a solution path that remains in the playing space is given by

$$\{(z_1, z_2) \mid z_1 = 0, -\sqrt{6} \leqq z_2 \leqq \sqrt{6}\}$$

(see Fig. 2).

Let us first consider terminal points on the subset

$$\{(z_1, z_2) \mid z_1 = 0, -\sqrt{2} \leqq z_2 \leqq \sqrt{6}\}$$

of the set of reachable terminal points. The control

$$u^*(t) = \begin{cases} -1 & \text{for } t \in [0, s], \\ 1 & \text{for } t \in (s, T] \end{cases}$$

renders a minimum of the transfer time T from (1, 0) to a given terminal point $(0, z_2)$, $-\sqrt{2} \leqq z_2 \leqq \sqrt{6}$, where the switching time s is a function of z_2. Conversely, for a given switching time s, the control $u^*(\cdot)$ results in a transfer time T and a corresponding terminal point $(0, z_2(T))$.

The solution path generated by $u^*(\cdot)$ is given by

$$z_1(t) = -\tfrac{1}{2}t^2 + 1, \qquad z_2(t) = -t, \qquad t \in [0, s],$$

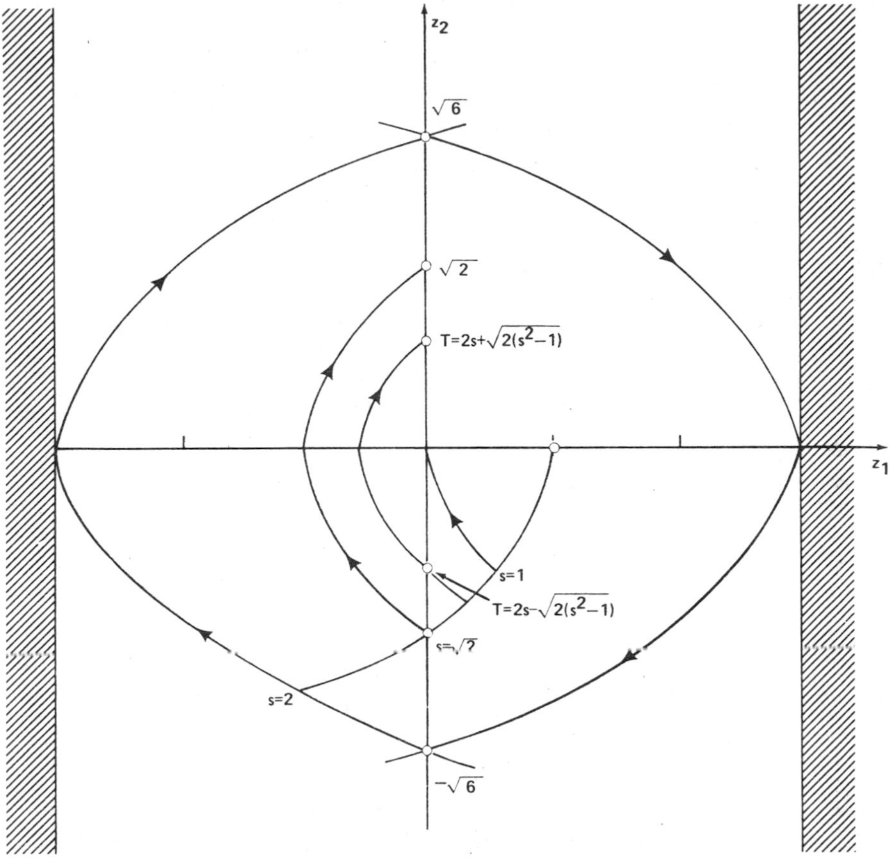

Fig. 2

and

$$z_1(\tau) = \tfrac{1}{2}\tau^2 - s\tau + 1 - s^2, \qquad z_2(\tau) = \tau - s, \qquad \tau \geqq 0, \tag{2}$$

where $\tau = t - s$. At termination, $z_1 = 0$ so that $\tau = s \pm \sqrt{[2(s^2 - 1)]}$, whence

$$T = 2s \pm \sqrt{[2(s^2 - 1)]}. \tag{3}$$

In order that T be defined, i.e., termination takes place, $s \geqq 1$.

Since, by (2), $dz_1/d\tau = 0$ for $\tau = s$, it follows that the minimum value of $z_1(\tau)$ is given by $1 - s^2$. But $z_1(\tau) \geqq -3$, so that $s \leqq 2$. Thus, $s \in [1, 2]$.

Since $\tau \geqq 0$,

$$\tau = s - \sqrt{[2(s^2 - 1)]} \geqq 0,$$

which is met iff $s \leqq \sqrt{2}$. Thus, for $s \in [1, \sqrt{2}]$, the portion of the solution path defined by (2) crosses the z_2-axis twice, the first time at

$$\tau = 2s - \sqrt{[2(s^2 - 1)]}$$

and the second time at

$$\tau = 2s + \sqrt{[2(s^2 - 1)]}.$$

Thus,

$$f_1 = 2s - \sqrt{[2(s^2 - 1)]}, \qquad f_2 = -\sqrt{[2(s^2 - 1)]}, \tag{4}$$

$$f_1 = 2s + \sqrt{[2(s^2 - 1)]}, \qquad f_2 = \sqrt{[2(s^2 - 1)]}. \tag{5}$$

For $s \in [\sqrt{2}, 2]$, the portion of the solution path given by (2) intercepts the z_2-axis once. Thus,

$$f_1 = 2s + \sqrt{[2(s^2 - 1)]}, \qquad f_2 = \sqrt{[2(s^2 - 1)]}. \tag{6}$$

These results are summarized in Table 1.

We shall now consider Curves 1 and 2 (see Fig. 3). Upon use of

$$df_2/df_1 = (df_2/ds)/(df_1/ds) \quad \text{and} \quad d^2f_2/df_1^2 = (d/ds)(df_2/df_1)(ds/df_1), \tag{7}$$

it is readily established that

$$df_2/df_1 > 0, \qquad d^2f_2/df_1^2 < 0 \tag{8}$$

for all points on Curves 1 and 2. Note also that Curve 1 joins Curve 2 smoothly at $(f_1, f_2) = (2, 0)$ corresponding to $s = 1$.

The domination cone for our problem is

$$\Lambda_1 = \{(d_1, d_2) \mid d_1 \geqq 0,\ d_2 \leqq 0\}.$$

It follows directly from Corollary 4.3 of Ref. 25 that a point $(f_1^0, f_2^0) \in \text{Ext}[Y \mid \Lambda_1]$ (i.e., is noninferior) iff f_1^0 is the unique minimum of f_1 for all $(f_1, f_2) \in Y$ and $f_2 \geqq f_2^0$.

Table 1

	f_1	f_2	s
Curve 1	$2s - \sqrt{[2(s^2 - 1)]}$	$-\sqrt{[2(s^2 - 1)]}$	$1 \leqq s \leqq \sqrt{2}$
Curve 2	$2s + \sqrt{[2(s^2 - 1)]}$	$\sqrt{[2(s^2 - 1)]}$	$\sqrt{2} \leqq s \leqq 2$

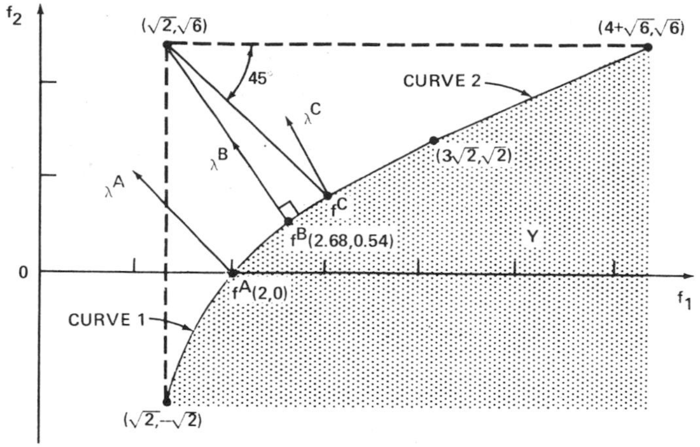

Fig. 3

Since $u^*(\cdot)$ renders the unique minimum of $f_1 = T$ for a given $f_2 = z_2(T) \in [-\sqrt{2}, \sqrt{6}]$, and $df_1/ds > 0$ on Curves 1 and 2, the points on these curves, and only these points, are noninferior among points (f_1, f_2) with $f_2 \in [-\sqrt{2}, \sqrt{6}]$.

On the other hand, consider $z_2(T) \in [-\sqrt{6}, -\sqrt{2})$. Among all (f_1, f_2) with $f_2 \geqslant z_2(T)$, the unique minimum of f_1 is rendered by the control $u^*(t) \equiv -1$. Thus, the point $(\sqrt{2}, -\sqrt{2})$ dominates all $(f_1, f_2) \in Y$ with $f_2 < -\sqrt{2}$. Consequently, the point on Curves 1 and 2 are the only noninferior points for all $(f_1, f_2) \in Y$.

Finally, we observe that, since $d^2f_2/df_1^2 < 0$ along Curves 1 and 2, f_2 is a strictly concave function of f_1. Since these curves represent $\mathrm{Ext}[Y \mid \Lambda_1]$, Y is Λ_1-convex. In order to see this point, we shall show that $Y + \Lambda_1$ is a convex set.

Let $y^1, y^2 \in Y + \Lambda_1$. Given α, $0 < \alpha < 1$, we show that

$$\alpha y^1 + (1 - \alpha)y^2 \in Y + \Lambda_1 .$$

By hypothesis, we can write

$$y^1 = f^1 + h_1 , \qquad y^2 = f^2 + h_2 ,$$

where $f^1, f^2 \in \mathrm{Ext}[Y \mid \Lambda_1]$ and $h_1, h_2 \in \Lambda_1$. Then,

$$\alpha y^1 + (1 - \alpha)y^2 = \alpha f^1 + (1 - \alpha)f^2 + \alpha h_1 + (1 - \alpha)h_2.$$

Since

$$\alpha h_1 + (1 - \alpha)h_2 \in \Lambda_1 ,$$

it suffices to show that

$$\alpha f^1 + (1 - \alpha)f^2 \in Y + \Lambda_1 .$$

Let

$$f^1 = (f_1^1, f_2(f_1^1)) \qquad \text{and} \qquad f^2 = (f_1^2, f_2(f_1^2)),$$

where $f_2(f_1)$ is specified as in Table 1. Then,

$$\begin{aligned} \alpha f^1 + (1-\alpha)f^2 &= (\alpha f_1^1 + (1-\alpha)f_1^2, \alpha f_2(f_1^1) + (1-\alpha)f_2(f_1^2)) \\ &= (\alpha f_1^1 + (1-\alpha)f_1^2, f_2(\alpha f_1^1 + (1-\alpha)f_1^2) - \beta) \\ &= (\alpha f_1^1 + (1-\alpha)f_1^2, f_2(\alpha f_1^1 + (1-\alpha)f_1^2)) + (0, -\beta) \\ &\in Y + \Lambda_1, \end{aligned}$$

where $\beta \geqslant 0$ and where the second equality is due to the fact that f_2 is a concave function of f_1.

Note that the Λ_1-convexity of Y is very desirable because all the properties listed in Section 2 are applicable. It also allows us to visualize the shape of Y as depicted in Fig. 3.

Referring to Fig. 3, let λ^A, λ^B, and λ^C be the normal vectors to ∂Y at the point f^A, f^B, and f^C. Observe that, at point f^A, the slope of the curve which represents $\text{Ext}[Y \mid \Lambda_1]$ is one. Thus, f^A is a compromise solution with $p = 1$. At f^B, the line from the utopia point to f^B is orthogonal to $\text{Ext}[Y \mid \Lambda_1]$. Thus, f^B is a compromise solution with $p = 2$. At f^C, both criteria suffer an equal regret, because $f_1^C - \sqrt{2} = \sqrt{6} - f_2^C$. In fact, f^C is the compromise solution with $p = \infty$.

The following may be worth noting.

(i) The solution suggested by Salukvadze is the point f^B which is the compromise solution with parameter $p = 2$ (see Remark 2.2).

(ii) All compromise solutions as functions of p vary continuously and monotonically from f^A to f^C. There is no special reason to pick f^B [see Properties (viii) and (ix) of Section 2].

(iii) The compromise solutions have the property that, when p is increased, the sum of the *resulting utility* [i.e., $(-f_1) + f_2$] decreases and the maximum of the individual regrets (i.e., $f_1 - \sqrt{2}$ and $\sqrt{6} - f_2$) decreases, too. How to select a p so that the group utility (the sum of the individual utilities) and the individual regret are best balanced remains to be answered [see Property (x)].

(iv) Point f^B also corresponds to the unique maximum of $\lambda^B \cdot f = \lambda_1^B f_1 + \lambda_2^B f_2$. For any weight vector $\lambda \neq \lambda^B$, f^B cannot correspond to the maximum of $\lambda \cdot f$, because f_2 is a strictly concave function of f_1 on $\text{Ext}[Y \mid \Lambda_1]$. By selecting $p = 2$, we determine implicitly the weight vector λ^B. Why should one not use another λ? Is the resulting solution desirable?

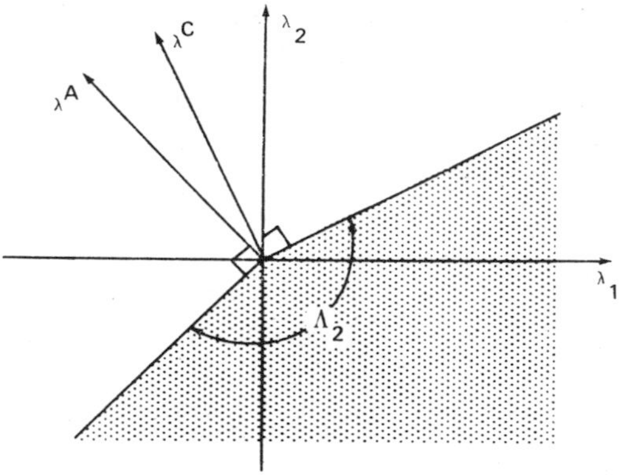

Fig. 4

(v) Let

$$\Lambda_2 = \{d \in R^2 \mid \lambda^A \cdot d \leqq 0, \lambda^C \cdot d \leqq 0\}$$

(see Fig. 4). One can show that

$$\text{Ext}[Y \mid \Lambda_2] = \text{curve}[f^A, f^C].$$

Thus, by restricting ourselves to compromise solutions, we assume implicitly that each domination cone $D(y)$, $y \in Y$, contains Λ_2. Is this assumption too strong?

References

1. Von Neumann, J., and Morgenstern, O., *Theory of Games and Economic Behavior*, Princeton University Press, Princeton, New Jersey, 1947.
2. Isaacs, R., *Differential Games*, John Wiley and Sons, New York, New York, 1965.
3. Isaacs, R., *Differential Games: Their Scope, Nature, and Future*, Journal of Optimization Theory and Applications, Vol. 3, pp. 283–295, 1969.
4. Blaquiere, A., Gerard, F., and Leitmann, G., *Quantitative and Qualitative Games*, Academic Press, New York, New York, 1969.
5. Ho, Y. C., *Final Report of the Frst International Conference on the Theory and Applications of Differential Games*, Amherst, Massachusetts, 1970.
6. Owen, G., *Game Theory*, W. B. Saunders Company, Philadelphia, Pennsylvania, 1968.
7. Rapoport, A., *N-Person Game Theory—Concepts and Applications*, The University of Michigan Press, Ann Arbor, Michigan, 1970.
8. Arrow, K. J., *Social Choice and Individual Values*, Cowles Commission Monograph No. 12, 1951.

9. Luce, R. D., and Raiffa, H., *Games and Decisions*, John Wiley and Sons, New York, New York, 1967.
10. Charnes, A., and Cooper, W. W., *Management Models and Industrial Applications of Linear Programming, Vols. I and II*, John Wiley and Sons, New York, New York, 1971.
11. Friedman, A., *Differential Games*, John Wiley and Sons (Interscience Publishers), New York, New York, 1971.
12. Shapley, L. S., *A Value for N-Person Games*, Contributions to the Theory of Games, II, Edited by H. W. Kuhn and A. W. Tucker, Princeton University Press, Princeton, New Jersey, 1953.
13. Blackwell, D., and Girschick, M. A., *Theory of Games and Statistical Decisions*, John Wiley and Sons, New York, New York, 1954.
14. Ferguson, T. S., *Mathematical Statistics, A Decision Theoretic Approach*, Academic Press, New York, New York, 1967.
15. Fishburn, P. C., *Utility Theory for Decision Making*, John Wiley and Sons, New York, New York, 1970.
16. Freimer, M., and Yu, P. L., *An Approach Toward Decision Problems with Multiobjectives*, University of Rochester, Center for System Science, Report No. 72-03, 1972.
17. Geoffrion, A. M., *Proper Efficiency and The Theory of Vector Maximization*, Journal of Mathematical Analysis and Applications, Vol. 22, pp. 618–630, 1968.
18. MacGrimmon, K. R., *Decision Making Among Multiple-Attribute Alternatives: A Survey and Consolidated Approach*, The Rand Corporation, Memorandum No. RM-4823-ARPA, 1963.
19. Raiffa, H., *Decision Analysis*, Addison-Wesley Publishing Company, Reading, Massachusetts, 1968.
20. Raiffa, H., *Preferences for Multi-Attributed Alternatives*, The Rand Corporation, Memorandum No. RM-5868-POT/RC, 1969.
21. Vincent, T. L., and Leitmann, G., *Control-Space Properties of Cooperative Games*, Journal of Optimization Theory and Applications, Vol. 6, pp. 91–113, 1970.
22. Leitmann, G., Rocklin, S., and Vincent, T. L., *A Note on Control Space Properties of Cooperative Games*, Journal of Optimization Theory and Applications, Vol. 9, pp. 379–390, 1972.
23. Stalford, H. L., *Criteria for Pareto-Optimality in Cooperative Differential Games*, Journal of Optimization Theory and Applications, Vol. 9, pp. 391–398, 1972.
24. Yu, P. L., *A Class of Solutions for Group Decision Problems*, Management Science, Vol. 19, pp. 936–946, 1973.
25. Yu, P. L., *Cone Convexity, Cone Extreme Points and Nondominated Solutions in Decision Problems with Multiobjectives*, Journal of Optimization Theory and Applications, Vol. 14, No. 3, 1974.

26. ZELENY, M., *Linear Multiobjective Programming*, The University of Rochester, Graduate School of Management, Ph.D. Thesis, 1972.
27. YU, P. L., *Introduction to Domination Structures in Multicriteria Decision Problems*, Multiple Criteria Decision Making, Edited by J. Cochrane and M. Zeleny, University of South Carolina Press, Columbia, South Carolina, 1973.
28. FREIMER, M., and YU, P. L., *The Application of Compromise Solutions to Reporting Games*, Game Theory as a Theory of Conflict Resolution, Edited by A. Rapoport, D. Reidel Publishing Company, Dordrecht, Holland, 1974.
29. YU, P. L., and ZELENY, M., *The Set of All Nondominated Solutions in the Linear Case and A Multicriteria Simplex Method*, Journal of Mathematical Analysis and Applications, Vol. 49, No. 2, 1975.
30. ZADEH, L. A., *Optimality and Non-Scalar-Valued Performance Criteria*, IEEE Transactions on Automatic Control, Vol. AC-8, pp. 59–60, 1963.
31. DACUNCHA, N. O., and POLAK, E., *Constrained Minimization Under Vector-Valued Criteria in Finite Dimensional Space*, Journal of Mathematical Analysis and Applications, Vol. 19, pp. 103–124, 1967.
32. SALUKVADZE, M. E., *Optimization of Vector Functionals, I, Programming of Optimal Trajectories* (in Russian), Avtomatika i Telemekhanika, No. 8, pp. 5–15, 1971.
33. SALUKVADZE, M. E., *Optimization of Vector Functionals, II, The Analytic Construction of Optimal Controls* (in Russian), Avtomatika i Telemekhanika, No. 9, pp. 5–15, 1971.
34. HUANG, S. C., *Note on the Mean-Square Strategy fot Vector-Valued Objective Functions*, Journal of Optimization Theory and Applications, Vol. 9, No. 5, pp. 364–366, 1972.
35. LEITMANN, G., *An Introduction to Optimal Control*, McGraw-Hill Book Company, New York, New York, 1966.
36. FREIMER, M., and YU, P. L., *Some New Results on Compromise Solutions*, Vol. 22, No. 6, 1976.
37. YU, P. L., *Nondominated Investment Policies in Stock Markets (Including an Empirical Study)*, University of Rochester, Graduate School of Management, Systems Analysis Program, Series No. F7222, 1973.
38. SHUBIK, M., *Readings in Game Theory and Political Behavior*, Doubleday and Company, New York, New York, 1954.

Additional Bibliography

39. SALUKVADZE, M., *On the Existence of Solutions in Problems of Optimization Under Vector-Valued Criteria*, Journal of Optimization Theory and Applications, Vol. 13, No. 2, 1974.

V

Multiple-Objective Optimization: Proper Equality Constraints (PEC) and Maximization of Index Vectors[1]

J. G. LIN

Abstract. In solving many practical problems we have to deal with conflicting multiple objectives (in performance, cost, gain, errors, risks, payoffs, etc.). Most multiple-objective problems do not have "supreme" solutions that can satisfy all of the objectives simultaneously. Many broad-sense optimality criteria, such as the Pareto optimum, the efficient point, the noninferior point, etc., have been introduced in various contexts, so that most multiple-objective problems have optimal solutions. However, such optimal solutions in general yield *nonunique vectors of optimal indexes* of the multiple objectives. In most cases, we have to make appropriate tradeoffs, compromises, or choices among those optimal solutions. To obtain the set of all such optimal solutions and, in particular, the set of all optimal index vectors, say for a comprehensive study on compromises, a usual practice is to optimize linear combinations of the multiple objectives for various weights. The success of such an approach relies heavily on a certain directional convexity condition; in other words, if such convexity is absent, this method will fail to obtain essential subsets. *The method of proper equality constraints* (PEC), however, relies on no convexity condition at all, and by it we can systematically obtain the *entire* set. In this paper, we attempt

[1] This work was supported by the National Science Foundation under Grant GK-32701; this paper is based on Ref. 18.

to lay the foundation for the method of PEC. We concentrate on the set of *all maximal index vectors*, for most of the broad definitions of optimal solutions are expressed in terms of maximal (index) vectors (see Ref. 1). We introduce the notion of *quasisupremal vector* as a substantially equivalent substitute for, but a rather practical and useful extension of, the notion of maximal vector. We then propose the method of PEC for computing the set of all quasisupremal (or maximal) index vectors. Various tests for quasisupremality (maximality, or optimality) are derived. An illustrative example in the allocation of funds is given. One of the important conclusions is that *optimizing the index of one objective* with the indexes of all other objectives equated to some arbitrary constants *may still result in inferior solutions*. The "sensitivity" to variations in these constants is also examined.

1. Introduction

1.1. In solving many system design or control problems, we have to optimize system performance under conflicting multiple performance objectives; in solving many planning or allocation problems, we have to deal with conflicting multiple economic objectives; in making most decisions or statistical hypotheses, we wish to minimize risks or errors of many kinds; and in most multiperson game problems, there are naturally conflicting multiple-payoff objectives to deal with. (See Refs. 1–12 and 17–21.) Suppose that in a typical problem there are N conflicting objectives ($N \geqslant 2$) and that the *maximization* of the index of each such objective is desired. (If the minimization of some of the indexes is desired instead, consider the maximization of the negative of the index.) Let z_i ($i = 1,..., N$) denote the *index* of the ith objective, and call $(z_1 ,..., z_N)$ an *index vector* for these multiple objectives. (For example, z_i may denote a value of the ith real-valued objective function.) Can all these indexes be maximized simultaneously? That is, does there exist an ideal, most desirable, optimal index vector $(z_1^*,..., z_N^*)$ *among those attainable* such that $z_i^* = \max z_i$ for all i? The answer generally is no, unfortunately. (If it is yes, these indexes will not be in real conflict and will be no more than one single index in different forms.)

Consider an alternative definition of optimality. Suppose a vector $z = (z_1 ,..., z_N)$ is considered to be *superior to* another vector $z' = (z_1',..., z_N')$ if $z_i \geqslant z_i'$ for all i and $z_j > z_j'$ for at least one j. Then we can consider an attainable index vector to which *no attainable index vector is superior* as an optimal index vector. Such an optimal index vector is called a *maximal vector* (Ref. 1). As pointed out in Ref. 1, maximal vector and weak-maximal vector (see Ref. 1 or Appendix A.2 for definition) are the two basic notions underlying various alternative definitions of optimality, such as Pareto's optimum (Refs. 2 and 3), Koopman's efficiency (Refs. 3 and 4), Zadeh's noninferiority (Refs. 5–10), and so forth, in different contexts.

In particular, consider a problem in which the multiple indexes are evaluated according to given multiple objective functions, say $J_1, \ldots, J_N$, at each "point" x as $z_i = J_i(x)$, $i = 1, \ldots, N$. A feasible solution x^0 is a *Pareto-optimal solution* (Refs. 1–3) if there exists no feasible solution x such that $J_i(x) \geqslant J_i(x^0)$ for all i and $J_j(x) > J_j(x^0)$ for at least one j. In this case $(J_1(x), \ldots, J_N(x))$ is an attainable index vector if x is a feasible solution. (A solution is feasible if it satisfies the constraints given in the problem.) It is clear from the above definitions that Pareto's optimum is defined via the notion of maximal vector.

1.2. Maximal vectors are generally not unique; in some problems there may be a few, while in others there may be an infinite number of them. Since none of the maximal vectors is to be superior to any others, we generally have to make appropriate tradeoffs, compromises, or choices according to some subjective or objective criteria. To obtain the set of *all maximal vectors* (or equivalently the set of all Pareto-optimal solutions), say for comprehensive studies on tradeoffs, compromises, or choices, it is a usual practice to maximize the family of linearly scalarized indexes (linear combinations of the multiple indexes) with weighting coefficients as the parameters (Refs. 5 and 6). The usefulness, or the reliability, of such a method depends on the convexity condition of *the set of all attainable index vectors*. It has been shown in Ref. 1 that if the latter set is *p-directionally convex for some nonzero positive vector p* in $\mathscr{R}^N$, *any* maximal vector can be obtained by maximizing some nonzero, positive linear combination of the indexes. This is the weakest condition so far available for assuring the reliability. In other words, if the assumption of directional convexity is invalid, the method of linear combinations will fail to include subsets of maximal vectors that are essential to making appropriate tradeoffs, compromises, or choices. (See Refs. 12 and 19–21 or Example 2.1 for illustration.)

Fortunately, the method of *proper equality constraints* (Refs. 10 and 18–21) and that of proper inequality constraints (Refs. 11, 18, and 21) make *no assumption on convexity or directional convexity* at all. This paper attempts to lay theoretical foundations for the former method and to offer some geometrical insights. Without loss of generality (but indeed making use of the several advantages pointed out in Ref. 1), we shall focus our attention mainly on *the maximization of vectors on the space of indexes*. In Refs. 19–21, we discuss Pareto-optimal solutions and the optimization under multiple objectives on the space of solutions.

1.3. In Section 2, we shall introduce the *quasisupremal vector* as a substantially equivalent, but rather useful, substitute for maximal vector. It will be seen there that a quasisupremal vector is practically as good as a maximal vector, and that with a fairly weak conditioning, the set of quasisupremal vectors is precisely that of maximal vectors. We shall then introduce the method of *proper equality constraints* (PEC) in Section 3 for computing the set of all quasisupremal vectors. It will be seen there that maximizing one index with all other indexes equated to some arbitrary constants *does not necessarily* produce quasisupremal (or maximal) vectors. Global tests for quasisupremality will be developed in Section 4, and local and pointwise tests in Section 5. Finally, an example of allocating funds to two activities will be given in Section 6 for illustrating the application of the method. Appendix A.1 contains some modifications for applications to the minimization of index vectors; Appendix A.2 discusses relations of quasisupremal vectors to weak-maximal vectors.

1.4. The following are some useful notations and preliminaries. We shall mainly be concerned with finite-dimensional Euclidean spaces $\mathscr{R}^k$. An element y of $\mathscr{R}^k$, denoted by a k-tuple $(y_1, \ldots, y_k)$ of real numbers, will be called *a vector or a point* depending on which is more suggestive or conventional in the context. We shall say that a vector y is positive if $y \geqslant 0$, definitely positive if $y \geqq 0$, and strictly positive if $y > 0$, where $\geqslant$, $\geqq$, and $>$ are partial orders defined as follows. For vectors $y = (y_1, \ldots, y_k)$ and $z = (z_1, \ldots, z_k)$,

$$y \geqslant z \quad \text{if and only if} \quad y_i \geqslant z_i \quad \text{for all } i;$$

$$y \geqq z \quad \text{if and only if} \quad y_i \geqslant z_i \quad \text{for all } i, \text{ and } y_j > z_j \quad \text{for at least one } j;$$

$$y > z \quad \text{if and only if} \quad y_i > z_i \quad \text{for all } i.$$

Negative, definitely negative, and strictly negative vectors are analogously defined by the inverses $\leqslant$, $\leqq$, and $<$ of $\geqslant$, $\geqq$, and $>$ respectively.

In matrix operations, a vector y will generally mean a column (vector). The superscript T will denote transpose. The scalar product of two vectors y and z in $\mathscr{R}^k$ will be denoted by $y^T z$; thus $y^T z = y_1 z_1 + \cdots + y_k z_k$. The norm or length $\| y \|$ of vector y is $\{y^T y\}^{1/2}$. A *unit vector* is one whose norm is unity. e_j will denote the unit vector $(0, \ldots, 0, 1, 0, \ldots, 0)$ in the direction of the jth coordinate axis.

The distance between two points y and z is the norm $\| y - z \|$ of the vector $y - z$. For a strictly positive real number δ, the *δ-neighborhood* $\mathcal{N}(z; \delta)$ of a point z is the set $\{y \in \mathcal{R}^k \mid \| y - z \| < \delta\}$.

2. Lineally Supremal Vectors; Quasisupremal and Maximal Vectors

2.1. Throughout the paper, $\mathcal{Z}$ denotes a given subset of the N-dimensional Euclidean space $\mathcal{R}^N$ (of indexes). Superiority defined by the strict partial order $\geqslant$ is considered. A vector z^0 is called a *maximal vector* (see Ref. 1) in $\mathcal{Z}$ if $z^0 \in \mathcal{Z}$ and there exists no z in $\mathcal{Z}$ such that $z \geqslant z^0$. We are concerned with obtaining the set of all maximal vectors in $\mathcal{Z}$. But we are actually interested in the set of "quasisupremal" vectors, which are substantially and realistically equivalent to maximal vectors.

Let $\bar{\mathcal{Z}}$ denote the closure of $\mathcal{Z}$. Given a nonempty subset $\mathcal{S}$ of $\bar{\mathcal{Z}}$ and a nonzero vector p in $\mathcal{R}^N$, we call a vector z^0 a *p-directional extremum* of $\mathcal{S}$ if $p^T z^0 = \sup\{p^T z \mid z \in \mathcal{S}\}$, i.e., if z^0 maximizes the projection $p^T z$ over $\mathcal{S}$. Of specific interest here are lineal subsets $\{z \in \bar{\mathcal{Z}} \mid z = z^0 + cp \text{ and } c \in \mathcal{R}\}$ given by the intersections of $\bar{\mathcal{Z}}$ with straight lines $\{z^0 + cp \mid c \in \mathcal{R}\}$. For a definitely positive (unit) vector p in $\mathcal{R}^N$, we specifically call z^0 a *p-lineally supremal vector* for $\mathcal{Z}$ if it is a p-directional extremum of the lineal subset $\{z \in \bar{\mathcal{Z}} \mid z = z^0 + cp \text{ and } c \in \mathcal{R}\}$ containing z^0.

Lemma 2.1. If z^0 is a p-lineally supremal vector for $\mathcal{Z}$ for some definitely positive (unit) vector p in $\mathcal{R}^N$, then z^0 is a point of the closure $\bar{\mathcal{Z}}$.

Proof. Let z^0 be p-lineally supremal for some definitely positive vector p but $z^0 \notin \bar{\mathcal{Z}}$. Then there is $\epsilon > 0$ such that $z \notin \bar{\mathcal{Z}}$ whenever $\| z - z^0 \| < \epsilon$. This implies that $c \leqslant -\epsilon$ whenever $z^0 + cp \in \bar{\mathcal{Z}}$. Consider the vector $z' \triangleq z^0 - \epsilon p$. Since $p^T(z^0 - \epsilon p) = p^T z^0 - \epsilon p^T p \geqslant p^T z^0 + c p^T p = p^T(z^0 + cp)$ for any c such that $z^0 + cp \in \bar{\mathcal{Z}}$, $p^T z'$ is an upper bound for $p^T z$ over the lineal subset $\{z \in \bar{\mathcal{Z}} \mid z = z^0 + cp \text{ and } c \in \mathcal{R}\}$. But since $p^T z' = p^T z^0 - \epsilon p^T p < p^T z^0$, $p^T z^0$ cannot be the least upper bound for $p^T z$ over the lineal subset, contradicting the definition of p-lineal supremality. Thus, $z^0 \in \bar{\mathcal{Z}}$. QED

Now, if z^0 is a p-lineally supremal vector of $\mathcal{Z}$ for every definitely positive (unit) vector p in $\mathcal{R}^N$, we call it a *quasisupremal vector* for $\mathcal{Z}$. Let $\mathcal{L}(p)$ denote the set of all p-lineally supremal vectors for $\mathcal{Z}$, and $\mathcal{Q}$

denote the set of all quasisupremal vectors. Then $\mathscr{Q} = \bigcap_{p>0} \mathscr{L}(p)$ by definition. Because a vector z^0 in $\bar{\mathscr{Z}}$ is p-lineally supremal for $\mathscr{Z}$ if and only if *there exists no strictly positive number* c such that $z^0 + cp \in \bar{\mathscr{Z}}$, $\mathscr{L}(p)$ geometrically represents those points of $\bar{\mathscr{Z}}$ that are "exposed" in the definitely positive direction p, and $\mathscr{Q}$ represents those that are "exposed" in all definitely positive directions.

Note that a quasisupremal vector is not necessarily a supremum of the entire set $\mathscr{Z}$. But we have:

Lemma 2.2. A quasisupremal vector for $\mathscr{Z}$ is a maximal vector in the closure $\bar{\mathscr{Z}}$, and vice versa.

Proof. Assume that z^0 is a quasisupremal vector not maximal in $\bar{\mathscr{Z}}$. Then there is z' in $\bar{\mathscr{Z}}$ such that $z' \gg z^0$. Consider the definitely positive vector p defined by $p = z' - z^0$. Then $z' = z^0 + p$ and $p^T z' = p^T z^0 + p^T p > p^T z^0$. Clearly z^0 is not even a p-lineally supremal vector for $\mathscr{Z}$, in contradiction to the assumption. On the other hand, consider a vector z'' that is not p-lineally supremal for some definitely positive vector p in $\mathscr{R}^N$. Then there is z''' in $\bar{\mathscr{Z}}$ such that $z''' = z'' + cp$ for some number c and that $p^T z''' > p^T z''$. Then c must be strictly positive. Consequently, $z''' \gg z''$ and z'' cannot be maximal in $\bar{\mathscr{Z}}$. QED

Observe from Lemma 2.2 that any quasisupremal vector possesses the essential characteristic of a maximal vector—namely, if z^0 is quasisupremal for $\mathscr{Z}$ then there exists no z in $\mathscr{Z}$ such that $z \gg z^0$. Observe also from Lemma 2.2 that a quasisupremal vector is actually a maximal vector if it is inside $\mathscr{Z}$, and is only infinitesimally distant from $\mathscr{Z}$ if otherwise. A quasisupremal vector is therefore practically as good as a maximal vector. In fact, a quasisupremal vector may even be superior to a maximal vector in $\mathscr{Z}$.

Example 2.1. Consider a subset $\mathscr{Z}_1$ of $\mathscr{R}^2$ given by $z_1 < 4.8$, $2z_2 \leqslant 5$, $10z_2 + 20 \geqslant z_1$, $(2z_1 - 9)^2 + (2z_2 - 1)^2 \geqslant 1$, $z_2 < 1 + (z_1 - 2.9)^2$, and $2z_2 \leqslant (6 - z_1) \cos 3.142 z_1$, as shown in Fig. 1. The set of *maximal* vectors in $\mathscr{Z}_1$ can be found (by definition or from the Contact Theorem—see Ref. 1) to consist of subsets *cd*, *hp* (without *h* and *p*), *rs*, and *tuv* (without *t* and *v*). (Subsets *gh* and *vw* are not contained in $\mathscr{Z}_1$.) The set of *quasisupremal* vectors for $\mathscr{Z}_1$ consists of subsets *cd* (without *d*), *ghp* (without *p*), *rs*, and *tuv* (without *t*). One can see that no vector in $\mathscr{Z}_1$ is superior to any (quasisupremal) vectors in the subset *gh*, and that any vector in *gh* is only of infinitesimal distance from

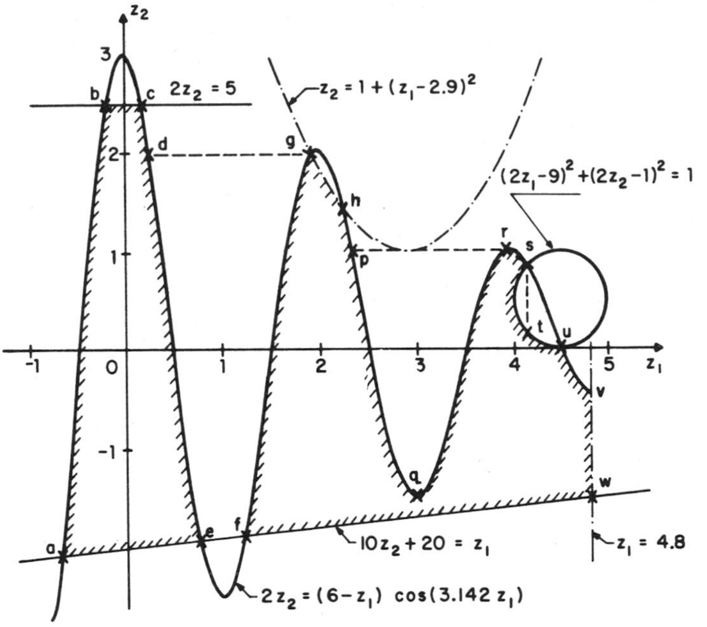

Fig. 1. The set $\mathscr{Z}_1$ in Example 2.1 given by

$$z_1 < 4.8,\ 2z_2 \leqslant 5,\ 10z_2 + 20 \geqslant z_1,\ (2z_1 - 9)^2 + (2z_2 - 1)^2 \geqslant 1,$$
$$z_2 < 1 + (z_1 - 2.9)^2, \text{ and } 2z_2 \leqslant (6 - z_1)\cos(3.142 z_1).$$

$\mathscr{Z}_1$. Any such vector practically is as good as a maximal vector in $\mathscr{Z}_1$. Now observe that vector g, a quasisupremal vector, can even be superior to vector d, which is a maximal vector in $\mathscr{Z}_1$.

Note that the maximization of *positive linear combinations* of z_1 and z_2 will yield *only* those in the subset bc and those in a portion of the subset rs as "solutions." Not only are not all of the "solutions" maximal vectors (for example only vector c in the entire subset bc is maximal), but also those "solutions" that are indeed maximal vectors constitute a very, very tiny portion of the whole set of maximal vectors.

2.2. Clearly from Lemma 2.2, the set $\mathscr{Q}$ of all quasisupremal vectors for $\mathscr{Z}$ is precisely the set of all maximal vectors in $\mathscr{Z}$ if $\mathscr{Z}$ is closed. Actually, $\mathscr{Z}$ need not be closed (so long as $\mathscr{Q}$ is contained in $\mathscr{Z}$). Call z^0 a *semisupremal vector for* $\mathscr{Z}$ if z^0 is an e_i-lineally supremal vector for every $i = 1,..., N$. Obviously, the set $\mathscr{S}$ of all semisupremal vectors for $\mathscr{Z}$ is given by $\bigcap_{i=1}^{i=N} \mathscr{L}(e_i)$, and $\mathscr{S}$ contains $\mathscr{Q}$ since $\mathscr{Q} = \bigcap_{p>0} \mathscr{L}(p)$. Thus we have

Lemma 2.3. The set of all quasisupremal vectors for $\mathscr{Z}$ is precisely the set of all maximal vectors in $\mathscr{Z}$ if (i) $\mathscr{Z}$ contains all semisupremal vectors for it, or (ii) for some definitely positive (unit) vector p, $\mathscr{Z}$ contains all p-lineally supremal vectors for it.

A point z^0 is said to be a *p-directional boundary point* (see Ref. 13) of $\mathscr{Z}$ if (i) for every $\epsilon > 0$, there exists z' in $\mathscr{Z}$ such that $\| z^0 - z' \| < \epsilon$ and (ii) for every $c > 0$, $z^0 + cp \notin \mathscr{Z}$. We say that $\mathscr{Z}$ is *p-directionally closed* if it contains all its p-directional boundary points. Evidently a p-directional boundary point of $\mathscr{Z}$ is a p-lineally supremal vector for $\mathscr{Z}$ (though not conversely). Accordingly, the set $\mathscr{L}(p)$ is a proper subset of $\mathscr{Z}$ if $\mathscr{Z}$ is p-directionally closed. As a consequence of Lemma 2.3, we have

Corollary 2.1. The set of all quasisupremal vectors for $\mathscr{Z}$ is precisely the set of all maximal vectors in $\mathscr{Z}$ if $\mathscr{Z}$ is (i) e_i-directionally closed for every $i = 1,..., N$, or (ii) p-directionally closed for some definitely positive (unit) vector p in $\mathscr{R}^N$.

2.3. Obviously from Lemma 2.2, *the existence of maximal vectors in $\mathscr{Z}$ is assured whenever there exist quasisupremal vectors inside $\mathscr{Z}$*. Example 2.2 shows that such an assurance does not require the boundedness of the set $\mathscr{Z}$ under consideration.

Example 2.2. Consider a subset $\mathscr{Z}_2$ of $\mathscr{R}^2$ given by $\{(z_1, z_2) \mid (z_1^2 + 4) z_2 \leqslant 8\}$ (the curve described by $z_1^2 z_2 = 4a^2(2a - z_2)$ is the witch of Agnesi). First of all, $\mathscr{Z}_2$ is not a bounded set. Second, the set of quasisupremal vectors for $\mathscr{Z}_2$ is given by $\{(z_1, z_2) \mid (z_1^2 + 4) z_2 = 8$ and $z_1 \geqslant 0\}$, which is nonempty and is contained in $\mathscr{Z}_2$. Thus the existence of maximal vectors in $\mathscr{Z}_2$ is assured. As a matter of fact, the above set of quasisupremal vectors is the set of *all* maximal vectors in $\mathscr{Z}_2$.

Therefore, we shall mainly be concerned with obtaining the set of *all quasisupremal vectors* for $\mathscr{Z}$. Simple computational methods can be developed for determining the set of *all* quasisupremal vectors.

3. The Method of Proper Equality Constraints

3.1. To obtain the set of all quasisupremal vectors for $\mathscr{Z}$ is by definition to find the intersection $\bigcap_{p \gg 0} \mathscr{L}(p)$ of all sets of p-lineally supremal vectors for $\mathscr{Z}$ with p ranging over the *infinite number* of definitely positive (unit) vectors in $\mathscr{R}^N$. This is generally a formidable, if not impossible, task. One may consider finding the set $\mathscr{S}$ of semisupremal vectors for $\mathscr{Z}$ since the determination of only a ***finite number*** of sets $\mathscr{L}(e_1),..., \mathscr{L}(e_N)$ is required. But notice that the set $\mathscr{S}$ may contain many very "inferior" vectors. For instance, vectors a, e, and f of $\mathscr{Z}_1$ (Fig. 1) are such very inferior semisupremal vectors. Nonquasisupremal vectors may constitute a rather large portion of the set $\mathscr{S}$. For example, the set of semisupremal vectors for $\{(z_1, z_2) \mid z_1 z_2 \geqslant 1$ and $z_1^2 + z_2^2 \leqslant 9\}$ consists of the hyperbolic segment $\{(z_1, z_2) \mid z_1 z_2 = 1$ and $-\sqrt{20} - \sqrt{7} \leqslant 2z_1 \leqslant -\sqrt{20} + \sqrt{7}\}$ and the circular segment $\{(z_1, z_2) \mid z_1^2 + z_2^2 = 9$ and $\sqrt{20} - \sqrt{7} \leqslant 2z_1 \leqslant \sqrt{20} + \sqrt{7}\}$, while none of the vectors in the entire hyperbolic segment is a quasisupremal vector. *Tests must be devised to "sift" out those semisupremal vectors that are not quasisupremal.*

We are in fact interested in extracting the set of quasisupremal vectors from even *only one* of the N sets: $\mathscr{L}(e_1),..., \mathscr{L}(e_N)$.

3.2. For each fixed integer j, $1 \leqslant j \leqslant N$, consider the following (single-objective) optimization problem:

$$\begin{aligned} &\text{maximize} \quad e_j^T z \\ &\text{subject to} \quad e_i^T z = \alpha_i, \quad i = 1,..., j-1, j+1,..., N, \quad \text{and} \quad z \in \bar{\mathscr{Z}}, \end{aligned} \tag{1}$$

where $\alpha_1,..., \alpha_{N-1}$ are real numbers. Let $\phi_j(\alpha^j)$ be the supremum of index $z_j = e_j^T z$,

$$\phi_j(\alpha^j) = \sup\{e_j^T z \mid z_i = \alpha_i \quad \text{for all} \quad i \neq j \quad \text{and} \quad z \in \bar{\mathscr{Z}}\},$$

where α^j denotes a vector $(\alpha_1,..., \alpha_{j-1}, \alpha_{j+1},..., \alpha_N)$ of $N - 1$ components. Note that if $\hat{z}(\alpha^j)$ is an optimal solution of problem (1) for vector α^j, then $\phi_j(\alpha^j)$ is the corresponding *optimal value* given by $e_j^T \hat{z}(\alpha^j)$. Also note that $e_i^T z \equiv z_i$ for each i. The following consequences are obvious.

Lemma 3.1. If $\hat{z}(\alpha^j)$ is an optimal solution of (1) for vector α^j, it is an e_j-lineally supremal vector for $\mathscr{Z}$ and is uniquely given by

$$\hat{z}(\alpha^j) = (\alpha_1,..., \alpha_{j-1}, \phi_j(\alpha^j), \alpha_{j+1},..., \alpha_N).$$

Lemma 3.2. A vector in $\mathscr{R}^N$ is an e_j-lineally supremal vector for $\mathscr{Z}$ if and only if it is given by $(\alpha_1,..., \alpha_{j-1}, \phi_j(\alpha^j), \alpha_{j+1},..., \alpha_N)$ for some vector α^j in $\mathscr{R}^{N-1}$ such that the set $\{z \in \bar{\mathscr{Z}} \mid z_i = \alpha_i$ for all $i \neq j\}$ is nonempty.

Lemmas 3.1 and 3.2 indicate that we can obtain e_j-lineally supremal vectors for $\mathscr{Z}$ by solving the optimization problem (1) for various vectors α^j in $\mathscr{R}^{N-1}$ and obtaining the corresponding optimal values $\phi_j(\alpha^j)$.

Call the family of optimization problems (1) with $\alpha_1, \ldots, \alpha_{j-1}, \alpha_{j+1}, \ldots, \alpha_N$ as $N - 1$ real-valued parameters the jth family of *associated parametric-equality-constrained single-objective (PECSO) optimization problems*. Then we can obtain the set $\mathscr{L}(e_j)$ of e_j-lineally supremal vectors for $\mathscr{Z}$ by solving the jth family of associated PECSO optimization problems. Consequently, we can obtain the set $\bigcap_{j=1}^{j=N} \mathscr{L}(e_j)$ of semisupremal vectors by solving the N families of associated PECSO optimization problems. In most cases, solving all these N families of optimization problems simultaneously or consecutively can still be a formidable task. Moreover, to be a semisupremal vector is not to be a quasisupremal vector unconditionally. Is it possible, then, to obtain the set of quasisupremal vectors by solving even *only one* of the N families of associated PECSO optimization problems? The method of proper equality constraints (PEC) will make it possible.

3.3. Suppose that the set $\mathscr{L}(e_N)$ of e_N-lineally supremal vectors are of specific interest. (Rearrange the components of the index vectors or relabel the coordinate axes of the space $\mathscr{R}^N$ if a direction other than e_N is actually of interest.) Consider the following associated PECSO optimization problems:

$$\begin{aligned} &\text{maximize} \quad z_N \\ &\text{subject to} \quad z_i = \alpha_i, \quad i = 1, \ldots, N-1, \quad \text{and} \quad (z_1, \ldots, z_N) \in \bar{\mathscr{Z}} \end{aligned} \tag{2}$$

for vectors $\alpha = (\alpha_1, \ldots, \alpha_{N-1})$ in $\mathscr{R}^{N-1}$. Let $\mathscr{A}$ denote the set of all vectors α for which the supremum

$$\phi(\alpha) = \sup\{z_N \mid (z_1, \ldots, z_{N-1}) = \alpha \quad \text{and} \quad (z_1, \ldots, z_N) \in \bar{\mathscr{Z}}\} \tag{3}$$

exists or $\phi(\alpha) = +\infty$. Then $\mathscr{L}(e_N) \subset \{(\alpha, \phi(\alpha)) \mid \alpha \in \mathscr{A}\}$. The following is a fundamentally important relation between quasisupremal vectors and e_N-lineally supremal vectors.

Theorem 3.1. For an e_N-lineally supremal vector $(\alpha^0, \phi(\alpha^0))$, $\alpha^0 \in \mathscr{A}$, to become a quasisupremal vector for $\mathscr{Z}$, it is necessary and sufficient that $\phi(\alpha) < \phi(\alpha^0)$ for any α in $\mathscr{A}$ such that $\alpha \gneqq \alpha^0$.

Proof. *Necessity*: Suppose that $\phi(\alpha') \geqslant \phi(\alpha^0)$ for some α' in $\mathscr{A}$ such that $\alpha' \gneqq \alpha^0$. Let p' be a definitely positive vector defined by $p' = (\alpha', \phi(\alpha')) - (\alpha^0, \phi(\alpha^0))$. Then $(\alpha', \phi(\alpha')) = (\alpha^0, \phi(\alpha^0)) + p'$ and the vector $(\alpha^0, \phi(\alpha^0))$ clearly cannot even be a p'-lineally supremal vector.

Sufficiency: Suppose that there is a z'' in $\bar{\mathscr{Z}}$ such that $z'' \gneqq (\alpha^0, \phi(\alpha^0))$. Let $\alpha'' = (z_1'', \ldots, z_{N-1}'')$. Then either (i) $\alpha'' = \alpha^0$ and $z_N'' > \phi(\alpha^0)$, or (ii) $\alpha'' \geqslant \alpha^0$ and $z''_N \geqslant \phi(\alpha^0)$. Case (i) is impossible by definition of $\phi(\alpha^0)$. But since $\phi(\alpha'') \geqslant z''_N$ by definition of $\phi(\alpha'')$, case (ii) is also impossible because of the hypothesis. Therefore, the existence of z'' is impossible, and $(\alpha^0, \phi(\alpha^0))$ must be maximal in $\bar{\mathscr{Z}}$. Thus $(\alpha^0, \phi(\alpha^0))$ is a quasisupremal vector for $\mathscr{Z}$ by Lemma 2.2. QED

Theorem 3.1 says that setting $z_1, \ldots, z_{N-1}$ equal to some arbitrary constants, say $\alpha_1^0, \ldots, \alpha_{N-1}^0$, and maximizing z_N subject to the equality constraints $z_i - \alpha_i^0 = 0$, $i = 1, \ldots, N-1$ [in addition to $(z_1, \ldots, z_N) \in \bar{\mathscr{Z}}$] *does not always yield a quasisupremal vector*; the resultant vector may be quite inferior even if it is e_N-lineally supremal. In other words, if the constants are not proper, even optimal solutions of the associated single-objective optimization problems will not produce quasisupremal vectors for $\mathscr{Z}$.

Let us call $z_i = \alpha_i^0$, $i = 1, \ldots, N-1$, *proper objective-converted equality constraints*, and call $\alpha^0 = (\alpha_1^0, \ldots, \alpha_{N-1}^0)$ a *proper vector of equality constants* for $(z_1, \ldots, z_{N-1})$, if (i) $\alpha^0 \in \mathscr{A}$ and if (ii) $\phi(\alpha) < \phi(\alpha^0)$ for any α in $\mathscr{A}$ such that $\alpha \geqslant \alpha^0$. Then, the *method of proper equality constraints* (PEC) involves in principle the determination of the set of quasisupremal vectors for $\mathscr{Z}$ by determining the set of optimal solutions of the associated single objective optimization problems (2) *with proper objective-converted equality constraints*. According to Theorem 3.1, an e_N-lineally supremal vector $(\alpha^0, \phi(\alpha^0))$ is a quasisupremal vector for $\mathscr{Z}$ *if and only if* α^0 *is a proper vector of equality constants for* $(z_1, \ldots, z_{N-1})$. We shall derive (in Sections 4 and 5) various useful tests for quasisupremality of e_N-lineally supremal vectors $(\alpha, \phi(\alpha))$ using Theorem 3.1. These tests are the foundations of various Pareto-optimality conditions and properness tests used in obtaining the set of all Pareto-optimal solutions (Refs. 19–21).

Before proceeding, note that $\mathscr{A}$ is a subset of the projection of $\bar{\mathscr{Z}}$ on the subspace $z_N = 0$ of $\mathscr{R}^N$ along the Nth coordinate axis. The set $\mathscr{A}$ may be as arbitrary as the set $\mathscr{Z}$ and may turn out to be an unconnected or a hollow subset of $\mathscr{R}^{N-1}$. That is, the parameter vector α may, depending on the nature of the set $\mathscr{Z}$, *vary discontinuously*. Since we consider ϕ as an extended-real-valued function defined on $\mathscr{A}$, the function can also be as arbitrary as $\mathscr{Z}$. That is, ϕ may be nondifferentiable, discontinuous, or even fragmentary.

4. Absolute Right-Decreasing Property: Global Tests for Quasisupremality and Properness

4.1. The condition for quasisupremality expressed in Theorem 3.1 requires that ϕ possess a certain pointwise directionally decreasing property. Let α^0 be a point in the set $\mathcal{A}$. We say that the function ϕ is *right-decreasing* on $\mathcal{A}$ *from* α^0 if $\phi(\alpha) \leqslant \phi(\alpha^0)$ for any α in $\mathcal{A}$ such that $\alpha \geqslant \alpha^0$, *strictly right-decreasing* if $\phi(\alpha) < \phi(\alpha^0)$ for any α in $\mathcal{A}$ such that $\alpha > \alpha^0$, and *absolutely right-decreasing* if $\phi(\alpha) < \phi(\alpha^0)$ for any α in $\mathcal{A}$ such that $\alpha \gneqq \alpha^0$.

Lemma 4.1. An e_N-lineally supremal vector $(\alpha^0, \phi(\alpha^0))$, $\alpha^0 \in \mathcal{A}$, is a quasisupremal vector for $\mathcal{L}$ if and only if ϕ is absolutely right-decreasing on $\mathcal{A}$ from α^0.

We can thus devise tests for quasisupremality by developing necessary and sufficient conditions for ϕ to be pointwise absolute right-decreasing.

The usual decreasing properties imply the pointwise right-decreasing properties: ϕ is decreasing, strictly decreasing, or absolutely decreasing on $\mathcal{A}$ if and only if it is respectively right-decreasing, strictly right-decreasing, or absolutely right-decreasing on $\mathcal{A}$ *from any point* in $\mathcal{A}$. As a corollary to Lemma 4.1 we observe that ϕ is absolutely decreasing on the subset $\{\alpha \in \mathcal{A} \mid (\alpha, \phi(\alpha))$ is a quasisupremal vector for $\mathcal{L}\}$.

The absolute right-decreasing property is stronger than the strict right-decreasing property, while both are stronger than the right-decreasing property. Consider, for example, functions ϕ_1, ϕ_2, and ϕ_3 defined on the rectangular subset $\mathcal{A} = \{(\alpha_1, \alpha_2) \mid -4 \leqslant \alpha_1 \leqslant 16$ and $-16 < \alpha_2 \leqslant 32\}$ of $\mathcal{R}^2$, respectively, by $\phi_1(\alpha) = (\alpha_1 + 1)^{-1} (\alpha_2 + 1)^{-1} \cos \alpha_2$, $\phi_2(\alpha) = (\alpha_1 + 1)^{-1} \cos \alpha_2$, and $\phi_3(\alpha) = \cos \alpha_1 \cos \alpha_2$. The function ϕ_1 is absolutely right-decreasing on $\mathcal{A}$ from (0, 0); ϕ_2 is strictly, but not absolutely, right-decreasing from (0, 0); ϕ_3 is right-decreasing, but neither absolutely nor strictly, from (0, 0). It is worth mentioning that such properties depend not only on the definition of the function but also on the *domain* of the function. For instance, if $\mathcal{A}' = \{(\alpha_1, \alpha_2) \mid 1/10 < \alpha_2/\alpha_1 < 10\}$ is the domain instead, the function ϕ_2 above is absolutely, as well as strictly, right-decreasing on $\mathcal{A}'$ from (0, 0).

4.2. Conditions on derivatives are natural candidates for the pointwise absolute right-decreasing property. Since the following conditions involve integrals of gradients and since a Lebesgue integral only requires its integrand to be well defined "almost everywhere" on the domain of integration, we shall also include (by parentheses) the case in which ϕ is differentiable "almost everywhere" but not necessarily everywhere on the domain of integration. Recall that a property is said to hold true *almost everywhere* on a set when it holds true everywhere on the set except a null subset (a set of measure zero). We denote by

$\partial\phi/\partial\alpha$ the row $[\partial\phi/\partial\alpha_1 \cdots \partial\phi/\partial\alpha_{N-1}]$ of partial derivatives and by $\nabla\phi$ the gradient vector of ϕ. Obviously $\partial\phi/\partial\alpha = (\nabla\phi)^T$.

Theorem 4.1. If $(\alpha^0, \phi(\alpha^0))$ is a quasisupremal vector for $\mathscr{X}$, then $\int_0^1 [\partial\phi(\alpha^0 + t\pi)/\partial\alpha]\pi\, dt < 0$ for any definitely positive vector π in $\mathscr{R}^{N-1}$ such that the function θ, defined by $\theta(t) = \phi(\alpha^0 + t\pi)$, is differentiable (or absolutely continuous) on interval $[0, 1]$.

Proof. Let π be any such definitely positive vector. Let θ be the function defined by $\theta(t) = \phi(\alpha^0 + t\pi)$ for t in $[0, 1]$. Then, by hypothesis, θ is differentiable (absolutely continuous) on $[0, 1]$. Hence the derivative $\theta'(t)$ exists (almost everywhere) on $[0, 1]$, and we have $\theta(1) - \theta(0) = \int_0^1 \theta'(t)\, dt$ (see Ref. 16, p. 188). But at each point $\alpha^0 + t\pi$ $(0 \leqslant t \leqslant 1)$ where ϕ is differentiable we have, by the chain rule of differentiation, $\theta'(t) = [\partial\phi(\alpha^0 + t\pi)/\partial\alpha]\pi$. A simple substitution yields

$$\phi(\alpha^0 + \pi) - \phi(\alpha^0) = \int_0^1 [\partial\phi(\alpha^0 + t\pi)/\partial\alpha]\pi\, dt.$$

The necessity of the condition in the theorem is thus obvious. QED

Let $\mathscr{B}$ be a subset of $\mathscr{A}$ and let $\alpha^0 \in \mathscr{B}$. We call α^0 a *radiant point* of $\mathscr{B}$ if for any α in $\mathscr{B}$ the closed line segment $\{\alpha^0 + t(\alpha - \alpha^0) \mid t \in [0, 1]\}$ is contained in $\mathscr{B}$. For example, α^0 is a radiant point of a convex set containing α^0 or of a star-shaped set with α^0 as its center. But if $\mathscr{B}$ consists of two or more unconnected subsets, no point in $\mathscr{B}$ is a radiant point of $\mathscr{B}$. We specifically consider subsets given by $\mathscr{A}(\alpha^0) = \{\alpha \in \mathscr{A} \mid \alpha \geqslant \alpha^0\}$ for $\alpha^0 \in \mathscr{A}$.

Corollary 4.1. If $(\alpha^0, \phi(\alpha^0))$ is a quasisupremal vector for $\mathscr{X}$ and if α^0 is a radiant point of $\mathscr{A}(\alpha^0)$, then

$$\int_0^1 [\partial\phi(\alpha^0 + t(\alpha - \alpha^0))/\partial\alpha](\alpha - \alpha^0)\, dt < 0$$

for any α in $\mathscr{A}(\alpha^0)$ with $\alpha \neq \alpha^0$ such that ϕ is differentiable at $\alpha^0 + t(\alpha - \alpha^0)$ for every t in $[0, 1]$.

Proof. Since α^0 is a radiant point of $\mathscr{A}(\alpha^0)$, any α in $\mathscr{A}(\alpha^0)$ with $\alpha \neq \alpha^0$ defines a definitely positive vector, $\pi = \alpha - \alpha^0$, so that ϕ is well defined at $\alpha^0 + t\pi$ for every t in $[0, 1]$. Consequently, the necessary condition in Theorem 4.1 must be satisfied by any such vector. QED

4.3. Necessary and sufficient conditions for the pointwise absolute right-decreasing property can be derived for differentiable functions at interior points. Recall that (Fréchet) differentiability is defined at interior points.

Theorem 4.2. Suppose that α^0 is an interior point of $\mathcal{A}$ and a radiant point of $\mathcal{A}(\alpha^0)$. Also suppose that $\phi(\alpha^0) \neq +\infty$ and ϕ is differentiable (or absolutely continuous) on $\mathcal{A}(\alpha^0)$. Then $(\alpha^0, \phi(\alpha^0))$ is a quasisupremal vector for $\mathscr{L}$ if and only if

$$\textstyle\int_0^1 [\partial\phi(\alpha^0 + t(\alpha - \alpha^0))/\partial\alpha](\alpha - \alpha^0)\, dt < 0$$

for every α in $\mathcal{A}(\alpha^0)$ such that $\alpha \neq \alpha^0$.

Proof. First, it follows from the hypotheses that for each α in $\mathcal{A}(\alpha^0)$ with $\alpha \neq \alpha^0$, ϕ is differentiable at $\alpha^0 + t(\alpha - \alpha^0)$ for (almost) every t in [0, 1]. Then, following the same arguments as in the proof of Theorem 4.1, we have

$$\begin{aligned}\phi(\alpha) - \phi(\alpha^0) &\equiv \phi(\alpha^0 + (\alpha - \alpha^0)) - \phi(\alpha^0)\\ &= \int_0^1 [\partial\phi(\alpha^0 + t(\alpha - \alpha^0))/\partial\alpha](\alpha - \alpha^0)\, dt\end{aligned}$$

for every α in $\mathcal{A}(\alpha^0)$ such that $\alpha \neq \alpha^0$. The necessity and sufficiency of the condition in the theorem follow immediately from Theorem 3.1. QED

If e_i-directional boundary points of $\mathcal{A}$ for $i = 1,..., N - 1$ can be explicitly determined, then for each α^0 in $\mathcal{A}$, the subset of those boundary points α of $\mathcal{A}$ such that $\alpha \geqslant \alpha^0$ can be easily determined. The following Corollaries 4.2 and 4.3 will offer useful for this case.

Corollary 4.2. Under the hypotheses of Theorem 4.2, $(\alpha^0, \phi(\alpha^0))$ is a quasisupremal vector for $\mathscr{L}$ if and only if

$$\textstyle\int_0^\tau [\partial\phi(\alpha^0 + t(\alpha - \alpha^0))/\partial\alpha](\alpha - \alpha^0)\, dt < 0$$

for every boundary point of $\mathcal{A}$ such that $\alpha \geqslant \alpha^0$ and every τ in (0, 1].

Proof. Let $\mathscr{B}(\alpha^0)$ denote the set of all boundary points α of $\mathcal{A}$ such that $\alpha \geqslant \alpha^0$, and $\mathscr{D}(\alpha^0)$ denote the subset $\{\alpha^0 + t(\alpha - \alpha^0) \mid \alpha \in \mathscr{B}(\alpha^0)$ and $t \in [0, 1]\}$. Then $\mathscr{D}(\alpha^0) \subset \mathcal{A}(\alpha^0)$, since α^0 is a radiant point of $\mathcal{A}(\alpha^0)$ and $\mathscr{B}(\alpha^0) \subset \mathcal{A}(\alpha^0)$. But for any α in $\mathcal{A}(\alpha^0)$, there obviously exists a boundary point α' in $\mathscr{B}(\alpha^0)$ such that $\alpha - \alpha^0 = t(\alpha' - \alpha^0)$ for some t in [0, 1]. This implies that $\mathcal{A}(\alpha^0) \subset \mathscr{D}(\alpha^0)$. Therefore, $\mathcal{A}(\alpha^0) = \mathscr{D}(\alpha^0)$. Now, to each α in $\mathcal{A}(\alpha^0)$ with $\alpha \neq \alpha^0$, let α' be the corresponding boundary point in $\mathscr{B}(\alpha^0)$ and τ be the corresponding number in (0, 1] such that $\alpha = \alpha^0 + \tau(\alpha' - \alpha^0)$. Then

$$\begin{aligned}\int_0^1 & [\partial\phi(\alpha^0 + t(\alpha - \alpha^0))/\partial\alpha](\alpha - \alpha^0)\,dt \\ &\equiv \int_0^1 [\partial\phi(\alpha^0 + t\tau(\alpha' - \alpha^0))/\partial\alpha]\tau\,(\alpha' - \alpha^0)\,dt \\ &\equiv \int_0^\tau [\partial\phi(\alpha^0 + t'(\alpha' - \alpha^0))/\partial\alpha](\alpha' - \alpha^0)\,dt,\end{aligned}$$

where $t' = t\tau$. The equivalence of the present condition to that in Theorem 4.1 is thus evident. QED

Corollary 4.3. Under the hypotheses of Theorem 4.2, $(\alpha^0, \phi(\alpha^0))$ is a quasisupremal vector for $\mathscr{Z}$ if and only if $\int_0^\tau [\partial\phi(\alpha^0 + t\pi)/\partial\alpha]\pi\,dt < 0$ for every definitely positive unit vector π in $\mathscr{R}^{N-1}$ and every number τ in $(0, \rho]$, where ρ is a strictly positive number, depending on π, such that $\alpha^0 + \rho\pi$ is a π-directional boundary point of $\mathscr{A}$.

Proof. First, each α in $\mathscr{A}(\alpha^0)$ with $\alpha \neq \alpha^0$ defines a definitely positive vector π by $\pi = (\alpha - \alpha^0)/\|\alpha - \alpha^0\|$ and a strictly positive number τ by $\tau = \|\alpha - \alpha^0\|$. Denote by ρ the corresponding strictly positive number such that $\alpha^0 + \rho\pi$ is a π-directional boundary point of $\mathscr{A}$. Obviously, $\tau \leqslant \rho$. Since α^0 is an interior point of $\mathscr{A}$, the set of definitely positive unit vectors thus defined contains all definitely positive unit vectors in $\mathscr{R}^{N-1}$. On the other hand, each definitely positive unit vector π in $\mathscr{R}^{N-1}$ and the corresponding strictly positive number ρ define a line segment $\{\alpha^0 + t\pi \mid t \in [0, \rho]\}$. Each such segment is contained entirely in $\mathscr{A}(\alpha^0)$ since α^0 is a radiant point of $\mathscr{A}(\alpha^0)$. The set of all such segments consequently contains the subset $\mathscr{A}(\alpha^0)$. The equivalence of the present condition to that in Theorem 4.2 is then evident from the following identity:

$$\int_0^1 [\partial\phi(\alpha^0 + t(\alpha - \alpha^0))/\partial\alpha](\alpha - \alpha)\,dt \equiv \int_0^\tau [\partial\phi(\alpha^0 + t'\pi)/\partial\alpha]\pi\,dt',$$

where $\alpha \in \mathscr{A}(\alpha^0)$, $\alpha \neq \alpha^0$, $\pi = (\alpha - \alpha^0)/\|\alpha - \alpha^0\|$, $\tau = \|\alpha - \alpha^0\|$, and $t' = t\tau$.

4.4. The following immediate corollaries to Corollary 4.3 give *sufficient* conditions in terms of the gradients $\nabla\phi$.

Corollary 4.4. Suppose that α^0 is an interior point of $\mathscr{A}$ and a radiant point of $\mathscr{A}(\alpha^0)$. Also suppose that $\phi(\alpha^0) \neq +\infty$ and ϕ is differentiable (or absolutely continuous) on $\mathscr{A}(\alpha^0)$. Then $(\alpha^0, \phi(\alpha^0))$ is a quasisupremal vector for $\mathscr{Z}$ if $\int_0^\tau \nabla\phi(\alpha^0 + t\pi)\, dt < 0$ for every definitely positive unit vector π in $\mathscr{R}^{N-1}$ and every number τ in $(0, \rho]$, where ρ is a strictly positive number (depending on π) such that $\alpha^0 + \rho\pi$ is a π-directional boundary point of $\mathscr{A}$.

Corollary 4.5. Under the hypotheses of Corollary 4.4, $(\alpha^0, \phi(\alpha^0))$ is a quasisupremal vector for $\mathscr{Z}$ if $\nabla\phi(\alpha^0 + t\pi) < 0$ for every definitely positive unit vector π in $\mathscr{R}^{N-1}$ and (almost) every t in $(0, \rho)$, where ρ is a strictly positive number, depending on π, such that $\alpha^0 + \rho\pi$ is a π-directional boundary point of $\mathscr{A}$.

4.5. Obviously, we can interpret $\nabla\phi(\alpha)$ as the *sensitivity* of the supremum $\phi(\alpha)$ of index z_N to variations in the equality constants $\alpha_1, \ldots, \alpha_{N-1}$, and $\int_0^\tau \nabla\phi(\alpha + t\pi)\, dt$ as the *average sensitivity* over the line segment $\{\alpha + t\pi \mid 0 \leqslant t \leqslant \tau\}$. In applying the method of PEC on the space of solutions, we can obtain the sensitivity $\nabla\phi(\alpha)$ without first evaluating the function ϕ (see Ref. 20).

5. Local Absolute Right-Decreasing Property: Local and Pointwise Tests for Quasisupremality and Properness

5.1. Preliminary tests for quasisupremality can be made on a local or pointwise base. Initial screening undoubtedly will save effort, especially when none of the foregoing global tests is applicable and the fundamental test expressed by Theorem 3.1 may have to be used.

We call a vector z^0 a *locally quasisupremal vector* (for $\mathscr{Z}$) if for some neighborhood $\mathscr{N}(z^0; \epsilon)$ in $\mathscr{R}^N$, z^0 is a quasisupremal vector for the subset $\mathscr{Z} \cap \mathscr{N}(z^0; \epsilon)$. For a point α^0 in $\mathscr{A}$, we say that ϕ is *locally absolutely right-decreasing* (on $\mathscr{A}$) from α^0 if, for some neighborhood $\mathscr{N}(\alpha^0; \delta)$ in $\mathscr{R}^{N-1}$, ϕ is absolutely right-decreasing on the subset $\mathscr{A} \cap \mathscr{N}(\alpha^0; \delta)$ from α^0, namely, if $\phi(\alpha) < \phi(\alpha^0)$ for any α in $\mathscr{A} \cap \mathscr{N}(\alpha^0; \delta)$ such that $\alpha \geqslant \alpha^0$.

Not all locally quasisupremal vectors are given by $(\alpha, \phi(\alpha))$ with ϕ defined by (3); but those that are not given by $(\alpha, \phi(\alpha))$ are of no interest since they are not even e_N-lineally supremal. Consider, for example, the subset of $\mathscr{R}^2$ given by the disk $\{z \mid 3 \leqslant \| z \| \leqslant 4\}$ containing the circle $\{z \mid \| z \| \leqslant 1\}$. The locally quasisupremal vectors are given by arc 1, $\{(z_1, z_2) \mid \sqrt{(z_1^2 + z_2^2)} = 4 \text{ and } 0 \leqslant z_1 \leqslant 4\}$, and arc 2, $\{(z_1, z_2) \mid \sqrt{(z_1^2 + z_2^2)} = 1 \text{ and } 0 \leqslant z_1 \leqslant 1\}$. Only those in arc 1 are given by

$(\alpha, \phi(\alpha))$, where $\phi(\alpha) = \sqrt{(16 - \alpha^2)}$; those in arc 2, however, are not even e_2-lineally supremal.

Lemma 5.1. An e_N-lineally supremal vector $(\alpha^0, \phi(\alpha^0))$, $\alpha^0 \in \mathcal{O}\!\!\mathcal{l}$, is a locally quasisupremal vector for $\mathscr{X}$ if and only if ϕ is locally absolutely right-decreasing on $\mathcal{O}\!\!\mathcal{l}$ from α^0.

Since local quasisupremality is obviously a necessary condition for (global) quasisupremality, we can localize Corollary 4.1 and obtain the following weaker necessary condition:

Corollary 5.1. If $(\alpha^0, \phi(\alpha^0))$ is a (locally) quasisupremal vector for $\mathscr{X}$ and if α^0 is a radiant point of $\mathcal{O}\!\!\mathcal{l}(\alpha^0) \cap \mathcal{N}(\alpha^0; \delta)$ for some $\delta > 0$, then for some δ' in $[0, \delta]$, $\int_0^1 [\partial\phi(\alpha^0 + t(\alpha - \alpha^0))/\partial\alpha](\alpha - \alpha^0)\, dt < 0$ for any α in $\mathcal{O}\!\!\mathcal{l}(\alpha^0) \cap \mathcal{N}(\alpha^0; \delta')$ with $\alpha \neq \alpha^0$ such that ϕ is differentiable at $\alpha^0 + t(\alpha - \alpha^0)$ for every t in $(0, 1]$.

Compared with Corollary 4.1, Corollary 5.1 is much more widely applicable, since a point α^0 is more likely to be a radiant point of subset $\mathcal{O}\!\!\mathcal{l}(\alpha^0) \cap \mathcal{N}(\alpha^0; \delta)$ for some neighborhood $\mathcal{N}(\alpha^0; \delta)$ than of the entire set $\mathcal{O}\!\!\mathcal{l}(\alpha^0)$.

By localizing Theorem 4.2 and Corollary 4.2, we immediately obtain the following *necessary and sufficient* conditions for *local* quasisupremality.

Corollary 5.2. Suppose that α^0 is an interior point of $\mathcal{O}\!\!\mathcal{l}$ and a radiant point of $\mathcal{O}\!\!\mathcal{l}(\alpha^0) \cap \mathcal{N}(\alpha^0; \delta)$ for some $\delta > 0$. Also suppose that $\phi(\alpha^0) \neq +\infty$ and ϕ is differentiable (or absolutely continuous) on $\mathcal{O}\!\!\mathcal{l}(\alpha^0) \cap \mathcal{N}(\alpha^0; \delta)$. Then $(\alpha^0, \phi(\alpha^0))$ is a locally quasisupremal vector for $\mathscr{X}$ if and only if, for some δ' in $(0, \delta]$,

$$\int_0^1 [\partial\phi(\alpha^0 + t(\alpha - \alpha^0))/\partial\alpha](\alpha - \alpha^0)\, dt < 0$$

for every α in $\mathcal{O}\!\!\mathcal{l}(\alpha^0) \cap \mathcal{N}(\alpha^0; \delta')$ such that $\alpha \neq \alpha^0$.

Corollary 5.3. Under the hypotheses of Corollary 5.2, $(\alpha^0, \phi(\alpha^0))$ is a locally quasisupremal vector for $\mathscr{X}$ if and only if for some ρ in $(0, \delta]$, $\int_0^\tau [\partial\phi(\alpha^0 + t\pi)/\partial\alpha]\pi\, dt < 0$ for every definitely positive unit vector π in $\mathscr{R}^{N-1}$ and every τ in $(0, \rho]$.

Immediately from Corollary 5.3, we have the following *sufficient* conditions for *local* quasisupremality.

Corollary 5.4. Under the hypotheses of Corollary 5.2, $(\alpha^0, \phi(\alpha^0))$ is a locally quasisupremal vector for $\mathscr{X}$ (a) if for some ρ in $(0, \delta]$,

$\int_0^\tau \nabla\phi(\alpha^0 + t\pi)\, dt < 0$ for every definitely positive unit vector π in $\mathscr{R}^{N-1}$ and every τ in $(0, \rho]$, or (b) if for some ρ in $(0, \delta]$, $\nabla\phi(\alpha^0 + t\pi) < 0$ for every definitely positive unit vector π in $\mathscr{R}^{N-1}$ and (almost) every t in $(0, \rho)$.

5.2. The existence of usual two-sided directional derivatives is not always necessary, nor is the interiority of the point α^0 in question. Given a nonzero vector ξ in $\mathscr{R}^{N-1}$, we call a point α^0 in $\mathscr{A}$ a *ξ-directional inner point* of $\mathscr{A}$ if there exists a sequence $\{t_k\}$ of strictly positive numbers converging to zero such that $\alpha^0 + t_k\xi \in \mathscr{A}$ for all such t_k.

Evidently, of the sequential set $\{(1 - 1/k)\alpha' + \alpha''/k \mid k \geqslant 1\}$, α' is an $(\alpha'' - \alpha')$-directional inner point; of the line segment $\{(1 - t)\alpha' + t\alpha'' \mid t \in [0, 1]\}$, α' is an $(\alpha'' - \alpha')$-directional inner point, so is any interior point. We call a point in $\mathscr{A}$ a *positive-directional inner point* of $\mathscr{A}$ if it is a ξ-directional inner point of $\mathscr{A}$ for *every definitely positive* (unit) vector ξ in $\mathscr{R}^{N-1}$.

Given a nonzero vector ξ in $\mathscr{R}^{N-1}$ and a ξ-directional inner point α^0 of $\mathscr{A}$, we define the *right-hand ξ-directional derivative* $\phi'(\alpha^0; \xi^+)$ of ϕ at α^0 by

$$\phi'(\alpha^0; \xi^+) = \lim_{t_k \downarrow 0}\{[\phi(\alpha^0 + t_k\xi) - \phi(\alpha^0)]/t_k\}, \tag{4}$$

where $\{t_k\}$ is a sequence of strictly positive numbers converging to zero such that $\alpha^0 + t_k\,\xi \in \mathscr{A}$ for small t_k. For each i $(i = 1,..., N - 1)$, the *right-hand partial derivative* $\partial^+\phi(\alpha^0)/\partial\alpha_i$ of ϕ at an e_i-directional inner point of $\mathscr{A}$ is defined to be the right-hand e_i-directional derivative $\phi'(\alpha^0; e_i{}^+)$. The $\partial^+\phi(\alpha^0)/\partial\alpha_i$ represents the *sensitivity* of the supremum $\phi(\alpha)$ to positive increments in α_i from α^0. In matrix operations it is convenient to denote the row vector $[\partial^+\phi(\alpha^0)/\partial\alpha_1 \cdots \partial^+\phi(\alpha^0)/\partial\alpha_{N-1}]$ simply by $\partial^+\phi(\alpha^0)/\partial\alpha$.

Left-hand ξ-directional derivatives $\phi'(\alpha^0; \xi^-)$ are analogously defined at $(-\xi)$-directional inner points α^0 of $\mathscr{A}$ by (4), except $t_k \downarrow 0$ is replaced by $t_k \uparrow 0$ and $\{t_k\}$ by a sequence of strictly *negative* numbers converging to zero. Trivially, when both right-hand and left-hand ξ-directional derivatives of ϕ are defined and identical at a point α^0 in $\mathscr{A}$, the usual two-sided ξ-directional derivative $\phi'(\alpha^0; \xi)$ of ϕ at α^0 exists, and consequently $\phi'(\alpha^0; \xi) = \phi'(\alpha^0; \xi^+) = \phi'(\alpha^0; \xi^-)$.

Now we are ready to present a necessary pointwise test for (local) quasisupremality, which is very useful in the application of the method of PEC.

Theorem 5.1. Suppose that π is a definitely positive vector in $\mathscr{R}^{N-1}$ and α^0 a π-directional inner point of $\mathscr{A}$. Then for $(\alpha^0, \phi(\alpha^0))$ to be a (locally) quasisupremal vector for $\mathscr{L}$, the right-hand π-directional derivative $\phi'(\alpha^0; \pi^+)$ of ϕ at α^0 must be negative, $\phi'(\alpha^0; \pi^+) \leqslant 0$, whenever it exists.

Proof. By the hypothesis on directional inner point, there exists a sequence $\{\alpha^k\}$ in $\mathscr{A}$ such that $\alpha^k = \alpha^0 + t_k\pi$ for some sequence $\{t_k\}$ of strictly positive numbers converging to zero. For $(\alpha^0, \phi(\alpha^0))$ to be a (locally) quasisupremal vector for $\mathscr{L}$, it is necessary by Lemma 5.1 that $\phi(\alpha^0 + t_k\pi) < \phi(\alpha^0)$ for any such t_k sufficiently small. Consequently,

$$\lim_{t_k \downarrow 0} \{[\phi(\alpha^0 + t_k\pi) - \phi(\alpha^0)]/t_k\}$$

must be negative whenever it exists. QED

5.3. When does a right-hand directional derivative exist? Let α^0 be a positive-directional inner point of $\mathscr{A}$. We say that ϕ is *right-differentiable* at α^0 if there is a vector γ in $\mathscr{R}^{N-1}$ such that

$$\lim_{\pi^k \downarrow 0} \{[\phi(\alpha^0 + \pi^k) - \phi(\alpha^0) - \gamma^T\pi^k]/\| \pi^k \|\} = 0$$

for any sequence $\{\pi^k\}$ of definitely positive vectors in $\mathscr{R}^{N-1}$ converging to the zero vector. Such a vector γ, if it exists, will be called the *right-hand gradient* of ϕ at α^0 and will be denoted by $\nabla^+\phi(\alpha^0)$. This weakened form of the usual (Fréchet) differentiability is sufficient to assure the existence of right-hand directional derivatives.

Lemma 5.2. Suppose that α^0 is a positive-directional inner point of $\mathscr{A}$, $\phi(\alpha^0) \neq +\infty$. Then the right-hand π-directional derivative $\phi'(\alpha^0; \pi^+)$ of ϕ at α^0 exists for every definitely positive vector π in $\mathscr{R}^{N-1}$ and $\nabla^+\phi(\alpha^0) = [\partial^+\phi(\alpha^0)/\partial\alpha]^T$ if ϕ is right-differentiable at α^0.

Proof. Let π be any definitely positive vector in $\mathscr{R}^{N-1}$ and $\{t_k\}$ be any sequence of strictly positive numbers converging to zero such that $\alpha^0 + t_k\pi \in \mathscr{A}$ for all k. Define a sequence $\{\pi^k\}$ of definitely positive vectors by $\pi^k = t_k\pi$. Note that $\| \pi^k \| = t_k \| \pi \|$ and $\pi^k \downarrow 0$. If ϕ is right-differentiable at α^0, then there is a vector γ in $\mathscr{R}^{N-1}$ such that

$$0 = \lim_{\pi^k \downarrow 0} \frac{\phi(\alpha^0 + \pi^k) - \phi(\alpha^0) - \gamma^T\pi^k}{\| \pi^k \|} = \left\{\lim_{t_k \downarrow 0} \frac{\phi(\alpha^0 + t_k\pi) - \phi(\alpha^0)}{t_k} - \gamma^T\pi\right\} \Big/ \| \pi \|.$$

Thus,

$$\lim_{t_k \downarrow 0} \{[\phi(\alpha^0 + t_k\pi) - \phi(\alpha^0)]/t_k\} = \gamma^T\pi;$$

hence the right-hand π-directional derivative $\phi'(\alpha^0; \pi^+)$ exists at α^0. By taking $\pi = e_i$, we have $\gamma_i = \partial^+\phi(\alpha^0)/\partial\alpha_i$, $i = 1,..., N-1$. QED

Immediately from Theorem 5.1 and Lemma 5.2 we have:

Theorem 5.2. Suppose that α^0 is a positive-directional inner point of $\mathcal{A}$, $\phi(\alpha^0) \neq +\infty$, and ϕ is right-differentiable at α^0. Then for $(\alpha^0, \phi(\alpha^0))$ to be a (locally) quasisupremal vector for $\mathcal{L}$, the right-hand gradient $\nabla^+\phi(\alpha^0)$ of ϕ at α^0 must be negative, $\nabla^+\phi(\alpha^0) \leqslant 0$.

5.4. An alternative approach to the existence problem of right-hand directional derivatives is to replace the limit in (4) by the limit superior, since limits superior always exists (see Ref. 14, e.g.). Let α^0 be a ξ-directional inner point of $\mathcal{A}$. We extend some notions similar to those of Refs. 14–16 and define the *upper right-hand ξ-directional derivate* of ϕ at α^0 by

$$\limsup_{t_k \downarrow 0} \{[\phi(\alpha^0 + t_k\xi) - \phi(\alpha^0)]/t_k\},$$

where $\{t_k\}$ is a sequence of strictly positive numbers converging to zero such that $\alpha_0 + t_k\xi \in \mathcal{A}$ for small t_k. The following generalizes Theorems 5.1 and 5.2.

Theorem 5.3. If $(\alpha^0, \phi(\alpha^0))$ is a (locally) quasisupremal vector for $\mathcal{L}$, then the upper right-hand π-directional derivate of ϕ at α^0 must be negative for any definitely positive unit vector π in $\mathcal{R}^{N-1}$ such that α^0 is a π-directional inner point of $\mathcal{A}$.

Proof. Suppose on the contrary that there exists a definitely positive unit vector π in $\mathcal{R}^{N-1}$ for which α^0 is a π-directional inner point of $\mathcal{A}$ such that

$$\beta \triangleq \limsup_{t_k \downarrow 0} \{[\phi(\alpha^0 + t_k\pi) - \phi(\alpha^0)]/t_k\} > 0,$$

where $\{t_k\}$ is a sequence of strictly positive numbers such that $\alpha^0 + t_k\pi \in \mathcal{A}$ for each such t_k and that $\lim_k t_k = 0$. Then by definition of limit superior, given a number ρ, say in interval $(0, \beta)$, there is a strictly positive number δ_0 such that

$$-\rho \leqslant \beta - \sup_{t_k < \delta} \{[\phi(\alpha^0 + t_k\pi) - \phi(\alpha^0)]/t_k\} \leqslant \rho$$

for every δ in $(0, \delta_0)$. Further, by definition of supremum, for $\epsilon = (\beta - \rho)/2$, there is a number τ in the sequence $\{t_k\}$ with $0 < \tau < \delta$ such that

$$\frac{\phi(\alpha^0 + \tau\pi) - \phi(\alpha^0)}{\tau} + \epsilon > \sup_{t_k < \delta} \frac{\phi(\alpha^0 + t_k\pi) - \phi(\alpha^0)}{t_k}.$$

Combining these inequalities yields $\phi(\alpha^0 + \tau\pi) - \phi(\alpha^0) > (\beta - \rho - \epsilon)\tau = (\beta - \rho)\,\tau/2 > 0$. Thus ϕ can not be locally absolutely right-decreasing on $\mathscr{A}$ from α^0, since for each δ in $(0, \delta_0)$ there exists a point $\alpha = \alpha^0 + \tau\pi$ in the subset $\mathscr{A} \cap \mathscr{N}(\alpha^0; \delta)$ such that $\alpha \geqslant \alpha^0$ and $\phi(\alpha) > \phi(\alpha^0)$. Consequently, $(\alpha^0, \phi(\alpha^0))$ cannot be a locally quasisupremal vector for $\mathscr{Z}$ by Lemma 5.1. QED

5.5. Finally, as mentioned at the end of Section 3, ϕ may be a discontinuous or even discrete function on $\mathscr{A}$. If none of the foregoing conditions is applicable at points α^0 of interest, we may have to test the fundamental relationship in Theorem 3.1 directly, or make appropriate extensions in the definition of the function ϕ, as illustrated by the following example.

6. An Illustrative Example

Consider a problem in allocating funds to two interdependent but indispensible activities. Let z_1 denote the units of amount to be allocated to activity 1 and z_2 the units to activity 2. The allocation has the following constraints: $z_1 < 5.8$, $2z_2 \leqslant 11$, $10z_2 \geqslant z_1 + 9$, $(2z_1 - 11)^2 + (2z_2 - 7)^2 \geqslant 1$, $z_2 < 4 + (z_1 - 3.9)^2$, $2(z_2 - 3) \leqslant (7 - z_1)\cos[3.142(z_1 - 1)]$. Obviously, both z_1 and z_2 are desired to be *maximized*.

Consider $z = (z_1, z_2)$ as an *index vector for the allocation* and let $\mathscr{Z}_3$ denote the set of all index vectors attainable within the constraints. The problem of primary interest is to determine those index vectors that are quasisupremal for $\mathscr{Z}_3$ (or maximal in $\mathscr{Z}_3$) so that comprehensive studies can be made for desirable tradeoffs or compromises.

We apply the method of PEC. First, we solve the family of associated PECSO optimization problems for e_2-lineally supremal vectors $(\alpha, \phi(\alpha))$—namely, maximize z_2 subject to $z_1 = \alpha$ in addition to the given constraints and obtain $\phi(\alpha) = (3.5 - 0.5\alpha)\cos[3.142(\alpha - 1)] + 3$ for $5.8 \geqslant \alpha \geqslant 5.5$, $5.138 \geqslant \alpha \geqslant 3.218$, $2.915 \geqslant \alpha \geqslant 2.213$, $1.763 \geqslant \alpha \geqslant 1.172$, or $0.798 \geqslant \alpha \geqslant 0.288$; $\phi(\alpha) = 3.5 - 0.5\,[1 - (2\alpha - 11)^2]^{1/2}$ for $5.5 \geqslant \alpha > 5.138$; $\phi(\alpha) = 4 + (\alpha - 3.9)^2$ for $3.218 \geqslant \alpha \geqslant 2.915$; $\phi(\alpha) = 5.5$ for $1.172 \geqslant \alpha \geqslant 0.798$. The domain $\mathscr{A}$ consists of two unconnected intervals, [0.288, 1.763] and [2.213, 5.8]. See Fig. 2. (Also compare with Fig. 1.)

Then, we test the quasisupremality of these e_2-lineally supremal vectors. First, local and pointwise tests are used to discard some of the unqualified ones. Since $\partial^+\phi(\alpha^0)/\partial\alpha = -3.142(3.5 - 0.5\alpha^0) \times \sin[3.142(\alpha^0 - 1)] - 0.5\cos[3.142(\alpha^0 - 1)] > 0$ for $4.951 > \alpha^0 > 3.967$,

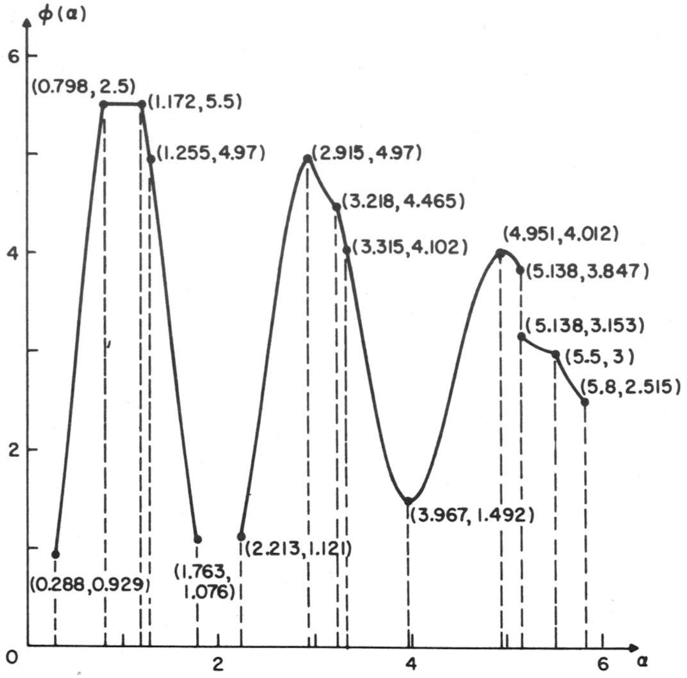

Fig. 2. The function ϕ for the set $\mathscr{X}_3$ in the illustrative example of Section 6.

$2.915 > \alpha^0 \geqslant 2.213$, or $0.798 > \alpha^0 \geqslant 0.288$, index vectors given by $(\alpha^0, \phi(\alpha^0))$ for any such α^0 are not even locally quasisupremal, by Theorem 5.2. (Note that 0.288 and 2.213 are not interior, but merely e_1-directional inner, points of $\mathscr{A}$, and that ϕ is not differentiable, but merely right-differentiable, at these points.) Since $\partial\phi(\alpha)/\partial\alpha = 0$ for $1.172 > \alpha > 0.798$, vectors given by $(\alpha^0, \phi(\alpha^0))$ for any α^0 in $[0.798, 1.172)$ are not even locally quasisupremal according to Corollaries 5.1, 5.2, or 5.3. [Note that Corollary 4.1 is not applicable at these α^0 since none is a radiant point of the entire subset $\mathscr{A}(\alpha^0)$.]

Now the vector (5.8, 2.515) given by $\alpha = 5.8$ is obviously a *quasisupremal vector*. ϕ is differentiable on $[4.951, 5.8)$ except points 5.138 and 5.5, and the Lebesgue integral

$$\int_0^\tau [\partial\phi(\alpha^0 + te_1)/\partial\alpha]\, e_1\, dt = \int_{\alpha^0}^{\alpha^0+\tau} (\partial\phi/\partial\alpha)\, d\alpha < 0$$

for any α^0 in $[4.951, 5.8)$ and any τ in $(0, \rho]$, where $\rho = 5.8 - \alpha^0$. Any vector $(\alpha^0, \phi(\alpha^0))$ for any such α^0 *is quasisupremal* by Corollary 4.3. ϕ is also differentiable on [3.315, 4.951], but for any α^0 in [3.315, 4.951),

$$\int_0^{4.951-\alpha^0} [\partial\phi(\alpha^0 + te_1)/\partial\alpha]\, e_1\, dt = 4.012 - \phi(\alpha^0) \geqslant 0;$$

vectors $(\alpha^0, \phi(\alpha^0))$ for any such α^0 are not quasisupremal, by Corollary 4.3. ϕ is also differentiable on [2.915, 3.315] and $\int_0^\tau [\partial\phi(\alpha^0 + te_1)/\partial\alpha]\, e_1\, dt < 0$ for any α^0 in [2.915, 3.315) and any τ in $(0, \rho]$, where $\rho = 5.8 - \alpha^0$. Any vector $(\alpha^0, \phi(\alpha^0))$ given by any such α^0 *is a quasisupremal vector* by Corollary 4.3.

Since ϕ is undefined on the interval (1.763, 2.213), those tests given in Sections 4 and 5 are not directly applicable. Consequently, Theorem 3.1 should be used. By this one can conclude that vectors given by $(\alpha^0, \phi(\alpha^0))$ are quasisupremal with $1.255 > \alpha^0 \geqslant 1.172$, but not with $1.763 \geqslant \alpha^0 \geqslant 1.255$. [Note that $\phi(2.915) \geqslant \phi(\alpha^0)$ for $1.763 \geqslant \alpha^0 \geqslant 1.255$.] We can, on the other hand, extend ϕ over this interval and apply Corollary 4.3 again. To do so, let $\phi(\alpha) = 1.076 + 0.1(\alpha - 1.763)$ for α in (1.763, 2.213). Then ϕ is also differentiable on (1.172, 2.213) except point 1.763. Now, for any α^0 in [1.255, 2.213],

$$\int_0^{2.915-\alpha^0} [\partial\phi(\alpha^0 + te_1)/\partial\alpha]\, e_1\, dt = 4.970 - \phi(\alpha^0) \geqslant 0;$$

vectors $(\alpha^0, \phi(\alpha^0))$ correspondingly given are not quasisupremal. For α^0 in [1.172, 1.255), $\int_0^\tau [\partial\phi(\alpha^0 + te_1)/\partial\alpha]\, e_1\, dt < 0$ for any τ in $(0, \rho]$, where $\rho = 5.8 - \alpha^0$; vectors $(\alpha^0, \phi(\alpha^0))$ correspondingly given *are quasisupremal.*

7. Appendix

7.1. Quasisupremality and the Minimization of Index Vectors. Suppose that the minimization of index vectors $z = (z_1, \ldots, z_N)$ is of interest instead, and that superiority is defined by the inverse $\leqslant$ of the partial order $\geqslant$, namely, z^0 is *superior to* z' if $z^0 \leqslant z'$. Again, let $\mathscr{Z}$ denote the set of all attainable index vectors under investigation. In this case, *minimal vectors* in $\mathscr{Z}$ are of interest. Quasisupremal vectors for $\mathscr{Z}$ are defined in the same manner as before except that lineally supremal vectors are modified for this case as follows. For a *definitely positive* (unit) vector p in $\mathscr{R}^N$, z^0 is called a p-lineally supremal vector for $\mathscr{Z}$ if it is a $(-p)$-directional extremum of the lineal subset $\{z \in \bar{\mathscr{Z}} \mid z = z^0 + cp \text{ and } c \in \mathscr{R}\}$.

Note that for a definitely positive vector p, z^0 is a $(-p)$-directional extremum of a subset $\mathscr{S}$ if and only if $p^T z^0 = \inf\{p^T z \mid z \in \mathscr{S}\}$. Thus the corresponding Nth family of associated PECSO optimization problems is to

minimize z_N

subject to $z_i = \alpha_i, \quad i = 1, ..., N-1, \quad$ and $\quad (z_1, ..., z_N) \in \bar{\mathscr{Z}}$

for vectors $(\alpha_1, ..., \alpha_{N-1})$ in $\mathscr{R}^{N-1}$. The corresponding optimal solutions are the corresponding e_N-lineally supremal vectors for $\mathscr{Z}$. Similarly let $\mathscr{A}$ denote the set of all vectors α for which the infimum $\phi(\alpha) = \inf\{z_N \mid (z_1, ..., z_{N-1}) = \alpha$ and $(z_1, ..., z_N) \in \bar{\mathscr{Z}}\}$ exists, or $\phi(\alpha) = -\infty$.

The fundamental relationship in Theorem 3.1 now reads: For an e_N-lineally supremal vector $(\alpha^0, \phi(\alpha^0))$, $\alpha^0 \in \mathscr{A}$, to be a quasisupremal vector for $\mathscr{Z}$ it is necessary and sufficient that $\phi(\alpha) > \phi(\alpha^0)$ for any α in $\mathscr{A}$ such that $\alpha \leqslant \alpha^0$. Consequently, the quasisupremality of $(\alpha^0, \phi(\alpha^0))$ requires the pointwise absolute "left-increasing" property of ϕ. ϕ is said to be *left-increasing* (*strictly left-increasing*, or *absolutely left-increasing*) on $\mathscr{A}$ from α^0 if all inequalities used in defining the property right-decreasing (strict right-decreasing, or absolute right-decreasing) are reversed. Pointwise left-increasing and right-decreasing properties are equivalent for decreasing functions; but for general functions, neither implies the other. For example, if ϕ is defined by $\phi(\alpha) = 1 - (\alpha - 1)^2$ on $[-1, 2]$, then ϕ is right-decreasing but not left-increasing from $\alpha^0 = 1$. (Instead, it is "left-decreasing" from $\alpha^0 = 1$.)

The tests for quasisupremality given in Sections 4 and 5 are all applicable to this case provided that appropriate, but generally obvious, modifications are made. For example, define $\mathscr{A}(\alpha^0)$ by $\{\alpha \in \mathscr{A} \mid \alpha \leqslant \alpha^0\}$, use the left-hand π-directional derivative $\phi'(\alpha^0; \pi^-)$ instead of the right-hand $\phi'(\alpha^0; \pi^+)$ (still for definitely positive vector π), assume left-differentiability instead of the right-differentiability, replace $\phi(\alpha^0) \neq +\infty$ by $\phi(\alpha^0) \neq -\infty$, and so forth.

7.2. Lineally Supremal Vectors and Weak-Maximal Vectors. We call a vector z^0 a *weak-maximal vector* in $\mathscr{Z}$ if there exists no vector z in $\mathscr{Z}$ such that $z > z^0$. One can easily see that a weak-maximal vector must be an e_i-lineally supremal vector for at least one i. But the converse is not true. Consider the set $\mathscr{Z}_1$ in Fig. 1, for example. Vector p is weak-maximal and e_2-lineally supremal; vector t is weak-maximal and e_1-lineally supremal; vector q is e_2-lineally supremal, vectors g, a, e, and f are both e_1- and e_2-lineally supremal, but none of these, q, g, a, e, or f, is a weak-maximal vector in $\mathscr{Z}_1$.

The vectors g, a, e, and f of $\mathscr{Z}_1$ indicate that a semisupremal vector is not necessarily a weak-maximal vector, while the vector $(1, 1)$ of the set $\{(z_1, z_2) \mid z_1 = 1$ or $z_2 = 1\}$ indicates that a weak-maximal vector is not necessarily semisupremal. Nevertheless, the union $\bigcup_{i=1}^{i=N} \mathscr{L}(e_i)$ contains *all weak-maximal vectors* in $\mathscr{X}$. Thus, in general, the set of all quasisupremal vectors for $\mathscr{X}$ is not a proper subset of the set of weak-maximal vectors in $\mathscr{X}$. But in case $\mathscr{X}$ contains the union $\bigcup_{i=1}^{i=N} \mathscr{L}(e_i)$, every quasisupremal vector is a weak-maximal vector.

References

1. Lin, J. G., *Maximal Vectors and Multi-Objective Optimization*, Journal of Optimization Theory and Applications, Vol. 18, No. 1, pp. 41–64, January 1976.
2. Pareto, V., *Manuale di Economia Politica*, Societa Editrice Libraria, Milan, Italy, 1906.
3. Karlin, S., *Mathematical Methods and Theory in Games, Programming, and Economics*, Vol. I, Addison-Wesley, Reading, Massachusetts, 1959.
4. Koopmans, T. C., *Analysis of Production as an Efficient Combination of Activities*, in *Activity Analysis of Production and Allocation* (Cowles Commission Monograph 13), Edited by T. C. Koopmans, Wiley, New York, 1951, pp. 33–97.
5. Zadeh, L. A., *Optimality and Non-Scalar-Valued Performance Criteria*, IEEE Transactions on Automatic Control, Vol. AC-8, No. 1, pp. 59–60, 1963.
6. Da Cunha, N. O., and Polak, E., *Constrained Minimization under Vector-Valued Criteria in Finite Dimensional Space*, Journal of Mathematical Analysis and Applications, Vol. 19, No. 1, pp. 103–124, July 1967.
7. Chu, K. C., *On the Noninferiority Set for the Systems with Vector-Valued Objective Function*, IEEE Transactions on Automatic Control, Vol. AC-15, pp. 591–593, October 1970.
8. Lin, J. G., Katopis, G., Patellis, S. T., and Shen, H. N., *On Relatively Optimal Controls of Dynamic Systems with Vector Performance Indices*, in *Proceedings of the Seventh Annual Princeton Conference on Information Sciences and Systems*, March 1973, pp. 579–584.
9. Schmitendorf, W. E., *Cooperative Games and Vector-Valued Criteria Problem*, IEEE Transactions on Automatic Control, Vol. AC-18, No. 2, pp. 139–144, April 1973.
10. Lin, J. G., *Circuit Design under Multiple Performance Objectives*, in *Proceedings of 1974 IEEE International Symposium on Circuits and Systems*, April 1974, pp. 549–552.
11. Lin, J. G., *Proper Inequality Constraints (PIC) and Maximization of Index Vectors*, Journal of Optimization Theory and Applications, to appear.
12. Katopis, G. A., and Lin, J. G., *Non-Inferiority of Controls under Double Performance Objectives: Minimal Time and Minimal Energy*, in *Proceedings*

of the Seventh Hawaii International Conference on System Sciences, January 1974, pp. 129–131.

13. HOLTZMAN, J. M., and HALKIN, H., *Directional Convexity and the Maximum Principle for Discrete Systems*, Journal of SIAM on Control, Vol. 4, No. 2, pp. 263–275, 1966.
14. TAYLOR, A. E., *General Theory of Functions and Integration*, Blaisdell, New York, 1965.
15. FLEMING, W. H., *Functions of Several Variables*, Addison-Wesley, Reading, Massachusetts, 1965.
16. MCSHANE, E. J., and BOTTS, T. A., *Real Analysis*, Van Nostrand, Princeton, New Jersey, 1959.
17. LIN, J. G., *Cooperative Control of Dynamic Systems with Multiple Performance Measures*, Proposal to National Science Foundation, November 1971.
18. LIN, J. G., *Multiple-Objective Optimization*, Columbia University, Department of Electrical Engineering and Computer Science, Systems Research Group Technical Reports, 1972.
19. LIN, J. G., *Multiple-Objective Problems: Pareto-Optimal Solutions by Method of Proper Equality Constraints (PEC)*, IEEE Transactions on Automatic Control, to appear.
20. LIN, J. G., *Multiple-Objective Programming: Lagrange Multipliers and Method of Proper Equality Constraints*, in Proceedings of 1976 Joint Automatic Control Conference, July 1976.
21. LIN, J. G., *Three Methods for Determining Pareto-Optimal Solutions of Multiple-Objective Problems*, presented at the Conference on Directions in Decentralized Control, Many-Person Optimization and Large-Scale Systems, September 1–3, 1975, Cambridge, Massachusetts.

VI

Sufficient Conditions for Preference Optimality

W. STADLER

Abstract. Preference optimality is an optimality concept in multicriteria problems. Formally, the objective there is to "optimize" N criterion functions $g_i(\cdot)$: $\mathscr{D} \to \mathbb{R}$, where $\mathscr{D}$ is a decision set, which may be a feasible set in programming or a set of admissible controls in control theory. Corresponsing to $d \in \mathscr{D}$ one obtains a set $\mathscr{A} \subset \mathbb{R}^N$ of attainable criterion values. On this set one introduces a preference $\lesssim$; that is, a binary relation with the purpose of providing a hierarchy among the elements of $\mathscr{A}$. A decision $d^* \in \mathscr{D}$ then is preference optimal iff either $g(d^*) \lesssim g(d)$ for every $d \in \mathscr{D}$ or $g(d) \lesssim g(d^*)$ implies $g(d) \sim g(d^*)$ for every comparable $d \in \mathscr{D}$. Here, sufficient conditions are given for preference-optimal decisions in general and for Pareto-optimal decisions in particular. An illustrative example is included.

1. Introduction

Preference optimality is a solution concept in problems where several criteria are to be "optimized" simultaneously subject to various side conditions. Many such solution concepts exist; a partial review and bibliography of such concepts may be found in Ref. 1. Obviously, there is no unique solution concept for these problems; the choice of an optimality concept depends upon the situation.

Preference optimality was introduced in Ref. 2 along with a discussion of existence, a statement of necessary conditions in terms of a maximum principle and a Fritz John condition, and a comparison with Pareto optimality. Here, sufficient conditions for preference optimality will be given. Although the concepts of preferences and orderings are well-established in mathematical economics, they have had little application in engineering; for this reason and in order to make the paper self-contained some of the basic discussion of Ref. 2 is necessarily repeated here.

Again, for definiteness the two basic problem formulations are given. Use is made of the following notation for points $x, y \in \mathbb{R}^n$:

(i) $x \leqslant y \Leftrightarrow x_i \leqslant y_i \quad \forall\, i \in I = \{1,..., n\}$.

(ii) $x < y \Leftrightarrow x_i \leqslant y_i \quad \forall\, i \in I$, and $x \neq y$.

(iii) $x \ll y \Leftrightarrow x_i < y_i \quad \forall\, i \in I$.

Subsequently it will be apparent that this is an example of a partial ordering of $\mathbb{R}^n$; it will be referred to as the natural order on $\mathbb{R}^n$. The notation differs somewhat from the usual in that inequality is emphasized ($<$, $\ll$) rather than equality ($\leqslant$, $\leqq$). Throughout, this notation in connection with vectors is to be interpreted in this light. It was thought desirable to use more or less conventional notation in the problem statements; this necessitates the use of the same symbol for different quantities. However, this should cause no difficulty, since the meaning will be clear from the context.

Problem I. The Programming Problem. Let Ω (open) $\subseteq \mathbb{R}^n$ and introduce the inequality constraints

$$f(\cdot): \quad \Omega \to \mathbb{R}^m$$

and the equality constraints

$$h(\cdot): \quad \Omega \to \mathbb{R}^k,$$

so that the functional constraint set is given by

$$X = \{x \in \Omega : f(x) \leqslant 0, \quad h(x) = 0\}.$$

The criterion functions are

$$g_i(\cdot): \quad X \to \mathbb{R}, \quad i = 1,..., N,$$

with corresponding criterion vector

$$g(\cdot) = (g_1(\cdot),..., g_N(\cdot)),$$

and values $g(x) \in \mathbb{R}^N$, the criteria space.

Definition 1.1. Attainable criteria set for Problem I. The criterion value $y \in \mathbb{R}^N$ is attainable iff there exists an $x \in X$ such that $g(x) = y$. The attainable criteria set consists of all such attainable criterion values; it is

$$Y = \{y \in \mathbb{R}^N: \quad y = g(x), \quad x \in X\} = g(X).$$

The basic multicriteria programming problem is: Obtain "optimal" decision(s) $x^* \in X$ for $g(x)$ subject to $x \in X$.

Problem II. The Control Problem. Let the state $x \in A$ (open) $\subset \mathbb{R}^n$ be controlled by means of a control $u(\cdot)$: $[t_0, t_1] \to U \subset \mathbb{R}^r$ in the system equations

$$\dot{x} = f(x, t, u) \tag{1}$$

with $x(t_0) \in \theta^0 \triangleq$ the initial set, and $x(t_1) \in \theta^1 \triangleq$ the terminal set. Furthermore, $f(\cdot)$: $A \times [t_0, t_1] \times U \to B$(open) $\subset \mathbb{R}^n$, and U, the control constraint set, is the set of possible values for $u(\cdot)$. It is usual to confine oneself to a set of admissible controls.

Definition 1.2. Admissible Controls. A control $u(\cdot)$: $[t_0, t_1] \to U$ is admissible iff

(i) U (bounded) $\subset \mathbb{R}^r$.

(ii) $u(\cdot)$ is Lebesgue-measurable.

(iii) $u(\cdot)$ generates a solution $x(\cdot)$: $[t_0, t_1] \to A$ of Eq. (1) such that $x(t_0) \in \theta^0$ and $x(t_1) \in \theta^1$.

The set of admissible controls is denoted by $\mathscr{F}$; it is assumed to be nonempty.

Strictly speaking, a solution of Eq. (1) is a function $s(\cdot)$ of the initial conditions x^0, the initial value t_0, and t; i.e., for a given set of such values, $x(t) = s(x_0, t_0; t)$. In addition, such a solution may be nonunique without some further assumptions on $f(\cdot)$. These dependences will be suppressed, since they are not relevant to the present discussion.

The criterion vector

$$g(\cdot): \quad \mathscr{F} \to \mathbb{R}^N$$

is defined in terms of the integrals

$$g_i(u(\cdot)) = \int_{t_0}^{t_1} f_{0i}(x(t), t, u(t))\, dt, \tag{2}$$

with $f_{0i}(\cdot)\colon A \times [t_0, t_1] \times U \to C_i$ (open) $\subset \mathbb{R}$, $i = 1,..., N$, where all dependences other than $u(\cdot)$ have again been suppressed.

The state space $\mathbb{R}^n$ is augmented by introducing a criterion response

$$\dot{y} = f^0(x, t, u) \qquad \text{with} \quad y(t_0) = 0, \tag{3}$$

where $y \in \mathbb{R}^N$, the criteria space, and where $f^0 = (f_{01}, ..., f_{0N})$.

Definition 1.3. Attainable criteria set for Problem II. Let $u(\cdot) \in \mathscr{F}$ and let $x(\cdot)$ be a corresponding solution of Eq. (1). For every $t \in [t_0, t_1]$ the attainable criteria set $K(t)$ is the set of all response points $y(t) \in \mathbb{R}^N$, where $y(\cdot)$ is a solution of (3). In particular, the set $K(t_1)$ is the set of all $y(t_1)$.

See Lee and Markus (Ref. 3) for a more thorough discussion of the set of attainability of a system of differential equations.

The multicriteria control problem is: Obtain "optimal" control(s) $u^*(\cdot) \in \mathscr{F}$ for $g(u(\cdot))$ subject to $u(\cdot) \in \mathscr{F}$.

This paper consists of a discussion of sufficient conditions for Problems I and II when "optimal" is "preference optimal." Briefly, without mathematical precision, preference optimality consists of the following basic ideas. Let M be an open set and $Y \subset M$. A preference relation $\precsim$ is introduced on M and thus induced on Y. In its most general form such a relation is nothing but a binary relation on M which is to be put to a particular use, namely it is used to indicate which elements of M are preferred to others. Thus, for $y^1, y^2 \in M$, $y^1 \prec y^2$ might be read as "y^2 is preferred to y^1." If then, a relation $\precsim$ induced on Y is such that it makes sense to talk about a greatest element among the elements of Y, that is, an element (or elements) $y^* \in Y$ such that $y \precsim y^* \; \forall y \in Y$, then one may take a decision x^* which yields $g(x^*) = y^*$ as preference optimal. This makes sense, for example, if $\precsim$ is a complete preordering of M. The calculation of such an element is eased considerably if there exists on M a function $\phi(\cdot)\colon M \to \mathbb{R}$ such that $\phi(y^1) \leqslant \phi(y^2) \Leftrightarrow y^1 \precsim y^2$, for the determination of a greatest element in Y then becomes equivalent to maximizing $\phi(\cdot)$ on Y. Most theorems concerning preference relations are linked to the existence of such a function along with some properties such as continuity or differentiability. It was precisely the existence of such a differentiable utility function which yielded esthetic necessary conditions for preference optimality. A complete rendition of preference optimality and with some examples of applications of Pareto optimality in mechanics are given in Ref. 4.

Here, only sufficiency theorems will be given. For the convenience of the reader some relevant mathematical preliminaries are given first.

2. Mathematical Preliminaries: Convexity

Most of the material and the notation of this section is based on Mangasarian (Ref. 5). All theorems deal with convex function theory; they are stated without proof. In the cited theorems and definitions $M \subseteq \mathbb{R}^n$ and $\phi(\cdot)$: $M \to \mathbb{R}$.

Definition 2.1. Convexity. The function $\phi(\cdot)$ is convex with respect to M at $\bar{x} \in M$ iff $x \in M$, $\theta \in [0, 1]$, and $(1 - \theta)\bar{x} + \theta x \in M$ imply

$$(1 - \theta)\,\phi(\bar{x}) + \theta\phi(x) \geqslant \phi((1 - \theta)\bar{x} + \theta x).$$

Definition 2.2. Quasiconvexity. The function $\phi(\cdot)$ is quasiconvex with respect to M at $\bar{x} \in M$ iff $x \in M$, $\phi(x) \leqslant \phi(\bar{x})$, $\theta \in [0, 1]$, and $(1 - \theta)\bar{x} + \theta x \in M$ imply

$$\phi((1 - \theta)\bar{x} + \theta x) \leqslant \phi(\bar{x}).$$

Definition 2.3. Pseudoconvexity. Let $\phi(\cdot)$ be differentiable. The function $\phi(\cdot)$ is pseudoconvex with respect to M at $\bar{x} \in M$ iff $x \in M$ and $\nabla\phi(\bar{x})(x - \bar{x}) \geqslant 0$ imply

$$\phi(\bar{x}) \leqslant \phi(x).$$

It follows that $\phi(\cdot)$ is convex, quasiconvex, or pseudoconvex on a convex set M iff $\phi(\cdot)$ is convex, quasiconvex, or pseudoconvex at every point of M.

Theorem 2.1. Let $g(\cdot)$: $M \to \mathbb{R}^N$. Assume that $g(\cdot)$ [that is, each of the component functions $g_i(\cdot)$] is convex at $\bar{x} \in M$ (convex on a convex set M), and let $c \geqslant 0$, $c \in \mathbb{R}^N$; then

$$\phi(\cdot) = cg(\cdot)$$

is convex at $\bar{x}$ (convex on a convex set M).

Note that there is no similar theorem for quasiconvex or pseudoconvex functions.

Theorem 2.2. Let M be open. Assume that $\phi(\cdot)$ is differentiable at $\bar{x} \in M$. If $\phi(\cdot)$ is convex at $\bar{x} \in M$, then

$$\phi(x) - \phi(\bar{x}) \geqslant \nabla\phi(\bar{x})(x - \bar{x})$$

for each $x \in M$.

Theorem 2.3. Let M be convex. Let $\Lambda_\nu = \{x \in M: \phi(x) \geqslant \nu\}$. Then $\phi(\cdot)$ is quasiconcave on M iff Λ_ν is convex for each $\nu \in \mathbb{R}$.

Theorem 2.4. Let M be open and convex and let $\phi(\cdot)$ be twice differentiable. Then $\phi(\cdot)$ is convex on M iff the quadratic form

$$\sum_{i,j=1}^{n} \phi_{ij}(x)\, \xi_i \xi_j\,, \qquad \phi_{ij}(x) = (\partial^2\phi/\partial x_i\, \partial x_j)(x),$$

is positive semidefinite in the variables $\xi_1, \ldots, \xi_n$ for every $x \in M$.

Theorem 2.5. Let $Q(\xi, \xi) = (\xi, Q\xi)$ be a quadratic form defined on $\mathbb{R}^n$, and let Q be the corresponding $n \times n$ matrix. Let $\alpha_1, \ldots, \alpha_n$ be the eigenvalues of Q. Then, the quadratic form is negative definite iff

$$\alpha_i < 0 \qquad \text{for all} \quad i \in \{1, \ldots, n\}.$$

3. Mathematical Preliminaries: Orderings and Preferences

The words "order" and "preference" both have acquired virtually fixed meanings in mathematics and economics, respectively. "Order" to a mathematician usually implies at least a transitive relation, termed a partial order by Kelley (Ref. 12), and based on the classic work of Debreu (Ref. 6) a "preference" is considered by economists to be at least a "partial preorder." This more or less colloquial usage of these terms makes it easy to lose sight of the fact that they simply represent binary relations which are tailored to a particular purpose—to introduce a hierarchy among the elements of a set.

Definition 3.1. Strict preference and indifference.

(i) Let R_1 be a binary relation on a set M. The relation R_1 is a strict preference on M iff R_1 serves to provide a hierarchy among elements of M. Then R_1 is denoted by $\prec$.

(ii) Let R_2 be a binary relation on a set M. The relation R_2 is an indifference on M iff R_2 serves to provide a sense of equality among elements of M. Then R_2 is denoted by $\sim$.

Definition 3.2. Preference. Let R be a binary relation on a set M. The relation R is a preference on M iff $R = R_1 \cup R_2$ is the disjoint union of a strict preference R_1 and an indifference R_2. Then R is denoted by $\precsim$.

An extensive treatment of preference and utility theory may be found in Fishburn (Ref. 7), who gives a thorough mathematical treatment of the subject, and in Debreu (Ref. 6), where both the mathematical aspects and applications in consumer theory are discussed. Some properties of preferences are in such frequent use that preferences that have them have been given special names. Only the essential properties and names have been cited here.

Definition 3.3. Properties of relations. Consider the following properties of a binary relation R defined on a set M with $x, y, z \in M$:

(i) xRx for every $x \in M$ (reflexivity).

(ii) "xRy and yRz" $\Rightarrow$ "xRz" (transitivity).

(iii) "xRy and yRx" $\Rightarrow$ "$x = y$" (antisymmetry).

(iv) For any $x, y \in M$ either xRy or yRx or both (connexity).

(v) "xRy" $\Rightarrow$ "$\neg yRx$" (asymmetry).

(vi) "xRy" $\Rightarrow$ "yRx" (symmetry).

Collectively one then defines:

(1) (i) and (ii) together as a partial preorder.

(2) (i)–(iii) together as a partial order.

(3) The inclusion of (iv) in (1) [resp. (2)] as a complete preorder [resp., linear order].

Definition 3.4. Properties of preferences. Let $\lesssim$ be a preference on a set M.

(i) *Continuity.* A preference $\lesssim$ on M is continuous iff the sets $I^+(y) = \{x \in M: x \gtrsim y\}$ and $I^-(y) = \{x \in M: x \lesssim y\}$ are closed in M for every $y \in M$.

(ii) *Monotonicity.* Let $M \subset \mathbb{R}^N$. The strict preference $\prec$ on M is: (1) weakly monotone iff for $x, y \in M$ one has $x \ll y \Rightarrow x \prec y$; (2) monotone iff for $x, y \in M$ one has $x < y \Rightarrow x \prec y$.

(iii) *Differentiability.* The preference $\lesssim$ on $M \subseteq \mathbb{R}^N$ is of class C^k on M iff $I = \{(x, y) \in M \times M: x \sim y\}$ is a C^k-hypersurface in $\mathbb{R}^{2N}$.

Most use of these latter properties for preferences is made in the derivation of necessary conditions for preference optimality. These properties are included here in order to give a meaningful quotation of the necessary conditions for Problem I. It will be needed in the presentation of the example in Section 5.

In the statement of sufficient conditions extensive use will be made of the explicit knowledge of a utility function, a preference-preserving mapping between $(\precsim, M)$ and $(\leq, \mathbb{R})$.

Definition 3.5. Utility function. Let $\precsim$ be a preference on a set M. A function $\phi(\cdot)$: $M \to \mathbb{R}$ is a utility function for $\precsim$ on M iff one has for every $x, y \in M$:

$$x \precsim y \Rightarrow \phi(x) \leq \phi(y),$$

$$x \sim y \Rightarrow \phi(x) = \phi(y),$$

$$x \prec y \Rightarrow \phi(x) < \phi(y).$$

There is nothing unique about such a function, since any strictly increasing function $F(\cdot)$: $\mathbb{R} \to \mathbb{R}$ defines another utility function $\psi(\cdot) = F \circ \phi(\cdot)$ for $\precsim$ on M. It is precisely this latter fact which will be useful for the derivation of the sufficient conditions. In particular, the following theorem will provide an answer to the question: Given $\phi(\cdot)$, whose level sets are the indifference surfaces of the preference relation on M—under what conditions on $\phi(\cdot)$ does there exist a strictly increasing, twice differentiable function $F(\cdot)$ such that $\psi(\cdot) = F \circ \phi(\cdot)$ is a convex utility function for $\precsim$ on M? Without differentiability assumptions this problem was first treated by de Finetti (Ref. 8) and subsequently by Fenchel (Ref. 9), who also derived further results with imposed differentiability assumptions (Ref. 10). The results obtained by Fenchel (Ref. 10) are summarized here in the form of a theorem. Most of the notation used by Fenchel has been retained.

In the following let M (open) $\subseteq \mathbb{R}^N$. Use

$$\phi_i(y) = (\partial\phi/\partial y_i)(y) \qquad \text{and} \qquad \phi_{ij}(y) = (\partial^2\phi/\partial y_i\, \partial y_j)(y),$$

with $\phi(\cdot)$: $M \to \mathbb{R}$, to define on M the quadratic form

$$Q(\xi, \xi) = \sum_{i,j=1}^{N} \phi_{ij}(y)\, \xi_i \xi_j + \sigma(y) \left[\sum_{i=1}^{N} \phi_i(y)\, \xi_i \right]^2,$$

where $\sigma(y)$ is a suitable Lagrange multiplier, which one may relate to $F(\cdot)$ by

$$\sigma(y) = F''(\phi(y))\, F'(\phi(y)),$$

for every $y \in M$. The primes denote differentiation with respect to the argument of $F(\cdot)$. For every $y \in M$ define

$$k^2(y) = \phi_1{}^2(y) + \phi_2{}^2(y) + \cdots + \phi_N{}^2(y)$$

along with the two characteristic polynomials

$$\Gamma(\alpha) = |\phi_{ij}(y) - \alpha\delta_{ij}| = S_N - S_{N-1}\alpha + \cdots + (-1)^N S_0 \alpha^N$$

$$\Gamma^*(\alpha) = -[1/k^2(y)] \begin{vmatrix} \phi_{ij}(y) - \alpha\delta_{ij} & \phi_i(y) \\ \phi_j(y) & 0 \end{vmatrix}$$

$$= S_{N-1}^* - S_{N-2}^*\alpha + \cdots + (-1)^{N-1} S_0^* \alpha^{N-1}.$$

The S_ν are the elementary symmetric functions of the characteristic polynomials. With roots $\alpha_1, \ldots, \alpha_N$ of $\Gamma(\alpha)$ and with $S_0 = 1$, these are given by

$$S_1 = \sum_{i=1}^{N} \alpha_i, \quad S_2 = \sum_{\substack{i,j=1 \\ i<j}}^{N} \alpha_i\alpha_j, \quad S_3 = \sum_{\substack{i,j,k=1 \\ i<j<k}}^{N} \alpha_i\alpha_j\alpha_k, \ldots, S_N = \alpha_1\alpha_2 \cdots \alpha_N.$$

With $S_0^* = 1$, the S_ν^* are similarly defined.

Furthermore, let

$$\epsilon = \inf\{\phi(y): \ y \in M\} \qquad \text{and} \qquad \beta = \sup\{\phi(y): \ y \in M\}$$

with $\pm\infty$ permitted. The notation $(=)$ in

$$\epsilon \underset{(=)}{<} t < \beta$$

is taken to mean that equality may hold iff $\phi(\cdot)$ has a minimum on M.

Theorem 3.1. (Fenchel). Assume that $\phi(\cdot)$ is a twice differentiable function on an open set M. The following conditions are necessary and sufficient for the existence of a twice differentiable, strictly increasing function $F(\cdot)\colon \mathbb{R} \to \mathbb{R}$ such that $\psi(\cdot) = F \circ \phi(\cdot)$ is a convex utility function on M:

(i) The function $\phi(\cdot)$ either has no stationary values or it has only an absolute minimum on M.

(ii) The quadratic form $Q(\xi, \xi)$ is positive semidefinite for every $y \in M$.

(iii) If the rank of the matrix of the quadratic form $Q(\xi, \xi)$ is $r(y) - 1$ at the point $y \in M$, then the rank of the matrix $[\phi_{ij}(y)]$ is at most $r(y)$.

(iv) For every fixed t with $\epsilon < t < \beta$

$$G(t) = \sup\{-S_r/k^2(y)\, S_{r-1}^*\colon \ \phi(y) = t, \ \ y \in M\} < \infty$$

where the subscripts r are to be identified with the rank in (iii).

(v) There exists a function $H(\cdot)$: $[\epsilon, \beta] \to \mathbb{R}$ such that $H(t) > 0$ on (ϵ, β), the derivative $H'(\cdot)$ exists for $\epsilon \underset{(=)}{<} t < \beta$, and such that

$$G(t) \leqslant H'(t)/H(t).$$

4. Sufficient Conditions

The necessary conditions (see Ref. 2) were given in an order which illustrated the increase in the required amount of information—no existence of a utility function, existence of a differentiable utility function, knowledge of a normal vector to the indifference surface at the optimal criterion value, and finally knowledge of a utility function itself. Ideally, of course, sufficient conditions should also make use only of the properties of the preferences; a recent result of Aumann (Ref. 11) would be in this direction, where no explicit knowledge of a utility function itself, but only of its derivatives, is required to deduce the existence of a concave utility representation for $\precsim$. However, the results given here will lean heavily on the knowledge of a utility function.

First a metatheorem is given which may be used to transform any sufficiency theorem in optimal control or programming into a corresponding one for preference optimality. This is followed by two illustrations of its possible use. Finally, some general sufficiency theorems for Pareto optimality are given in view of the popularity of this concept in multicriteria problems.

In all general discussions of a multicriteria decision problem the decisions are $d \in \mathscr{D}$, a decision set with attainable criterion values $g(d) \in \mathscr{A}$, the attainable criteria set.

Definition 4.1. Preference optimal decision. Let $\precsim$ be a preference on $\mathscr{A}$. A decision $d^* \in \mathscr{D}$ is preference optimal iff it results in a $g(d^*) \in \mathscr{A}$ such that either:

(i) $g(d^*) \precsim g(d) \quad \forall d \in \mathscr{D}$; or

(ii) $g(d) \precsim g(d^*) \Rightarrow g(d) \sim g(d^*) \quad \forall d^*$-comparable $d \in \mathscr{D}$.

Now let $G(\cdot)$: $\mathscr{D} \to \mathbb{R}$ be a decision problem in the usual sense, that is, a standard programming or a control problem with the objective: Minimize $G(d)$ subject to $d \in \mathscr{D}$. Let *Th* be any sufficiency theorem that guarantees that a decision d^* satisfies $G(d^*) \leq G(d)$ $\forall d \in \mathscr{D}$. Then the hypotheses of *Th*, together with suitable assumptions for the preference relation on $\mathscr{A}$, may be used to construct the following metasufficiency theorem for preference optimality.

Theorem 4.1. Let $\lesssim$ be a preference on $\mathcal{A}^0$ (open) $\supset \mathcal{A}$ and let $a^* = g(d^*)$ with $d^* \in \mathcal{D}$. Assume:

(i) It is known that there exists a differentiable utility function $\phi(\cdot)$ for $\lesssim$ on $\mathcal{A}^0$ such that $\phi(\cdot)$ is pseudoconvex with respect to $\mathcal{A}$ at $a^* \in \mathcal{A}$ and such that $\nabla\phi(a^*) \neq 0$.

(ii) A normal vector $n(a^*)$ to the indifference set $I(a^*) = \{b \in \mathcal{A}^0: a^* \sim b\}$ at a^* is known and it has the same orientation as $\nabla\phi(a^*)$.

(iii) All the hypotheses of *Th* are satisfied with $G(\cdot) = n(a^*)g(\cdot)$.

Then d^* is a preference-optimal decision.

Proof. As a consequence of (iii),

$$n(a^*)\, g(d^*) \leq n(a^*)\, g(d) \quad \forall d \in \mathcal{D}.$$

From (ii) it follows that $n(a^*) = k\, \nabla\phi(a^*)$, $k > 0$, with the result

$$\nabla\phi(a^*)(a - a^*) = \nabla\phi(a^*)(g(d) - g(d^*)) \geq 0 \quad \forall d \in \mathcal{D}.$$

But since $\phi(\cdot)$ is pseudoconvex at a^*, this implies

$$\phi(a^*) \leq \phi(a) \quad \forall a \in \mathcal{A}$$

and consequently

$$g(d^*) \lesssim g(d) \quad \forall d \in \mathcal{D}.$$

Since $d^* \in \mathcal{D}$, d^* is a preference-optimal decision. QED

In essence the theorem provides for the existence of a separating hyperplane between $\mathcal{A}$ and the set $I^-(a^*) = \{b \in \mathcal{A}^0: b \lesssim a^*\}$. The three fundamental influences in preference optimality with differentiability are clearly apparent from this theorem; namely, the normal vectors $n(a)$ characterizing the indifference surfaces, the utility function $\phi(\cdot)$ characterizing preference, and the functions $g_i(\cdot)$, which determine the attainable criteria set.

The following two sufficiency theorems for the programming problem and the control problem respectively will illustrate the use of the metatheorem. For the programming theorem a sufficiency theorem from Mangasarian (Ref. 5) was modified and for the control theorem one of Lee and Markus (Ref. 3) was used as a foundation. The theorems also make use of the existence of a convexifying transformation for the utility function. The attempt is to provide the potential user with a checklist. The notation $f_I(\cdot)$ simply denotes that subvector of a mapping

$f(\cdot)$ whose components are $f_i(\cdot)$, $i \in I$, where I is an appropriate index set. Furthermore, let $n(y)$ denote a normal vector to $I(y)$ at y.

Theorem 4.2. Let $\lesssim$ be a preference on Y^0 (open) $\supset Y$ and let $y^* = g(x^*)$ with $x^* \in X$. Assume:

(i) A utility function $\phi(\cdot)$ is known for $\lesssim$ on Y^0 with $\nabla\phi(y^*) \neq 0$ and such that $\phi(\cdot)$ satisfies the hypotheses of Theorem 3.1.

(ii) A normal vector $n(y^*)$ to $I(y^*)$ at y^* is known and has the same orientation as $\nabla\phi(y^*)$.

(iii) The criterion functions $g_i(\cdot)$, $i = 1,..., N$, are convex with respect to X at $x^* \in X$.

(iv) There exist vectors $c \in \mathbb{R}^N$, $\lambda \in \mathbb{R}^m$, and $\mu \in \mathbb{R}^k$ such that

$$c\,\nabla g(x^*) + \lambda\,\nabla f(x^*) + \mu\,\nabla h(x^*) = 0, \qquad x^* \in X,$$

and $\lambda f(x^*) = 0$, $\lambda \geqslant 0$, $c > 0$, where c satisfies the compatibility condition $c = n(y^*)$.

(v) Let $J = \{1,..., m\}$. With $I = \{i \in J: f_i(x^*) = 0\}$, $f_I(\cdot)$ is differentiable and quasiconvex at x^*.

(vi) The equality constraints $h_i(\cdot)$ are differentiable and both quasiconcave and quasiconvex at x^*.

Then x^* is a preference-optimal decision.

Proof. For the first part one need only modify the proof of the programming theorem in Mangasarian (Ref. 5) to include equality constraints. The introduction of $\tilde{f}(\cdot) = (f(\cdot), h(\cdot), -h(\cdot))$ and $\mu = \eta - \nu$ reduces the hypothesis (iv) to one without equality constraints; hypothesis (iii) and $c > 0$ in conjunction with Theorem 2.1 suffice to conclude that $G(\cdot) = cg(\cdot)$ is convex. One may then follow Ref. 5 to show that

$$cg(x^*) \leq cg(x) \quad \forall x \in X.$$

As a next step, choose $c = n(y^*) = k\,\nabla\phi(y^*)$ to obtain

$$\nabla\phi(y^*)\,g(x^*) \leq \nabla\phi(y^*)\,g(x) \quad \forall x \in X,$$

or

$$0 \leq \nabla\phi(y^*)(y - y^*) \quad \forall y \in Y.$$

By hypothesis (i) there exists an $F(\cdot)$ such that $\psi(\cdot) = F \circ \phi(\cdot)$ is convex. Since $F(\cdot)$ is a strictly increasing function, $F'(\xi) \geq 0 \; \forall \xi \in \mathbb{R}$ and

$$\begin{aligned} 0 &\leq F'(\phi(y^*))\, \nabla\phi(y^*)(y - y^*) \\ &= \nabla\psi(y^*)(y - y^*) \leq \psi(y) - \psi(y^*). \end{aligned}$$

Consequently,

$$\psi(y^*) \leq \psi(y) \Rightarrow y^* \lesssim y \Rightarrow g(x^*) \lesssim g(x)$$

for every $x \in X$, and with $x^* \in X$ the result is established. QED

Sufficient conditions for Problem II are given next. For the statement of the theorem, introduce the mappings

$$\begin{aligned} &k^0(\cdot)\colon\ A \times [t_0, t_1] \to K^0(\text{open}) \subset \mathbb{R}^N, \quad k^0 = (k_{01}, \ldots, k_{0N}), \\ &h^0(\cdot)\colon\ U \times [t_0, t_1] \to H^0(\text{open}) \subset \mathbb{R}^N, \quad h^0 = (h_{01}, \ldots, h_{0N}), \\ &h(\cdot)\colon\ U \times [t_0, t_1] \to H(\text{open}) \subset \mathbb{R}^n, \quad h = (h_1, \ldots, h_n), \end{aligned}$$

and an $n \times n$ matrix $A(t)$ to define

$$f^0(x(t), t, u(t)) = k^0(x(t), t) + h^0(u(t), t)$$

and

$$f(x, t, u) = A(t)x + h(u, t).$$

Theorem 4.3. Let $\lesssim$ be a preference on P^0 (open) $\supset K(t_1)$ and let $u^*(\cdot) \in \mathscr{F}$, with $z^*(\cdot) = (x^*(\cdot), y^*(\cdot))$ as the corresponding augmented response. Assume:

(i) A utility function $\phi(\cdot)$ is known for $\lesssim$ on P^0 with $\nabla\phi(y^*) \neq 0$, $y^* = y^*(t_1)$, and such that $\phi(\cdot)$ satisfies the hypotheses of Theorem 3.1.

(ii) A normal vector $n(y^*)$ to $I(y^*)$ at y^* is known and has the same orientation as $\nabla\phi(y^*)$.

(iii) $k_{0i}(\cdot)$, $\nabla k_{0i}(\cdot)$, and $h_{0i}(\cdot)$, $i = 1, \ldots, N$, along with $A(\cdot)$ and $h(\cdot)$ are continuous on their respective domains.

(iv) The $k_{0i}(\cdot)$, $i = 1, \ldots, N$, are convex in x for each fixed t in the interval $[t_0, t_1]$.

(v) $\theta^0 = \{x^0\}$, a single point, and $\theta^1 \subseteq \mathbb{R}^n$ is a closed convex set.

(vi) The control $u^*(\cdot)$ satisfies the maximum principle

$$-ch^0(u^*(t), t) + \lambda(t)\, h(u^*(t), t) = \max_{u \in U}(-ch^0(u, t) + \lambda(t)\, h(u, t)),$$

where $c > 0$ and satisfies the compatibility condition $c = n(y^*)$ and where $\lambda(t)$ is any nontrivial solution of

$$\dot{\lambda}(t) = c \, \nabla k^0(x^*(t), t) - \lambda(t) \, A(t)$$

satisfying the transversality condition: $\lambda(t_1)$ is an inward normal to θ^1 at the boundary point $x(t_1)$.

Then $u^*(\cdot) \in \mathscr{F}$ is a preference-optimal control.

Proof. With $G(\cdot) = ck^0(\cdot)$ convex in x due to $c > 0$ and hypothesis (iv), one may use the procedure in Ref. 3 to show that

$$cy^* \leq cy(t_1) \quad \forall y(t_1) \in K(t_1).$$

The remaining steps are the same as those in the preceding theorem. QED

Rather general preferences and preference-optimal decisions with respect to them may be used to generate Pareto-optimal decisions; that is, to provide sufficient conditions for Pareto optimality.

Theorem 4.4. Let $\precsim$ be a preference on $\mathscr{A}^0$ (open) $\supset \mathscr{A}$ with $\mathscr{A}^0 \subset \mathbb{R}^N$ and assume that the derived strict preference $\prec$ is monotone and asymmetric on $\mathscr{A}^0$ and that the derived indifference $\sim$ is symmetric on $\mathscr{A}^0$. Then a preference-optimal decision $d^* \in \mathscr{D}$ is also a Pareto-optimal decision.

Proof. Let $d^* \in \mathscr{D}$ be a preference-optimal decision and assume that d^* is not Pareto optimal. Then there exists a decision $\bar{d} \in \mathscr{D}$ such that $g(\bar{d}) < g(d^*)$. Since the derived relation $\prec$ is monotone on $\mathscr{A}^0$, it follows that $g(\bar{d}) \prec g(d^*)$. Thus, $g(d^*)$ cannot be a minimal element of $\mathscr{A}$ with respect to $\precsim$, for $g(\bar{d}) \prec g(d^*) \Leftrightarrow g(\bar{d}) \precsim g(d^*)$ and $\neg g(\bar{d}) \sim g(d^*)$, by definition. It cannot be a least element, since $g(\bar{d}) \prec g(d^*) \Rightarrow \neg g(\bar{d}) \sim g(d^*)$ by definition and $\neg g(d^*) \sim g(\bar{d})$ by the symmetry of $\sim$. From the asymmetry of $\prec$ follows $g(d^*) \nprec g(\bar{d})$. Together these statements imply $\neg g(d^*) \precsim g(\bar{d})$, all of which collectively yields a contradiction to the preference optimality of d^*. QED

Corollary 4.1. Let $\precsim$ be a monotone preference on $\mathscr{A}^0$ (open)$\supset\mathscr{A}$. Assume that a utility function $\phi(\cdot)$ is known for $\precsim$ and that $a^* = g(d^*)$, $d^* \in \mathscr{D}$, satisfies $\phi(a^*) \leq \phi(a)$ $\forall a \in \mathscr{A}$. Then d^* is a Pareto-optimal decision.

Note that the utility function mentioned in the corollary need not even be continuous. Finally, one may state a sufficient condition that makes use only of a C^1 utility function, or rather any C^1 function with the indicated properties will do.

Corollary 4.2. Let $\mathcal{A}^0$ (open) $\supset \mathcal{A}$ and let $\phi(\cdot)$: $\mathcal{A}^0 \to \mathbb{R}$ be a C^1-function that satisfies $\nabla\phi(a) \gg 0$ on $\mathcal{A}^0$. Then any decision $d^* \in \mathcal{D}$ with $a^* = g(d^*)$ that satisfies $\phi(a^*) \leq \phi(a)$ $\forall a \in \mathcal{A}$ is a Pareto-optimal decision.

Proof. The function $\phi(\cdot)$ generates a complete preorder $\lesssim$ on $\mathcal{A}^0$; furthermore, $\lesssim$ is continuous and of class C^1 on $\mathcal{A}^0$. The preorder is also monotone. By the mean value theorem on $\mathbb{R}^N$ one may write, for any a^1, $a^2 \in \mathcal{A}^0$,

$$\phi(a^2) - \phi(a^1) = \phi(\theta a^1 + (1 - \theta)\, a^2)(a^2 - a^1), \qquad 0 < \theta < 1.$$

Now, assume $a^1 < a^2$; then $\nabla\phi(a) \gg 0$ for every $a \in \mathcal{A}^0$ implies $\phi(a)^2 - \phi(a^1) > 0$ and consequently $a^2 \succ a^1$. The conclusion then follows from Theorem 4.4. QED

5. Illustrative Example

The example is given in a programming framework since a control-theoretic example was given in Ref. 2. The geometry of the example is such that maximization is more suitable as the basic objective. Since there will be a need for the necessary conditions, they are quoted here for a maximizer of preference.

Theorem 5.1. Let $h(\cdot)$ and $g(\cdot)$ have continuous first partial derivatives at x^* and let $f(\cdot)$ be differentiable at x^*. Let $\lesssim$ be a complete preordering, continuous, weakly monotone, and of class C^2 on P(open)$\supset X$. If $x^* \in X$ is a preference-optimal decision, then there exist vectors $c \in \mathbb{R}^N$, $(\lambda_0, \lambda) \in \mathbb{R}^{1+m}$, and $\mu \in \mathbb{R}^k$ such that

$$-\lambda_0 c\, \nabla g(x^*) + \lambda\, \nabla f(x^*) + \mu\, \nabla h(x^*) = 0, \qquad x^* \in X,$$

$$\lambda f(x^*) = 0, \qquad (\lambda_0, \lambda) \geq 0, \qquad (\lambda_0, \lambda, \mu) \neq 0, \qquad \text{and} \qquad c > 0.$$

The mathematics of the example is purposely kept simple, but to give an indication of the possible applicability of the present concept, the problem is stated in an agricultural context. Assume that a farmer wishes to raise two different species of livestock, designated 1 and 2. Let x_1

represent the amount of time spent in grooming and caring for the species and x_2 the amount of allocated fodder. Let $g_1(x)$ and $g_2(x)$ be the respective numbers of the species capable of thriving for given amounts x. Now the farmer may control the amounts x and, because of the possibly varying benefit of the species to him, he may consider some combinations of these species more desirable than others, thus giving rise in a natural way to a preference over $g_1 g_2$ space.

Let the preference be described in the following manner. With $P = \{y \in \mathbb{R}^2 \colon y \gg 0\}$, consider the family of hyperbolas $y_2 = a/y_1$, $a > 0$, $y = (y_1, y_2) \in P$. Let the indifference surfaces of the relation be the members of this family; that is, if $z = (z_1, z_2)$ is any point in P and a is such that z may be written as $z = (z_1, a/z_1)$, then

$$I_a(z) \triangleq \{x \in P\colon \quad x_2 = a/x_1\} = \{x \in P\colon \quad x \sim z\}$$

is the indifference class of z. With

$$I_a^+(z) \triangleq \{x \in P\colon \quad x_2 \geqslant a/x_1\}$$

the statement

$$\forall z \in P, \quad x \gtrsim z \text{ iff } x \in I_a^+(z),$$

along with an equivalent one for $x \lesssim z$, defines a monotone and continuous preference of class C^2 (at least) on P which is a complete preordering of P. In addition, it follows from Theorem 2.3 that any utility function on P will be quasiconcave, since the set $\{x \in P : x \gtrsim y\} = \{x \in P : \phi(x) \geqslant \phi(y)\}$ is convex for every $y \in P$.

In order to apply Theorem 4.2, a normal vector must be known for the necessity part and a utility function for the sufficiency part. For every $y \in P$, $a > 0$, the following choice is made for these:

$$n(y) = (a/y_1^2, 1) \qquad \text{and} \qquad \phi(y) = y_1 y_2 .$$

This takes care of the preference on P.

With maximization as the basic objective, consider then the following specific problem: Obtain preference-optimal decisions for

$$g_1(x) = (x_1 x_2)^{1/2} \qquad \text{and} \qquad g_2(x) = x_1 - x_2$$

subject to the inequality constraints

$$f_1(x) = x_2 - x_1 < 0, \qquad f_2(x) = x_1 - 1 \leqslant 0, \qquad f_3(x) = -x_2 < 0.$$

The sets X and Y are defined by

$$X = \{x \in \mathbb{R}^2\colon \quad f(x) < 0\} \qquad \text{and} \qquad Y = g(X).$$

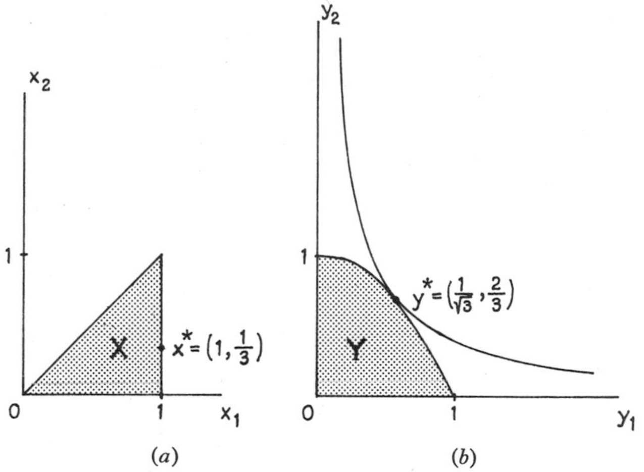

Fig. 1. (a) Functional constraint set X and preference-optimal decision $x^* = (1, \frac{1}{3})$.
(b) Attainable criteria set Y and preference-optimal criterion value $y^* = (1/\sqrt{3}, \frac{2}{3})$.

These sets, together with the preference optimal decision and corresponding criterion value, are shown in Fig. 1.

Solution: The solution consists of two parts, application of the necessary condition Theorem 5.1 to obtain a candidate (or candidates) and a check of the sufficiency conditions to assure preference optimality.

Part I. With three inequality constraints the necessary conditions become

$$-(\tfrac{1}{2}c_1 \sqrt{(x_2/x_1)} + c_2\,,\, \tfrac{1}{2}c_1 \sqrt{(x_1/x_2)} - c_2) + (-\lambda_1 + \lambda_2\,,\, \lambda_1 - \lambda_3) = 0.$$

However, $f_1(\cdot)$ and $f_3(\cdot)$ are strict inequalities for every $x \in \mathbb{R}^2$; hence they are never active, so that $\lambda_1 = \lambda_3 = 0$. An interior point of X cannot be a solution, so that only the possibility with $f_2(\cdot)$ active remains. With $f_2(\cdot)$ as the active constraint, $x_1 = 1$, $\lambda_2 \neq 0$, and

$$-\tfrac{1}{2}c_1 \sqrt{x_2} - c_2 = -\lambda_2\,, \qquad -\tfrac{1}{2}c_1(1/\sqrt{x_2}) + c_2 = 0$$

are all that ramain of the necessary conditions. The solution of the system yields $x_2 = \frac{1}{4}(c_1/c_2)^2$.

To apply the compatibility condition, y is needed in terms of c; that is,

$$y_1 = \tfrac{1}{2}c_1/c_2\,, \qquad y_2 = 1 - (c_1/2c_2)^2.$$

Since (y_1, y_2) must belong to $I_a(y)$, $a = (c_1/2c_2)\,[1 - (c_1/2c_2)^2]$, and with the above choice of normal, $n(y)$ in terms of c becomes

$$\tilde{n}(c) = ((2c_2/c_1)[1 - (c_1/2c_2)^2], 1).$$

Thus,

$$c_1 = (2c_2/c_1)[1 - (c_1/2c_2)^2], \qquad c_2 = 1,$$

with the result

$$c_1 = 2/\sqrt{3}, \qquad c_2 = 1,$$

along with $x_2 = \frac{1}{3}$. Note that the specific values for the c_i are illusory, since all the results remain unchanged with $c_1 = 2\delta/\sqrt{3}$, $c_2 = \delta$, $\delta > 0$.

Part II. Claim: $x^* = (1, \frac{1}{3})$ is a preference optimal decision with $y^* = (1/\sqrt{3}, \frac{2}{3})$.

Note first that $\phi(\cdot)$ is not concave. Essentially, this leaves two approaches, either to show that $\phi(\cdot)$ is pseudoconcave and to make implicit use of Theorem 4.1 or to show that $\phi(\cdot)$ satisfies the hypotheses of Theorem 4.2 and, in particular, those that concern Theorem 3.1.

The function $\phi(\cdot)$ is indeed pseudoconcave since $y \in P$, $\nabla\phi(y^*)(y - y^*) \leqslant 0$, implies $\phi(y) \leqslant \phi(y^*)$; i.e., $y_2 y_1{}^* + y_2{}^* y_1 \leq 2y_2{}^* y_1{}^*$ implies $y_1 y_2 \leqslant y_1{}^* y_2{}^*$. This is most easily seen from the fact that $y_1{}^* y_2{}^*$ is the largest possible area of an inscribed rectangle.

However, it may usually be quite difficult to show pseudoconcavity or concavity directly, so that it is instructive to illustrate the application of Theorem 4.2. The satisfaction of the hypotheses of Theorem 3.1 is of particular interest, since all other conditions are obviously satisfied. Naturally, the conditions (i)–(v) of the theorem are modified to conform to the basic objective of maximization.

(i) Since $\nabla\phi(y) = (y_2, y_1) \gg 0\ \forall y \in P$, $\phi(\cdot)$ has neither a stationary nor a maximum point on P.

(ii) The quadratic form Q is given by

$$Q(\xi, \xi) = (\xi_1, \xi_2)\begin{bmatrix} \phi_{11}(y) + \sigma(y)\,\phi_1{}^2(y) & \phi_{12}(y) + \sigma(y)\,\phi_1(y)\,\phi_2(y) \\ \phi_{21}(y) + \sigma(y)\,\phi_1(y)\,\phi_2(y) & \phi_{22}(y) + \sigma(y)\,\phi_2{}^2(y) \end{bmatrix}\begin{pmatrix} \xi_1 \\ \xi_2 \end{pmatrix}$$

with characteristic polynomial

$$\Gamma_Q(\alpha) = \alpha^2 - \alpha\sigma(y)(y_1{}^2 + y_2{}^2) - (1 + 2\sigma(y)\,y_1 y_2),$$

where

$$\phi_1(y) = y_2, \quad \phi_2(y) = y_1, \quad \phi_{12}(y) = \phi_{21}(y) = 1, \quad \phi_{11}(y) = \phi_{22}(y) = 0$$

have been used. A choice of

$$\sigma(y) = -1/2y_1 y_2$$

yields

$$\alpha_1 = 0, \qquad \alpha_2 = -(y_1^2 + y_2^2)\, 2y_1 y_2$$

assuring the negative semidefiniteness of $Q(\cdot)$ (by a simple extension of Theorem 2.5). One may now use Theorem 3.1 in a constructive manner, for, with

$$\sigma(y) = F''(\phi(y))\, F'(\phi(y)),$$

the desired transformation $F(\cdot)$ may be obtained from

$$F''(\phi) + (1/2\phi) F'(\phi) = 0$$

as

$$F(\phi) = \tfrac{1}{2} d_1 \sqrt{\phi} + d_2 .$$

The d_i are irrelevant and a concave utility function $\psi(\cdot)$ for $\lesssim$ on P is defined by

$$\psi(y) = F \circ \phi(y) = \sqrt{(y_1 y_2)}.$$

(iii) The matrix associated with $Q(\xi, \xi)$ has rank 1, and $[\phi_{ij}(y)]$ has rank 2 for every $y \in P$.

(iv) Here

$$\Gamma(\alpha) = \alpha^2 - 1 \qquad \text{and} \qquad \Gamma^*(\alpha) = -[2y_1 y_2/(y_1^2 + y_2^2)] - \alpha$$

so that

$$S_0 = 1, \quad S_1 = 0, \quad S_2 = -1; \qquad S_0{}^* = 1, \quad S_1{}^* = 2y_1 y_2/(y_1^2 + y_2^2).$$

Since maximization is the objective,

$$G(t) = \sup\{S_2/k^2(y)\, S_1{}^*: \ \phi(y) = t\} = 1/2t < \infty$$

for every fixed $t \in (\epsilon, \beta) = (0, \infty)$.

(v) Let a minorant function $\gamma(\cdot)$: $[\epsilon, \beta) \to \mathbb{R}$ be defined by

$$\gamma(t) = H'(t)/H(t).$$

An obvious choice for $H(\cdot)$ is $H(t) = \sqrt{t}$. Then,

$$G(t) = 1/2t \geqslant 1/2t$$

follows.

Hence, $x^* = (1, \frac{1}{3})$ is a preference-optimal decision with resultant criterion value $y^* = (1/\sqrt{3}, \frac{2}{3})$.

Naturally, all the variables are to be thought of as dimensionless variables characterizing the physical quantities. With this caveat in mind let x_1 be equivalent to the time spent in hours per day, let x_2 be equivalent to hundreds of pounds of food, and let y_1 and y_2 be representative of hundreds of members of the respective species. Then 1 hr of grooming and care together with $33\frac{1}{3}$ lb of food will result in approximately 58 members of the first species and 67 of the second.

References

1. Yu, P. L., and Leitmann, G., *Compromise Solutions, Domination Structures and Salukvadze's Solution*, Journal of Optimization Theory and Applications, Vol. 13, No. 3, 1974.
2. Stadler, W., *Preference Optimality*, to be published.
3. Lee, E. B., and Markus, L., *Foundations of Optimal Control Theory*, John Wiley and Sons, New York, 1967.
4. Stadler, W., *Preference Optimality and Applications of Pareto Optimality*, in *Multicriteria Decision Making*, edited by A. Marzollo and G. Leitmann, *CISM Courses and Lectures*, Springer Verlag, New York, 1976.
5. Mangasarian, O. L., *Nonlinear Programming*, McGraw-Hill, New York, 1969.
6. Debreu, G., *Theory of Value*, John Wiley and Sons, New York, 1959.
7. Fishburn, P. C., *Utility Theory for Decision Making*, John Wiley and Sons, New York, 1970.
8. De Finetti, B., *Sulle stratifiazioni convesse*, Annali di Matematica Pura ed Applicata, Vol. 4, No. 30, 1949.
9. Fenchel, W., *Convex Cones, Sets and Functions*, mimeographed notes, Department of Mathematics, Princeton University, 1953.
10. Fenchel, W., *Über konvexe Funktionen mit vorgeschriebenen Niveaumannigfaltigkeiten*, Mathematische Zeitschrift, Vol. 63, pp. 496–506, 1956.
11. Aumann, R. J., *Values of Markets with a Continuum of Traders, Technical Report No. 121, The Economic Series*, Institute for Mathematical Studies in the Social Sciences, Stanford University, Stanford, California, January 1974.
12. Kelley, J. L., *General Topology*, van Nostrand, Princeton, New Jersey, 1955.

VII

Individual and Collective Rationality in a Dynamic Pareto Equilibrium[1]

A. HAURIE AND M. C. DELFOUR

Abstract. This paper deals with the problem of establishing the conditions for individual and collective rationality when a set of players cooperate in a Pareto equilibrium. To derive such conditions one follows the approach of the theory of reachability of perturbed systems. Open-loop and closed-loop concepts are discussed and are shown to be nonequivalent.

1. Introduction

The notion of a Pareto equilibrium is commonly used for the study of cooperative dynamic games (see Refs. 1–6). The object of this paper is to study the individual and collective rationality underlying the participation of decision-makers in a Pareto equilibrium for a multistage system where coalitions are permitted.

More precisely, a Pareto equilibrium is said to be *individually optimal* (I-optimal) when no decision-maker, at any stage, in the situation where all other decision-makers decide to cooperate in minimizing his gain, can provide himself a higher gain than what he obtains in the

[1] The research of the first author was supported in part by Canada Council Grant No. S-701-491 and has benefited from collaboration with the Laboratoire d'Automatique Théorique de l'Université de Paris VII, Paris, France.

Pareto equilibrium. Analogously, a Pareto equilibrium is *collectively optimal* (C-optimal) when, at any stage, no coalition of a subset of the decision-makers can *assure* each of its members a higher gain than what he can get by full cooperation with all the other decision-makers.

Cooperative games where coalitions are permitted are commonly studied in characteristic form (see Ref. 7) and important notions of optimality are the *set of imputations* and the *core of the game*. I-optimality and C-optimality make it possible to extend these notions to dynamic multistage games. This is not a straightforward generalization because of the closed-loop definitions which are natural to adopt in dynamic games with perfect information.

Section 2 contains the main definitions. In Section 3, we introduce the concepts of I- and C-optimality and discuss the information structure of the game. In Section 4, we characterize I- and C-optimality in the space of gains and give constructive definitions of the set of imputations and the core of the game. In Section 5, we specialize the system of Section 2 and present a formal algorithm to characterize C-optimal decision sequences. In Section 6, a simple example is worked out, to show the difference between open-loop and closed-loop concepts.

Results on qualitative multistage games coalitions have been reported by Blaquiere and Wiese (Ref. 8). However, this paper is technically closer to the reachability theory for perturbed systems, since, given a coalition of decision-makers, the action of all other decision-makers is considered as a perturbation (see Refs. 9–11). The reader will notice that all the results presented can be applied to sequential machines.

Notation. Given any set S, $\mathscr{P}(S)$ will denote the set of all subsets of S, and $\mathbb{R}$ the set of real numbers.

We use indifferently the words *decision-maker* and *player*.

2. Definition of the System, Gain Functionals, and Pareto Equilibria

Let F be an arbitrary set and let $K \geqslant 1$, $m \geqslant 1$ be integers. We define the time set $\mathbb{K} \triangleq \{0, 1,..., K\}$, the *set of players* $M \triangleq \{1, 2,..., m\}$, and a dynamic multistage system

$$x(k + 1) = f^k(x(k), u_M(k)), \qquad k = 0, 1,..., K - 1, \tag{1-1}$$

$$x(0) = x^0, \tag{1-2}$$

where

(i) for each $j \in M$ and for each $k \in \mathbb{K} - K$, $U_j(k) \neq \varnothing$ is the *decision set* of Player j at stage k and $u_j(k) \in U_j(k)$ is an *admissible decision* of Player j at stage k;

(ii) $U_M(k) \triangleq \prod_{j \in M} U_j(k)$ and $u_M(k) \triangleq (u_j(k))_{j \in M} \in U_M(k)$

are the decision set and the admissible decision of the set of all players at stage k;

(iii) for all $k \in \mathbb{K}$, $x(k)$ is the value of the state variable at stage k, and $x^0 \in F$ is the given state at stage 0;

(iv) $f^k : F \times U(k) \to F$, $k = 0, 1, \ldots, K - 1$, is a sequence of given maps.

An *admissible decision sequence* will be defined as an element

$$\tilde{u}_M \triangleq (u_M(k))_{k \in \mathbb{K} - K} \in \prod_{k \in \mathbb{K} - K} U_M(k).$$

Given the initial state x^0 at stage 0 and an admissible decision sequence, we denote by

$$\tilde{x}(\cdot\,; x^0, \tilde{u}_M) : \mathbb{K} \to F$$

the trajectory emanating from $(0, x^0) \in \mathbb{K} \times F$ generated by $\tilde{u}_M$. For $i \in \mathbb{K}$ and $x^i \in F$, we adopt the notations

$$\mathbb{K}^i \triangleq \{i, i + 1, \ldots, K\}, \qquad \tilde{u}_M{}^i \in \prod_{k \in \mathbb{K}^i - K} U_M(k);$$

then,

$$\tilde{x}^i(\cdot\,; x^i, \tilde{u}_M{}^i) : \mathbb{K}^i \to F$$

is the *trajectory* emanating from x^i at stage i, generated by $\tilde{u}_M{}^i$.

A *coalition* S is defined as a nonempty subset of M. For a coalition, we shall use the products

$$U_S(k) \triangleq \prod_{j \in S} U_j(k), \qquad U_{M-S}(k) \triangleq \prod_{j \in M-S} U_j(k)$$

and the notations

$$\tilde{u}_S \in \prod_{k \in \mathbb{K} - K} U_S(k), \qquad \tilde{u}_{M-S} \in \prod_{k \in \mathbb{K} - K} U_{M-S}(k)$$

for the admissible decision sequences of the coalition S and of the set $M - S$ of all other players respectively.

We associate with each player a *gain functional* $g_j : F \to \mathbb{R}$. For the system (1) in state x^i in F at stage i, we define the gain

$$V_j(i, x^i, \tilde{u}_M{}^i) \triangleq g_j(\tilde{x}^i(K; x^i, \tilde{u}_M{}^i))$$

of the jth player along the trajectory $\tilde{x}^i(\cdot; x^i, \tilde{u}_M{}^i)$.

First we shall recall the definition of a Pareto equilibrium.

Definition 2.1. Given an initial state x^i at stage i, an admissible decision sequence $(\tilde{u}_M{}^i)^*$ defines a *Pareto equilibrium* at (i, x^i) if, for any admissible decision sequence $\tilde{u}_M{}^i$ with the property that

$$V_j(i, x^i, \tilde{u}_M{}^i) \geqslant V_j(i, x^i, (\tilde{u}_M{}^i)^*) \qquad \forall\, j \in M,$$

we necessarily obtain

$$V_j(i, x^i, \tilde{u}_M{}^i) = V_j(i, x^i, (\tilde{u}_M{}^i)^*) \qquad \forall\, j \in M.$$

Such a sequence is said to be *Pareto-optimal* at (i, x^i).

Remark 2.1. Strictly speaking, a decision sequence $(\tilde{u}_M{}^i)^*$ which defines a Pareto equilibrium is dependent on the initial state x^i and should be written as $(u_M{}^i(\cdot; x^i))^*$.

Remark 2.2. Pareto equilibrium is a full-cooperation notion of optimality. The absence of perturbations due to conflicts between players makes unnecessary the use of a closed-loop definition.

3. I-Optimality, C-Optimality, and the Sets $X^k(S, \omega^S)$

3.1. Definitions. Let S be a coalition. An ordered sequence $\{\omega_j\}_{j \in S}$ of real numbers ω_j will be called a *gain vector* for S and written as ω_S.

Definition 3.1. Given a coalition S and two gain vectors $\omega_S{}^1$ and $\omega_S{}^2$, we say that $\omega_S{}^2$ *dominates* $\omega_S{}^1$ if $\omega_j{}^2 \geqslant \omega_j{}^1\, \forall\, j \in S$ and there exists $l \in S$ such that $\omega_l{}^2 > \omega_l{}^1$.

Definition 3.2. Given an initial state x^0 at stage 0, a Pareto-optimal decision sequence $\tilde{u}_M{}^*$ generating a trajectory $\tilde{x}^*$,

$$\tilde{x}^*(k) \triangleq \tilde{x}(k; x^0, \tilde{u}_M{}^*), \qquad \forall\, k \in \mathbb{K},$$

is said to be *C-optimal* at $(0, x^0)$ if, for any coalition S and any initial stage i in $\mathbb{K} - K$, the following property is true:

$$\forall u_S(i)\, \exists u_{M-S}(i) \text{ s.t. } \forall u_S(i+1)\, \exists u_{M-S}(i+1) \ldots \text{ s.t.}$$
$$\forall u_S(K-1)\, \exists u_{M-S}(K-1)$$

such that the admissible decision sequences $\tilde{u}_S{}^i$ and $\tilde{u}^i_{M-S}$ thus defined generate a gain vector

$$\omega_S \triangleq \{V_j(i, \tilde{x}^*(i), (\tilde{u}_S{}^i, \tilde{u}^i_{M-S}))\}_{j\in S}$$

which does not dominate the optimal gain vector

$$\omega_S{}^* \triangleq \{V_j(0, x^0, \tilde{u}_M{}^*)\}_{j\in S}\,.$$

Remark 3.1. C-optimality corresponds to the case where there is no way for a coalition S to *assure* its members a gain vector dominating the gain vector obtained through full cooperation, from any state of the Pareto-optimal trajectory. In this case, there is *collective rationality* in the Pareto equilibrium.

Definition 3.3. If we restrict the class of coalitions considered to be constituted only of one decision-maker subsets

$$S \triangleq \{j\}, \qquad \forall j \in M,$$

and if the characteristic property of Definition 3.2 is true for this subclass of coalitions, we shall say that the Pareto-optimal sequence is *I-optimal at* $(0, x^0)$.

Remark 3.2. Obviously, I-optimality is less restrictive than C-optimality. A Pareto equilibrium is *individually rational* if there is no way for each player to *assure* himself more than what he will obtain in the Pareto equilibrium, and this from any state of the Pareto-optimal trajectory.

3.2. Links with Reachability Theory. Closed-loop reachability of target sets can be used to characterize I- and C-optimality.

Proposition 3.1. Given an initial state x^0 at stage 0 and a decision sequence $\tilde{u}_M{}^*$, let

$$\omega_M{}^* \triangleq \{V_j(0; x^0, \tilde{u}_M{}^*)\}_{j\in M}\,.$$

For all coalitions S, one constructs the sequence

$$\{X^k(S, \omega_S^*)\}_{k \in \mathbb{K}}$$

of subsets of F as follows:

(i) $X^K(S, \omega_S^*) \triangleq \{x \in F : \omega_S \triangleq \{g_j(x)\}_{j \in S}$ dominates $\omega_S^*\}$;

(ii) for all $k \in \mathbb{K} - K$, $X^k(S, \omega_S^*)$ is defined in terms of $X^{k+1}(S, \omega_S^*)$ as the set of all x in F for which there exists an admissible decision $u_S(k)$ such that, for all admissible decisions $u_{M-S}(k)$, the following property is true:

$$f^k(x, (u_S(k), u_{M-S}(k))) \in X^{k+1}(S, \omega_S^*).$$

Then, the following assertions hold:

(a) $\tilde{u}_M^*$ is I-optimal at $(0, x^0)$ iff it is Pareto-optimal at $(0, x^0)$ and generates a trajectory $\tilde{x}^*$ such that

$$\forall j \in M, \qquad \forall k \in \mathbb{K} - K, \; \tilde{x}^*(k) \notin X^k(\{j\}, \omega_j^*);$$

(b) $\tilde{u}_M^*$ is C-optimal at $(0, x^0)$ iff

$$\forall S \subset M, \qquad \forall k \in \mathbb{K} - K, \; \tilde{x}^*(k) \notin X^k(S, \omega_S^*).$$

Proof. By construction, the state x^k is in $X^k(S, \omega^*)$ iff the following is true:

$$\exists u_S(k) \text{ s.t.} \quad \forall u_{M-S}(k) \;\; \exists u_S(k+1) \text{ s.t.} \quad \forall u_{M-S}(k+1)$$

$$\ldots \exists u_S(K-1) \text{ s.t.} \quad \forall u_{M-S}(K-1)$$

the gain vector

$$\omega_S \triangleq \{V_j(k; x^k, (\tilde{u}_S^{\,k}, \tilde{u}_{M-S}^k))\}_{j \in S}$$

dominates ω_S^*. Then, the proof is by definition.

Remark 3.3. With the preceding notations, if x^0 is not in $X^0(M, \omega_M^*)$, then $\tilde{u}^*$ is Pareto-optimal at $(0, x^0)$.

The set $X^K(S, \omega_S^*)$ may be interpreted as a *target at stage* K for coalition S.

If full cooperation is not achieved, then coalition S may consider the decisions of the other decision-makers as unknown perturbations of the system.

$X^k(S, \omega_S^*)$ is the set of initial states at stage k from which the target $X^K(S, \omega_S^*)$ is reachable at stage K by S for any decision sequence $\tilde{u}_{M-S}^k$ of $M - S$.

3.3. Closed-Loop, Open-Loop, and Information Structure. In the construction of the sets $X^k(S, \omega_S^*)$, we have implicitly assumed that a coalition S can observe without error the state of the system at each stage. If such an information is not allowed to the decision-makers, it is possible to adopt an open-loop point of view.

Definition 3.4. Given an initial state x^0 at stage 0, a Pareto-optimal decision sequence $\tilde{u}_M^*$ generating a trajectory $\tilde{x}^*$ is said to be *open-loop C-optimal at* $(0, x^0)$ if, for any coalition S, any initial stage i in $\mathbb{K} - K$, and any admissible decision sequence $\tilde{u}_S{}^i$, there exists an admissible decision sequence $\tilde{u}^i_{M-S}$ such that the gain vector

$$\omega_S \triangleq \{V_j(i; \tilde{x}^*(i), (\tilde{u}_S{}^i, \tilde{u}^i_{M-S}))\}_{j\in S}$$

does not dominate

$$\omega_S{}^* \triangleq \{V_j(0; x^0, \tilde{u}_M{}^*)\}_{j\in S}\,.$$

Open-loop I-optimality is defined similarly.

A straightforward consequence of Definitions 3.2 and 3.4 is the following.

Proposition 3.2. If $\tilde{u}_M{}^*$ is C-optimal at $(0, x^0)$, then it is *open-loop* C-optimal at $(0, x^0)$.

The converse is not true as will be seen in Section 6.

For multistage games with perfect information, the closed-loop notions are the natural ones when we do not consider full cooperation of all decision-makers.

4. Imputations and Core of the Game

In this section, we show that the concept of I- and C-optimality can also be described in the *space of gains* (see Refs. 7 and 12). The reader will notice that there are obvious differences between the definitions which we shall introduce below and the standard definitions in m-person game theory. However, they are conceptually similar.

Given a coalition S, we denote by $\mathbb{R}^S$ the product $\prod_{j\in S} \mathbb{R}$.

Definition 4.1. Given a coalition S, the map

$$(k, x) \mapsto v^k(x, S) : \mathbb{K} \times F \to \mathscr{P}(\mathbb{R}^S),$$

defined as

$$v^k(x, S) \triangleq \{\omega_S \in \mathbb{R}^S : x \in X^k(S, \omega_S)\},$$

is called the *characteristic function of the game.*

When the system (1) is in state x at stage k, then the coalition S can obtain a gain vector dominating any ω_S in $v^k(x, S)$ whatever be the action of the other players.

Definition 4.2. Let $(k, x) \in (\mathbb{K} - K) \times F^n$ We denote by $\mathscr{V}^k(x)$ the *set of imputations at* (k, x). An *imputation at* (k, x) is a gain vector $\omega_M \in \mathbb{R}^M$ such that

(i) there exists an admissible decision sequence $(\tilde{u}_M{}^k)^*$ which defines a Pareto equilibrium at (k, x) such that

$$\omega_j = V_j(k, x, (\tilde{u}_M{}^k)^*) \qquad \forall j \in M;$$

(ii) $\forall j \in M, \qquad \omega_j \notin v^k(x, \{j\}).$

Remark 4.1. Property (ii) is commonly referred to as *individual rationality.*

Proposition 4.1. Let $\tilde{u}^*$ be I-optimal at $(0, x^0)$, and let $\omega_j{}^* = V_j(0, x^0, \tilde{u}_M{}^*)$ $\forall j \in M$. Then, at each stage $k \in \mathbb{K} - K$, $\omega_M{}^* \in \mathscr{V}^k(\tilde{x}(k; x^0, \tilde{u}_M{}^*))$.

Proof. It is well known (see Refs. 1–6), that, if $\tilde{u}_M{}^*$ is a Pareto-optimal decision sequence at $(0, x^0)$, then, for all k in $\mathbb{K}$, $(\tilde{u}_M{}^*)^k$ is a *Pareto-optimal decision sequence* at $(k, \tilde{x}^*(k))$, where $\tilde{x}^*(k) \triangleq \tilde{x}(k; x^0, \tilde{u}_M{}^*)$. Then, the proof is by definition.

Definition 4.3. Let $(k, x) \in (\mathbb{K} - K) \times F$. We denote by $\mathscr{C}^k(x)$ the *core of the game* at (k, x), which is defined as the set of all gain vectors $\omega_M \in \mathbb{R}^M$ such that

(i) there exists an admissible decision sequence $(\tilde{u}_M{}^*)^k$ which defines a Pareto equilibrium at (k, x) such that

$$\omega_j \triangleq V_j(k, x, (\tilde{u}_M{}^*)^k) \qquad \forall j \in M;$$

(ii) for all coalitions S, $\omega_S \notin v^k(x, S)$.

Remark 4.2. By definition, $\mathscr{C}^k(x) \subset \mathscr{V}^k(x)$. Property (ii) will also be referred to as *collective rationality.*

Proposition 4.2. Let $\tilde{u}_M{}^*$ be C-optimal at $(0, x^0)$ and let

$$\omega_j{}^* \triangleq V_j(0, x^0, \tilde{u}_M{}^*) \qquad \forall j \in M.$$

Then, at each stage $k \in \mathbb{K} - K$,

$$\omega_M{}^* \in \mathscr{C}^k(\tilde{x}(k; x, \tilde{u}_M{}^*)).$$

Proof. The proof is similar to the proof of Proposition 4.1.

5. Algorithm for Finding C-Optimal Decision Sequences

This section is concerned with a backward algorithm for the construction of the set of all initial states for which there exists a C-optimal decision sequence generating a given gain vector. However, this construction requires a specialization to the case where F is a semigroup and system (1) is of the following form:

$$x(k+1) = q^k(x(k)) + \sum_{j \in M} h_j{}^k(u_j(k)), \tag{2-1}$$

$$x(0) = x^0, \tag{2-2}$$

where

$$q^k : F \to F, \qquad k = 0, \ldots, K-1,$$

$$h_j{}^k : U_j(k) \to F, \qquad \forall j \in M,\ k - 0, \ldots, K-1,$$

are given maps.

For this problem we can now adapt some techniques developed in Refs. 9–11. More precisely, consider the following steps.

Step (1). Choose $\omega_M \in \mathbb{R}^m$.

Step (2). Define $C^K(\omega_M) \triangleq \{x : g_j(x) = \omega_j\ \forall\, j \in M\}$.

Step (3). Define, for all $S \subset M$, $S \neq \varnothing$,

$$X^K(S, \omega_S) \triangleq \{x : g_j(x) \geqslant \omega_j\ \forall j \in S, g_l(x) > \omega_l \quad \text{for some} \quad l \in S\}.$$

Step (4). Take $k = K$.

Step (5). Define

$$T^{k-1}(\omega_M) \triangleq \{x : \exists u_M(k) \in U_M(k) \text{ s.t. } q^{k-1}(x) + \sum_{j \in M} h_j^{k-1}(u_j(k-1)) \in C^k(\omega_M)\}.$$

Step (6). Define, for all $S \subset M$, $S \neq \varnothing$, the set

$$M^{k-1}(S, \omega_S) \triangleq \{x : x + \sum_{j \in M-S} h_j^{k-1}(u_j(k)) \in X^k(S, \omega_S)\ \forall u_{M-S}(k) \in U_{M-S}(k)\}$$

and the set

$$X^{k-1}(S, \omega_S) \triangleq \{x : \exists u_S(k) \in U_S(k) \text{ s.t. } q^{k-1}(x) + \sum_{j \in S} h_j^{k-1}(u_j(k)) \in M^{k-1}(S, \omega_S)\}.$$

Step (7). Form

$$C^{k-1}(\omega_M) \triangleq T^{k-1}(\omega_M) \bigcap \Bigg(\bigcap_{\substack{S \subset M \\ S \neq \varnothing}} \bar{X}^{k-1}(S, \omega_S) \Bigg),$$

where the bar denotes the complementary set w.r.t. F.

Step (8). If $C^{k-1}(\omega_M) \neq \varnothing$, take $k - 1$ instead of k and go to Step (5). If $C^{k-1}(\omega_M) = \varnothing$, stop. Then, $C^i(\omega_M) = \varnothing$ for all $i \leqslant k - 1$.

Proposition 5.1. With the preceding notations, $C^0(\omega_M)$ is the set of all states x at stage 0 such that there exists a decision sequence $\tilde{u}_M{}^*$ which is C-optimal at $(0, x)$ and such that

$$V_j(0, x^0, \tilde{u}_M{}^*) = \omega_j \qquad \forall j \in M.$$

Proof. *Necessity.* If $x \in C^0(\omega_M)$ then, $\forall\ S \subset M$, $S \neq \varnothing$, $x \notin X^0$ (S, ω_S). Then, there exists $u_M(0) \in U_M(0)$ such that

$$q^0(x) + \sum_{j \in M} h_j{}^0(u_j(k)) \in C^1(\omega_M).$$

By repeating this construction, one defines a control sequence $\tilde{u}_M$ generating a trajectory $\tilde{x}(\cdot; x, \tilde{u}_M)$ emanating from $(0, x)$, avoiding all sets

$$X^l(S, \omega_S), \qquad \forall S \subset M, S \neq \varnothing, \qquad \forall l = 0, 1, \ldots, K,$$

and such that

$$\tilde{x}^k(K; x, \tilde{u}_M{}^k) \in C^K(\omega_M).$$

Therefore $\tilde{u}_M$ is C-optimal at $(0, x)$ and

$$V_j(0, x; \tilde{u}_M) = \omega_j \qquad \forall j \in S.$$

Sufficiency. If $\tilde{u}_M$ is C-optimal at $(0, x)$ generating a trajectory with gains $V_j(0, x, \tilde{u}_M) = \omega_j$, then the terminal state $\tilde{x}(K; x, \tilde{u}_M)$ is in $C^K(\omega_M)$ and, following Proposition 3.1 on C-optimality, it is easy to show that $x \in C^0(\omega_M)$.

6. Example

To show the nonequivalence of open-loop and closed-loop concepts consider the following system:

$$x(k+1) = x(k) + \sum_{j=1}^{3} u_j(k),$$

where

$$k \in \mathbb{K} \triangleq \{0, 1, 2\}, \qquad j \in M \triangleq \{1, 2, 3\},$$
$$x = (x_1, x_2) \in \mathbb{R}^2,$$
$$u_1 \in U_1 \triangleq \{(0, 2), (-1, 2)\},$$
$$u_2 \in U_2 \triangleq \{(0, 2), (1, 2)\},$$
$$u_3 \in U_3 \triangleq \{y \in \mathbb{R}^2 : \| y \| = 1\},$$

with the gain functionals defined by

$$g_1(x) = -| x_1 |, \qquad g_2(x) = x_2$$

for decision-makers 1 and 2, respectively. We are interested in the construction of sets

$$X^k(\{1, 2\}, \omega_{\{1,2\}});$$

therefore, we do not need to specify the gain functional of the third decision-maker.

Consider the gain vector

$$\omega_{\{1,2\}} \triangleq (-2, 8).$$

Then, we have

$$X^2(\{1, 2\}, \omega_{\{1,2\}}) = \{x : | x_1 | \leqslant 2 \text{ and } x_2 > 8 \text{ or } | x_1 | < 2 \text{ and } x_2 \geqslant 8\}.$$

Following Step (6) of the preceding algorithm, we shall construct the sets

$$M^1(\{1, 2\}, \omega_{\{1,2\}}) = \{x : | x_1 | \leqslant 1 \text{ and } x_2 > 9 \text{ or } | x_1 | < 1 \text{ and } x_2 \geqslant 9\},$$

$$X^1(\{1, 2\}, \omega_{\{1,2\}}) \triangleq \{x : x + u_1 + u_2 \in M^1(\{1, 2\}, \omega_{\{1,2\}})\}$$
$$= \{x : | x_1 | \leqslant 2 \text{ and } x_2 > 5 \text{ or } | x_1 | < 2 \text{ and } x_2 \geqslant 5\}.$$

Now, we construct the sets

$$M^0(\{1, 2\}, \omega_{\{1,2\}}) = \{x : | x_1 | \leqslant 1 \text{ and } x_2 > 6 \text{ or } | x_1 | < 1 \text{ and } x_2 \geqslant 6\},$$
$$X^0(\{1, 2\}, \omega_{\{1,2\}}) = \{x : | x_1 | \leqslant 2 \text{ and } x_2 > 2 \text{ or } | x_1 | \leqslant 2 \text{ and } x_2 \geqslant 2\}.$$

In Fig. 1, the construction of these sets has been illustrated.

The point $(-1/2, 5/2)$ is in

$$X^0(\{1, 2\}, \omega_{\{1,2\}}),$$

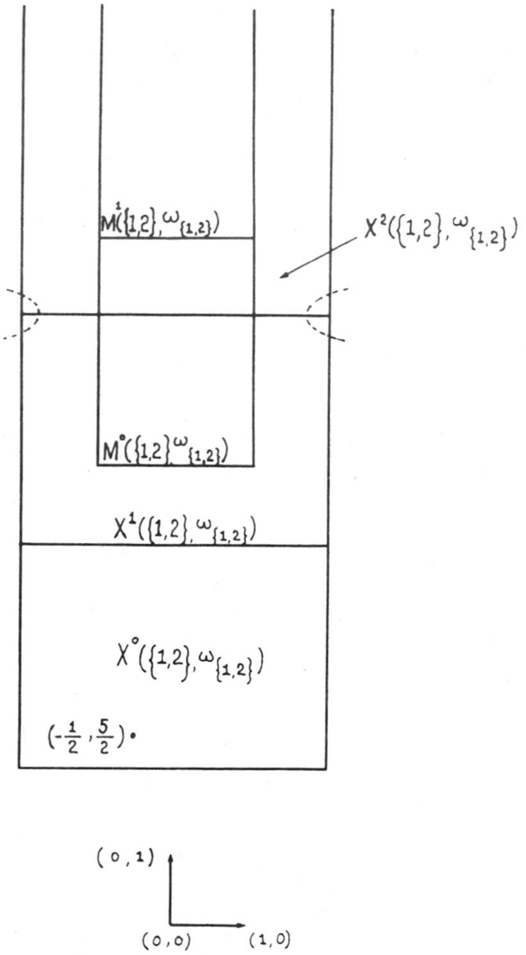

Fig. 1

but it is readily seen that there exists no decision sequence

$$\tilde{u}_{\{1,2\}} = (u_{\{1,2\}}(0),\ u_{\{1,2\}}(1))$$

such that, for all decision sequences $\tilde{u}_3 = (u_3(0),\ u_3(1))$, the trajectory emanating from $(-1/2,\ 5/2)$ and generated by $(\tilde{u}_{\{1,2\}}\ ,\ \tilde{u}_3)$ reaches the target

$$X^2(\{1,\ 2\},\ \omega_{\{1,2\}})$$

at Stage (2). Therefore, *open-loop* C-optimality is less restrictive than C-optimality.

References

1. Starr, A. W., and Ho, Y. C., *Nonzero-Sum Differential Games*, Journal of Optimization Theory and Applications, Vol. 3, No. 3, 1969.
2. Starr, A. W., and Ho, Y. C., *Further Properties of Nonzero-Sum Differential Games*, Journal of Optimization Theory and Applications, Vol. 3, No. 4, 1969.
3. Vincent, T. L., and Leitmann, G., *Control Space Properties of Cooperative Games*, Journal of Optimization Theory and Applications, Vol. 6, No. 2, 1970.
4. Blaquiere, A., *Sur la Géométrie des Surfaces de Pareto d'un Jeu Différentiel à N Joueurs*, Comptes Rendus des Séances de l'Académie des Sciences de Paris, Séries A, Vol. 271, No. 15, 1970.
5. Blaquiere, A., Juricek, L., and Wiese, K. E., *Sur la Géométrie des Surfaces de Pareto d'un Jeu Différentiel à N Joueurs; Théorème du Maximum*, Comptes Rendus des Séances de l'Académie des Sciences de Paris, Séries A, Vol. 271, No. 20, 1970.
6. Haurie, A., *Jeux Quantitatifs à M Joueurs*, Université de Paris, Paris, France, Doctoral Dissertation, 1970.
7. Aumann, R. J., *A Survey of Cooperative Games without Side Payments*, Essays in Mathematical Economics, Edited by M. Shubik, Princeton University Press, Princeton, New Jersey, 1969.
8. Blaquiere, A., and Wiese, K. E., *Jeux Qualitatifs Multi-Étages à N Personnes, Coalitions*, Comptes Rendus des Séances de l'Académie des Sciences de Paris, Séries A, Vol. 270, No. 19, 1970.
9 Delfour, M. C., and Mitter, S. K., *Reachability of Perturbed Linear Systems and Min Sup Problems*, SIAM Journal on Control, Vol. 7, No. 4, 1969.
10. Bertsekas, D. P., and Rhodes, I. B., *On the Minimax Reachability of Targets and Target Tubes*, Automatica, Vol. 7, No. 2, 1971.
11. Glover, J. D., and Schweppe, F. C., *Control of Linear Dynamic Systems with Set Constrained Disturbances*, IEEE Transactions on Automatic Control, Vol. AC-16, No. 5, 1971.
12. Lucas, W. F., *Some Recent Developments in n-Person Game Theory*, SIAM Review, Vol. 13, No. 4, 1971.
13. Kuhn, H. W., and Szego, G. P., *Differential Games and Related Topics*, North Holland, Amsterdam, Holland, 1971.

VIII

A Sufficiency Condition for Coalitive Pareto-Optimal Solutions

W. E. SCHMITENDORF AND G. MORIARTY

Abstract. In a k-player, nonzero-sum differential game there exists the possibility that a group of players will form a coalition and work together. If all k players form the coalition, the criterion usually chosen is Pareto optimality, whereas if the coalition consists of only one player, a minimax or Nash equilibrium solution is sought. In this paper, games with coalitions of more than one but less than k players are considered. Coalitive Pareto optimality is chosen as the criterion. Sufficient conditions are presented for coalitive Pareto-optimal solutions and the results are illustrated with an example.

1. Problem Formulation

Consider a k-player differential game. The dynamical system is described by

$$\dot{x}(t) = f(x(t), u_1(t),..., u_k(t)), \qquad x(t_0) = x_0, \tag{1}$$

where the state $x(t) \in R^n$ and Player i's control variable $u_i(t) \in R^{m_i}$, $i = 1,..., k$. We assume f is C^1 on $R^n \times R^{m_1} \times \cdots R^{m_k}$. The game starts at specified initial point (x_0, t_0) and terminates at a prescribed t_f.

Let $\mathcal{N}_i$, $i = 1,..., k$, denote the set of all piecewise C^1 functions from $[t_0, t_f]$ into R^{m_i}. A control function u_i is admissible if, and only if, $u_i \in \mathcal{N}_i$ and, for all $t \in [t_0, t_f]$, $u_i(t) \in U_i$, where U_i is a fixed subset of R^{m_i}. The set $\mathcal{M}_i$ is the set of admissible u_i.

Assumption 1. We assume that for each admissible k-tuple $(u_1, ..., u_k)$, there exists a unique solution of (1) on $[t_0, t_f]$ passing through (x_0, t_0).

Player i's cost is

$$J_i(u_1, ..., u_k) = g_i(x(t_f)) + \int_{t_0}^{t_f} L_i(x(t), u_1(t), ..., u_k(t))\, dt, \tag{2}$$

where $x: [t_0, t_f] \to R^n$ is the trajectory corresponding to $(u_1, ..., u_k)$ from (x_0, t_0). It is assumed that each L_i is C^1 on $R^n \times R^{m_i} \times \cdots R^{m_k}$ and each g_i is C^1 on R^n.

2. Noncooperative Moods of Play

2.1. Minmax Solution. In this case, Player i assumes all the other players are collectively playing against him by attempting to maximize his cost $J_i(u_1, ..., u_s)$. Let u_{K-I} denote the control $(k-1)$-tuple $(u_1, ..., u_{i-1}, u_{i+1}, ..., u_k)$ and $\mathscr{M}_{K-I}$ the set of admissible u_{K-I}. Then Player i considers the two-player, zero-sum game with cost $J_i(u_i, u_{K-I})$ and seeks a minmax solution $u_i{}^*$ for this game. The control $u_i{}^*$ is the minmax strategy for Player i. By using $u_i{}^*$, Player i is assured that his performance index will not exceed $\sup_{u_{K-I} \in \mathscr{M}_{K-I}} J_i(u_i{}^*, u_{K-I})$ no matter what the other players do and this is the smallest cost he can guarantee. In fact, he will probably do better than this since it is unlikely that all the other players will be collectively playing directly against him. Sufficient conditions for a minmax strategy are presented in Ref. 1.

2.2. Nash Equilibrium Solutions. A control k-tuple $(u_1{}^*, ..., u_k{}^*)$ is a Nash equilibrium control if for all $i \in \{1, ..., k\}$ the equilibrium conditions $J_i(u_1{}^*, ..., u_k{}^*) \leq J_i(u_1{}^*, ..., u_i, ..., u_k{}^*)$ are satisfied for all admissible $(u_1{}^*, ..., u_i, ..., u_k{}^*)$.

This type of solution is secure against any attempt by one player to unilaterally alter his control, since he cannot improve his performance by such a change. In Ref. 2, necessary conditions for open-loop Nash equilibrium solutions are treated, and in Ref. 3, sufficient conditions for linear quadratic games are derived.

3. k-Player Coalitions

In some situations all k players may agree to cooperate. Then each player may be able to achieve a lower cost than he would if all the players played minmax or Nash. For this case, an attractive solution concept is the concept of Pareto optimality (see Refs. 4–9). An admissible control k-tuple $(u_1^*,..., u_k^*)$ is Pareto optimal if, and only if, for every admissible k-tuple

$$\Delta J_i = J_i(u_1,..., u_k) - J_i(u_1^*,..., u_k^*) = 0$$

for all $i \in \{1, 2,..., k\}$ or there exists at least one $i \in \{1,..., k\}$ such that $\Delta J_i > 0$. In general, there is an infinite number of Pareto-optimal solutions.

When all the players agree to cooperate, they should play one of the control k-tuples that is Pareto optimal. If they play a control that is not Pareto optimal, then there is another control k-tuple with which at least one player improves his performance while none do worse.

Necessary conditions and sufficient conditions for Pareto optimality have been presented in terms of a scalar cost $J = \sum_{i=1}^{k} \alpha_i J_i$, where $\alpha_i \geq 0$ or $\alpha_i > 0$. A solution of an optimal control problem with cost $J(u_1,..., u_s, \alpha)$ is a Pareto-optimal solution. As the α's vary, an infinite number of Pareto optimal solutions may be obtained. Before a particular one can be agreed upon, further rules of negotiation must be specified.

4. Coalitions with Less Than k Players

Coalitions need not be restricted to those involving all k players. A subset of the players may form a coalition. For a criterion of optimality, we shall use coalitive Pareto optimality, first introduced for differential games by Haurie (Ref. 10).

Let Players 1,..., s $(s \geq 1)$ form a coalition.[1] Denote the control s-tuple $(u_1,..., u_s)$ by u_S and the $(k - s)$-tuple $(u_{s+1},..., u_k)$ by u_{K-S}. Let $\mathscr{M}_{K-S}$ be the set of admissible u_{K-S}, i.e., $u_{K-S} \in \mathscr{M}_{K-S}$ if, and only if, $u_j \in \mathscr{M}_j$, $j = s + 1,..., k$. Similarly, $\mathscr{M}_S$ is the set of admissible u_S.

Definition 4.1. The control s-tuple $u_S^* = (u_1^*,..., u_s^*)$ is *coalitive Pareto-optimal* (CPO) *at* (x_0, t_0) if, and only if, for all admissible u_S either

$$\Delta J_i \triangleq \max_{u_{K-S} \in \mathscr{M}_{K-S}} J_i(u_S, u_{K-S}) - \max_{u_{K-S} \in \mathscr{M}_{K-S}} J_i(u_S^*, u_{K-S}) = 0 \qquad (3)$$

for all $i \in \{1,..., s\}$ or there is at least one $i \in \{1,..., s\}$ such that $\Delta J_i > 0$.[2]

[1] Any coalition can be represented this way by simply renumbering the players.

[2] We assume that $\max_{u_{K-S} \in \mathscr{M}_{K-S}} J_i(u_S, u_{K-S})$ exists for all $u_S \in \mathscr{M}_S$.

For a coalition of all k players, the definition becomes that of Pareto optimality, whereas if the coalition consists of one player, the definition becomes the definition of a minmax control for the player in the coalition. The following lemma is a sufficient condition for $u_S{}^*$ to be CPO at (x_0, t_0).

Lemma 4.1. Let the control s-tuple $u_S{}^* = (u_1{}^*, \ldots, u_s{}^*)$ be admissible. If there exists an $\alpha \in E^s$ with $\alpha_i > 0$, $i = 1, \ldots, s$, and $\sum_{i=1}^{s} \alpha_i = 1$ such that

$$\sum_{i=1}^{s} \alpha_i \max_{u_{K-S} \in \mathscr{M}_{K-S}} J_i(u_S{}^*, u_{K-S}) \leq \sum_{i=1}^{s} \alpha_i \max_{u_{K-S} \in \mathscr{M}_{K-S}} J_i(u_S, u_{K-S}) \tag{4}$$

for all $u_S \in \mathscr{M}_S$, then $u_S{}^*$ is CPO at (x_0, t_0).

Proof. Suppose $u_S{}^*$ satisfies (4) but is not CPO. Then there exists a $\bar{u}_S \in \mathscr{M}_S$ such that

$$\max_{u_{K-S} \in \mathscr{M}_{K-S}} J_i(\bar{u}_S, u_{K-S}) \leq \max_{u_{K-S} \in \mathscr{M}_{K-S}} J_i(u_S{}^*, u_{K-S}), \qquad i = 1, \ldots, s, \tag{5}$$

with the strict inequality holding for at least one i. Multiplying the ith equation in (5) by α_i and adding leads to

$$\sum_{i=1}^{s} \alpha_i \max_{u_{K-S} \in \mathscr{M}_{K-S}} J_i(\bar{u}_S, u_{K-S}) < \sum_{i=1}^{s} \alpha_i \max_{u_{K-S} \in \mathscr{M}_{K-S}} J_i(u_S{}^*, u_{K-S}),$$

which contradicts (4). QED

A direct application of this lemma suggests that one consider the scalar problem of minimizing with respect to u_S,

$$J(u_S, \alpha) = \sum_{i=1}^{s} \alpha_i \max_{u_{K-S} \in \mathscr{M}_{K-S}} J_i(u_S, u_{K-S}),$$

in an analogous fashion to the scalar problem for Pareto optimality. Unfortunately, solving this scalar problem is quite difficult. It requires finding, for all admissible u_S, the function $u_{K-S}^i(t) = \gamma_{K-S}^i(u_S(t), t)$ which maximizes $J_i(u_S, u_{K-S})$. Even if these functions could be found, the resulting $J_i(u_S, \gamma_{K-S}^i)$ would most likely not have the requisite differentiability to be treated by standard techniques.

We have been able to obtain a sufficient condition in terms of a related two-player, zero-sum game. With this result, verifying CPO sufficiency is no more difficult than using sufficient conditions for a two-player, zero-sum game.

Consider the following two-player, zero-sum differential game. The system is described by

$$\begin{aligned} \dot{x}_1 &= f(x_1, u_S, u^1_{K-S}), & x_1(t_0) &= x_0, \\ \dot{x}_2 &= f(x_2, u_S, u^2_{K-S}), & x_2(t_0) &= x_0, \\ &\vdots & &\vdots \\ \dot{x}_s &= f(x_s, u_S, u^s_{K-S}), & x_s(t_0) &= x_0 \end{aligned} \tag{6}$$

$$u_S \in \mathscr{M}_S, \qquad u^i_{K-S} \in \mathscr{M}_{K-S}, \qquad i = 1,\ldots, s, \tag{7}$$

$$J(u_S, u^1_{K-S}, \ldots, u^s_{K-S}) = \sum_{i=1}^{s} \alpha_i J_i(u_S, u^i_{K-S}), \tag{8}$$

where

$$J_i(u_S, u^i_{K-S}) = g_i(x_i(t_f)) + \int_{t_0}^{t_f} L_i(x_i(t), u_S(t), u^i_{K-S}(t))\, dt.$$

In this two-player, zero-sum game, one player controls the s-tuple u_S while the other chooses the $s(k - s)$-tuple $(u^1_{K-S}, \ldots, u^s_{K-S})$, which will be denoted by u_M.

A further understanding of this game can be obtained by looking at the specific case of a four-player game where Players 1 and 2 form a coalition. The differential equations are

$$\dot{x} = f(x, u_1, u_2, u_3, u_4), \qquad x(t_0) = x_0$$

and the costs for Players 1 and 2 are

$$J_i(u_1, u_2, u_3, u_4) = g_i(x(t_f)) + \int_{t_0}^{t_f} L_i(x(t), u_1(t), \ldots, u_4(t))\, dt, \qquad i = 1, 2.$$

The game corresponding to (6)–(8) is

$$\begin{aligned} \dot{x}_1 &= f(x_1, u_1, u_2, u_3{}^1, u_4{}^1), & x_1(t_0) &= x_0, \\ \dot{x}_2 &= f(x_2, u_1, u_2, u_3{}^2, u_4{}^2), & x_2(t_0) &= x_0, \end{aligned}$$

with cost

$$\begin{aligned} J(u_1, u_2, u_3{}^1, u_4{}^1, u_3{}^2, u_4{}^2) &= \alpha_1 g_1(x_1(t_f)) + \alpha_1 \int_{t_0}^{t_f} L_1(x_1(t), u_1(t), u_2(t), u_3{}^1(t), u_4{}^1(t))\, dt \\ &\quad + \alpha_2 g_2(x_2(t_f)) + \alpha_2 \int_{t_0}^{t_f} L_2(x_2(t), u_1(t), u_2(t), u_3{}^2(t), u_4{}^2(t))\, dt. \end{aligned}$$

The dimension of the state is now $2n$ and we have, in a sense, added two fictitious players who choose u_3^2 and u_4^2. To state it another way, one complementary coalition has been added, which chooses (u_3^2, u_4^2). Note that Player 1's portion of the new cost is only affected by x_1 and Player 2's only by x_2. Also, the complementary coalition (u_3^1, u_4^1) only influences x_1, while the fictitious complementary coalition only influences x_2.

In the zero-sum game, one player chooses (u_1, u_2) while the other chooses $(u_3^1, u_4^1, u_3^2, u_4^2)$. Thus Players 1 and 2 act as one player, while Players 3 and 4 and the two fictitious players act as the other.

We now present a sufficient condition for the general case.

Theorem 4.1. Let (u_S^*, u_M^*) be an open-loop saddle-point solution for the two-player, zero-sum game defined by (6)–(8) with $\alpha_i > 0$, $\sum_{i=1}^{s} \alpha_i = 1$. Then u_S is a CPO solution at (x_0, t_0) for the coalition consisting of Players 1,..., s.

Proof. Since (u_S^*, u_M^*) is an open-loop saddle-point solution,

$$J(u_S^*, u_M) \leq J(u_S^*, u_M^*) \leq J(u_S, u_M^*) \qquad \forall u_S \in \mathscr{M}_S \text{ and } \forall u_M \in \mathscr{M}_M. \tag{9}$$

With the assumption that for every admissible k-tuple $(u_1, ..., u_k)$ there exists a unique solution of (1) on $[t_0, t_f]$ passing through (x_0, t_0), (9) implies

$$J(u_S^*, u_M^*) = \max_{u_M \in \mathscr{M}_M} J(u_S^*, u_M) \tag{10}$$

and

$$\max_{u_M \in \mathscr{M}_M} J(u_S^*, u_M) \leq \max_{u_M \in \mathscr{M}_M} J(u_S, u_M) \tag{11}$$

for all $u_S \in \mathscr{M}_S$. Also, since u_{K-S}^i only affects the ith term of $J(u_S, u_{K-S}^1, ..., u_{K-S}^s)$,

$$\max_{u_M \in \mathscr{M}_M} J(u_S, u_M) = \sum_{i=1}^{s} \alpha_i \max_{u_{K-S}^i \in \mathscr{M}_{K-S}} J_i(u_S, u_{K-S}^i)$$

$$= \sum_{i=1}^{s} \alpha_i \max_{u_{K-S} \in \mathscr{M}_{K-S}} J_i(u_S, u_{K-S}). \tag{12}$$

Combining (11) and (12), we have

$$\sum_{i=1}^{s} \alpha_i \max_{u_{K-S} \in \mathscr{M}_{K-S}} J_i(u_S^*, u_{K-S}) \leq \sum_{i=1}^{s} \alpha_i \max_{u_{K-S} \in \mathscr{M}_{K-S}} J_i(u_S, u_{K-S}) \tag{13}$$

for all $u_S \in \mathscr{M}_S$ and, from Lemma 1, the theorem is proved. QED

Sufficient conditions for open-loop saddle-point solutions can be found in Refs. 11 and 12. These results can be used for the game defined by (6)–(8) and, hence, for coalitive Pareto optimality.

The assumption that for each admissible k-tuple $(u_1, ..., u_k)$ there exists a unique solution through (x_0, t_0) on $[t_0, t_f]$ is a strong one. It is satisfied if the system (1) is linear. Additional conditions assuring that this assumption is met are contained in Theorem 1.2.1 of Ref. 13. For a particular problem one may be able to verify Assumption 1 without resorting to a general theorem.

It is possible to derive sufficient conditions for CPO solutions when the game (6)–(8) does not have a saddle-point solution. This result is contained in the following corollary.

Corollary 4.1. If $u_S{}^*$ is an open-loop minmax control for the game defined by (6)–(8) with $\alpha_i > 0$ and $\sum_{i=1}^{s} \alpha_i = 1$, then $u_S{}^*$ is a CPO solution at (x_0, t_0) for the coalition consisting of players $1, ..., s$.

If $u_S{}^*$ is a minmax control for the game (6)–(8), then it satisfies (11) and the proof follows in exactly the same manner.

If $\max_{u_{K-S} \in \mathscr{M}_{K-S}} J_i(u_S, u_{K-S})$ does not exist for all $u_S \in \mathscr{M}_S$, this corollary still applies if "sup" replaces "max" in Definition 1. The corollary is also applicable even if the assumption that all admissible k-tuples are playable is not met, since Ref. 1 contains a sufficient condition for a minmax control without imposing this assumption. Instead a weaker playability condition must be satisfied.

For games where the describing differential equations are linear,

$$\dot{x}(t) = F(t)\, x(t) + \sum_{j=1}^{k} G_j(t)\, u_j(t), \qquad u_j(t) \in U^j = R^{m_j},$$

and the costs are quadratic,

$$J_i(u_1, ..., u_k) = \tfrac{1}{2} x(t_f)\, S_i x(t_f) + \tfrac{1}{2} \int_{j_0}^{t_f} \sum_{j=1}^{k} u_j(t)\, R_{ij}(t)\, u_j(t)\, dt, \qquad i = 1, 2, ..., k,$$

it is relatively easy to find coalitive Pareto-optimal solutions for any coalition S by using Theorem 1 along with the results for open-loop saddle-point solutions for two-player, zero-sum games presented in Ref. 12.

In the next section, an open-loop sufficiency result for a two-player, zero-sum game is presented. This result is then used in conjunction with Theorem 1 to determine coalitive Pareto-optimal solutions for an example problem.

5. An Example

For the example of this section, the following open-loop sufficiency result for two-player, zero-sum games will be used. Consider the two-player, zero-sum game analogous to the game defined in Section 1. [We call the controls (w_1, w_2) rather than (u_1, u_2) to avoid confusion later in this section].

$$\dot{x} = f(x, w_1, w_2), \qquad x(t_0) = x_0,$$

$$\begin{aligned} J_1(w_1, w_2) &= -J_2(w_1, w_2) \triangleq J(w_1, w_2) \\ &= g(x(t_f)) + \int_{t_0}^{t_f} L(x(t), w_1(t), w_2(t))\, dt. \end{aligned}$$

Let $[w_1^*, w_2^*]$ be continuous and, for all $t \in [t_0, t_f]$, satisfy $w_i(t) \in W_i$. If there exist two C^1 functions $V_1 : R^n \times [t_0, t_f] \to R^1$ and $V_2 : R^n \times [t_0, t_f] \to R^1$ satisfying $V_1(x, t_f) = V_2(x, t_f) = g(x)$ such that

(a) $V_{1t}(x, t) + L(x, w_1, w_2^*(t)) + V_{1x}(x, t) f(x, w_1, w_2^*(t)) \geq 0,$

(b) $V_{1t}(x^*(t), t) + L(x^*(t), w_1^*(t), w_2^*(t))$
$\qquad + V_{1x}(x^*(t), t) f(x^*(t), w_1^*(t), w_2^*(t)) = 0,$

(c) $V_{2t}(x, t) + L(x, w_1^*(t), w_2) + V_{2x}(x, t) f(x, w_1^*(t), w_2) \leq 0,$

(d) $V_{2t}(x^*(t), t) + L(x^*(t), w_1^*(t), w_2^*(t))$
$\qquad + V_{2x}(x^*(t), t) f(x^*(t), w_1^*(t), w_2^*(t)) = 0,$

for all $(x, t) \in R^n \times [t_0, t_f]$, for all $w_1 \in W_1$ and $w_2 \in W_2$, where $x^*(t)$ is the trajectory corresponding to (w_1^*, w_2^*) starting at (x_0, t_0), then (w_1^*, w_2^*) is an open-loop saddle-point solution at (x_0, t_0).

The proof follows that of Theorem 1 of Ref. 14. Conditions (a) and (b) lead to $J(w_1^*, w_2^*) \leqslant J(w_1, w_2^*)$, while (c) and (d) lead to $J(w_1^*, w_2^*) \geqslant J(w_1^*, w_2)$, verifying the saddle-point inequalities at (x_0, t_0).[3]

The example problem is a three-player game where Players 1 and 2 form a coalition:

$$\dot{x} = u_1 + u_2 + u_3, \qquad x(0) = 0,$$

$$| u_i(t) | \leq 1, \qquad i = 1, 2, 3,$$

$$J_1 = -x(1) + \tfrac{1}{2} \int_0^1 u_1^2(t)\, dt, \qquad J_2 = x(1) + \tfrac{1}{2} \int_0^1 u_2^2(t)\, dt.$$

---

[3] The assumption that $[w_1, w_2]$ is continuous can be relaxed to piecewise continuous by using the results of Ref. 15.

The game corresponding to (6)–(8) is

$$\dot{x}_1 = u_1 + u_2 + u_3^1, \qquad x_1(0) = 0, \tag{14a}$$

$$\dot{x}_2 = u_1 + u_2 + u_3^2, \qquad x_2(0) = 0, \tag{14b}$$

$$|u_1(t)| \leq 1, \qquad |u_2(t)| \leq 1, \qquad |u_3^1(t)| \leq 1, \qquad |u_3^2(t)| \leq 1, \tag{15}$$

$$J = -\alpha_1 x_1(1) + \alpha_2 x_2(1) + \tfrac{1}{2}\int_0^1 [\alpha_1 u_1^2(t) + \alpha_2 u_2^2(t)]\, dt, \tag{16}$$

where one player chooses $w_1 \triangleq (u_1, u_2)$ and the other $w_2 \triangleq (u_3^1, u_3^2)$. If $\alpha_2 \geq 2\alpha_1$, then

$$w_1^* = (-1, (\alpha_1 - \alpha_2)/\alpha_2), \qquad w_2^* = (-1, +1)$$

is an open-loop saddle-point solution for the game (14)–(16). This is easily verified using $V_1(x, t) = V_2(x, t) = +\frac{7}{2}\alpha_1(1 - t) - \alpha_1 x_1 + \alpha_2 x_2 - \frac{1}{2}(\alpha_1^2/\alpha_2)(1 - t) - \frac{1}{2}\alpha_2(1 - t)$ in the above sufficient conditions. Thus, from Theorem 1, $u_1^*(t) = -1$, $u_2^*(t) = (\alpha_1 - \alpha_2)/\alpha_2$ is coalitive Pareto optimal for any α_1, α_2 satisfying $\alpha_1 > 0$, $\alpha_2 > 0$, $\alpha_1 + \alpha_2 = 1$, and $\alpha_2 \geq 2\alpha_1$.

Similarly $u_1^*(t) = (\alpha_1 - \alpha_2)/\alpha_1$, $u_2^*(t) = (\alpha_1 - \alpha_2)/\alpha_2$ is CPO for any α_1, α_2 satisfying $\alpha_1 > 0$, $\alpha_2 > 0$, $\alpha_1 + \alpha_2 = 1$, and $\alpha_2/2 < \alpha_1 < 2\alpha_2$, while $u_1^*(t) = (\alpha_1 - \alpha_2)/\alpha_1$, $u_2^*(t) = +1$ is CPO for any $\alpha_1 > 0$, $\alpha_2 > 0$, $\alpha_1 + \alpha_2 = 1$, and $2\alpha_2 \leq \alpha_1$.

References

1. Schmitendorf, W. E., *Differential Games Without Pure Strategy Saddle Point Solutions*, Journal of Optimization Theory and Applications, Vol. 18, No. 1, 1976.
2. Starr, A. W., and Ho, Y. C., *On Some Further Properties of Nonzero Sum Differential Games*, Journal of Optimization Theory and Applications, Vol. 3, No. 4, 1969.
3. Schmitendorf, W. E., and Foley, M. H., *On a Class of Nonzero Sum Linear Quadratic Differential Games*, Journal of Optimization Theory and Applications, Vol. 7, No. 5, 1971.
4. Starr, W. S., and Ho, Y. C., *Nonzero Sum Differential Games*, Journal of Optimization Theory and Applications, Vol. 3, No. 3, 1969.
5. Da Cunha, N. O., and Polak, E., *Constrained Minimization Under Vector-Valued Criteria in Linear Topological Spaces*, in *Mathematical Techniques of Optimization*, Academic Press, New York, 1967.

6. VINCENT, T. L., and LEITMANN, G., *Control Space Properties of Cooperative Games*, Journal of Optimization Theory and Applications, Vol. 9, No. 5, 1972.
7. LEITMANN, G., ROCKLIN, S., and VINCENT, T. L., *A Note on Control Space Properties of Cooperative Games*, Journal of Optimization Theory and Applications, Vol. 9, No. 6, 1972.
8. LEITMANN, G., and SCHMITENDORF, W. E., *Some Sufficiency Conditions for Pareto-Optimal Control*, ASME Journal of Dynamic Systems, Measurement, and Control, Vol. 95, No. 4, December 1973.
9. SCHMITENDORF, W. E., and LEITMANN, G., *A Simple Derivation of Necessary Conditions for Pareto Optimality*, IEEE Transactions on Automatic Control, Vol. AC-19, No. 5, October 1974.
10. HAURIE, A., *On Pareto-Optimal Decisions for Coalition of a Subset of Players*, IEEE Transactions on Automatic Control, Vol. AC-18, No. 2, April 1973.
11. REKASIUS, Z. V., *On Closed-Loop and Open-Loop Solutions of Differential Games*, paper presented at the Purdue Centennial Symposium of Information Processing, Lafayette, Indiana, 1969.
12. SCHMITENDORF, W. E., *Existence of Optimal Open-Loop Strategies for a Class of Differential Games*, Journal of Optimization Theory and Applications, Vol. 5, No. 5, 1970.
13. FRIEDMAN, A., *Differential Games*, Wiley–Interscience, New York, 1971.
14. LEITMANN, G., *Sufficiency Theorems for Optimal Control*, Journal of Optimization Theory and Applications, Vol. 2, No. 5, 1968.
15. STALFORD, H., *Sufficient Conditions for Optimal Controls with State and Control Constraints*, Journal of Optimization Theory and Applications, Vol. 7, No. 2, 1971.

IX

On the Stackelberg Strategy in Nonzero-Sum Games[1]

M. Simaan and J. B. Cruz, Jr.

Abstract. The properties of the Stackelberg solution in static and dynamic nonzero-sum two-player games are investigated, and necessary and sufficient conditions for its existence are derived. Several game problems, such as games where one of the two players does not know the other's performance criterion or games with different speeds in computing the strategies, are best modeled and solved within this solution concept. In the case of dynamic games, linear-quadratic problems are formulated and solved in a Hilbert space setting. As a special case, nonzero-sum linear-quadratic differential games are treated in detail, and the open-loop Stackelberg solution is obtained in terms of Riccati-like matrix differential equations. The results are applied to a simple nonzero-sum pursuit–evasion problem.

1. Introduction

The solution of a nonzero-sum game is generally defined in terms of the rationale that each player adopts as a means of describing optimality. One of the most commonly known rationales is the Nash strategy (Ref. 1), first introduced in dynamic games in Ref. 2–3. The Nash strategy safeguards each player against attempts by any one player to further improve on his individual performance criterion. This solution

[1] This work was supported in part by the US Air Force under Grant No. AFOSR-68-1579D, in part by NSF under Grant No. GK-36276, and in part by the Joint Services Electronics Program under Contract No. DAAB-07-72-C-0259 with the Coordinated Science Laboratory, University of Illinois, Urbana, Illinois.

generally assumes that the players know each other's performance functions and that, when the strategies have been calculated, they are announced at the same instant of time. However, because these assumptions may not always hold, many games cannot be modeled and solved in this manner.

For example, in a two-player game, if the first assumption does not hold and one player does not have information about the other's performance function, then it is no longer possible for this player to calculate his Nash strategy. Instead, allowing for the worst possible behavior on his rival's part, he may choose to play a minimax strategy whose calculation requires only knowledge of his own performance function. Or, instead of risking such a pessimistic strategy, he may select to play the game passively, that is, by waiting until the other player's strategy is announced and then solving an ordinary optimization problem for his corresponding strategy. Similarly, the same situation arises in games where, due to faster means of information processing, one player is capable of announcing his strategy before the other. These cases are only a few examples of a class of games that are formulated in such a way that the strategies are announced sequentially. The main question that this paper is concerned with is the following: for the player that has to announce his strategy first, what will be the best strategy to choose? Assuming that the sole objective of the players is to minimize their respective cost functions, a solution concept most reasonable for games of this nature is known as the Stackelberg strategy. This strategy is well known in static competitive economics (Refs. 4–6) and was recently introduced in dynamic games (Ref. 7). It will be shown that, if the player that has to announce his strategy first follows a Stackelberg strategy, he will do no worse, in terms of obtaining lower cost, than the corresponding Nash solution.

As an example of the type of problems considered, let us examine the following simple matrix game. Assume that a government G wants to select a tax rate from the following set of allowable rates: $\{a_1\%, a_2\%, a_3\%\}$ for taxing a certain firm F, which in turn has to decide on manufacturing one out of three possible varieties $\{v_1, v_2, v_3\}$ of the products that it can manufacture. Let the objectives (for example, representing net income for the firm and a combination of income and price stability for the government) of the government and the firm be measured quantitatively for every pair of tax rate and product variety by the entries in Fig. 1.

Here, the first entries correspond to the firm and the second entries correspond to the government. It is assumed that the only desire of F and G is to maximize their individual objective measures. If, due to the

		G		
		a_1	a_2	a_3
F	v_1	8,10	5,10	8,11
	v_2	7,5	8,6	11,7
	v_3	5,6	9,9	12,6

Fig. 1

nature of the game,[2] the government were to fix the tax rate before the firm decides on its product, then, by choosing a Stackelberg strategy, the government will actually be selecting the most advantageous tax rate as well as imposing some influence in the selection of the variety of product to be manufactured by the firm.

The purpose of this paper is to study some of the important characteristics of the Stackelberg strategy and derive necessary and sufficient conditions under which its existence is guaranteed in static as well as dynamic two-player nonzero-sum games. In the case of dynamic problems, linear-quadratic games are formulated and solved in a Hilbert space setting, thus including continuous-time, discrete-time, distributed-parameter, and delay-differential systems. A continuous-time differential game problem is then treated in detail as a special case, and the solution for the open loop Stackelberg strategies is obtained in terms of Riccati-like differential equations. Finally, a pursuit–evasion differential game is solved as an illustrative example.

2. Definition and Properties of the Stackelberg Strategy

Let U_1 and U_2 be the sets of admissible strategies for players 1 and 2, respectively. Let the cost functions $J_1(u_1, u_2)$ and $J_2(u_1, u_2)$ be two functions mapping $U_1 \times U_2$ into the real line such that Player 1 wishes to minimize J_1 and Player 2 wishes to minimize J_2. Following the terminology in Ref. 5, the player that selects his strategy first is called *the leader* and the player that selects his strategy second is called *the follower*. Unless otherwise stated, for the rest of this paper a Stackelberg strategy will always refer to a Stackelberg strategy with Player 2 as leader.

Definition 2.1. If there exists a mapping $T\colon U_2 \to U_1$ such that, for any fixed $u_2 \in U_2$, $J_1(Tu_2, u_2) \leqslant J_1(u_1, u_2)$ for all $u_1 \in U_1$, and if

[2] For instance, the firm might only know the first entries of the table in Fig. 1, corresponding to its own objectives, and thus chooses to wait until a tax rate is fixed before deciding on its product.

there exists a $u_{2s2} \in U_2$ such that $J_2(Tu_{2s2}, u_{2s2}) \leqslant J_2(Tu_2, u_2)$ for all $u_2 \in U_2$, then the pair $(u_{1s2}, u_{2s2}) \in U_1 \times U_2$, where $u_{1s2} = Tu_{2s2}$, is called a Stackelberg strategy pair with Player 2 as leader and Player 1 as follower.

In other words, the Stackelberg strategy is the optimal strategy for the leader when the follower reacts by playing optimally. A Stackelberg strategy with Player 1 as leader is also defined in a similar way. Let the graph $D_1 = \{(u_1, u_2) \in U_1 \times U_2 : u_1 = Tu_2\}$ of the mapping T be called *the rational reaction set* of Player 1. This set represents the collection of strategy pairs in $U_1 \times U_2$ according to which Player 1 reacts to every strategy $u_2 \in U_2$ that Player 2 may choose. By playing according to the set D_1, Player 1 is referred to as being a rational player. In the Stackelberg strategy, the follower is always assumed to be rational. Similarly, let D_2 denote the rational reaction set of Player 2 when Player 1 is the leader. The sets D_1 and D_2 have significant importance in characterizing both the Stackelberg and the Nash strategies as demonstrated in the following two propositions.

Proposition 2.1. A strategy pair (u_{1s2}, u_{2s2}) is a Stackelberg strategy with Player 2 as leader iff $(u_{1s2}, u_{2s2}) \in D_1$ and

$$J_2(u_{1s2}, u_{2s2}) \leqslant J_2(u_1, u_2), \qquad \forall (u_1, u_2) \in D_1. \tag{1}$$

Proposition 2.2. A strategy pair (u_{1N}, u_{2N}) is a Nash strategy pair iff $(u_{1N}, u_{2N}) \in D_1 \cap D_2$.

The proofs of these propositions are straightforward and follow directly from the definitions of the Nash and Stackelberg strategies and the sets D_1 and D_2.

Several interesting properties relating the Nash and Stackelberg strategies can be derived from these propositions. From (1) and Proposition 2.2, it is seen that

$$J_2(u_{1s2}, u_{2s2}) \leqslant J_2(u_{1N}, u_{2N}), \tag{2}$$

which means that the leader in the Stackelberg solution achieves at least as good (possibly better) a cost function as the corresponding Nash solution. This is so because, by choosing a Stackelberg strategy, the leader is actually imposing a solution which is favorable to himself. If the Stackelberg strategies with either player as leader coincide, then they both coincide with the Nash strategy and, clearly, in this case the leader loses its advantage. It is also evident that, in zero-sum games with saddle point, the Nash strategy, the Stackelberg strategy with either player as leader, and the minimax strategy coincide. Similarly, in identical goal games, the Nash and the Stackelberg strategies are the same. The following examples are presented to illustrate the basic idea and related properties.

Example 2.1. In the matrix game of Fig. 1, both F and G want to maximize J_1 and J_2. The rational reaction set of F when G is the leader is the set of pairs $D = \{(a_1, v_1), (a_2, v_3), (a_3, v_3)\}$, and the Stackelberg strategy with G as leader is the element of D that maximizes J_2. This is achieved by the pair (a_1, v_1). Thus, by selecting a tax rate of $a_1\%$, the government leaves no choice for the firm but to manufacture the product v_1; hence, the resulting J_2 is more than what it would have been, had a_2 or a_3 been chosen instead. The Stackelberg strategy with F as leader (a_3, v_2) and the Nash strategy (a_2, v_3) are also easily computed. Note that, in this example, both leaders in the Stackelberg solution obtain better results than in the Nash solution and that the followers are worse off.

Example 2.2. Consider the following single-state two-stage matrix game. The 2×2 matrix games shown in Fig. 2 are to be played consecutively. The first player controls u_1 and u_2 in game (a) and p_1 and p_2 in game (b), while the second player controls v_1 and v_2 in game (a) and q_1 and q_2 in game (b). The first entries in the tables are the costs borne by Player 1 and the second entries are those borne by Player 2. The game may be played starting with either (a) or (b), and the costs to every player, as shown in Fig. 3, are the sum of the costs borne in (a) and (b).

The Stackelberg strategy with Player 2 as leader as obtained from Fig. 3 is $\{(u_1 p_2), (v_2 q_1)\}$. The Stackelberg strategies with Player 2 as leader of the subgames (a) and (b) are (u_1, v_2) and (p_2, q_1), respectively. Thus, it is seen that the Stackelberg strategies of the individual subgames are components of the Stackelberg strategies of the composite game. Stated in more general terms, we have the following proposition.

Proposition 2.3. If a game is composed of N simultaneous separate subgames, where the cost functions are the sum of the corre-

	v_1	v_2
u_1	3,0	3,1
u_2	2,3	5,4

(a)

	q_1	q_2
p_1	5,0	9,9
p_2	4,4	0,5

(b)

Fig. 2

	$v_1 q_1$	$v_1 q_2$	$v_2 q_1$	$v_2 q_2$
$u_1 p_1$	8,0	12,9	8,1	12,10
$u_1 p_2$	7,4	3,5	7,5	3,6
$u_2 p_1$	7,3	11,12	10,4	14,13
$u_2 p_2$	6,7	2,8	9,8	5,9

Fig. 3

sponding cost functions of the subgames, then a composite strategy is a Stackelberg strategy pair for the composite game iff its components are Stackelberg strategy pairs for the component subgames.

Proof. Let the ith subgame be defined by $u_j^{(i)} \in U_j^{(i)}$ and $J_j^{(i)}(u_1^{(i)}, u_2^{(i)})$, $j = 1, 2$. Then,

$$J_j(u_1, u_2) = \sum_{i=1}^{N} J_j^{(i)}(u_1^{(i)}, u_2^{(i)}), \qquad j = 1, 2, \tag{3}$$

where

$$u_j = \{u_j^{(i)}\} = \{u_j^{(1)},..., u_j^{(i)},..., u_j^{(N)}\}, \qquad U_j = \prod_{i=1}^{N} U_j^{(i)}, \qquad j = 1, 2.$$

(a) Let $(u_{1s2}^{(i)}, u_{2s2}^{(i)})$ be a Stackelberg strategy for the ith subgame $\forall i = 1,..., N$. Then, there exists $T^{(i)}$ such that $u_{1s2}^{(i)} = T^{(i)}u_{2s2}^{(i)}$ and, for $u_2^{(i)} \in U_2^{(i)}$,

$$J_1^{(i)}(T^{(i)}u_2^{(i)}, u_2^{(i)}) \leqslant J_1^{(i)}(u_1^{(i)}, u_2^{(i)}), \qquad \forall u_1^{(i)} \in U_1^{(i)}, \qquad \forall i = 1,..., N, \tag{4}$$

$$J_2^{(i)}(T^{(i)}u_{2s2}^{(i)}, u_{2s2}^{(i)}) \leqslant J_2^{(i)}(T^{(i)}u_2^{(i)}, u_2^{(i)}), \qquad \forall u_2^{(i)} \in U_2^{(i)}, \qquad \forall i = 1,..., N. \tag{5}$$

Let $Tu_2 = \{T^{(i)}u_2^{(i)}\}$. By summing (4) and (5) for $i = 1,..., N$, we obtain

$$J_1(Tu_2, u_2) \leqslant J_1(u_1, u_2), \qquad u_2 \in U_2, \qquad \forall u_1 \in U_1, \tag{6}$$

$$J_2(Tu_{2s2}, u_{2s2}) \leqslant J_2(Tu_2, u_2), \qquad \forall u_2 \in U_2. \tag{7}$$

$(u_{1s2} = \{u_{1s2}^{(i)}\}, u_{2s2} = \{u_{2s2}^{(i)}\})$ is therefore a Stackelberg strategy for the composite game.

(b) Let (u_{1s2}, u_{2s2}) be a Stackelberg solution for the composite game. Then, there exists T such that $u_{1s2} = Tu_{2s2}$ and (6)–(7) are satisfied. Fix $u_2 = \{u_2^{(i)}\} \in U_2$, and let $u_1 = Tu_2$ or $u_1^{(i)} = T^{(i)}u_2^{(i)}$. Now, select $u_1 \in U_1$ such that $u_1 = \{T^{(1)}u_2^{(1)},..., T^{(i-1)} u_2^{(i-1)}, u_1^{(i)}, T^{(i+1)}u_2^{(i+1)},..., T^{(N)}u_2^{(N)}\}$, where $u_1^{(i)} \in U_1^{(i)}$; then (6) reduces to

$$J_1^{(i)}(T^{(i)}u_2^{(i)}, u_2^{(i)}) \leqslant J_1^{(i)}(u_1^{(i)}, u_2^{(i)}), \qquad \forall u_1^{(i)} \in U_1^{(i)}, \qquad i = 1,..., N. \tag{8}$$

Similarly, if

$$u_2 = \{u_{2s2}^{(1)},..., u_{2s2}^{(i-1)}, u_2^{(i)}, u_{2s2}^{(i+1)},..., u_{2s2}^{(N)}\}, \qquad u_2^{(i)} \in U_2^{(i)},$$

then (7) reduces to

$$J_2^{(i)}(T^{(i)}u_{2s2}^{(i)}, u_{2s2}^{(i)}) \leqslant J_2^{(i)}(T^{(i)}u_2^{(i)}, u_2^{(i)}), \qquad \forall u_2^{(i)} \in U_2^{(i)}, \qquad i = 1,..., N. \tag{9}$$

Therefore, $(u_{1s2}^{(i)}, u_{2s2}^{(i)})$ is a Stackelberg strategy for the ith subgame.

The significance of this proposition lies in the fact that, if a set of N games are played consecutively such that the outcome of every game does not affect the outcome of the following games, the players need only calculate the Stackelberg strategies for every subgame in order to obtain the Stackelberg strategies for the composite games. Note that this property does not hold for the noninferior solutions as shown in Ref. 3.

3. Static Games

Static games are games that do not evolve over time. In this section, a class of static games in which the cost functions $J_1(u_1, u_2)$ and $J_2(u_1, u_2)$ are real-valued continuous functions defined over a subset or all of the Euclidean space $R^{m_1} \times R^{m_2}$, where m_1 and m_2 are positive integers, will be considered. Unlike matrix games, the Stackelberg solution in static games need not always exist. In these games, the Nash solution may exist, but the Stackelberg solution may not exist (and *vice versa*), as demonstrated by the following examples.

Example 3.1. Let the cost functions of the two players be

$$J_1(u_1, u_2) = -u_1u_2 + \tfrac{1}{2}u_1^2 + u_1,$$
$$J_2(u_1, u_2) = -(u_1^2 + 1)\, u_2 + \tfrac{1}{2}(u_2^2 - u_1^2) - 2u_1,$$

where $u_1 \in R^1$ and $u_2 \in R^1$. We have, respectively,

$$\partial J_1/\partial u_1 = -u_2 + u_1 + 1, \qquad \partial^2 J_1/\partial u_1^2 = 1$$
$$\partial J_2/\partial u_2 = -(u_1^2 + 1) + u_2, \qquad \partial^2 J_2/\partial u_2^2 = 1.$$

The rational reaction sets D_1 and D_2 are therefore the line $u_1 = u_2 - 1$ and the curve $u_2 = u_1^2 + 1$. The solution of these equations ($u_1 = 0, u_2 = 1$) and ($u_1 = 1, u_2 = 2$) are the Nash strategies for this game. On the other hand, the Stackelberg strategy with Player 2 as leader is obtained by minimizing J_2 subject to the constraint $u_1 = u_2 - 1$. This reduces to minimizing the function

$$J_2 = -u_2^3 + 2u_2^2 - 3u_2 - 1.5,$$

which has no minimum with respect to u_2, thus implying that a Stackelberg strategy does not exist.

In order to guarantee the existence of Stackelberg strategies, one generally requires compactness of the spaces U_1 and U_2. The following proposition gives sufficient conditions for the existence of the Stackelberg strategies in static games.

Proposition 3.1. If U_1 and U_2 are compact sets, $U_1 \subset R^{m_1}$ and $U_2 \subset R^{m_2}$, and if J_1 and J_2 are real-valued continuous functions on $U_1 \times U_2$, then Stackelberg strategies with either player as leader exist.

Proof. The existence of a Stackelberg strategy with Player 2 as leader will be proved. The proof for the case where Player 1 is the leader is analogous. Since D_1 is a subset of the compact set $U_1 \times U_2$, we need only show that it is closed. Let (u_1^0, u_2^0) be a point in $\bar{D}_1$, the closure of D_1, and let (u_1^n, u_2^n) be a sequence of points in D_1 converging to (u_1^0, u_2^0). We will show that $(u_1^0, u_2^0) \in D_1$. Suppose that $(u_1^0, u_2^0) \notin D_1$, then $\exists (u_1^*, u_2^0) \in D_1$ such that $J_1(u_1^0, u_2^0) > J_1(u_1^*, u_2^0)$. Let $\mathscr{E} = J_1(u_1^0, u_2^0) - J_1(u_1^*, u_2^0)$. Since J_1 is continuous, $\exists \delta_1$ and $\delta_2 > 0$ such that

$$| J_1(u_1, u_2) - J_1(u_1^0, u_2^0)| < \mathscr{E}/3, \qquad \forall (u_1, u_2) \in A,$$

where

$$A = \{(u_1, u_2) \in U_1 \times U_2 : |(u_1, u_2) - (u_1^0, u_2^0)| < \delta_1\},$$

and

$$| J_1(u_1, u_2) - J_1(u_1^*, u_2^0)| < \mathscr{E}/3, \qquad \forall (u_1, u_2) \in B,$$

where

$$B = \{(u_1, u_2) \in U_1 \times U_2 : |(u_1, u_2) - (u_1^*, u_2^0)| < \delta_2\},$$

and $A \cap B = \varnothing$, where $|\cdot|$ denotes the Euclidean norm. Since $(u_1^n, u_2^n) \to (u_1^0, u_2^0)$, $\exists N_1$ such that $(u_1^n, u_2^n) \in A$, $\forall n > N_1$, and also $\exists$ a sequence $u_1^{*n} \to u_1^*$ and N_2 such that $(u_1^{*n}, u_2^n) \in B$, $\forall n > N_2$. Now, pick $N = \max\{N_1, N_2\}$; then, $| J_1(u_1^n, u_2^n) - J_1(u_1^0, u_2^0)| < \mathscr{E}/3$ and $| J_1(u_1^{*n}, u_2^n) - J_1(u_1^*, u_2^0)| < \mathscr{E}/3$, $\forall n > N$. This means that $J_1(u_1^n, u_2^n) > J_1(u_1^{*n}, u_2^n)$, $\forall n > N$. This is a contradiction, since (u_1^n, u_2^n) is a sequence in D_1. Hence, $(u_1^0, u_2^0) \in D_1$, and D_1 is closed. By the continuity of J_2, $\exists (u_{1s2}, u_{2s2}) \in D_1$ such that (1) is satisfied.

When the Stackelberg strategies happen to be in the interior of $U_1 \times U_2$, or when $U_1 \times U_2$ is the whole Euclidean space, necessary conditions for the existence of a solution can be derived easily. If $U_1 = R^{m_1}$, $U_2 = R^{m_2}$ and $J_1(u_1, u_2)$, $J_2(u_1, u_2)$ are twice differentiable on $U_1 \times U_2$, then, if a Stackelberg solution (u_{1s2}, u_{2s2}) with Player 2 as leader exists, it must satisfy the following set of equations:

(i) $$\nabla_{u_1} J_1(u_{1s2}, u_{2s2}) = 0, \tag{10}$$

(ii) $$\nabla_{u_1} J_2(u_{1s2}, u_{2s2}) + J_{1u_1u_1}(u_{1s2}, u_{2s2})\lambda = 0, \tag{11}$$

(iii) $$\nabla_{u_2} J_2(u_{1s2}, u_{2s2}) + J_{1u_2u_1}(u_{1s2}, u_{2s2})\lambda = 0, \tag{12}$$

where λ is an m_1-dimensional Lagrange multiplier. The notation $\nabla_{u_1} J_1$ denotes the gradient vector of J_1 with respect to u_1 and $J_{1u_1u_1}$ and $J_{1u_2u_1}$ represent the $m_1 \times m_1$ and $m_2 \times m_1$ matrices of second partial derivatives whose ijth elements are $\partial^2 J_1/\partial u_1^{(j)}\, \partial u_1^{(i)}$ and $\partial^2 J_1/\partial u_2^{(j)}\, \partial u_1^{(i)}$, respectively.

A simple example of planar games illustrating some of the basic properties of the solutions is presented below.

Example 3.2. The static minimization game considered in Refs. 6–8 is reproduced in Fig. 4. The cost functions J_1 and J_2, defined on $R^1 \times R^1$, are assumed to be convex and twice differentiable with respect to u_1 and u_2 and having contour lines as shown in Fig. 4. The rational reaction sets D_1 and D_2 are obtained by joining the points of tangency between the contour lines and the lines of constant u_2 and u_1, respectively. It is clear that J_2 achieves its minimum over D_1 at the point S_2 whose coordinates (u_{1s2}, u_{2s2}) are the Stackelberg strategies when Player 2 is the leader. In other words, if Player 2 is to select his strategy first, he has no better choice than u_{2s2} as long as Player 1 reacts according to the curve D_1. Similarly, point S_1 is the Stackelberg solution when Player 1 is the leader and point N is the Nash solution. An interesting feature of these strategies, illustrated in this example, is that the leader is not necessarily always the only player that benefits. In fact, in this example, both Stackelberg solutions give lower costs for both players than the Nash solution. Thus, by playing Stackelberg (that is, by agreeing that one player will lead and the other will follow), the players will be playing an enforceable solution from which both can benefit over the Nash solution. Furthermore, the Stackelberg solution has great

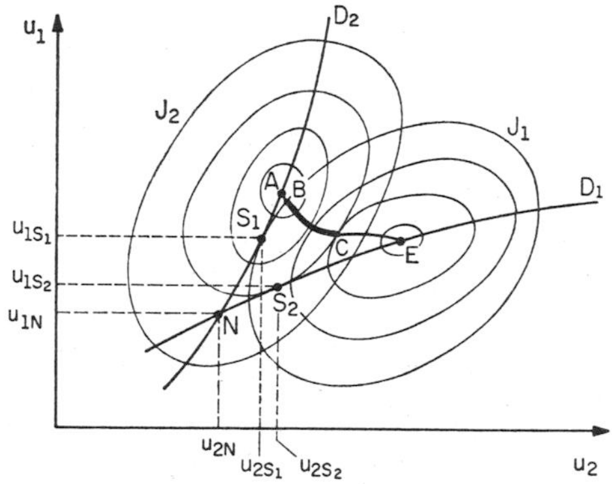

Fig. 4. A game with Nash and Stackelberg solutions.

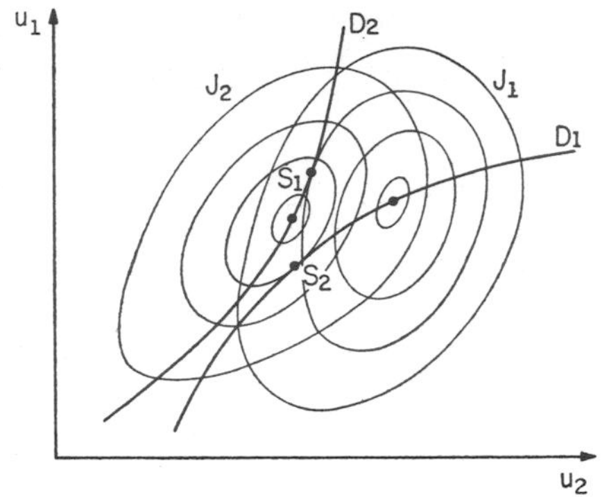

Fig. 5. A game without Nash but with Stackelberg solutions.

impact on the two players in the process of making negotiation.[3] Finally, a slight modification of this game shown in Fig. 5 illustrates a situation where Stackelberg strategies exist while a Nash strategy does not exist. In this case, the Stackelberg strategies are potential substitutes for the Nash strategies.

4. Dynamic Games

Dynamic games are games that evolve over time. Their description is usually done in terms of a dynamic equation that describes the evolution of the state of the game in response to control variables selected by the players from sets of allowable controls. Linear-quadratic games are generally represented by a linear state equation and quadratic cost functions. In this section, linear-quadratic games defined over real Hilbert spaces are treated. This formulation includes several dynamic games, such as continuous-time, discrete-time, etc., that are of interest to control engineers.

[3] The authors would like to thank one of the reviewers for bringing this fact to their attention. If $P \subset U_1 \times U_2$ is the noninferior set, then a negotiation set N_e can be defined as follows:

$$N_e = \{(u_1, u_2) \in P;\ J_i(u_1, u_2) \leqslant \underline{J_i(u_{1si}, u_{2si}) \leqslant J_i(u_{1N}, u_{2N})},\ i = 1, 2\},$$

where (u_{1si}, u_{2si}) is the Stackelberg solution when player i is the leader. The underlined relation is always true if the Stackelberg and Nash strategies exist. If any of them do not exist, the corresponding part of the inequality can be ignored. In Fig. 4, P is the curve AE, and N_e is the curve BC.

Let H, H_1, H_2 be real Hilbert spaces, and let the state equation of the game be of the form

$$x = \phi x_0 + L_1 u_1 + L_2 u_2 , \tag{13}$$

where the state variable x and the initial state x_0 are in H, and where the control variables u_1 and u_2 of Players 1 and 2 are selected from H_1 and H_2, respectively. $\phi: H \to H$, $L_1: H_1 \to H$, and $L_2: H_2 \to H$ are bounded linear transformations. The cost functionals that the players seek to minimize are of the form

$$J_1(u_1, u_2) = \tfrac{1}{2}(\langle x, Q_1 x\rangle + \langle u_1, R_{11} u_1\rangle + \langle u_2, R_{12} u_2\rangle), \tag{14}$$

$$J_2(u_1, u_2) = \tfrac{1}{2}(\langle x, Q_2 x\rangle + \langle u_1, R_{21} u_1\rangle + \langle u_2, R_{22} u_2\rangle), \tag{15}$$

where Q_1 and Q_2 are bounded linear self-adjoint (BLSA) operators on H, R_{11} and R_{21} are BLSA operators on H_1, and R_{12} and R_{22} are BLSA operators on H_2. The inner products in (14)–(15) are taken over the underlying spaces. Necessary and sufficient conditions for the existence of open-loop and closed-loop Nash controls for this game have been obtained in Refs. 9–10. In the following analysis, necessary and sufficient conditions for the existence of an open-loop Stackelberg control pair $(u_{1s2}, u_{2s2}) \in H_1 \times H_2$ are obtained. Because of difficulties encountered with resulting nonlinear equations, closed-loop Stackelberg controls will not be considered here.

If $u_2 \in H_2$ is fixed, Player 1 can calculate his corresponding optimal strategy by minimizing $J_1(u_1, u_2)$. This minimization, when repeated for all $u_2 \in H_2$, will lead to a description of the rational reaction set D_1. When (13) is substituted into (14), $J_1(u_1, u_2)$ becomes[4]

$$J_1(u_1, u_2) = \tfrac{1}{2}(\langle u_1, (R_{11} + L_1{}^* Q_1 L_1)\, u_1\rangle + 2\langle u_1, L_1{}^* Q_1(\phi x_0 + L_2 u_2)\rangle + J_{10}), \tag{16}$$

where

$$J_{10} = (\langle \phi x_0 + L_2 u_2, Q_1(\phi x_0 + L_2 u_2)\rangle + \langle u_2, R_{12} u_2\rangle).$$

A necessary condition for u_1 to minimize (16) is obtained by setting

$$[dJ_1(u_1 + \alpha h_1, u_2)/d\alpha]_{\alpha=0} = 0,$$

where $h_1 \in H_1$ and α is a real number. This gives

$$(R_{11} + L_1{}^* Q_1 L_1)\, u_1 + L_1{}^* Q_1(\phi x_0 + L_2 u_2) = 0.$$

[4] $L_1{}^*$ denotes the adjoint of the operator L_1. For $i = 1, 2$, if $L_i: H_i \to H$, then $L_i{}^*: H \to H_i$ and is defined by $\langle x, L_i u_i\rangle = \langle L_i{}^* x, u_i\rangle$, $x \in H$, $u_i \in H_i$.

A sufficient condition (Ref. 11) for u_1 to minimize J_1 is that the operator

$$S_1 = R_{11} + L_1{}^*Q_1L_1$$

be strongly positive. That is, it must satisfy

$$\alpha \| u_1 \|^2 \leqslant \langle S_1u_1, u_1\rangle \leqslant \beta \| u_1 \|^2, \qquad 0 < \alpha \leqslant \beta, \qquad \forall u_1 \in H_1, \qquad (17)$$

where α and β are real numbers defined by

$$\alpha = \inf_{\|u_1\|=1} \langle S_1u_1, u_1\rangle \qquad \text{and} \qquad \beta = \sup_{\|u_1\|=1} \langle S_1u_1, u_1\rangle.$$

If (17) holds, S_1^{-1} exists and

$$u_1 = -S_1^{-1}L_1{}^*Q_1(\phi x_0 + L_2u_2). \qquad (18)$$

This relationship defines the mapping T, and the collection of pairs (u_1, u_2) such that (18) is satisfied constitutes the rational reaction set D_1.

Now, for u_1 as in (18), Player 2 (the leader) can find his optimal control by solving an optimization problem for $\tilde{J}_2(u_2) = J_2(u_1, u_2)$ when (18) and (13) are substituted for u_1 and x. Following the same procedure as above, by setting

$$[d\tilde{J}_2(u_2 + \alpha h_2)/d\alpha]_{\alpha=0} = 0,$$

$h_2 \in H_2$ and α a real number, necessary and sufficient conditions for a minimizing u_{2s2} are obtained by

$$S_2u_{2s2} + (L_2{}^*M_1{}^*Q_2M_1\phi + L_2{}^*Q_1L_1S_1^{-1}R_{21}S_1^{-1}L_1{}^*Q_1\phi)\, x_0 = 0, \qquad (19)$$

where

$$S_2 = L_2{}^*M_1{}^*Q_2M_1L_2 + L_2{}^*Q_1L_1S_1^{-1}R_{21}S_1^{-1}L_1{}^*Q_1L_2 + R_{22}, \qquad (20)$$

$$M_1 = I - L_1S_1^{-1}L_1{}^*Q_1, \qquad (21)$$

I being the identity operator on H. If S_2 is strongly positive, S_2^{-1} exists and

$$u_{2s2} = -S_2^{-1}(L_2{}^*M_1{}^*Q_2M_1\phi + L_2{}^*Q_1L_1S_1^{-1}R_{21}S_1^{-1}L_1{}^*Q_1\phi)\, x_0. \qquad (22)$$

u_{1s2} is then obtained by substituting (22) into (18). These results are summarized in the following proposition:

Proposition 4.1. If the operators S_1 and S_2 are strongly positive, then the game defined by (13)–(15) has a unique pair (u_{1s2}, u_{2s2}) of open-loop Stackelberg controls satisfying the relations

$$u_{1s2} = -S_1^{-1}L_1{}^*Q_1(\phi - L_2S_2^{-1}L_2{}^*(M_1{}^*Q_2M_1\phi + Q_1L_1S_1^{-1}R_{21}S_1^{-1}L_1{}^*Q_1\phi))\,x_0, \tag{23}$$

$$u_{2s2} = -S_2^{-1}L_2{}^*(M_1{}^*Q_2M_1\phi + Q_1L_1S_1^{-1}R_{21}S_1^{-1}L_1{}^*Q_1\phi)\,x_0. \tag{24}$$

By properly selecting the various operators in (14)–(15), S_1 and S_2 can be made strongly positive in order to guarantee the existence of a solution. One such possible selection is given in the following proposition.

Proposition 4.2. If R_{11} and R_{22} are strongly positive, and Q_1, Q_2, and R_{21} are positive semidefinite,[5] then an open-loop Stackelberg solution (u_{1s1}, u_{2s2}) exists.

Proof. If R_{11} is strongly positive and $Q_1 \geqslant 0$, then clearly S_1 is strongly positive. Similarly, if R_{22} is strongly positive, $Q_2 \geqslant 0$, and $R_{21} \geqslant 0$, then S_2 is strongly positive. By Proposition 4.1, a Stackelberg solution exists and satisfies (23)–(24).

Note that the existence of a Stackelberg solution with Player 2 as leader does not generally imply the existence of a Stackelberg solution with Player 1 as leader. The form in which the Stackelberg strategies (23)–(24) are obtained is most convenient for solving games defined over finite-dimensional Euclidean spaces. However, when infinite-dimensional spaces are considered, an alternative form that does not require inverting the operators S_1 and S_2 is preferable. One such representation, helpful in the numerical computation of the strategies as functions of time only, will be to express the open-loop controls (23)–(24) as a function of x, rather than x_0, and then solve for x as a function of x_0 separately from the state equation. After some algebraic manipulations, (23)–(24) reduce to

$$u_{1s2} = -R_{11}^{-1}L_1{}^*Q_1x, \tag{25}$$

$$u_{2s2} = -R_{22}^{-1}L_2{}^*(Q_2 - Q_1P)x, \tag{26}$$

where the operator P and the state x satisfy the relations

$$P - L_1R_{11}^{-1}L_1{}^*(Q_2 - Q_1P) + L_1R_{11}^{-1}R_{21}R_{11}^{-1}L_1{}^*Q_1 = 0, \tag{27}$$

$$(I + L_1R_{11}^{-1}L_1{}^*Q_1 + L_2R_{22}^{-1}L_2{}^*(Q_2 - Q_1P))x = \phi x_0. \tag{28}$$

The state vector x is obtained in terms of x_0 from (28) and then substituted

[5] If H is a Hilbert space and R is a BLSA operator on H, then R is positive definite (semidefinite), denoted by $R > 0$ ($\geqslant 0$) if $\langle h, Rh\rangle > 0$ ($\geqslant 0$), $\forall h \neq 0 \in H$.

in (25)–(26). It is important to note that (25)–(26) are only used for generating the open-loop controls and are not intended for closed-loop control implementation. As a special case of the above analysis, linear-quadratic differential games will be considered in the following section.

5. Linear-Quadratic Differential Games

Recently, linear-quadratic differential games have received considerable interest in the differential games literature (Refs. 8–13). These games have significant importance in studying the local behavior of corresponding nonlinear differential games. The dynamics of the game considered here are assumed to obey the linear differential equation

$$\dot{x} = Ax + B_1u_1 + B_2u_2 , \qquad x(t_0) = x_0 , \tag{29}$$

and the performance criteria are of the form

$$J_1(u_1 , u_2) = \tfrac{1}{2}x_f'K_{1f}x_f + \tfrac{1}{2}\int_{t_0}^{t_f} (x'Q_1x + u_1'R_{11}u_1 + u_2'R_{12}u_2)\, dt, \tag{30}$$

$$J_2(u_1 , u_2) = \tfrac{1}{2}x_f'K_{2f}x_f + \tfrac{1}{2}\int_{t_0}^{t_f} (x'Q_2x + u_1'R_{21}u_1 + u_2'R_{22}u_2)\, dt. \tag{31}$$

In these equations, the initial time t_0 and the final time t_f are finite and fixed, the state x is an n-dimensional vector of continuous functions defined on $[t_0 , t_f]$ with $x_f = x(t_f)$, and the controls u_1 and u_2 are square (Lebesgue)-integrable m_1-dimensional and m_2-dimensional vector functions defined on $[t_0 , t_f]$. The various matrices in (29)–(31) are of proper dimensions and with elements continuous functions on $[t_0 , t_f]$. In order to guarantee the existence of an open-loop Stackelberg solution (Propositions 4.1 or 4.2), the matrices in (30)–(31) are assumed to be symmetric and to satisfy the conditions $K_{1f} \geqslant 0$, $K_{2f} \geqslant 0$, $Q_1(t) \geqslant 0$, $Q_2(t) \geqslant 0$, $R_{11}(t) > 0$, $R_{22}(t) > 0$, $R_{21}(t) \geqslant 0$. These are only sufficient conditions and, in games where these conditions are not satisfied (for example, zero-sum or almost zero-sum games), it must be insured that S_1 and S_2 are strongly positive. We note that the matrices $R_{11}(t)$ and $R_{22}(t)$, being positive definite, are in fact strongly positive.[6] The players are seeking an

[6] Since $R_{11}(t) > 0$, $\forall t \in [t_0 , t_f]$, then there exists a $\lambda > 0$ such that, $\forall t \in [t_0 , t_f]$, $u_1'(t)\, R_{11}(t)\, u_1(t) \geqslant \lambda u_1'(t)\, u_1(t)$, $\forall u_1 \in R^{m_1}$. Therefore,

$$\int_{t_0}^{t_f} u_1'(t)\, R_{11}(t)\, u_1(t)\, dt \geqslant \lambda \int_{t_0}^{t_f} u_1'(t)\, u_1(t)\, dt,$$

which implies that $\langle u_1 , R_{11}u_1\rangle \geqslant \lambda \| u_1 \|^2$. Similar results hold for $R_{22}(t)$.

open-loop Stackelberg solution. That is, the leader is seeking a strategy $u_{2s2}(t)$, a function of time only, that he announces before the game starts. The follower will then calculate his strategy $u_{1s2}(t)$ also as a function of time only.

In order to formulate this problem into the Hilbert-space structure of (13)–(15), the solution of (29) is first obtained. Using the variation of parameters formula, we have

$$x(t) = \phi(t, t_0)\, x_0 + \int_{t_0}^{t} \phi(t, \tau)\, B_1(\tau)\, u_1(\tau)\, d\tau + \int_{t_0}^{t} \phi(t, \tau)\, B_2(\tau)\, u_2(\tau)\, d\tau, \quad (32)$$

where $\phi(t, t_0)$ satisfies the relations

$$\dot{\phi}(t, t_0) = A\phi(t, t_0), \qquad \phi(t, t) = I. \quad (33)$$

Let H, H_1, H_2 be the following spaces:

$$H = \mathscr{L}_2^{n}[t_0, t_f] \times R^n, \qquad H_1 = \mathscr{L}_2^{m_1}[t_0, t_f], \qquad H_2 = \mathscr{L}_2^{m_2}[t_0, t_f],$$

where $\mathscr{L}_2^{j}[t_0, t_f]$ is the set of all j-dimensional real-valued square-integrable functions $v(t)$ satisfying the inequality

$$\int_{t_0}^{t_f} v'(t)\, v(t)\, dt < \infty$$

and accompanied with the inner product

$$\langle v_1(t), v_2(t)\rangle = \int_{t_0}^{t_f} v_1'(t)\, v_2(t)\, dt. \quad (34)$$

Let

$$\tilde{x} = \begin{bmatrix} x(t) \\ x_f \end{bmatrix} \in H, \qquad \tilde{x}_0 = \begin{bmatrix} x_0 \\ x_0 \end{bmatrix} \in H, \qquad u_1 \in H_1 \quad \text{and} \quad u_2 \in H_2.$$

Equation (32), when evaluated at t and t_f, can be written in the form

$$\tilde{x} = \phi\tilde{x}_0 + L_1 u_1 + L_2 u_2, \quad (35)$$

where

$$\phi = \begin{bmatrix} \phi(t, t_0) & 0 \\ 0 & \phi(t_f, t_0) \end{bmatrix} \quad (36)$$

$$L_i u_i = \begin{bmatrix} \displaystyle\int_{t_0}^{t} \phi(t, \tau)\, B_i(\tau)\, u_i(\tau)\, d\tau \\ \displaystyle\int_{t_0}^{t_f} \phi(t_f, \tau)\, B_i(\tau)\, u_i(\tau)\, d\tau \end{bmatrix}, \qquad i = 1, 2. \quad (37)$$

The performance criteria (30)–(31) reduce to[7]

$$J_1(u_1, u_2) = \tfrac{1}{2}(\langle \tilde{x}, \tilde{Q}_1\tilde{x}\rangle + \langle u_1, R_{11}u_1\rangle + \langle u_2, R_{12}u_2\rangle), \tag{38}$$

$$J_2(u_1, u_2) = \tfrac{1}{2}(\langle \tilde{x}, \tilde{Q}_2\tilde{x}\rangle + \langle u_1, R_{21}u_1\rangle + \langle u_2, R_{22}u_2\rangle), \tag{39}$$

where

$$\tilde{Q}_i\tilde{x} = \begin{bmatrix} Q_i(t)\, x(t) \\ K_{if}x_f \end{bmatrix}, \qquad i = 1, 2. \tag{40}$$

The next step is to determine the adjoints of L_1 and L_2. Consider the inner product of L_iu_i with an arbitrary vector

$$\tilde{w} = \begin{bmatrix} w \\ w_f \end{bmatrix} \in H,$$

that is,

$$\langle w, L_iu_i\rangle = \int_{t_0}^{t_f} w'(t) \int_{t_0}^{t} \phi(t, \tau)\, B_i(\tau)\, u_i(\tau)\, d\tau\, dt + w_f' \int_{t_0}^{t_f} \phi(t_f, \tau)\, B_i(\tau)\, u_i(\tau)\, d\tau.$$

By interchanging the order of integration, one easily obtains the relation

$$\langle w, L_iu_i\rangle = \int_{t_0}^{t_f} u_i'(\tau)\, B_i'(\tau) \int_{\tau}^{t_f} \phi'(t, \tau)\, w(t)\, dt\, d\tau + \int_{t_0}^{t_f} u_i'(\tau)\, B_i'(\tau)\, \phi'(t_f, \tau)\, w_f\, d\tau,$$

from which we conclude that $L_i^*\colon H \to H_i$ is defined by

$$L_i^*w = B_i'(t) \int_{t}^{t_f} \phi'(\sigma, t)\, w(\sigma)\, d\sigma + B_i'(t)\, \phi'(t_f, t)\, w_f, \qquad i = 1, 2. \tag{41}$$

Using these results, and omitting a few algebraic manipulations, (25)–(26) reduce to

$$u_{1s2} = -R_{11}^{-1}B_1'K_1x, \tag{42}$$

$$u_{2s2} = -R_{22}^{-1}B_2'K_2x, \tag{43}$$

[7] The inner product on H is defined by

$$\langle \tilde{x}_1, \tilde{x}_2\rangle = \int_{t_0}^{t_f} x_1'x_2\, dt + x_{1f}'x_{2f}.$$

where x, $K_1(t)$, $K_2(t)$ satisfy the relations

$$\dot{x} = (A - B_1R_{11}^{-1}B_1'K_1 - B_2R_{22}^{-1}B_2'K_2)x, \qquad x(t_0) = x_0, \tag{44}$$

$$K_1(t)\,x(t) = \int_t^{t_f} \phi'(\sigma, t)\,Q_1(\sigma)\,x(\sigma)\,d\sigma + \phi'(t_f, t)\,K_{1f}x_f, \tag{45}$$

$$K_2(t)\,x(t) = \int_t^{t_f} \phi'(\sigma, t)[Q_2(\sigma) - Q_1(\sigma)\,P(\sigma)]\,x(\sigma)\,d\sigma + \phi'(t_f, t)[K_{2f} - K_{1f}P(t_f)]\,x_f, \tag{46}$$

where $P(t)$ is obtained from (27) and satisfies the relation

$$P(t)\,x(t) = \int_{t_0}^{t} \phi(t, \tau)\,B_1(\tau)\,R_{11}^{-1}(\tau)[B_1'(\tau)\,K_2(\tau) - R_{21}(\tau)\,R_{11}^{-1}(\tau)\,B_1'(\tau)\,K_1(\tau)]\,x(\tau)\,d\tau. \tag{47}$$

Differentiating (45)–(47) with respect to t, we see that they reduce to

$$\dot{K}_1 = -A'K_1 - K_1A - Q_1 + K_1B_1R_{11}^{-1}B_1'K_1 + K_1B_2R_{22}^{-1}B_2'K_2, \qquad K_1(t_f) = K_{1f}, \tag{48}$$

$$\dot{K}_2 = -A'K_2 - K_2A - Q_2 + Q_1P + K_2B_1R_{11}^{-1}B_1'K_1 + K_2B_2R_{22}^{-1}B_2'K_2, \qquad K_2(t_f) = K_{2f} - K_{1f}P(t_f), \tag{49}$$

$$\dot{P} = AP - PA + PB_1R_{11}^{-1}B_1'K_1 + PB_2R_{22}^{-1}B_2'K_2 - B_1R_{11}^{-1}R_{21}R_{11}^{-1}B_1'K_1 + B_1R_{11}^{-1}B_1'K_2, \qquad P(t_0) = 0, \tag{50}$$

and the open-loop Stackelberg controls are

$$u_{1s2} = -R_{11}^{-1}B_1'K_1\xi(t, t_0)\,x_0, \tag{51}$$

$$u_{2s2} = -R_{22}^{-1}B_2'K_2\xi(t, t_0)\,x_0, \tag{52}$$

where

$$\dot{\xi}(t, t_0) = (A - B_1R_{11}^{-1}B_1'K_1 - B_2R_{22}^{-1}B_2'K_2)\,\xi(t, t_0), \qquad \xi(t, t) = I. \tag{53}$$

Equations (48)–(49) are identical to the Riccati equations obtained in the corresponding open-loop Nash solution (Ref. 8), except for the terms containing P in (49). These terms account for the fact that the leader is now minimizing his cost on the rational reaction set [Eq. (18)] of the follower. Equation (50), however, is not of the Riccati type, and its solution must be done forward in time in contrast to (48)–(49), whose solution is obtained backward in time.

This two-point boundary-value problem is generally not easy to solve. It is possible, however, to obtain its solution from the solution of a single-point boundary-value problem, as follows. If there exists a $2n \times 2n$ matrix $F(t)$ satisfying

$$\dot{F} = -\tilde{A}'F - F\tilde{A} - \tilde{Q} + F\tilde{B}F, \qquad F(t_f) = F_f, \tag{54}$$

where

$$\tilde{A} = \begin{bmatrix} A & 0 \\ 0 & A \end{bmatrix}, \quad \tilde{Q} = \begin{bmatrix} Q_1 & 0 \\ Q_2 & -Q_1 \end{bmatrix}, \quad \tilde{B} = \begin{bmatrix} B_1R_{11}^{-1}B_1' & B_2R_{22}^{-1}B_2' \\ B_1R_{11}^{-1}R_{21}R_{11}^{-1}B_1' & -B_1R_{11}^{-1}B_1' \end{bmatrix},$$

$$F = \begin{bmatrix} F_{11} & F_{12} \\ F_{21} & F_{22} \end{bmatrix}, \quad F_f = \begin{bmatrix} K_{1f} & 0 \\ K_{2f} & -K_{1f} \end{bmatrix},$$

and if (50) has a solution when K_1 and K_2 are of the form

$$\begin{bmatrix} K_1 \\ K_2 \end{bmatrix} = \begin{bmatrix} F_{11} & F_{12} \\ F_{21} & -F_{11} \end{bmatrix} \begin{bmatrix} I \\ P \end{bmatrix}, \tag{55}$$

then the open-loop Stackelberg strategies (51)–(52) can be written as

$$u_{1s2} = -R_{11}^{-1}B_1'(F_{11} + F_{12}P)\,\xi(t, t_0)\,x_0, \tag{56}$$

$$u_{2s2} = -R_{22}^{-1}B_2'(F_{21} - F_{11}P)\,\xi(t, t_0)\,x_0. \tag{57}$$

In (56)–(57), $\xi(t, t_0)$ is the solution of (53) with K_1 and K_2 obtained from (55). It is shown in the appendix that the construction of K_1, K_2, P, as obtained above, satisfies (48)–(50).

It will now be shown that, in the special case of zero-sum games and identical goal games, the Stackelberg solution reduces to the familiar saddle-point and cooperative solutions.

(i) *Zero-Sum Games with Saddle Point.* In zero-sum games, $J_1 = -J_2$. That is, $R_{11} = -R_{21} = R_1$, $R_{22} = -R_{12} = R_2$, $Q_1 = -Q_2 = Q$, $K_{1f} = -K_{2f} = K_f$. The sufficient conditions of Proposition 4.2 are not satisfied. However, assuming that these matrices are selected in such a way that S_1 and S_2 are strongly positive, then a Stackelberg solution exists; and, upon substituting in (48)–(50), we conclude that $P(t) \equiv 0$, $\forall t \in [t_0, t_f]$ and $K_1(t) = -K_2(t) = K(t)$ and satisfying

$$\dot{K} = -A'K - KA - Q + K(B_1R_1^{-1}B_1' - B_2R_2^{-1}B_2')K, \qquad K(t_f) = K_f,$$

in agreement with the saddle-point solution (Ref. 12).

(ii) *Identical Goal Games.* If the two players are cooperating in minimizing the same performance function $J_1 = J_2$, the game is called an identical goal game. This problem can be formulated as an optimal control problem, and its solution obtained in terms of the Riccati equation of the regulator theory. When $R_{11} = R_{21} = R_1$, $R_{12} = R_{22} = R_2$, $Q_1 = Q_2 = Q$, $K_{1f} = K_{2f} = K_f$ are substituted in (48)–(50), it is easily concluded that $P(t) \equiv 0$, $\forall t \in [t_0, t_f]$, and $K_1(t) = K_2(t) = K(t)$ and satisfying

$$\dot{K} = -A'K - KA - Q + K(B_1R_1^{-1}B_1' + B_2R_2^{-1}B_2')K, \qquad K(t_f) = K_f,$$

in agreement with the Riccati equation of regulator theory.

Except for those two special cases, the Stackelberg solution is generally different from the Nash solution. In what follows, a simple pursuit–evasion problem will be considered.

Example 5.1. Consider the nonzero-sum velocity-controlled pursuit–evasion game studied in Refs. 2 and 13. The dynamics of the game are described by the equations

$$\dot{x} = u_1 - u_2, \qquad x(t_0) = x_0,$$

where u_1 and u_2 are the velocities of the pursuer and evader, respectively, and x is their relative position. The performance criteria are

$$J_1 = \tfrac{1}{2}x_f^2 + (1/2c_p)\int_0^1 u_1^2\, dt, \qquad J_2 = -\tfrac{1}{2}x_f^2 + (1/2c_e)\int_0^1 u_2^2\, dt,$$

where

$$c_p > 0, \qquad c_e > 0, \qquad c_pc_e = 1, \qquad c_p/c_e = \omega^2.$$

Assume that Player 2 decides to evade before Player 1 decides to pursue. Naturally, in this case, his best choice will be to announce a Stackelberg strategy u_{2s2} with himself as leader.

Applying (48)–(53), we obtain the open-loop Stackelberg solution with the evader as leader as

$$u_{1s2} = [-c_p/(c_p - \sigma c_e + 1)]\, x_0, \qquad u_{2s2} = [-\sigma c_e/(c_p - \sigma c_e + 1)]\, x_0,$$

where

$$\sigma = 1/(1 + c_p).$$

Sufficient conditions for the existence of a solution are obtained from Propositions 4.1. With the conditions that $c_p > 0$, $c_e > 0$, $c_pc_e = 1$, the operators S_1 and S_2 will be strongly positive if $\sigma^2 c_e < 1$. These conditions imply that $c_e < 2.116$. That is, if $c_e < 2.116$, an open-loop

Stackelberg solution with the evader as leader exists. On the other hand (Ref. 13), the open-loop Nash solution exists if $c_e < 1$, it does not exist if $c_e > 1$, and nothing can be said about its existence if $c_e = 1$. Therefore, in this example, the existence of a Stackelberg strategy with Player 2 as leader is guaranteed over a wider range of parameters.

The performance functions as calculated when the Stackelberg strategy is used are

$$J_1(u_{1s2}, u_{2s2}) = J_{1s2} = \tfrac{1}{2}[(1 + c_p)/(1 - \sigma c_e + c_p)^2]\, x_0^2,$$
$$J_2(u_{1s2}, u_{2s2}) = J_{2s2} = \tfrac{1}{2}[(\sigma^2 c_e - 1)/(1 - \sigma c_e + c_p)^2]\, x_0^2.$$

For the sake of comparison, the open-loop Nash solution for this problem (Ref. 13) is shown below as follows:

$$u_{1N} = [-c_p/(1 + c_p - c_e)]\, x_0,$$
$$u_{2N} = [-c_e/(1 + c_p - c_e)]\, x_0,$$
$$J_1(u_{1N}, u_{2N}) = J_{1N} = \tfrac{1}{2}[(1 + c_p)/(1 - c_e + c_p)^2]\, x_0^2,$$
$$J_2(u_{1N}, u_{2N}) = J_{2N} = \tfrac{1}{2}[(c_e - 1)/(1 - c_e + c_p)^2]\, x_0^2.$$

For the range where both the Stackelberg and the Nash solutions exist (i.e., $c_e < 1$), a comparison of the above quantities will yield $J_{1s2} < J_{1N}$ and $J_{2s2} < J_{2N}$. That is, not only the leader will benefit by using a Stackelberg solution but also the follower will benefit as well. Thus, in this special case, the Stackelberg solution can be looked at as an enforceable negotiated solution that is preferred by both players over the Nash solution.

6. Conclusions

A class of nonzero-sum games in which the strategies are announced sequentially has been investigated, and it was shown that, if the players' sole objective is to minimize their cost functions, the Stackelberg strategy is the most natural way of defining optimality. Games where one player does not know the other's cost function while the other player knows both cost functions and games where one player is faster than the other in computing his strategy are best modeled and solved within this solution concept. In this strategy, the roles of the players, whether leader or follower, must be properly defined and, when compared to the Nash solution, it was shown that it is advantageous to the leader. Conditions for the existence of the Stackelberg strategies have been obtained. It was shown that, in general, one cannot conclude the existence of a Stackelberg solution from the existence of a Nash solution nor *vice versa*. Several examples were considered in order to illustrate the properties of this solution concept. In dynamic games, an abstract formulation in Hilbert spaces has been considered, and necessary and sufficient conditions for the existence of an open-loop Stackelberg solution were obtained. Linear-quadratic differential games were treated as a special case, and the solution was expressed in terms of Riccati-like differential equations. A simple pursuit–evasion problem was solved, and the results were compared to the Nash solution.

7. Appendix

Write Eqs. (48)–(50) in the following form, using the notation in Ref. 14:

$$\dot{K} = -\tilde{A}'K - K\tilde{A} + KNK + K\bar{N}\bar{K} - Q + Q_0\tilde{P}, \qquad K(t_f) = K_f, \tag{58}$$

$$\dot{\tilde{P}} = \tilde{A}'\tilde{P} - \tilde{P}\tilde{A} + \tilde{P}NK + \tilde{P}\bar{N}\bar{K} + RK + \bar{R}\bar{K}, \qquad P(t_0) = 0, \tag{59}$$

where all matrices in (58)–(59) are $2n \times 2n$ and where

$$K = \begin{bmatrix} K_1 & 0 \\ 0 & K_2 \end{bmatrix}, \quad \tilde{P} = \begin{bmatrix} P & 0 \\ 0 & P \end{bmatrix}, \quad N = \begin{bmatrix} B_1R_{11}^{-1}B_1' & 0 \\ 0 & B_2R_{22}^{-1}B_2' \end{bmatrix},$$

$$Q = \begin{bmatrix} Q_1 & 0 \\ 0 & Q_2 \end{bmatrix}, \quad Q_0 = \begin{bmatrix} 0 & 0 \\ 0 & Q_1 \end{bmatrix},$$

$$R = \begin{bmatrix} -B_1R_{11}^{-1}R_{21}R_{11}^{-1}B_1' & 0 \\ 0 & B_1R_{11}^{-1}B_1' \end{bmatrix}, \quad K_f = \begin{bmatrix} K_{1f} & 0 \\ 0 & K_{2f} \end{bmatrix},$$

and the operation $\bar{K}$ is as follows:

$$\bar{K} = \begin{bmatrix} K_2 & 0 \\ 0 & K_1 \end{bmatrix}.$$

Let K_1 and K_2 be related to P by (55). In the above notation, this is written as

$$K = F_1 + F_2\tilde{P},$$

where

$$F_1 = \begin{bmatrix} F_{11} & 0 \\ 0 & F_{21} \end{bmatrix}, \quad F_2 = \begin{bmatrix} F_{12} & 0 \\ 0 & F_{22} \end{bmatrix}. \tag{60}$$

Upon differentiating (60) and substituting in it (58)–(59), and after several steps involving algebraic manipulation, (54) is obtained. Furthermore, from the symmetry of (54), it is clear that $F_{22} = -F_{11}$. If (54) has a solution, then K_1 and K_2 can be written as functions of P as in (55) or (60). P and ξ are then obtained by solving (50) and (53) forward in time.

References

1. NASH, J. F., *Non-Cooperative Games*, Annals of Mathematics, Vol. 54, No. 2, 1951.
2. STARR, A. W., and HO, Y. C., *Nonzero-Sum Differential Games*, Journal of Optimization Theory and Applications, Vol. 3, No. 3, 1969.
3. STARR, A. W., and HO, Y. C., *Further Properties of Nonzero-Sum Differential Games*, Journal of Optimization Theory and Applications, Vol. 3, No. 4, 1969.
4. VON STACKELBERG, H., *The Theory of the Market Economy*, Oxford University, Press, Oxford, England, 1952.
5. COHEN, K. J., and CYERT, R. M., *Theory of the Firm: Resource Allocation in a Market Economy*, Prentice Hall, Englewood Cliffs, New Jersey, 1965.
6. INTRILLIGATOR, M. D., *Mathematical Optimization and Economic Theory*, Prentice-Hall, Englewood Cliffs, New Jersey, 1971.
7. CHEN, C. I., and CRUZ, J. B., JR., *Stackelberg Solution for Two-Person Games with Biased Information Patterns*, IEEE Transactions on Automatic Control, Vol. AC-17, No. 5, 1972.
8. STARR, A. W., *Nonzero-Sum Differential Games: Concepts and Models*, Harvard University, Division of Engineering and Applied Physics, TR No. 590, 1969.
9. LUKES, D. L., and RUSSEL, D. L., *A Global Theory for Linear Quadratic Differential Games*, Journal of Mathematical Analysis and Applications, Vol. 33, No. 1, 1971.
10. LUKES, D. L., *Equilibrium Feedback Control in Linear Games with Quadratic Costs*, SIAM Journal on Control, Vol. 9, No. 2, May 1971.
11. VULIKH, B. Z., *Introduction to Functional Analysis*, Addison Wesley Publishing Company, Reading, Massachusetts, 1963.
12. HO, Y. C., BRYSON, A. E., JR., and BARON, S., *Differential Games and Optimal Pursuit–Evasion Strategies*, IEEE Transactions on Automatic Control, Vol. AC-10, No. 4, 1965.
13. FOLEY, M. H., and SCHMITENDORF, W. E., *On a Class of Nonzero-Sum Linear Quadratic Differential Games*, Journal of Optimization Theory and Applications, Vol. 7, No. 5, 1971.

14. Krikelis, N. J., and Rekasius, Z. V., *On the Solution of the Optimal Linear Control Problems Under Conflict of Interest*, IEEE Transactions on Automatic Control, Vol. AC-16, No. 2, 1971.

Additional Bibliography

15. Luce, R., and Raiffa, H., *Games and Decision*, John Wiley and Sons, New York, 1957.

X

Additional Aspects of the Stackelberg Strategy in Nonzero-Sum Games[1]

M. Simaan and J. B. Cruz, Jr.

Abstract. The Stackelberg strategy in nonzero-sum games is a reasonable solution concept for games where, either due to lack of information on the part of one player about the performance function of the other, or due to different speeds in computing the strategies, or due to differences in size or strength, one player dominates the entire game by imposing a solution which is favorable to himself. This paper discusses some properties of this solution concept when the players use controls that are functions of the state variables of the game in addition to time. The difficulties in determining such controls are also pointed out. A simple two-stage finite state discrete game is used to illustrate these properties.

1. Introduction

The Stackelberg solution of a two-player nonzero-sum game (Refs. 1–3) assumes that the roles of the players are different. There is a leader and there is a follower. The follower conforms to the policies of the leader by allowing him to determine his strategy first. The leader foresees this and, in effect, controls the entire system.

Let U_1 and U_2 be the sets of admissible controls for Players 1 and 2,

[1] This work was supported in part by the U.S. Air Force under Grant No. AFOSR-68-1579D, in part by NSF under Grant No. GK-36276, and in part by the Joint Services Electronics Program under Contract No. DAAB-07-72-C-0259 with the Coordinated Science Laboratory, University of Illinois, Urbana, Illinois.

respectively, and let $J_1(u_1, u_2)$ and $J_2(u_1, u_2)$ be their corresponding cost functions. If there exists a mapping $T: U_2 \to U_1$ such that[2]

$$J_1(Tu_2, u_2) \leqslant J_1(u_1, u_2) \qquad \forall u_1 \in U_1 \tag{1}$$

for every $u_2 \in U_2$, then the set

$$D_1 = \{(u_1, u_2) \in U_1 \times U_2 \colon u_1 = Tu_2 \ \forall u_2 \in U_2\} \tag{2}$$

is called the rational reaction set for Player 1 when Player 2 is the leader. Furthermore, if there is a $(u_{1s2}, u_{2s2}) \in D_1$ such that

$$J_2(u_{1s2}, u_{2s2}) \leqslant J_2(u_1, u_2) \qquad \forall (u_1, u_2) \in D_1, \tag{3}$$

then (u_{1s2}, u_{2s2}) is called a Stackelberg strategy pair when Player 2 is the leader. When Player 1 is the leader, the rational reaction set for Player 2 and the Stackelberg solution are denoted by D_2 and (u_{1s1}, u_{2s1}), respectively. It is clear that, if D_1 and D_2 intersect, then their common element (u_{1N}, u_{2N}) is the Nash solution of the game. In this case, it follows (Ref. 2) that

$$J_2(u_{1s2}, u_{2s2}) \leqslant J_2(u_{1N}, u_{2N})$$

when Player 2 is the leader and that

$$J_1(u_{1s1}, u_{2s1}) \leqslant J_1(u_{1N}, u_{2N})$$

when Player 1 is the leader.

The properties of the open-loop Stackelberg solution for a class of linear quadratic games were discussed in Ref. 2. In this paper, additional properties of this solution are obtained. It is shown that, unlike the case of closed-loop Nash controls (Refs. 4–5), dynamic programming cannot be used to calculate the closed-loop Stackelberg controls. To differentiate closed-loop Stackelberg controls from controls obtained via dynamic programming, the latter are called Stackelberg feedback strategies. Both closed-loop Stackelberg controls and Stackelberg feedback strategies have attractive properties that are discussed via a simple two-stage finite state game. The difficulties in deriving the necessary conditions for the existence of the closed-loop Stackelberg controls and feedback Stackelberg strategies are pointed out and, finally, necessary conditions for the existence of feedback Stackelberg strategies for a class of discrete multistage games are derived.

[2] It is clear, from the definition of T, that only problems where, for every $u_2 \in U_2$, there corresponds only one element $Tu_2 \in U_1$ such that (1) is satisfied are considered in this paper.

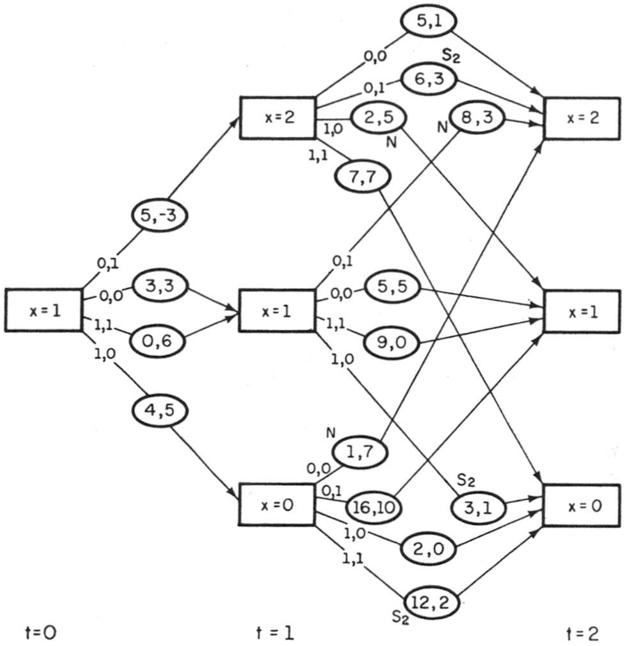

Fig. 1. A two-stage discrete game.

2. Illustrative Example

There are several forms according to which controls in dynamic games can be selected. There are controls that are functions of the time only and known as open-loop controls, and there are controls that are functions of the time and the state of the game as well, and these are known as closed-loop controls. In order to illustrate how the selection of such controls is done, let us consider the following simple two-stage finite state game.[3]

Example 2.1. Consider the game shown in Fig. 1. At every stage and from every state, each player has a choice between two possible controls, 0 and 1. After decisions have been made, the transition and costs borne by the players are shown in Fig. 1, where the first entries in the encircled quantities are the costs borne by Player 1 and the second entries are those borne by Player 2. The subscripts or superscripts

[3] A similar example was considered in Ref. 5 for the Nash solution.

Table 1

	o_{i1}	o_{i2}	o_{i3}	o_{i4}
$u_i(0)$	0	0	1	1
$u_i(1)$	0	1	0	1

o and c will be used to denote open-loop and closed-loop quantities. Consider first the open-loop controls for this game.

(i) *Open-loop controls.* Assume that, before the start of the game, the players have to commit themselves to controls $u_1(t)$ and $u_2(t)$ that are functions of time only; then, each player has four such possible functions to choose from. In other words, the sets of admissible open-loop controls are $U_1^o = \{o_{1j}\,;\, j = 1,..., 4\}$ and $U_2^o = \{o_{2j}\,;\, j = 1,..., 4\}$, where o_{ij} are obtained from Table 1 for $i = 1, 2$.

The game with these admissible controls can be represented as a bimatrix game, as shown in Fig. 2. Suppose that Player 2 is the leader; then, for every $o_{2j} \in U_2^o$, Player 1 will choose a control in U_1^o that minimizes J_1. Thus, the rational reaction set D_{1o} can be easily determined as follows:

$$D_{1o} = \{(o_{13}\,, o_{21}), (o_{11}\,, o_{22}), (o_{14}\,, o_{23}), (o_{13}\,, o_{24})\},$$

and the Stackelberg strategy is $(o_{11}\,, o_{22})$. The corresponding trajectory is $x(1) = 1$ and $x(2) = 2$, and the costs are $J^o_{1s2} = 11$ and $J^o_{2s2} = 6$. Similarly, the rational reaction set D_{2o} is

$$D_{2o} = \{(o_{11}\,, o_{23}), (o_{12}\,, o_{23}), (o_{13}\,, o_{24}), (o_{14}\,, o_{21})\},$$

Open-Loop Stackelberg with 2 as Leader

		Player 2			
		o_{21}	o_{22}	o_{23}	o_{24}
Player 1	o_{11}	8,8	(11,6)	10,-2	11,0
	o_{12}	6,4	12,3	7,2	12,4
	o_{13}	5,12	20,15	5,11	(8,9)
	o_{14}	(6,5)	16,7	3,7	9,6

Open-Loop Stackelberg with 1 as Leader — Open-Loop Nash

Fig. 2. Open-loop bimatrix game for Example 2.1.

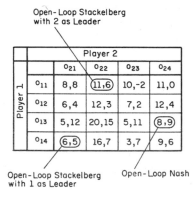

Table 2

	c_{i1}	c_{i2}	c_{i3}	c_{i4}	c_{i5}	c_{i6}	c_{i7}	c_{i8}	c_{i9}	c_{i10}	c_{i11}	c_{i12}	c_{i13}	c_{i14}	c_{i15}	c_{i16}
$u_i(0,1)$	0	0	0	0	0	0	0	0	1	1	1	1	1	1	1	1
$u_i(1,2)$	0	0	0	0	1	1	1	1	0	0	0	0	1	1	1	1
$u_i(1,1)$	0	0	1	1	0	0	1	1	0	0	1	1	0	0	1	1
$u_i(1,0)$	0	1	0	1	0	1	0	1	0	1	0	1	0	1	0	1

and the Stackelberg strategy when Player 1 is the leader is (o_{14}, o_{21}), leading to the trajectory $x(1) = 0$ and $x(2) = 0$ and the costs $J^o_{1s1} = 6$ and $J^o_{2s1} = 5$. Furthermore, the Nash solution, which is the common element in D_{1o} and D_{2o}, is (o_{13}, o_{24}); its trajectory is $x(1) = 1$ and $x(2) = 2$, and the costs are $J^o_{1N} = 8$ and $J^o_{2N} = 9$. Consider, next, the closed-loop controls.

(ii) *Closed-loop controls.* Assume that the players are restricted to announce, before the start of the game, control laws $u_1(t, x)$ and $u_2(t, x)$ that are functions of the time and the state of the game. The actual values of their controls can then be determined only while the game is played, once the actual value of the state at each time is known. Such controls are called closed-loop controls. In this game, there are 16 such choices for each player, and the sets of closed-loop admissible controls are $U_1^c = \{c_{1j}\,;\, j = 1,..., 16\}$ and $U_2^c = \{c_{2j}\,;\, j = 1,..., 16\}$, where c_{ij} are as in Table 2 for $i = 1, 2$.

For every c_{2j} that Player 2 may choose, Player 1 will have to solve an optimization problem, and his corresponding optimal closed-loop control can be easily obtained (e.g., via dynamic programming). For example, if Player 2 chooses c_{26}, then the optimization problem for Player 1 is shown in Fig. 3, and it is easily seen that c_{14} is his optimal

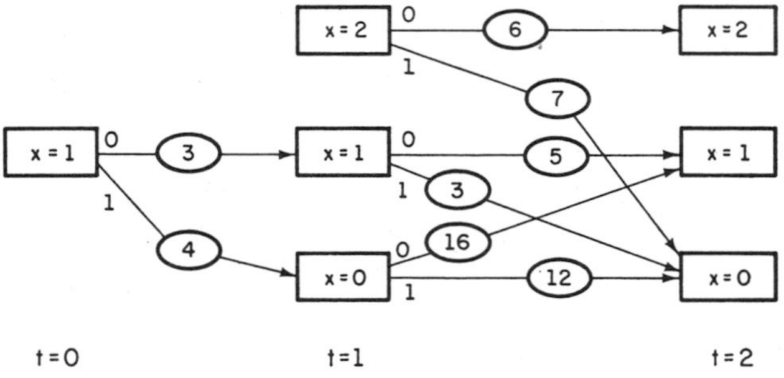

Fig. 3. Optimization problem for Player 1 when $u_2(t, x) = c_{26}$.

closed-loop control. If this procedure is repeated for all $u_2(x, t) \in U_2{}^c$, the rational reaction set D_{1c} for Player 1 is obtained as follows:

$$D_{1c} = \{(c_{115}, c_{21}), (c_{18}, c_{22}), (c_{113}, c_{23}), (c_{16}, c_{24}), (c_{111}, c_{25}), (c_{14}, c_{26}), (c_{19}, c_{27}), (c_{12}, c_{28}), (c_{115}, c_{29}), (c_{116}, c_{210}), (c_{15}, c_{211}), (c_{16}, c_{212}), (c_{111}, c_{213}), (c_{112}, c_{214}), (c_{11}, c_{215}), (c_{12}, c_{216})\}.$$

Following the same procedure, the set D_{2c} is obtained as follows:

$$D_{2c} = \{(c_{11}, c_{211}), (c_{12}, c_{211}), (c_{13}, c_{211}), (c_{14}, c_{211}), (c_{15}, c_{211}), (c_{16}, c_{211}), (c_{17}, c_{211}), (c_{18}, c_{211}), (c_{19}, c_{211}), (c_{110}, c_{23}), (c_{111}, c_{211}), (c_{112}, c_{23}), (c_{113}, c_{211}), (c_{114}, c_{23}), (c_{115}, c_{211}), (c_{116}, c_{23})\}.$$

There are two closed-loop Stackelberg controls with Player 2 as leader, (c_{15}, c_{211}) and (c_{16}, c_{212}), both leading to the same costs $J_{1s2}^c = 7$ and $J_{2s2}^c = 2$ and to the same trajectory $x(1) = 2$ and $x(2) = 1$. Taking $D_{1c} \cap D_{2c}$, there is only one closed-loop Nash control pair, (c_{15}, c_{211}), giving the same costs $J_{1N}^c = 7$ and $J_{2N}^c = 2$ and the same[4] trajectory $x(1) = 2$ and $x(2) = 1$.

Thus, as in the Nash solution, open-loop and closed-loop Stackelberg solutions are generally different. However, since the Stackelberg and Nash controls are selected from the same spaces, whether $U_1{}^o \times U_2{}^o$ or $U_1{}^c \times U_2{}^c$, it is always true that the leader is better off in the Stackelberg solution than in the Nash solution.

(iii) *Feedback strategies.* The procedure described above for calculating the closed-loop Stackelberg controls is quite lengthy, especially in games with a large number of states, stages and controls, as is usually the case. One approach for simplifying this computation is to use the following dynamic programming technique, which is similar to the one described in Ref. 5 for the Nash solution. At time $t = 1$, there are three possible states. From every state, the transition to the next stage $t = 2$ is a 2×2 matrix game whose Nash and Stackelberg controls (with 2 as leader) are calculated and marked by N and S_2 in Fig. 1. Eliminating all other trajectories except for those marked by S_2, we see that, at stage $t = 0$, there are two choices of controls for each player, resulting in the 2×2 matrix game shown

[4] The coincidence of one closed-loop Stackelberg solution with Player 2 as leader and the closed-loop Nash solution in this example is only accidental. In fact, it can be easily checked that, when Player 1 is the leader, the closed-loop Stackelberg solution is different from the closed-loop Nash solution.

		Player 2	
		0	1
Player 1	0	(6,4)	11,0
	1	16,7	3,7

(a) Stackelberg

		Player 2	
		0	1
Player 1	0	11,6	(7,2)
	1	5,12	8,9

(b) Nash

Fig. 4. Stackelberg and Nash feedback strategies via dynamic programming.

in Fig. 4a. The Stackelberg strategy computed from this bimatrix game is encircled on Fig. 4a and is identified as (c_{14}, c_{26}). Similarly (Ref. 5), eliminating all trajectories except for those marked by N, we see that the Nash controls calculated from the resulting bimatrix game at $t = 0$ and shown in Fig. 4b are identified as (c_{15}, c_{211}). Therefore, we readily note that this dynamic programming technique, when used for calculating the Stackelberg strategies, did not lead to the closed-loop Stackelberg controls obtained in (ii), while when used for the Nash controls it did indeed lead to the closed-loop Nash controls. It is therefore necessary to differentiate between these two types of controls. For this reason, we shall call the controls obtained via dynamic programming approach feedback strategies; and the subscripts or superscript f will be used to denote related quantities. Thus, the Stackelberg feedback strategies for this example are (c_{14}, c_{26}) with trajectory $x(1) = 1$, $x(2) = 0$ and costs $J^f_{1s2} = 6$ and $J^f_{2s2} = 4$, and the Nash feedback strategies are (c_{15}, c_{211}) with trajectory $x(1) = 2$ and $x(2) = 1$ and costs $J^f_{1N} = 7$ and $J^f_{2N} = 2$ which are identical to the closed-loop Nash controls. Another conclusion which can be drawn from the above computations is that the property that the leader is better off in the Stackelberg solution than in the Nash solution is no longer true when feedback strategies are considered (note that, in this example, $J^f_{2s2} > J^f_{2N}$). Naturally, this is due to the fact that the trajectories eliminated at $t = 1$ are different in both cases.

Although this example is quite simple, it illustrates the difficulties encountered in determining the closed-loop Stackelberg controls in nonzero-sum dynamic games. The principle of optimality (Ref. 6), which in effect is the tool that simplifies the computational procedures, does not generalize to the Stackelberg solution as it does to the Nash solution. In fact, it can be easily checked that neither the open-loop nor the closed-loop Stackelberg solutions in the previous example have the Stackelberg property for the game resulting from the transition from $t = 1$ to $t = 2$ (i.e., starting at $t = 1$). At $x(1) = 2$, which is on the trajectory of the closed-loop Stackelberg solution for the game starting at $t = 0$, the remaining Stackelberg solution is $(1, 0)$, leading to $J_1 = 2$ and $J_2 = 5$, while the closed-loop Stackelberg control for the game starting at $t = 1$ gives the controls (0, 1), leading to $J_1 = 6$ and $J_2 = 3$. These results will now be generalized, and the equivalence between the closed-loop Nash controls and the Nash feedback strategies will be proven.

3. Stackelberg Solutions in Differential Games

The differences among the open-loop, closed-loop, and feedback controls are due to the fact that they are selected from different sets of admissible controls. A definition of these sets in differential games is therefore necessary. Let $[t_0, t_f]$ be the interval of time over which the game is defined. Let $\tau \in [t_0, t_f)$ and let $U^o_{1\tau}$ and $U^o_{2\tau}$ denote the sets of all admissible open-loop (for example, measurable functions) controls on $[\tau, t_f]$ for Players 1 and 2, respectively. Now, define the set

$$X_{\xi,\tau} = \{x(t) \colon \dot{x} = f(x, t, u_1, u_2),\ t \in [\tau, t_f],\ x(\tau) = \xi,\ u_1 \in U^o_{1\tau},\ u_2 \in U^o_{2\tau}\},$$

and consider the following performance indices defined on $[\tau, t_f]$:

$$J_i(\xi, \tau, u_1, u_2) = K_i(x(t_f)) + \int_\tau^{t_f} L_i(x, t, u_1, u_2)\, dt, \qquad i = 1, 2, \tag{4}$$

with $u_1 \in U^o_{1\tau}$, $u_2 \in U^o_{2\tau}$ and $x(t)$ satisfying

$$\dot{x} = f(x, t, u_1, u_2) \qquad x(\tau) = \xi, \qquad t \in [\tau, t_f]. \tag{5}$$

Clearly, the open-loop admissible control sets for the game starting at t_0 are $U^o_{1t_0}$ and $U^o_{2t_0}$, and the necessary conditions for the existence of Stackelberg strategies in these sets are easily obtained as follows. The rational reaction set D_{1o}, when Player 2 is the leader, is composed of $(u_1(t), u_2(t)) \in U^o_{1t_0} \times U^o_{2t_0}$ satisfying the following necessary conditions:

$$\dot{x} = f(x, t, u_1(t), u_2(t)), \qquad x(t_0) = x_0, \tag{6}$$

$$\dot{p}' = -\partial H_1/\partial x, \qquad p'(t_f) = \partial K_1(x(t_f))/\partial x(t_f), \tag{7}$$

$$0 = \partial H_1/\partial u_1, \tag{8}$$

where

$$H_1(x, t, u_1, u_2, p) \triangleq L_1(x, t, u_1, u_2) + p'f(x, t, u_1, u_2), \tag{9}$$

and the pair $(u_{1s2}(t), u_{2s2}(t)) \in D_{1o}$ that minimizes $J_2(x_0, t_0, u_1, u_2)$

subject to (6)–(9) as constraints can be shown, simply by using variational techniques, to satisfy the following necessary conditions[5]:

$$\dot{\lambda}_1' = -\partial H_2/\partial x, \quad \lambda_1'(t_f) = \partial K_2(x(t_f))/\partial x(t_f) - \lambda_2'(t_f)[\partial^2 K_1(x(t_f))/\partial x(t_f)^2], \tag{10}$$

$$\dot{\lambda}_2' = -\partial H_2/\partial p, \qquad \lambda_2(t_0) = 0, \tag{11}$$

$$0 = \partial H_2/\partial u_1 = \partial H_2/\partial u_2, \tag{12}$$

where

$$H_2(x, t, p_1, u_1, u_2, \lambda_1, \lambda_2, \lambda_3) \triangleq L_2(x, t, u_1, u_2) + \lambda_1' f(x, t, u_1, u_2) + \lambda_2'(-\partial H_1/\partial x)' + \lambda_3'(\partial H_1/\partial u_1)'. \tag{13}$$

When $K_1(x(t_f)) = 0$ and $K_2(x(t_f)) = 0$, these conditions become similar to those obtained in Ref. 3.

The closed-loop admissible control sets $U_{1t_0}^c$ and $U_{2t_0}^c$ at t_0, from which the players select their closed-loop controls before the start of the game are defined as follows:

$$U_{it_0}^c = \{u_i(t, x(t)) \colon [t_0, t_f] \times X_{\xi, t_0} \to U_{it_0}^o\}, \qquad i = 1, 2.$$

These are the sets of all functions of t and $x(t)$ defined on $[t_0, t_f]$ such that the corresponding solution $x(t)$ of (5) with τ replaced by t_0, when substituted back in $u_i(t, x(t))$, produces a function of time which belongs to the space $U_{it_0}^o$. The necessary conditions for the existence of a closed-loop Stackelberg solution are not easily obtained by using variational techniques as in the case of the open-loop solution. The closed-loop rational reaction set D_{1c} when Player 2 is the leader is composed of the pairs $(u_1(t, x(t)), u_2(t, x(t))) \in U_{1t_0}^c \times U_{2t_0}^c$ satisfying the following necessary conditions:

$$\dot{x} = f(x, t, u_1(t, x(t)), u_2(t, x(t))), \qquad x(t_0) = x_0, \tag{14}$$

$$\dot{p}' = -\partial H_1/\partial x - \partial H_1/\partial u_2[\partial u_2(t, x(t))/\partial x(t)], \qquad p'(t_f) = \partial K_1(x(t_f))/\partial x(t_f), \tag{15}$$

$$0 = \partial H_1/\partial u_1, \tag{16}$$

where H_1 is as defined in (9). Because of the term including $\partial u_2(t, x(t))/\partial x(t)$ in (15), it can be easily shown that variational techniques fail to produce a candidate in $U_{1t_0}^c \times U_{2t_0}^c$ which minimizes $J_2(x_0, t_0, u_1, u_2)$ when subject to (14)–(16) as constraints. Other

---

[5] Note that $\partial H_2/\partial x = (\nabla_x H_2)'$ is a row vector. The same notation is used for all partial derivatives of a scalar function with respect to a vector.

possible approaches for performing this minimization are not investigated in this paper and remain unexplored for future research.

We now consider the case of feedback strategies. Let the interval $[t_0, t_f]$ be divided into N equal subintervals of time. Consider an arbitrary interval $[\tau_j, \tau_{j+1}]$, and assume that the Stackelberg feedback strategies (in the sense that is being defined) $u_{1s2}^f(t, x(t))$ and $u_{2s2}^f(t, x(t))$ for the game defined over the interval $[\tau_{j+1}, t_f]$ have been obtained. Let $J_i(x(\tau_{j+1}), \tau_{j+1}, u_{1s2}^f, u_{2s2}^f)$, $i = 1, 2$, be the costs corresponding to these strategies. Furthermore, let $U_{1\tau_j}^f$ and $U_{2\tau_j}^f$ denote the subsets of $U_{1\tau_j}^c$ and $U_{2\tau_j}^c$ obtained by eliminating all controls that do not coincide with $u_{1s2}^f(t, x(t))$ and $u_{2s2}^f(t, x(t))$, respectively, over $[\tau_{j+1}, t_f]$. Now, consider the game defined over $[\tau_j, t_f]$, where the state equation is as in (5) and the performance indices (4) are reduced to

$$J_i(x(\tau_j), \tau_j, u_1, u_2) = J_i(x(\tau_{j+1}), \tau_{j+1}, u_{1s2}^f, u_{2s2}^f) + \int_{\tau_j}^{\tau_{j+1}} L_i(x, t, u_1, u_2)\, dt, \qquad i = 1, 2, \tag{17}$$

where $u_1 \in U_{1\tau_j}^f$ and $u_2 \in U_{2\tau_j}^f$. Let $(u_{1s2}^f(t, x(t)), u_{2s2}^f(t, x(t))) \in U_{1\tau_j}^f \times U_{2\tau_j}^f$ be the Stackelberg strategies[6] for the game defined by (17) and (5) with τ replaced by τ_j. Now, if there exist such strategies for all $\tau_j \in [t_0, t_f)$ when this procedure is repeated backward in time until $U_{1t_0}^f$ and $U_{2t_0}^f$ are obtained, and if their limit as $|\tau_{j+1} - \tau_j| \to 0$ for all j (or as $N \to \infty$) exist, then the resulting strategies in $U_{1t_0}^f \times U_{2t_0}^f$ are called Stackelberg feedback strategies.

In a similar way (simply by replacing the word Stackelberg by the word Nash) as above, the Nash feedback strategies $u_{1N}^f(t, x(t))$ and $u_{2N}^f(t, x(t))$ are defined. We now prove the equivalence between the closed-loop Nash controls and the Nash feedback strategies.[7]

Proposition 3.1. The closed-loop Nash control pair $(u_{1N}^c(t, x(t)), u_{2N}^c(t, x(t)))$ are equal to the Nash feedback strategies $(u_{1N}^f(t, x(t)), u_{2N}^f(t, x(t)))$.

Proof. The proof of this proposition is straightforward. Let (ξ, τ)

[6] It should be clear that these strategies defined over $[\tau_j, t_f]$ and those obtained originally for $[\tau_{j+1}, t_f]$ coincide over $[\tau_{j+1}, t_f]$; hence, in order to avoid proliferation of notation, they are denoted by the same expressions $u_{is2}^f(t, x(t))$.

[7] In this proposition, we assume that the Nash closed-loop and feedback strategies exist and are unique.

be any point in the state space. Every pair $(\hat{u}_1(t, x(t)), \hat{u}_2(t, x(t))) \in D_{1c}$ has the property that, if $\hat{u}_2(t, x(t))$ is restricted to the time interval $[\tau, t_f]$, then $\hat{u}_1(t, x(t))$, which is obtained by solving an ordinary optimization (rather than a game) problem where it is known that dynamic programming or any other method lead to the same optimal control, when restricted to $[\tau, t_f]$, is also optimal for the optimization problem resulting from (4)–(5) when $u_2 = \hat{u}_2(t, x(t))$. Similarly, every pair $(\hat{u}_1(t, x(t)), \hat{u}_2(t, x(t)) \in D_{2c}$ has the same property for $\hat{u}_2(t, x(t))$ when $\hat{u}_1(t, x(t))$ is restricted to $[\tau, t_f]$. Therefore, since $(u_{1N}^c(t, x(t)), u_{2N}^c(t, x(t))) \in D_{1c} \cap D_{2c}$, it follows that these properties hold simultaneously for both controls and, hence, it is also a Nash feedback strategy.

This proposition justifies the simultaneous use in Ref. 5 of the closed-loop Nash controls and the Nash feedback strategies as being identical solutions. However, because the closed-loop Stackelberg control pair lies on the rational reaction set D_{1c} of the follower and not generally in the intersection of D_{1c} and D_{2c}, it cannot be concluded that it coincides with the Stackelberg feedback strategies. In other words, Proposition 3.1 simply says that, at any time during the course of play and from any allowable state at that instant, if the players recalculate their closed-loop Nash controls, these controls will be the same as the remaining part of the controls calculated initially. This, however, is not true in the case of the closed-loop Stackelberg controls. In fact, when dynamic programming is used, several controls in closed-loop form are eliminated from consideration at t_0 because they do not possess this optimality (Nash or Stackelberg) property from all other possible starting points (ξ, τ). Thus, because of Proposition 3.1, in the case of the Nash solution, the closed-loop Nash controls will not be among those controls that are eliminated at t_0; while, in the case of the Stackelberg solution, the closed-loop Stackelberg controls most likely will. Furthermore, the closed-loop controls eliminated at t_0 in the Nash solution are not the same as those eliminated at t_0 in the Stackelberg solution (i.e., $U_{1t_0}^f$ is not the same in both cases; for example, see Fig. 4) and, hence, it is no longer possible to conclude that the Stackelberg feedback solution is more beneficial to the leader than the Nash feedback solution.

In order to illustrate the dynamic programming technique described earlier, the necessary conditions for the existence of Stackelberg feedback strategies for a class of discrete games, where dynamic programming is more easily applied, are obtained in the following section.

4. Stackelberg Feedback Strategies in Discrete Games

Consider the multistage discrete game defined by

$$x(l+1) = f(x(l), l, u_1(l), u_2(l)), \qquad x(0) = x_0, \qquad l = 0, \ldots, N-1, \quad (18)$$

where the state $x(l)$ and the decision (control) variables $u_1(l)$ and $u_2(l)$ are n-dimensional, m_1-dimensional, and m_2-dimensional vectors of real numbers for all $l = 0, 1, \ldots, N-1$. Let the cost functionals defined over stages $k, \ldots, N$ be of the form

$$J_i(x(k), k, u_1, u_2) = K_i(x(N)) + \sum_{l=k}^{N-1} L_i(x(l), l, u_1(l), u_2(l)), \quad (19)$$

where $u_i = (u_i(k), \ldots, u_i(N-1))$, $i = 1, 2$. Suppose that Player 2 is the leader, and assume that the transition from the kth to the $(k+1)$th stage is under consideration. Let u_{1s2}^f and u_{2s2}^f be the Stackelberg feedback strategies for the game starting at stage $k+1$ and ending at stage N, and let $V_i(x(k+1), k+1) = J_i(x(k+1), k+1, u_{1s2}^f, u_{2s2}^f)$, $i = 1, 2$, be the costs corresponding to these strategies and obtained from

$$V_i(x(k+1), k+1) = K_i(x(N)) + \sum_{l=k+1}^{N-1} L_i(x(l), l, u_{1s2}^f(l, x(l)), u_{2s2}^f(l, x(l))), \quad i = 1, 2, \quad (20)$$

where $x(l)$ is obtained from

$$x(l+1) = f(x(l), l, u_{1s2}^f(l, x(l)), u_{2s2}^f(l, x(l))), \qquad l = k+1, \ldots, N-1. \quad (21)$$

The cost functionals for the game defined over stages $k, \ldots, N$ can therefore be written as

$$J_i(x(k), k, u_1(k), u_2(k)) = V_i(x(k+1), k+1) + L_i(x(k), k, u_1(k), u_2(k)). \quad (22)$$

Assuming that no constraints exist on the controls, we see that, for a fixed $u_2(k)$, the follower (Player 1) determines his optimal $u_1(k)$ (assuming that it exists) as a function of $u_2(k)$ and $x(k)$ from

$$\partial J_1(x(k), k, u_1(k), u_2(k))/\partial u_1(k)$$
$$= [\partial V_1(x(k+1), k+1)/\partial x(k+1)][\partial f/\partial u_1(k)] + \partial L_1/\partial u_1(k) = 0. \quad (23)$$

The leader, therefore, must minimize $J_2(x(k), k, u_1(k), u_2(k))$ subject to

(23) as constraint. The necessary conditions for this minimization are

$$\partial L_2/\partial u_i(k) + [\partial V_2(x(k+1), k+1)/\partial x(k+1)][\partial f/\partial u_i(k)] + (\partial/\partial u_i(k)) \times [\lambda'(k)[\partial L_1/\partial u_1(k)]' + \lambda'(k)[\partial f/\partial u_1(k)]'[\partial V_1(x(k+1), k+1)/\partial x(k+1)]'] = 0, \quad i = 1, 2, \tag{24}$$

where $\lambda(k)$ is an m_1-dimensional Lagrange multiplier. When (23)–(24) are solved, $u^f_{is_2}(k, x(k))$ and $V_i(x(k), k)$ for the game from k to N are obtained. This procedure is then repeated until the starting stage is reached.[8] The boundary conditions for (23)–(24) are given at the terminal stage by $V_i(x(N), N) = K_i(x(N))$; $i = 1, 2$. Note that, when stage $k = 0$ is reached, all feedback strategies defined over the stages 1 to $N - 1$ will have been eliminated except for those that are feedback Stackelberg strategies for the game defined on stages 1 to $N - 1$.

5. Conclusions

Several properties of the closed-loop Stackelberg controls have been investigated and the difficulties (conceptual and computational) encountered in their determination have been pointed out. Unlike the closed-loop Nash controls, it has been shown that the closed-loop Stackelberg controls cannot be obtained by applying dynamic programming techniques. The solution obtained via dynamic programming, called Stackelberg feedback strategies, has the property that, at any instant of time during the course of play and from any allowable state at that instant of time, it provides the leader with the best choice of control (in the sense of Stackelberg), regardless of previous decisions and with the assumption that Stackelberg feedback strategies will be used for the remainder of the interval of play. On the other hand, if the starting time is fixed, the leader's closed-loop Stackelberg control is the best control law (among all other admissible closed-loop controls) that he can announce prior to the start of the game, but it does not have this same desirable property from any other starting time. Since the leader is virtually the only decision maker in the Stackelberg solution, the decision of choosing between a closed-loop or a feedback strategy depends generally on whether t_0 is known or not and whether the system parameters are completely certain or not.

[8] In discrete games, the Stackelberg feedback strategies also have the property that the players may announce their controls (the leader first) stage by stage, once the current value of the state vector at each stage is known. Such development of information is called successive (Ref. 7).

References

1. VON STACKELBERG, H., *The Theory of the Market Economy*, Oxford University, Oxford, England, 1952.
2. SIMAAN, M., and CRUZ, J. B., JR., *On the Stackelberg Strategy in Nonzero-Sum Games*, Journal of Optimization Theory and Applications, Vol. 11, No. 5, 1973.
3. CHEN, C. I., and CRUZ, J. B., JR., *Stackelberg Solution for Two-Person Games with Biased Information Patterns*, IEEE Transactions on Automatic Control, Vol. AC-17, No. 6, 1972.
4. STARR, A. W., and HO, Y. C., *Nonzero-Sum Differential Games*, Journal of Optimization Theory and Applications, Vol. 3, No. 3, 1969.
5. STARR, A. W., and HO, Y. C., *Further Properties of Nonzero-Sum Differential Games*, Journal of Optimization Theory and Applications, Vol. 3, No. 4, 1969.
6. BELLMAN, R., *Dynamic Programming*, Princeton University Press, Princeton, New Jersey, 1957.
7. PROPOI, A. I., *Minimax Problems of Control Under Successively Acquired Information*, Automation and Remote Control, Vol. 31, No. 1, 1970.

XI

Switching Surfaces in N-Person Differential Games

I. G. SARMA AND U. R. PRASAD

Abstract. Switching surfaces in N-person differential games are essentially similar to those encountered in optimal control and two-person, zero-sum differential games. The differences between the Nash noncooperative solution and the saddle-point solution are reflected in the dispersal surfaces. These are discussed through classification and construction procedures for switching surfaces. The noncooperative solution of a simple two-person, nonzero sum game under perfect information to the players is presented to illustrate certain special features of the theory.

1. Introduction

Differential games with more than two players have been receiving attention lately (Refs. 2–6). In an important class of these problems, in which the players are permitted to use discontinuous strategies, the discontinuities in optimal strategies of the players lie on certain surfaces in the *playing space* of the game. Furthermore, the solution of the game might often exhibit surfaces containing singular and abnormal solutions. All such surfaces are termed here as the switching surfaces, and this paper is devoted to their study.

Since the strategy of any player with perfect information is a feedback control law, the solution of the game requires (loosely speaking) the mandatory solution of the synthesis problem for all those players endowed with perfect information. Perhaps, one could also solve the synthesis problem for the remaining players,[1] for convenience in repre-

[1] The misleading term *optimal open-loop feedback control* is used by some authors for this solution for players with no observations.

senting their control laws and the optimal paths in the playing space of the game. The construction of switching surfaces holds in either case with suitable interpretation.

The solution of the game between the switching surfaces is obtained in a routine fashion by integrating the canonical equations obtained by the appropriate necessary conditions (Refs. 2–6). This is referred to as the *solution in the small* (Ref. 7). In contrast to this, the construction of the switching surfaces itself constitutes the *solution in the large*. This latter aspect is on an uneasy terrain in the literature and mostly is example oriented.

In Section 2, we formulate the problem and discuss the solution aspects, restricting our attention to the noncooperative situation. This is followed in Section 3 with a general classification and construction of the switching surfaces in optimal control and differential game problems. In Section 4, we give simple examples illustrating the method of construction of the switching surfaces. The complete solution of the theme example of Section 4 is presented in Section 5 under perfect information to the players.

2. Formulation of *N*-Person Differential Games

The state of an N-person differential game, x of dimension n, satisfies the vector differential equation

$$\dot{x} = f(x, \mathbf{u}, t), \tag{1}$$

where

$$\mathbf{u} = (u^1, \ldots, u^p, \ldots, u^N). \tag{2}$$

In (2), u^p is the r^p-dimensional control action vector of the pth player. The region $\mathscr{R}$ of interest in the state–time space in which the game evolves is known as the playing space.

The state of the game is to be transferred by the control actions of the players from an initial state $x(t_0) = x_0$ to a final state contained in a terminal surface of dimension n given by

$$x_f = X(\sigma), \qquad t_f = T(\sigma), \tag{3}$$

where σ ranges over an n-dimensional cube. The terminal surface forms a part of the boundary of $\mathscr{R}$.

The pth player chooses his control action $u^p \in \Omega^p$ so as to minimize his payoff functional

$$J^p[x_0, t_0, \mathbf{u}] = \phi^p(x_f, t_f) + \int_{t_0}^{t_f} L^p(x, \mathbf{u}, t)\, dt. \tag{4}$$

The strategy of any player is a function of his own information of the state of the game. No notational difference will be made between the control action of a player and his strategy.

The functions f and L^p are assumed to belong to the class $C^{(1)}$. The functions T and X in (3) are assumed piecewise $C^{(1)}$, and ϕ^p is of class $C^{(1)}$ on each smooth section of the terminal surface. These are some of the usual assumptions made in the case of optimal control problems as well.

The noncooperative solution,[2] in which unrestricted communication and binding agreements are not allowed between the players, is given in terms of the equilibrium and minmax points of the game (Ref. 18). The game formulated in (1)–(4) with the initial condition $x(t_0) = x_0$ has an equilibrium point, if a joint strategy $\mathbf{u}^* = (u^{1*},..., u^{p*},..., u^{N*})$ exists such that, for $p = 1, 2,..., N$, we have

$$J^p[x_0, t_0, (\mathbf{u}^*; u^p)] \geqslant J^p[x_0, t_0, \mathbf{u}^*], \tag{5}$$

where

$$(\mathbf{u}^*; u^p) = (u^{1*},..., u^{p-1*}, u^p, u^{p+1*},..., u^{N*}). \tag{6}$$

Thus, no player can unilaterally deviate from such a strategy to improve his payoff. If, for any (x_0, t_0), there exists a unique equilibrium strategy, then it is the noncooperative solution. If not, additional concepts are introduced to define the solution.

Two equilibrium strategies $\mathbf{u}^*$ and $\hat{\mathbf{u}}$ are equivalent if, for $p = 1, 2,..., N$, we have

$$J^p[x_0, t_0, \mathbf{u}^*] = J^p[x_0, t_0, \hat{\mathbf{u}}]. \tag{7}$$

Two equilibrium strategies $\mathbf{u}^*$ and $\hat{\mathbf{u}}$ are said to be interchangeable if any recombined strategy $\tilde{\mathbf{u}}$ (every $\tilde{\mathbf{u}}^p$ is either u^{p*} or $\hat{u}^p$) is also an equilibrium strategy. If all the equilibrium strategies of a game are interchangeable, then they constitute the Nash noncooperative solution.[3] The above concepts are readily applicable to the null-information case, in which the players have no information of the state of the game, except the initial condition.

---

[2] The solution concepts to be discussed are strictly applicable for the normal form of the game. The construction of the maximum nonvoid class of playable strategies representing the normal form for a perfect information differential game is given in Ref. 6.

[3] If not, the solution is defined through certain reduction procedures on the game. Minmax strategies also may have to be considered for solution in this case. These concepts are discussed with reference to differential games by means of an example in Ref. 1.

For the perfect information game, we have to consider the game situation for all the initial conditions in the playing space $\mathscr{R}$. Here, we consider a class of games having the Nash noncooperative solution. We further assume that the solution $\mathbf{u}^*$ induces a regular decomposition on $\mathscr{R}$ (Ref. 6). Thus, discontinuities in $\mathbf{u}^*$ lie on certain well-defined surfaces in $\mathscr{R}$. The corners of the optimal paths lie on the transition surfaces. Starting points from which multiple paths arise lie on the dispersal surfaces. Those multiple paths arising due to the discontinuities in u^{p*} should obviously yield the same payoff to the pth player. But this need not be true for players whose strategies are continuous at the dispersal surfaces. Such a situation has no parallel in two-person, zero-sum games, since saddle points are automatically both equivalent and interchangeable. We state below the necessary conditions to be satisfied for such a solution $\mathbf{u}^*$. These are stated as generalizations of the earlier results (Refs. 2–6).

Theorem 2.1. If $\mathbf{u}^*$ is a Nash noncooperative solution of the perfect-information game defined by (1)–(4), then there exist adjoint variables $(\lambda_0{}^p, \lambda^p)$, $\lambda_0{}^p \geqslant 0$ for $p = 1, 2,..., N$, one set for each player, such that each set is nonzero along the trajectory and satisfies the following conditions.

(i) *Euler–Lagrange Equations.* Between the corners of the optimal trajectory x^*, the variables x and λ^p satisfy the equations[4]

$$\dot{x} = f(x, \mathbf{u}^*, t), \tag{8}$$

$$\dot{\lambda}^p = -H_x^{p*} - \sum_j H_{u^{j*}}^{p*} u_x^{j*}, \tag{9}$$

where

$$H^p(x, \lambda^p, \mathbf{u}, t) = \lambda_0{}^p L(x, \mathbf{u}, t) + \langle \lambda^p, f(x, \mathbf{u}, t) \rangle. \tag{10}$$

(ii) *Corner Conditions.* At corners of x^*, corresponding to the discontinuities in the strategies of one or more players, the following condition is satisfied across the manifold of discontinuity (or switching surface):

$$(H^{p+} - H^{p-})\, dt - (\lambda^{p+} - \lambda^{p-})\, dx = 0, \tag{11}$$

where dt and dx are arbitrary displacements along the manifold at the corner point.

[4] Subscripts on H, u, and W indicate partial derivatives.

Also, if the trajectories on either side of the switching surface are not tangential to it, and if all except one player, say the pth player, employ continuous strategies across the manifold, then his λ^p is continuous.

(iii) *Transversality Condition.* For $p = 1, 2,..., N$, at the terminal time t_f, we have[5]

$$\lambda_0^p \phi_\sigma^p + H^p T_\sigma - \lambda^p X_\sigma = 0. \tag{12}$$

(iv) *Equilibrium Point Principle.* For every t, $u^*(t)$ corresponds to an equilibrium point in the Hamiltonians; that is, for $p = 1, 2,..., N$,

$$H^p(x, \lambda^p, (\mathbf{u}^*; u^p), t) \geqslant H^p(x, \lambda^p, \mathbf{u}^*, t) = \min_{u^p \in \Omega^p} H^p(x, \lambda^p, (\mathbf{u}^*; u^p), t). \tag{13}$$

If u^p is interior to Ω^p, (13) implies that

$$\partial H^p / \partial u^p = 0, \tag{14}$$

$$\partial^2 H^p / \partial u^{p^2} \geqslant 0. \tag{15}$$

Inequality (15), is known as the Legendre–Clebsch condition.

(v) *Hamilton–Jacobi Equation.* The noncooperative value function vector $\mathbf{W} = (W^1,..., W^p,..., W^N)$ of the game satisfies a Hamilton–Jacobi equation; that is, for $p = 1, 2,..., N$,

$$\begin{aligned} -W_t^p(x, t) &= \min_{u^p \in \Omega^p} \{\langle W_x^p(x, t), f(x, (\mathbf{u}^*; u^p), t)\rangle + \lambda_0^p L^p(x, (\mathbf{u}^*; u^p), t)\} \\ &= H^p(x, W_x^p, \mathbf{u}^*, t), \end{aligned} \tag{16}$$

where

$$W^p(x, t) = J^p[x, t, \mathbf{u}^*]. \tag{17}$$

Equation (16) has the final condition

$$W^p(x_f, t_f) = \phi^p(x_f, t_f). \tag{18}$$

The results follow from the one-sided optimal control problems of the players. The corresponding results for the null-information case appear in Refs. 2, 4, 5, and the essential differences between these two forms of information are discussed by Starr and Ho (Refs. 4–5). For the null-information case, the adjoint equation (9) is modified with $u_x^{p*} = 0$, and the Hamilton–Jacobi equations (16) need not be satisfied.

The optimal paths on which the theorem holds with $\lambda_0^p = 1$ are called the normal paths. The canonical equations (8), (9) and the

[5] Equation (11) can also be expressed in this form if the switching surface has a parametric representation similar to Eq. (3).

minimum principle (13) are satisfied between the switching surfaces. So, for obtaining the solution, one should construct the Hamiltonians for the players and find the equilibrium point in them. If, on an optimal trajectory, the minimum principle fails to determine u^*, then we call such paths singular paths.

In the next section, we classify the switching surfaces encountered in differential games and optimal control problems.

3. Classification and Construction of Switching Surfaces

Switching surfaces can be exhaustively classified by considering the nature of the optimal paths on the surface and its neighborhood, i.e., whether the paths enter, leave, or are parallel to the surface (not necessarily of the same nature on either side). The universal and dispersal surfaces of Isaacs (Ref. 7) derive their names on this basis. He also showed that some of the surfaces under this classification are unrealizable.

A different method is to label a switching surface as belonging to the player or players whose strategies are discontinuous across the surface. Yet another classification is based on the method of construction or the condition to be satisfied on the switching surface. Thus, transition surfaces are constructed by applying the corner conditions (11). The other candidates in this classification are the singular, dispersal, and abnormal surfaces. Before going into the construction of these surfaces, we present a general discussion on the construction of the switching surfaces in optimal control problems and differential games.

The construction of switching surfaces in optimal control problems appears in contemporary literature. The familiar problems have the Hamiltonians linear or sectionally linear in the control variables on which bounds are specified. The resulting bang–bang, three-level, and other controls are given in terms of signum, dead zone, or similar functions of a suitable switching function with the state and adjoint variables (x, λ) as its arguments. Under the usual smoothness assumptions on the formulation functions [f and L are assumed of class $C^{(1)}$], the switching function as well as λ are continuous functions of their respective arguments. In these problems, the construction of switching surfaces, which are mainly of the transition type, is straight forward.

On the other hand, in differential games, the switching function as well as the adjoint variables λ^p need not be continuous in spite of similar smoothness assumptions on f and L^p. This situation arises because any discontinuities in the optimal strategies of the other players are reflected in the dynamic equations of the one-sided optimal control problem of the remaining player p. The preceding result is true whatever be the information patterns to the players. The corner conditions stated by Berkovitz (Ref. 8) for such a problem are equivalent to the results stated as corner conditions (11) in Theorem 2.1.

The possible discontinuities in the adjoint variables, together with the simultaneity involved in obtaining the strategies of all the players, makes the construction of switching surfaces more difficult in differential games. This explains to a large extent the occurrence of certain unusual transition surfaces, termed jocularly as bang–bang–bang surfaces by Isaacs (Ref. 9). In what follows, we consider one-by-one the singular, dispersal, and abnormal surfaces.

3.1. Singular Surfaces. Singular surfaces contain singular optimal paths. Singular extremals in optimal control have been studied in the literature (Refs. 10–12). A definition of Robbins (Ref. 12) is generalized here. An extremal arc is singular if, at each point of the arc, there is some allowable first-order weak control variation for at least one player p, which leaves his Hamiltonian H^p unchanged to second order. If u^p is interior to the restraint set Ω^p, then this condition reduces to $H^p_{u^p u^p}$ being singular or that the Legendre–Clebsch condition (15) is satisfied in its weak form only. As in optimal control, the most common examples are the linear singular arcs. These arise when at least one of the Hamiltonians, say H^p, is linear or sectionally linear in some components of the corresponding u^p. Most universal surfaces of Isaacs (Ref. 7) fall in this category.

Of the various methods of obtaining the singular arcs, we present here the Robbins' second variation method. The linear singular control variables cannot be determined from (14) of the minimum principle. However, differentiation of $H^p_{u^p}$ a sufficient number of times equal to the order of singularity will determine these variables after suitable manipulation. This technique consists in substituting the canonical equations and the expressions for the nonsingular control variables after each differentiation.

The generalized Legendre–Clebsch condition is stated as a test for the optimality of singular extremals. For player p, $\Phi_k{}^p$ is formed for different values of k, where

$$\Phi_k{}^p = \{(d/dt)^k H^p_{u^p}\}_{u^p} . \tag{19}$$

The first time $\Phi_k{}^p$ is not equal to a null matrix, k should be even, that is, $k = 2l$. The matrix $(-1)^l \Phi^p_{2l}$ must be positive semidefinite. If this condition fails, the singular extremal is not optimal. The two-person, zero-sum version of this condition is given by Anderson (Ref. 13) along with a few junction conditions. Such junction conditions are applicable only when all players, except one, use continuous strategies across the junction.

The analytical difficulties in the actual construction mount with several players having singular control variables and with different control variables having different orders of singularity.

3.2. Dispersal Surfaces. Dispersal surfaces contain starting points in $\mathscr{R}$ where multiple optimal paths arise. Multiple optimal paths result on account of the discontinuities in the strategies of one or more players. These discontinuities can be interpreted as different strategy choices for the concerned player corresponding to each different path at the starting point. If we freeze the strategies of the other players at this point, then the payoffs to the player on the different paths, resulting from his different strategy choices, should be equal. In a general N-person differential game, the payoff to the other players on these paths need not be equal.

In optimal control problems with the usual smoothness assumptions on f and L and the terminal surface, dispersal surfaces are rarely met with. Isaacs discusses thoroughly the construction of these surfaces in two-person, zero-sum differential games. Dispersal surfaces of any player are constructed based on the property that the payoff to this player is independent of the multiple optimal paths when the strategies of the others are frozen at the surface, as described earlier. However, on the dispersal surface $N^p_{i_1,i_2,\ldots,i_k}$ of player p, the following condition is satisfied for any variations dx, dt on the surface (see Ref. 14 for a similar result):

$$\begin{aligned} H^p(x, \lambda^p_{i_1}, (\mathbf{u}^*; u^*_{i_1}), t) - \lambda^p_{i_1}\, dx &= H^p(x, \lambda^p_{i_2}, (\mathbf{u}^*; u^*_{i_2}), t) - \lambda^p_{i_2}\, dx \\ &\;\;\vdots \\ &= H^p(x, \lambda^p_{i_k}, (\mathbf{u}^*; u^*_{i_k}) - \lambda^p_{i_k}\, dx. \end{aligned} \tag{20}$$

The subscripts $i_1, i_2, \ldots, i_k$ denote the different optimal paths corresponding to the different strategy choices.

3.3. Abnormal Surfaces. Abnormal surfaces are surfaces containing abnormal solutions. On the abnormal paths, the minimum principle of Theorem 2.1 is satisfied with $\lambda_0{}^p = 0$. Thus, $\lambda_0{}^p = 0$ in the expression (10) for the Hamiltonian and the transversality and corner conditions (11)–(12). The semipermeable surfaces of Isaacs (Ref. 7) follow the same construction and constitute examples of abnormal surfaces.

Results on the optimality of abnormal solutions appear in optimal control literature (see for example, Ref. 15, Theorem 12, p. 364, and Ref. 16, pp. 37f and 53). Similar considerations arise in differential games as well. The abnormal surfaces in problems with time as payoff have a special significance in two-person, zero-sum games as is evidenced by the concept of barrier (Ref. 7). This surface does not exhibit time optimality but corresponds to the neutral outcome associated with the game of kind. In a general problem, the roles of the players with regard to the termination of the game have to be specified to determine the optimality of abnormal solutions. Thus, Isaacs specifies a drastic penalty for nontermination to both the players in a game (see Appendix of Ref. 9) and, under this assumption, the abnormal solutions of this game were shown to be optimal.

The construction of switching surfaces across which more than one player employ discontinuous strategies is much more difficult than the construction of one-player switching surfaces. Examples are the bang-bang–bang surfaces (Ref. 9) and the equivocal surfaces (Ref. 7) of Isaacs and the switch envelope of Breakwell and Merz (Ref. 17). It is in particular the construction of these surfaces that is highly example oriented.

In the next section, we present a few simple examples. Included also is an example of a new type of dispersal surface for nonzero-sum games which has no counterpart in two-person, zero-sum differential games.

4. Examples of Switching Surfaces

In this section, we consider the game described by a double integral plant defined by

$$\dot{x}_1 = x_2, \qquad \dot{x}_2 = u^1 + cu^2, \tag{21}$$

with the control variables constrained as follows:

$$|u^1| \leqslant 1, \qquad |u^2| \leqslant 1. \tag{22}$$

The terminal state is specified as the origin at the free terminal time t_f. The payoffs to the players are given by

$$\begin{aligned} J^1[x_0, u^1, u^2] &= \int_0^{t_f} dt, \\ J^2[x_0, u^1, u^2] &= \int_0^{t_f} \{|u^1| + b\,|u^2|\}\, dt. \end{aligned} \tag{23}$$

This problem originally arose in a bicriterion optimal control problem: a satellite attitude controller represented by a double-integral plant with time and fuel minimization as the twin objectives. This problem was solved fully for the case $c > b$ (Ref. 1). Here, we obtain the singular and abnormal solutions of the noncooperative solution. We also obtain the complete noncooperative solution of the game for the case $2 < c < b$ under a modified playing space and terminal specification. This exhibits a dispersal surface for the second player.

The Hamiltonians for the players are given by

$$\begin{aligned} H^1 &= 1 + \lambda_1^1 x_2 + \lambda_2^1(u^1 + cu^2), \\ H^2 &= | u^1 | + b | u^2 | + \lambda_1^2 x_2 + \lambda_2^2(u^1 + cu^2). \end{aligned} \tag{24}$$

For obtaining the equilibrium strategies, H^1 and H^2 are to be minimized with respect to u^1 and u^2, respectively. Thus, we have

$$u^{1*} = -\text{sign}(\lambda_2^1), \qquad u^{2*} = -\text{dez}(\lambda_2^2 c/b), \tag{25}$$

where we define

$$\text{sign}\, z = \begin{cases} +1, & z > 0, \\ -1, & z < 0, \end{cases} \tag{26}$$

$$\text{dez}(z) = \begin{cases} +1, & z > 1, \\ 0, & -1 < z < 1, \\ -1, & z < -1. \end{cases} \tag{27}$$

When neither player has any observations, the adjoint equations are given for $p = 1, 2$ as

$$\dot{\lambda}_1^p = 0, \qquad \dot{\lambda}_2^p = -\lambda_1^p. \tag{28}$$

However, since u^{1*} and u^{2*} assume only the values ± 1 and 0, in view of (25)–(27), the expressions $u_{x_l}^{p*}$ for $p, l = 1, 2$ will be zero. Hence, (28) is valid when the players have perfect information as well.

Since the terminal specification violates the dimensionality requirement, we choose the terminal surface as $x_1 = a \cos \theta$ and $x_2 = a \sin \theta$, and then apply the transversality condition (12) with $\sigma = (t_f, \theta)$. We have

$$[\lambda_1^1(t_f), \lambda_2^1(t_f)] \begin{bmatrix} a \sin \theta, & -a \sin \theta \\ u^1 + cu^2, & a \cos \theta \end{bmatrix} = [-1, 0], \tag{29}$$

$$[\lambda_1^2(t_f), \lambda_2^2(t_f)] \begin{bmatrix} a \sin \theta, & -a \sin \theta \\ u^1 + cu^2, & a \cos \theta \end{bmatrix} = [-\{| u^1 | + b | u^2 |\}, 0]. \tag{30}$$

Solving (29)–(30) and letting $a \to 0$, we have

$$\lambda_1^p(t_f) = \lambda_2^p(t_f) \cot \theta, \qquad p = 1, 2, \tag{31}$$

$$\lambda_2^2(t_f) = \{| u^1 | + b | u^2 |\} \lambda_2^1(t_f) = -(| u^1 | + b | u^2 |)/(u^1 + cu^2). \tag{32}$$

4.1. Singular Solutions. A singular u^1 requires that $\lambda_2^1(t) = 0$ and $\dot{\lambda}_2^1(t) = -\lambda_1^1(t) = 0$; this violates the condition $H^1 = 0$ on the trajectory and, hence, does not arise. On the other hand, a singular u^2 requires that

$$\lambda_2^2(t)c/b = \pm 1 \qquad \text{or} \qquad \lambda_2^2(t) = \pm b/c, \tag{33}$$

and

$$\dot{\lambda}_2^2(t) = -\lambda_1^2(t) = 0. \tag{34}$$

Thus, for the terminal sequence $\left[\begin{smallmatrix}-1\\-\epsilon\end{smallmatrix}\right]$, $0 \leqslant \epsilon \leqslant 1$, to be optimal, we should have from the transversality condition (12) and (33) (considering the upper values) and (34)

$$\lambda_1^1(t) = \lambda_1^2(t) = 0, \tag{35}$$

$$\lambda_2^2(t) = (1 + b\epsilon) \lambda_2^1(t) = -(1 + b\epsilon)/(1 + c\epsilon) = b/c. \tag{36}$$

Equation (36) requires that $b = c$ and that ϵ is a constant on the trajectory. It can be seen that this sequence does not violate the generalized Legendre–Clebsch condition. Considering the lower values of (33), we can show a similar result for the terminal sequence $\left[\begin{smallmatrix}+1\\+\epsilon\end{smallmatrix}\right]$. The resulting trajectories for $0 \leqslant \epsilon \leqslant 1$ are shown in regions G_5 and G_6 of Fig. 1. The equation of the trajectory $\gamma_{1\epsilon}$ along which the state reaches the origin with the control law $\left[\begin{smallmatrix}\pm 1\\\pm\epsilon\end{smallmatrix}\right]$, $0 \leqslant \epsilon \leqslant 1$, is obtained as follows.

On integrating (21) with the initial condition $x_1(0) = s_1$ and $x_2(0) = s_2$, we have

$$x_2 = s_2 + (u^1 + cu^2)t, \qquad x_1 = s_1 + s_2 t + \tfrac{1}{2}(u^1 + cu^2)t^2. \tag{37}$$

For the system to reach the origin, there must exist some $t_f > 0$ such that $x_1(t_f) = x_2(t_f) = 0$ in (37). Hence, we get

$$t_f = -s_2/(u^1 + cu^2), \qquad \operatorname{sign} s_2 = -\operatorname{sign}(u^1 + cu^2), \tag{38}$$

and

$$s_1 = -s_2^2/2(u^1 + cu^2). \tag{39}$$

Thus, $\gamma_{1\epsilon}$ is given by

$$\gamma_{1\epsilon} = \{(x_1, x_2) : x_1 = -x_2^2 \operatorname{sign} x_2/2(1 + c\epsilon)\}. \tag{40}$$

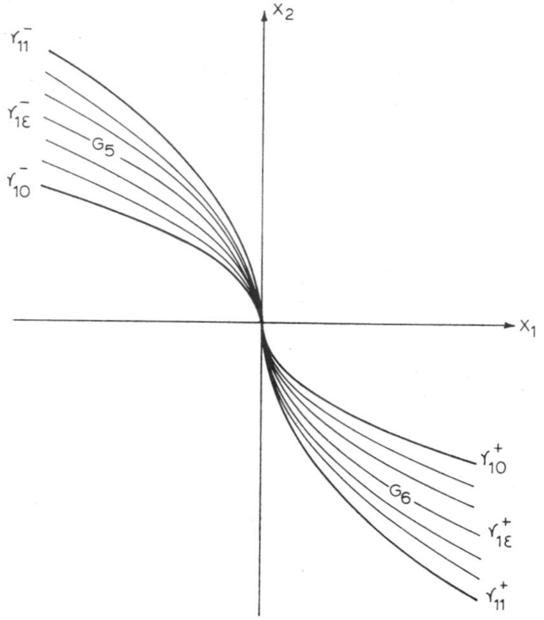

Fig. 1. Singular solutions for the example of the double integral plant.

Now, we can determine whether these trajectories are optimal for the perfect-information case by verifying the Hamilton–Jacobi equation (16). Expressing the control laws as feedback policies in the region G_5, we have, in view of (38)–(40),

$$u^1 = -1, \qquad u^2 = -\epsilon = (x_2^2 + 2x_1)/2x_1c, \tag{41}$$

$$W^1(x_1, x_2) = x_2/(1 + c\epsilon) = -2x_1/x_2, \qquad W^2(x_1, x_2) = x_2. \tag{42}$$

Now, for Player 1, we have the Hamilton–Jacobi equation

$$0 = \min_{|u^1|\leqslant 1} \{1 + (-2/x_2)\, x_2 + (2x_1/x_2^2)(u^1 + cu^2)\}$$

or

$$u^{1*} = -\text{sign}(2x_1/x_2^2) = +1. \tag{43}$$

Since (43) contradicts (41), the cluster of singular solutions in G_5 (and similarly in G_6) are not optimal for Player 1.

However, by writing the Hamilton–Jacobi equation for Player 2, it can be easily seen that the cluster of singular solutions in G_5 and G_6 are optimal for this player. Which of the singular solutions in (G_5, G_6) are optimal under the null and perfect information cases can only be answered by a complete solution of the problem, which will be given in Section 5.

4.2. Abnormal Surfaces. We obtain the abnormal solutions of the same problem now. Since $\lambda_0^1 = \lambda_0^2 = 0$ in this case, (24)–(25) are modified as follows:

$$H^1 = \lambda_1^1 x_2 + \lambda_2^1(u^1 + cu^2), \qquad H^2 = \lambda_1^2 x_2 + \lambda_2^2(u^1 + cu^2), \tag{44}$$

and

$$u^{1*} = -\text{sign}(\lambda_2^1), \qquad u^{2*} = -\text{sign}(\lambda_2^2). \tag{45}$$

The adjoint equation (28) are not modified, obviously.

The curves γ_{11}^+, γ_{11}^-, $\gamma_{1,-1}$, $\gamma_{-1,1}$ are the abnormal curves with the corresponding control sequences

$$\begin{bmatrix}+1\\+1\end{bmatrix}, \quad \begin{bmatrix}-1\\-1\end{bmatrix}, \quad \begin{bmatrix}+1\\-1\end{bmatrix}, \quad \begin{bmatrix}-1\\+1\end{bmatrix}.$$

We show this below for γ_{11}^- by showing that it satisfies the required necessary conditions. The results for the other curves follow similarly.

From the transversality conditions (14),[6] we have λ_1^1 and λ_1^2 arbitrary and

$$\lambda_2^1(t_f) = \lambda_2^2(t_f) = 0. \tag{46}$$

By (38), we have for any initial state (x_1, x_2) on γ_{11}^-

$$t_f = x_2/(1 + c). \tag{47}$$

From (28), (46), (47), on integration, we have

$$\lambda_2^p(0) = \lambda_1^p\, x_2/(1 + c). \tag{48}$$

Equations (45) and (48) yield

$$u^1 = u^2 = -1, \tag{49}$$

with λ_1^p being any negative number for $p = 1, 2$.

The optimality of the different abnormal solutions can be decided in the context of the full solution of the game.

4.3. Dispersal Surfaces. Now, we modify the playing space, the terminal specification, and the payoff functions of the game formulated in (21)–(23). The terminal surface $T_1 U T_2$ is shown in Fig. 2. The space between T_1 and T_2 is the playing space. We have

$$\begin{aligned} T_1 &= \{(x_1, x_2) : x_1 = -x_2^2/2(1 + c)\}, \\ T_2 &= \{(x_1, x_2) : x_1 = -\tfrac{1}{2}x_2^2\}. \end{aligned} \tag{50}$$

[6] Equations (29)–(30) are modified with null vectors on the right-hand side in this case.

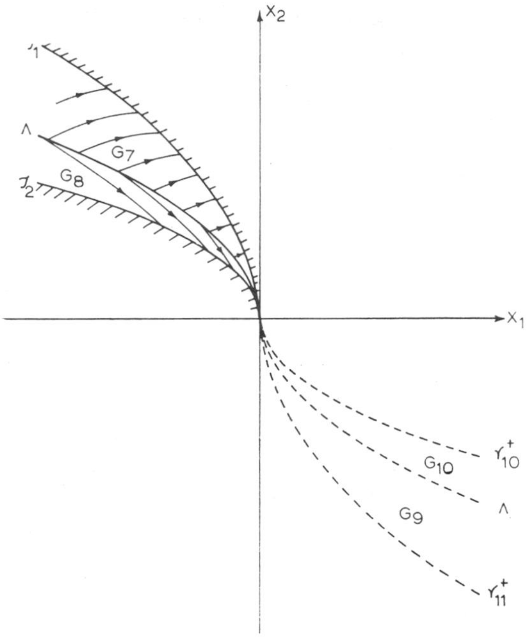

Fig. 2. An example of a game with a dispersal surface.

Comparing (40) and (50), we know that T_1 and T_2 are identical with $\gamma_{\bar{1}\bar{1}}$ and $\gamma_{\bar{1}\bar{0}}$, respectively.

The payoff functionals are of the composite type given by

$$\begin{aligned} J^1[x_0, \mathbf{u}] &= \phi^1(x_f) + \int_0^{t_f} dt, \\ J^2[x_0, \mathbf{u}] &= \phi^2(x_f) + \int_0^{t_f} \{| u^1 | + b \,| u^2 |\}\, dt, \end{aligned} \tag{51}$$

where x_f is the terminal state at the free terminal time t_f. The functions ϕ^1 and ϕ^2 are given below as

$$\phi^1(x_f) = \begin{cases} x_{2f}/(1 + c), & x_f \in T_1, \\ x_{2f}, & x_f \in T_2, \end{cases} \tag{52}$$

$$\phi^2(x_f) = \begin{cases} x_{2_f}(1 + b)/(1 + c), & x_f \in T_1, \\ x_{2_f}, & x_f \in T_2. \end{cases} \tag{53}$$

The Hamiltonians for the players, the adjoint equations, and the optimal control actions are given again as in (24), (28), and (25), respectively.

Applying the transversality condition (12) at T_1, with $\sigma = (x_{2_f}, t_f)$, we have

$$
\begin{aligned}
1/(1+c) - \lambda_1^1[-x_{2_f}/(1+c)] - \lambda_2^1 &= 0, \\
1 + \lambda_1^1 x_{2_f} + \lambda_2^1(u^1 + cu^2) &= 0, \\
(1+b)/(1+c) - \lambda_1^2[-x_{2_f}/(1+c)] - \lambda_2^2 &= 0, \\
|u^1| + b|u^2| + \lambda_1^2 x_{2_f} + \lambda_2^2(u^1 + cu^2) &= 0.
\end{aligned}
\tag{54}
$$

The quantities in (54) refer to time t_f. Solving (54) to be consistent with (25) and (28) yields

$$
\begin{aligned}
&\lambda_1^1(t_f) = -1/x_{2_f}, \qquad \lambda_2^1(t_f) = 0, \\
&\lambda_1^2(t_f) = -(2+c+b)/(2+c)\,x_{2f}, \qquad \lambda_2^2(t_f) = b/(2+c),
\end{aligned}
\tag{55}
$$

and

$$
u^1(t) = 1, \qquad u^2(t) = 0 \qquad \text{for} \quad t < t_f. \tag{56}
$$

By a similar application of T_2, we have

$$
\begin{aligned}
1 - \lambda_1^1(-x_{2_f}) - \lambda_2^1 &= 0, \\
1 + \lambda_1^1 x_{2_f} + \lambda_2^1(u^1 + cu^2) &= 0, \\
1 - \lambda_1^2(-x_{2_f}) - \lambda_2^2 &= 0, \\
|u^1| + b|u^2| + \lambda_1^2 x_{2_f} + \lambda_2^2(u^1 + cu^2) &= 0.
\end{aligned}
\tag{57}
$$

Solving (57), (25), (28) consistently, we have

$$
\begin{aligned}
&\lambda_1^1(t_f) = -1/x_{2_f}, \qquad \lambda_2^1(t_f) = 0, \\
&\lambda_1^2(t_f) = [-1 - b - b(1-c)/(c-2)]/x_{2_f}, \qquad \lambda_2^2(t_f) = b/(c-2),
\end{aligned}
\tag{58}
$$

and

$$
u^1(t) = 1, \qquad u^2(t) = -1 \qquad \text{for} \quad t < t_f. \tag{59}
$$

The optimal paths resulting from the control laws (56) and (59) are made to flood the playing space. From (56) and (59), it is clear that the strategy of Player 1 is continuous in the playing space and that of the second player is discontinuous. Hence, there arises a switching surface which, in this case, is the second player's dispersal surface Λ. For starting points on Λ, W^2 must be the same whether the optimal paths reach T_1 or T_2. By actually equating the values, it is straightforward to show that Λ is given by

$$
\Lambda = \{(x_1, x_2) : x_1 = -\tfrac{1}{2}\eta x_2^2, x_2 \geqslant 0\}, \tag{60}
$$

where η is given by

$$(b+1)/(c-1) - [(2+b-c)/(c-1)]\sqrt{[(-1+(c-1)\eta)/(c-2)]}$$
$$= -1 + (2+c+b)\sqrt{(1+\eta)}/\sqrt{[(1+c)(2+c)]}. \tag{61}$$

The optimality of this solution is conclusively shown by verifying the Hamilton–Jacobi equation. A similar construction between γ_{11}^{+} and γ_{10}^{+} shown by broken lines in Fig. 2 will be used in Section 5.

From (52) and (38), one can easily see that $\phi^1(x_f)$ is the time taken for the system state x_f to reach the origin along either T_1 or T_2 as the case may be. Hence, W^1 can be interpreted as the total time taken for the system state to reach the origin along the optimal path and the terminal surface. Thus, for any point A on Λ, the value of W^1 for either optimal path can be written as the line integral

$$W^1 = \int_A^0 dt = \int_A^0 dx_1/x_2\,. \tag{62}$$

As the trajectory reaching T_2 is below that reaching T_1, it follows from (62) that the time taken for the former path is larger than that of the latter. Hence, the two equilibrium points are nonequivalent for Player 1. There is no counterpart of this result in two-person, zero-sum games, for obvious reasons.

5. A Two-Person Differential Game with No Nash Solution

The two-person differential game introduced at the beginning of Section 4 will be shown to have a unique equilibrium point only for the case $c > b$, which then is its noncooperative solution. On the other hand, for $c \leqslant b$, it exhibits multiple noninterchangeable equilibrium points and hence does not have a Nash solution. The solution is then obtained by evaluating the risks to the players when each of them enforces his preferred strategy.

5.1. Solution for the Case $c > b$. The solution is constructed by the flooding technique to fill the state space with optimal trajectories. After satisfying the terminal transversality conditions (31) and (32), the canonical equations (21) and (28) are integrated backward in time and the transition surfaces are constructed at the corners of the resulting trajectories. Unlike in similar problems of optimal control, the adjoint variables here exhibit discontinuities at these surfaces, which have to be determined by the application of (11). We summarize the solution below, since the procedure is straightforward.

The only control sequences that are equilibrium optimal in this case are

$$\begin{bmatrix} u^1 \\ u^2 \end{bmatrix} = \begin{bmatrix} \pm 1 & \pm 1 & \mp 1 \\ \pm 1 & 0 & \mp 1 \end{bmatrix}. \tag{63}$$

The corresponding decomposition of the state space is shown in Fig. 3. The transition surface Γ is given by

$$\Gamma = \{(x_1, x_2): \quad x_1 = -\tfrac{1}{2}\alpha x^2 \operatorname{sign} x_2\}, \tag{64}$$

where

$$\alpha = \frac{(2+b+c)^2c^2 + 4c(1+c)^2(2+c)(c-b)b + 4(1+c)^3b^3}{(1+c)(2+c)^2(c-b)^2} \tag{65}$$

The control law is stated in accordance with Fig. 3:

$$\begin{aligned}
&(x_1, x_2) \in G_1, && (u^1, u^2) = (1, 0),\\
&(x_1, x_2) \in G_2 \cup \gamma_{11}^+, && (u^1, u^2) = (1, 1),\\
&(x_1, x_2) \in G_3, && (u^1, u^2) = (-1, 0),\\
&(x_1, x_2) \in G_4 \cup \gamma_{11}^-, && (u^1, u^2) = (-1, -1).
\end{aligned} \tag{66}$$

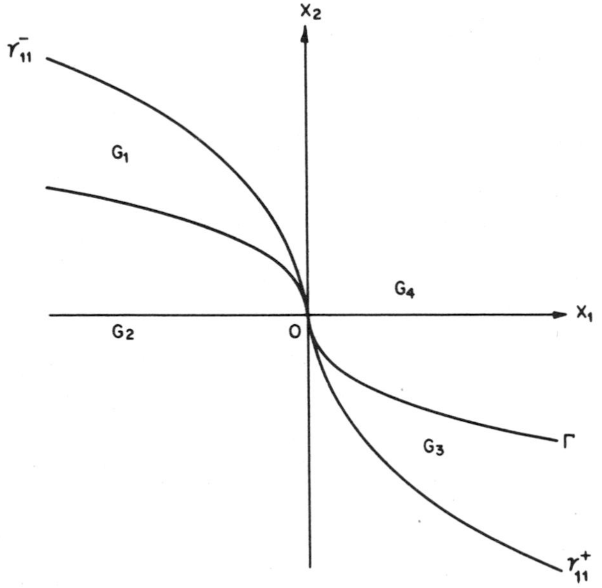

Fig. 3. Switching surfaces for the noncooperative solution. Case $c > b$.

The solution (66) can indeed be shown to be optimal by verifying the Hamilton–Jacobi equation (12) for the value function.

5.2. Solution for the Case $c < b$ and $2c - b^2 + c^2 + 2bc > 0$. From (25) and (32), the terminal sequences $\left[\begin{smallmatrix}\pm 1\\ 0\end{smallmatrix}\right]$ are optimal and the state of the game reaches the origin along γ_{10} given by (40) with $\epsilon = 0$. This generates the equilibrium optimal control sequences

$$\begin{bmatrix} u^1 \\ u^2 \end{bmatrix} = \begin{bmatrix} \pm 1 & \pm 1 & \mp 1 \\ \pm 1 & 0 & 0 \end{bmatrix}, \tag{67}$$

and the corresponding decomposition of the state space shown in Fig. 4(ii). The transition surface Γ' is given by

$$\Gamma' = \{(x_1, x_2):\quad x_1 = -\alpha' x_2{}^2 \operatorname{sign} x_2/2\}, \tag{68}$$

where

$$\alpha' = (b^2 - c^2 - 2bc)/(b - c)^2. \tag{69}$$

The control law is expressed as

$$\begin{aligned}
&(x_1, x_2) \in G_1' \cup \gamma_{10}^+, && (u^1, u^2) = (1, 0),\\
&(x_1, x_2) \in G_2', && (u^1, u^2) = (1, 1),\\
&(x_1, x_2) \in G_3' \cup \gamma_{10}^-, && (u^1, u^2) = (-1, 0),\\
&(x_1, x_2) \in G_4', && (u^1, u^2) = (-1, -1).
\end{aligned} \tag{70}$$

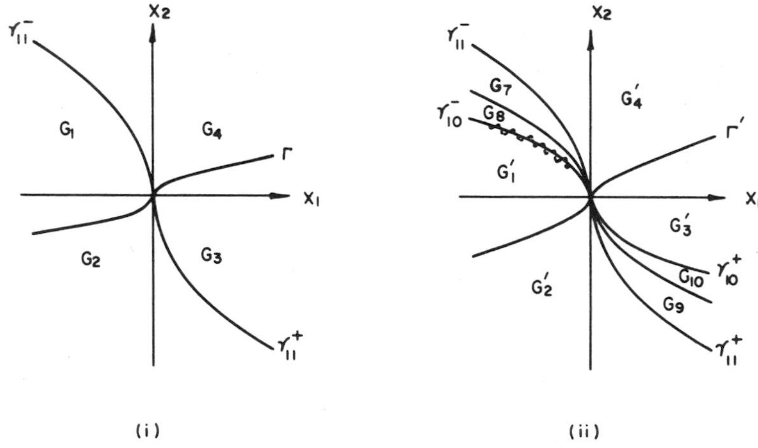

Fig. 4. Switching curves for noncooperative solution (preferences of Players 1 and 2). Case $c < b$ and $2c - b^2 + c^2 + 2bc > 0$.

Since $\gamma_{11}^{\pm}$ forms the boundary to G_2' and G_4', these abnormal surfaces (see Section 4.2) satisfy Isaacs' envelope principle and are optimal.

Since γ_{10} and γ_{11} are both optimal, we can construct the dispersal surface Λ for the second player when $c > 2$ (see Section 4.3, identifying T_1 and T_2 with γ_{11} and γ_{10}). The control laws in regions G_7, G_8, G_9, and G_{10} are given by

$$\begin{aligned} (x_1, x_2) &\in G_7, & (u^1, u^2) &= (1, 0), \\ (x_1, x_2) &\in G_8, & (u^1, u^2) &= (1, -1), \\ (x_1, x_2) &\in G_9, & (u^1, u^2) &= (-1, 0), \\ (x_1, x_2) &\in G_{10}, & (u^1, u^2) &= (-1, 1). \end{aligned} \tag{71}$$

It also follows from the optimality of γ_{11} that the sequence (63) with the corresponding decomposition of Fig. 4(i) in this case is equilibrium optimal. The transition surface Γ and the control law are given by (64)–(66).

The equilibrium control sequences and decomposition corresponding to Fig. 4(i) are preferred by Player 1 and those corresponding to Fig. 4(ii) are preferred by Player 2. This can easily be verified by evaluating the payoffs to the players in both cases. Under perfect information to the players, the recombined strategies when each player adheres to his preferred strategy are playable. This results in trajectories which chatter along γ_{10} to the origin for all starting points in the state space except $G_7 \cup G_8$, where the players' preferences match. A payoff analysis clearly shows that this results in the less preferred equilibrium payoff to Player 1. But Player 2 gets a payoff inferior to even his less preferred equilibrium payoff. Because of this risk, Player 2 has to yield to his opponent's preferences and the noncooperative solution is given by (64)–(66) and Fig. 4(i).

5.3. Solution for the Case $c < b$ and $2c - b^2 + c^2 + 2bc < 0$. In this case the transition curve Γ' of Fig. 4(ii) goes into the space between γ_{11} and γ_{10} and the abnormal solution γ_{11} no longer is optimal. In the region between Γ' and γ_{10}, there are no equilibrium trajectories reaching Γ'. However, for the case $c > 2$, we have $\left[\begin{smallmatrix}\pm 1\\ \mp 1\end{smallmatrix}\right]$ as equilibrium sequences corresponding to trajectories reaching γ_{10} as in regions G_8 and G_{10}.

An interesting strategy available to Player 2 to reach Γ' when the game is between γ_{10} and Γ' is to use $u^2 = 0$. Since Player 2 thus leaves the optimization problem entirely to his opponent, we call this his abstaining strategy. Under the perfect observation assumption, Player 1 can detect this and play $u^1 = \pm 1$, depending upon whether the state of the game is below or above γ_{10}. When $c > 2$ one can fit a dispersal surface Λ' for Player 2 separating his abstaining strategy and the strategy $\left[\begin{smallmatrix}\pm 1\\ \mp 1\end{smallmatrix}\right]$ leading to γ_{10}. However, this strategy is risky to Player 2 in the same way as discussed in Section 5.2. Thus the solution consists of Player 2 completely abstaining from the play. It may be noted that except in the regions between γ_{10} and Γ' where the game has no equilibrium point, the rest of the strategies are in equilibrium.

5.4. Case $c < b$ and $2c - b^2 + c^2 + 2bc = 0$. This separates the previous two cases. Considering in the limit the solutions to them, it is obvious that the abstaining strategy is suited best for Player 2.

5.5. Solution for the Case $c = b$. It was shown in Section 4.1 that the sequences $\left[\begin{smallmatrix}\pm 1\\ \pm\epsilon\end{smallmatrix}\right]$ with $0 \leqslant \epsilon \leqslant 1$ qualify as equilibrium terminal sequences since they do not violate the generalized Legendre–Clebsch condition. Thus one encounters an infinite number of equilibrium optimum optimal sequences

$$\begin{bmatrix}\pm 1 & \pm 1 & \mp 1\\ \pm 1 & 0 & \mp\epsilon\end{bmatrix}, \qquad 0 \leqslant \epsilon \leqslant 1, \tag{72}$$

and

$$\begin{bmatrix}\pm 1 & \mp 1\\ \mp 1 & \mp\epsilon\end{bmatrix}, \qquad \epsilon < (c-2)/c. \tag{73}$$

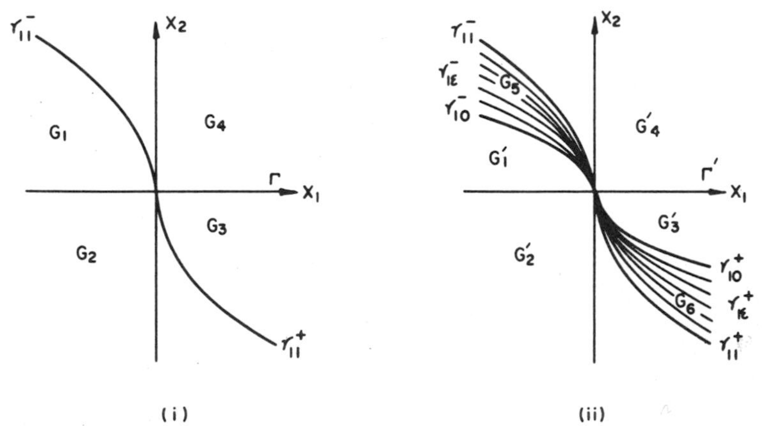

Fig. 5. Switching curves for noncooperative solution (preferences of Players 1 and 2). Case $c = b$.

Payoff analysis shows that Player 1 prefers the control sequence (63) with the corresponding decomposition shown in Fig. 5(i). Similarly for Player 2 the preferred decomposition is represented by Fig. 5(ii). The transition curve Γ coincides with the x_1 axis. The control law is given by (70) and (41).

The recombined preference strategies of the players are playable under the perfect-information assumption. Once again it is risky for Player 2 to enforce his preferred strategy against his opponent enforcing his own preferences. A repetition of the previous arguments gives that the noncooperative solution is as given by (64)–(66) and Fig. 5(i).

6. Conclusions

The Nash noncooperative solution of an N-person differential game is described for a class of games with null and perfect information to the players. Switching surfaces encountered in optimal control and differential games are classified, and the method of construction of these surfaces is reviewed with reference to the N-person differential games. An example brings out the distinct nature of dispersal surfaces in the Nash noncooperative solution. The optimality of singular and abnormal solutions of the example can be established by the complete solution of the game.

References

1. PRASAD, U. R., *N-Person Differential Games and Multicriterion Optimal Control Problems*, Ph.D. Thesis, Department of Electrical Engineering, Indian Institute of Technology, Kanpur, India, 1969.
2. KARVOVSKIY, G. S., and KUZNETSOV, A. D., *A Maximum Principle for* N *Players*, Engineering Cybernetics, No. 6, 1966.
3. CASE, J. H., *Toward a Theory of Many Player Differential Games*, SIAM Journal on Control, Vol. 7, No. 1, 1969.
4. STARR, A. W., and HO, Y. C., *Nonzero-Sum Differential Games*, Journal of Optimization Theory and Applications, Vol. 3, No. 3, 1969.
5. STARR, A. W., and HO, Y. C., *Further Properties of Nonzero-Sum Differential Games*, Journal of Optimization Theory and Applications, Vol. 3, No. 4, 1969.
6. SARMA, I. G., RAGADE, R. K., and PRASAD, U. R., *Necessary Conditions for Optimal Strategies in a Class of Noncooperative* N-*Person Differential Games*, SIAM Journal on Control, Vol. 7, No. 4, 1969.

7. Isaacs, R., *Differential Games*, John Wiley and Sons, New York, 1965.
8. Berkovitz, L. D., *Variational Methods in Problems of Control and Programming*, Journal of Mathematical Analysis and Applications, Vol. 3, No. 1, 1961.
9. Isaacs, R., *Differential Games: Their Scope, Nature, and Future*, Journal of Optimization Theory and Applications, Vol. 3, No. 5, 1969.
10. Johnson, C. D., *Singular Solutions in Problems of Automatic Control*, Advances in Control Systems, Vol. 2, Edited by C. T. Leondes, Academic Press, New York, 1965.
11. Kelley, H. J., Kopp, R. E., and Moyer, H. G., *Singular Extremals*, Topics in Optimization, Edited by G. Leitmann, Academic Press, New York, 1966.
12. Robbins, H. M., *A Generalized Legendre–Clebsch Condition for the Singular Cases in Optimal Control*, IBM Journal of Research and Development, Vol. 11, No. 4, 1967.
13. Anderson, G. M., *Necessary Conditions for Singular Solutions in Differential Games with Controls Appearing Linearly*, Paper presented at the First International Conference of the Theory and Applications of Differential Games, Amherst, Massachusetts, 1969.
14. Berkovitz, L. D., *A Variational Approach to Differential Games*, Advances in Game Theory, Edited by M. Dresher, L. S. Shapley, and A. W. Tucker, Annals of Mathematical Studies, No. 52, Princeton University Press, Princeton, New Jersey, 1964.
15. Leitmann, G., Editor, *Topics in Optimization*, Academic Press, New York, 1967.
16. Leitmann, G., *An Introduction to Optimal Control*, McGraw-Hill Book Company, New York, 1966.
17. Breakwell, J. V., and Merz, A. W., *Toward a Complete Solution of the Homicidal Chauffeur Game*, Paper presented at the First International Conference on the Theory and Applications of Differential Games, Amherst, Massachusetts, 1969.
18. Luce, R. D., and Raiffa, H., *Games and Decisions*, John Wiley and Sons, New York, 1957.

XII

Some Fundamental Qualifications of Optimal Strategies and Transition Surfaces in Differential Games[1]

P. L. Yu

Abstract. At each point of a regular region of a differential game, there are two tangent cones which are complementary to each other. The two vectograms defined by the pair of pure optimal strategies must be contained separately in the two tangent cones. The velocity vector resulting from the selection of the optimal strategies consequently must represent a semipermeable direction. These conditions, which reveal a fundamental separating property of the optimal velocity vector, are weaker than that of Isaac's main equation. Moreover they hold even on singular surfaces in the regular regions. The conditions also reveal a necessary condition for a differential game to have a regular region. Each isovalued surface in the regular region is essentially a semipermeable surface. A transition surface arises only when the resulting directed isovalued surface of the previous optimal strategies cannot have a smooth semipermeable extension at the singular surface. This observation yields a necessary condition for a transition surface to occur. The localization of transition surfaces is then possible. Finally, a jump and smooth condition of the isovalued surfaces is given.

[1] This research was supported in part by the Systems Analysis Program, The University of Rochester, under the Bureau of Naval Personnel Contract No. N00022-70-C-0076. It was also supported in part by the Center for Naval Analyses of the University of Rochester. The author is grateful to Professors R. Isaacs, M. Freimer, and H. Gerber for their helpful discussion during the research. He is especially grateful to Professor G. Leitmann, University of California at Berkeley, for his very helpful comments and suggestions.

1. Introduction

Let us consider a class of differential games described by the following elements:

(i) The playing space is $\mathscr{E} = \{E^1 \times \mathscr{E}'\}$, where $\mathscr{E}'$ is an open set in E^n (the Euclidean n-space), with its element denoted by

$$X = (x_0, x_1, \dots, x_n), \qquad x_0 \in E^1, \qquad (x_1, \dots, x_n) \in \mathscr{E}'.$$

(ii) The terminal surface is $\mathscr{C} = \{E^1 \times \mathscr{C}'\}$, $\mathscr{C}' \subset \partial\mathscr{E}'$, where $\partial\mathscr{E}'$ is the boundary of $\mathscr{E}'$.

(iii) The kinematic equation (time-forward derivative) is

$$\dot{X} = f(X, \phi, \psi), \qquad f = (f_0, f_1, \dots, f_n),$$
$$\phi \in \Phi, \qquad \psi \in \Psi, \qquad \Phi \subset E^p, \qquad \Psi \subset E^q,$$

where f has continuous first derivatives with respect to its arguments throughout $\mathscr{E}$.

(iv) The payoff $H(X)$, which is defined over $\mathscr{C}$, is strictly increasing in x_0.[2] If the game cannot terminate, we assume that the payoff is arbitrarily large.[3] The first player, who controls the parameter ϕ, called the ϕ-player, tries to minimize the payoff, while his opponent, the ψ-player, who controls the parameter ψ, tries to maximize it.

Remark 1.1. By introducing a new state variable, we see that a game with integral payoff can be transformed into the form described above (see Refs. 1–2). A very vivid geometric account of this transformation has also been provided by Blaquière and Leitmann (see Refs. 3–7).

By a pair of strategies, we shall mean a pair of functions $\mu(X)$ and $\nu(X)$ such that $\mu(X) \in \Phi$ and $\nu(X) \in \Psi$ for all $X \in \mathscr{E}$. Namely, $\mu(X)$ and $\nu(X)$ are strategies selected by the ϕ-player and ψ-player, respectively.

A pair of strategies $\mu(X)$ and $\nu(X)$ is said to be admissible if, given $X_0 \in \mathscr{E}$, there is a path $\{X(t) \mid 0 \leqslant t \leqslant t_1\} \subset \mathscr{E}$ which satisfies $\dot{X} = f(X, \mu(X), \nu(X))$ and $X(0) = X_0$, and $\mu(X(t))$ and $\nu(X(t))$ are right-hand continuous along the path.[4] For convenience, a strategy $\mu(X)$ [or $\nu(X)$] is said to be admissible if there is a strategy $\nu(X)$ [or $\mu(X)$]

---

[2] That is, $H(X'') > H(X')$, if $X' = (x_0', x_1, \dots, x_n)$ and $X'' = (x'', x_1, \dots, x_n)$ with $x_0'' > x_0'$.

[3] This assumption is used to make the problems well defined.

[4] We assume right-hand continuity in order to be consistent with the kinematic equations which are in terms of time-forward derivatives. However, if we assume left-hand continuity, our main results still hold. In order to make our presentation simple, we have made more assumptions than needed. See Ref. 8 for its extension.

such that $\{\mu(X), \nu(X)\}$ form a pair of admissible strategies. We introduce the following concepts.

Definition 1.1. Given $X \in \mathscr{E}$, $\phi^0 \in \Phi$, and $\psi^0 \in \Psi$, then at X the ϕ-vectogram when $\psi = \psi^0$ is defined by

$$f(X, \Phi, \psi^0) = \{f(X, \phi, \psi^0) \mid \phi \in \Phi\};$$

the ψ-vectogram when $\phi = \phi^0$ is defined by

$$f(X, \phi^0, \Psi) = \{f(X, \phi^0, \psi) \mid \psi \in \Psi\}.$$

Remark 1.2. The ϕ-vectogram when $\psi = \psi^0$ denotes the totality of the vectors $\dot{X}$ under the control of the minimizing player when his opponent fixes his control at ψ^0. An analogous interpretation holds for the ψ-vectogram when $\phi = \phi^0$.

Remark 1.3. Suppose that, at each point X of $\mathscr{E}$, we can successfully select a *balanced pair* (or saddle-point pair) of controls (ϕ^*, ψ^*) such that the velocity vector $\dot{X} = f(X, \phi^*, \psi^*)$ is the *best* direction of the motion of X from both players' viewpoint when their choices of the velocity vectors are restricted to $f(X, \Phi, \psi^*)$ and $f(X, \phi^*, \Psi)$, respectively.[5] Then, the differential game is essentially solved; for, once we know the correct value of ϕ and ψ for optimality, the problem reduces to one of ordinary differential equations. This observation has served as the guideline of our research.

2. Tangent Cones[6]

A sequence $\{X_n\}$ in E^{n+1} is said to converge to X_0 in the direction h, denoted by $\{X_n\} \xrightarrow{h} X_0$, if h is a unit vector, $X_n \neq X_0$, and

$$\lim_{n\to\infty} | X_n - X_0 | = 0, \qquad \lim_{n\to\infty} \frac{(X_n - X_0)}{| X_n - X_0 |} = h.$$

Given h, $\{\lambda h \mid 0 \leqslant \lambda < \infty\}$ is called the halfline generated by h.

[5] If the *payoff* at X is given by $P(X, \phi, \psi)$, then (ϕ^*, ψ^*) being a *balanced pair* (or saddle-point pair) means that the following inequalities hold: $P(X, \phi^*, \psi) \leqslant P(X, \phi^*, \psi^*) \leqslant P(X, \phi, \psi^*)$ for all $\psi \in f(X, \phi^*, \Psi)$ and $\phi \in f(X, \Phi, \psi^*)$. See Ref. 16 for further discussions.

[6] The definition follows Hestenes' book (Ref. 2). Blaquière and Leitmann have supplied another definition (Ref. 4). Under suitable assumptions (such as Assumptions 1 and 2 of Ref. 4), one can show some equivalent relations among the different definitions. The definition used here may not be as geometrically vivid as those of Ref. 4. Nevertheless, it yields more direct proofs of the results which we are looking for.

Definition 2.1. The *tangent cone* of a set S in E^{n+1} at X_0, denoted by $\Lambda(S, X_0)$, is the collection of the halflines that are generated by those unit vectors h for which there exists a sequence $\{X_n\}$ in S converging to X_0 in the direction h. That is,

$$\Lambda(S, X_0) = \{\lambda h, 0 \leqslant \lambda < \infty \mid \text{all } h \text{ such that } \exists \{X_n\} \subset S : \{X_n\} \xrightarrow{h} X_0\}.$$

We state the following results.

Lemma 2.1. The following relations hold[7]:

$$\Lambda(S, X_0) = \phi \qquad \text{if} \qquad X_0 \cap \bar{S} = \phi, \tag{1}$$

$$\Lambda(S, X_0) = E^{n+1} \qquad \text{if} \qquad X_0 \in \text{int } S, \tag{2}$$

$$\Lambda(S_1, X_0) \cap \Lambda(S_2, X_0) \supset \Lambda(S_1 \cap S_2, X_0), \tag{3}$$

$$\Lambda(S_1, X_0) \cup \Lambda(S_2, X_0) = \Lambda(S_1 \cup S_2, X_0), \tag{4}$$

$$\Lambda(S, X_0) \text{ is closed}, \tag{5}$$

$$\Lambda(S, X_0) = \Lambda(\bar{S}, X_0). \tag{6}$$

Proof. Statements (1)–(3) are obvious from definition.

Statement (4). It is clear from definition that

$$\Lambda(S_1, X_0) \cup \Lambda(S_2, X_0) \subset \Lambda(S_1 \cup S_2, X_0).$$

In order to show that

$$\Lambda(S_1, X_0) \cup \Lambda(S_2, X_0) \supset \Lambda(S_1 \cup S_2, X_0),$$

let h be a unit generating vector of $\Lambda(S_1 \cup S_2, X_0)$. We would like to show that $h \in \Lambda(S_1, X_0) \cup \Lambda(S_2, X_0)$. By hypothesis, there exists a sequence $\{X_n\}$ in $S_1 \cup S_2$ such that $\{X_n\} \xrightarrow{h} X_0$. Let $\{X_{n1}\}$ and $\{X_{n2}\}$ be the subsequences of $\{X_n\}$ such that $\{X_{n1}\} \subset S_1$ and $\{X_{n2}\} \subset S_2$. It is seen that at least one of these two subsequences contains infinite terms of $\{X_n\}$ and converges to X_0 in the direction of h.

Statement (5). Let $\{h^k\}$ be a sequence of unit generating vectors of $\Lambda(S, X_0)$ such that $\{h^k\}$ converge to h^0. Hence, for each k, there exists a sequence $\{X_n{}^k\}$ in S such that $\{X_n{}^k\} \xrightarrow{h^k} X_0$. We would like to show that h^0 is also a unit generating vector of $\Lambda(S, X_0)$. Select a sequence $\{X_k'\}$ in such a way that

$$X_k' \in \{X_n{}^k\}, \qquad | X_k' - X_0 | < \frac{1}{k}, \qquad \left| \frac{(X_k' - X_0)}{| X_k' - X_0 |} - h^k \right| < \frac{1}{k}.$$

[7] The symbol $\bar{S}$ denotes the closure of S.

It is seen that h^0 is a unit vector, $\{X_k'\}$ converges to X_0, and

$$\left|\frac{(X_k' - X_0)}{|X_k' - X_0|} - h^0\right| \leqslant \left|\frac{(X_k' - X_0)}{|X_k' - X_0|} - h^k\right| + |h^k - h^0|.$$

Therefore,

$$\{X_k'\} \xrightarrow{h^0} X_0 .$$

Statement (6). The fact that $\Lambda(S, X_0) \subset \Lambda(\bar{S}, X_0)$ is obvious. To show that $\Lambda(S, X_0) \supset \Lambda(\bar{S}, X_0)$, it suffices to show that, by (4), $\Lambda(\partial S, X_0) \subset \Lambda(S, X_0)$. Let h be a unit generating vector of $\Lambda(\partial S, X_0)$. So, there exists a sequence $\{X_n\} \subset \partial S$ such that $\{X_n\} \xrightarrow{h} X_0$. Select an infinite subsequence $\{X_k\}$ of $\{X_n\}$ such that $1/(k + 1) < | X_k - X_0 | < 1/k$, k a positive integer. One can show that the selection is possible by a contradiction proof. Since $X_k \in \partial S$, we can select $X_k' \in S$ such that $| X_k' - X_k | < 1/k^2$. It is seen that $\{X_k'\} \xrightarrow{h} X_0$, because

$$| X_k' - X_0 | \leqslant | X_k' - X_k | + | X_k - X_0 |$$

and

$$\begin{aligned}
&\left|\frac{(X_k' - X_0)}{|X_k' - X_0|} - h\right| \\
&\quad = \left|\frac{(X_k' - X_k)}{|X_k' - X_0|} + \frac{(X_k - X_0)}{|X_k' - X_0|} - h\right| \\
&\quad \leqslant \left|\frac{(X_k' - X_k)}{|X_k' - X_0|}\right| + \left|\frac{(X_k - X_0)}{|X_k - X_0|} + 0\left(\frac{1}{k^2}\right)(X_k - X_0) - h\right| \\
&\quad \leqslant \left|\frac{(k+1)}{k^2}\right| + \left|\frac{(X_k - X_0)}{|X_k - X_0|} - h\right| + 0\left(\frac{1}{k^2}\right) | X_k - X_0 | .
\end{aligned}$$

3. Regular Region

Let R be a subregion of $\mathscr{E}$.[8] We say that $\mathscr{D} = \{R_i \mid i = 1, 2,..., m\}$ is a decomposition of R provided

$$\text{each } R_i \text{ is a region in } E^{n+1}, \tag{7}$$

$$R_i \subset R, \tag{8}$$

$$R_i \cap R_j = \phi \qquad \text{if} \qquad i \neq j, \tag{9}$$

$$R \subset \bigcup_{i=1}^{m} \bar{R}_i \text{ where } \bar{R}_i \text{ is the closure of } R_i . \tag{10}$$

[8] By a region, we mean an open and connected set in E^{n+1}.

A function defined over R is piecewise continuous with respect to $\mathscr{D}$ if it is continuous in each R_i, $i = 1, 2,..., m$.

Definition 3.1. Let Σ be a subset of R. By Σ being a piecewise smooth surface with respect to the decomposition

$$\mathscr{D} = \{R_i \mid i = 1, 2,..., m\}$$

of R, we mean that $\Sigma \cap R_i$ is an n-dimensional manifold such that the tangent plane of Σ and its normal vector exist and vary continuously throughout $\Sigma \cap R_i$; furthermore, if $\mathscr{Q}$ is a continuous path of positive length contained in $\Sigma \cap R_i$, except for its initial point X_0 which is at $\Sigma \cap \partial R_i$, then the tangent plane of Σ and its normal vector have limits at X_0 along the path.

Definition 3.2. A subset Σ of a region R is said to *nicely separate* R into two parts if there exist two nonempty subregions R_1 and R_2 of R such that each continuous arc in R, which connects a point of R_1 and a point of R_2, will intersect Σ; furthermore, when R is treated as a subspace with the relative topology, $\Sigma = \partial R_1 = \partial R_2$.

Definition 3.3. Let $\mathscr{R} \subset \mathscr{E}$. We say that $\mathscr{R}$ is a regular region of the game if the following assumptions are satisfied.

Assumption 3.1. The value function[9] of the game $V(X)$ is continuous in $\mathscr{R}$ and strictly increasing[10] in x_0.

Remark 3.1. Clearly, from this assumption, an isovalued surface $\Sigma(a) = \{X \mid X \in \mathscr{R}, V(X) = a\}$ in R nicely separates $\mathscr{R}$ into two parts, namely,

$$A/\Sigma(a) = \{X \mid X \in \mathscr{R}, V(X) > a\}, \qquad B/\Sigma(a) = \{X \mid X \in \mathscr{R}, V(X) < a\}.$$

Note that, if we treat x_0 as the vertical axis, $A/\Sigma(a)$ denotes the points of $\mathscr{R}$ *above* $\Sigma(a)$, and $B/\Sigma(a)$ denotes those below $\Sigma(a)$. This notation has been used by Leitmann and his associates (Refs. 3–7). We shall adopt it.

Remark 3.2. Given $X_0 \in \mathscr{R}$, there is a unique isovalued surface

[9] For the definition of the value of a game, we refer to Refs. 1 and 7.

[10] See Footnote 2 for the *strictly increasing* assumption. Note that the assumption implies that, under the optimal play, the game can always terminate.

$\Sigma(V(X_0))$ containing X_0. The surface separates $\mathscr{R}$ into two parts, and the tangent cones

$$\Lambda(\Sigma(V(X_0)), X_0), \qquad \Lambda(A/\Sigma(V(X_0)), X_0), \qquad \Lambda(B/\Sigma(V(X_0)), X_0)$$

are then defined at X_0. For notational convenience, we shall set

$$\Lambda_\Sigma(X_0) = \Lambda(\Sigma(V(X_0)), X_0), \qquad \Lambda_A(X_0) = \Lambda(A/\Sigma(V(X_0)), X_0),$$
$$\Lambda_B(X_0) = \Lambda(B/\Sigma(V(X_0)), X_0).$$

Assumption 3.2.[11] The tangent cone $\Lambda_\Sigma(X_0)$ nicely separates E^{n+1} into two parts. Both $\Lambda_A(X_0)$ and $\Lambda_B(X_0)$ have interiors, and $\Lambda_A(X_0) \cap \Lambda_B(X_0) \subset \Lambda_\Sigma(X_0)$.

Assumption 3.3. There exists a decomposition

$$\mathscr{D} = \{R_i \mid i = 1, 2, \ldots, m\}$$

such that the optimal strategies $\mu^*(X)$ and $\nu^*(X)$ are piecewise continuous with respect to $\mathscr{D}$ and such that all isovalued surfaces in $\mathscr{R}$ are piecewise smooth with respect to $\mathscr{D}$.

Assumption 3.4. Throughout $\mathscr{R}$, an optimal path

$$\mathscr{Q} = \{X(t) \mid t_1 \leqslant t \leqslant t_2\}$$

resulting from the optimal strategies μ^* and ν^* can only have a finite number of common points with the boundaries of $\mathscr{R}_i$, $i = 1, 2, \ldots, m$; and, along $\mathscr{Q}$,

$$\lim_{\epsilon \to 0} f(X(t \pm \epsilon), \mu^*(X(t \pm \epsilon)), \nu^*(X(t \pm \epsilon))) \neq 0 \qquad \text{for any } t \in (t_1, t_2).$$

From now on, unless otherwise stated, $\mathscr{R}$ will be used to denote a regular region. We first show the following relations among the tangent cones in $\mathscr{R}$.

Lemma 3.1. At each point X_0 of a regular region $\mathscr{R}$, the following relations hold:

$$\Lambda_A(X_0) \cup \Lambda_B(X_0) = E^{n+1}, \tag{11}$$
$$\Lambda_A(X_0) \cap \Lambda_B(X_0) = \Lambda_\Sigma(X_0), \tag{12}$$
$$\operatorname{comp} \Lambda_A(X_0) = \operatorname{int} \Lambda_B(X_0), \tag{13}$$
$$\operatorname{comp} \Lambda_B(X_0) = \operatorname{int} \Lambda_A(X_0). \tag{14}$$

[11] Blaquière and Leitmann (Ref. 4), by using different assumptions and definitions, have worked out some quite detailed relations among cones.

Proof. *Statement* (11). By (2), (4), (6) of Lemma 2.1, we have that

$$\Lambda_A(X_0) \cup \Lambda_B(X_0) = \Lambda(\mathscr{R}, X_0) = E^{n+1},$$

because $X_0 \in \operatorname{int} \mathscr{R}$.

Statement (12). By Assumption 3.2 on $\mathscr{R}$,

$$\Lambda_A(X_0) \cap \Lambda_B(X_0) \subset \Lambda_\Sigma(X_0).$$

To show the inverse inclusion, note that

$$\overline{A/\Sigma(V(X_0))} \cap \overline{B/\Sigma(V(X_0))} = \Sigma(V(X_0)).$$

By applying (3) and (6) of Lemma 2.1, we have that

$$\Lambda_A(X_0) \cap \Lambda_B(X_0)$$
$$= \Lambda(\overline{A/\Sigma(V(X_0))}, X_0) \cap \Lambda(\overline{B/\Sigma(V(X_0))}, X_0) \supset \Lambda(\Sigma(V(X_0)), X_0) = \Lambda_\Sigma(X_0).$$

Statements (13)–(14). From Assumption 3.2 that $\Lambda_\Sigma(X_0)$ nicely separates E^{n+1} into two parts, there exist two connected nonempty regions $\mathscr{A}$ and $\mathscr{B}$ of E^{n+1} such that $\Lambda_\Sigma(X_0)$, $\mathscr{A}$, $\mathscr{B}$ form a partition of E^{n+1}, and $\Lambda_\Sigma(X_0) = \partial\mathscr{A} = \partial\mathscr{B}$. We like to show that comp $\Lambda_A(X_0)$ and int $\Lambda_B(X_0)$ are equal to one of $\{\mathscr{A}, \mathscr{B}\}$ and comp $\Lambda_B(X_0)$ and int $\Lambda_A(X_0)$ are equal to the other of $\{\mathscr{A}, \mathscr{B}\}$.

First, let us treat $\mathscr{A}$ as a subspace. As $\Lambda_A(X_0)$ and $\Lambda_B(X_0)$ are closed in E^{n+1} [see (5) of Lemma 2.1], so are $\mathscr{A} \cap \Lambda_A(X_0)$ and $\mathscr{A} \cap \Lambda_B$ in the relative topology.

Next, from (11)–(12), we have that $\Lambda_A(X_0) \cup \Lambda_B(X_0) = E^{n+1}$ and $\Lambda_A(X_0) \cap \Lambda_B(X_0) = \Lambda_\Sigma(X_0)$. Since $\mathscr{A} \subset E^{n+1} - \Lambda_\Sigma(X_0)$, $\mathscr{A} \cap \Lambda_A(X_0)$ and $\mathscr{A} \cap \Lambda_B(X_0)$ form a partition of $\mathscr{A}$. Since $\mathscr{A} \cap \Lambda_A(X_0)$ and $\mathscr{A} \cap \Lambda_B(X_0)$ are closed with respect to $\mathscr{A}$ and form a partition of $\mathscr{A}$, $\mathscr{A} \cap \Lambda_A(X_0)$ and $\mathscr{A} \cap \Lambda_B(X_0)$ are also open with respect to $\mathscr{A}$. However, as $\mathscr{A}$ is connected, $\mathscr{A} \cap \Lambda_A(X_0)$ and $\mathscr{A} \cap \Lambda_B(X_0)$ must be either $\mathscr{A}$ or ϕ. From forming a partition, if $\mathscr{A} \cap \Lambda_A(X_0) = \mathscr{A}$, then $\mathscr{A} \cap \Lambda_B(X_0) = \phi$ and vice versa. That is, either

Case (1): $\mathscr{A} \subset \Lambda_A(X_0)$ and $\mathscr{A} \cap \Lambda_B(X_0) = \phi$,

or

Case (2): $\mathscr{A} \cap \Lambda_A(X_0) = \phi$ and $\mathscr{A} \subset \Lambda_B(X_0)$.

Similarly, if instead of $\mathscr{A}$ we consider $\mathscr{B}$, we have either

Case (3): $\mathscr{B} \subset \Lambda_A(X_0)$ and $\mathscr{B} \cap \Lambda_B(X_0) = \phi$, or

Case (4): $\mathscr{B} \cap \Lambda_A(X_0) = \phi$ and $\mathscr{B} \subset \Lambda_B(X_0)$.

There are four possible combinations.

Case (1) and Case (3) imply that $\Lambda_B(X_0) \subset \text{comp}(\mathscr{A} \cup \mathscr{B}) = \Lambda_\Sigma(X_0)$. So, int $\Lambda_B(X_0) = \phi$ [because, by nice separation, int $\Lambda_\Sigma(X_0) = \phi$]. This contradicts Assumption 3.2 of a regular region.

Case (1) and Case (4) imply that

$$\mathscr{A} \subset \Lambda_A(X_0) \subset \text{comp}(\mathscr{B}) = \mathscr{A} \cup \Lambda_\Sigma(X_0) = \bar{\mathscr{A}},$$
$$\mathscr{B} \subset \Lambda_B(X_0) \subset \text{comp}(\mathscr{A}) = \mathscr{B} \cup \Lambda_\Sigma(X_0) = \bar{\mathscr{B}}.$$

Since $\Lambda_A(X_0)$ and $\Lambda_B(X_0)$ are closed [see (5) of Lemma 2.1],

$$\Lambda_A(X_0) = \bar{\mathscr{A}} = \mathscr{A} \cup \Lambda_\Sigma(X_0), \qquad \Lambda_B(X_0) = \bar{\mathscr{B}} = \mathscr{B} \cup \Lambda_\Sigma(X_0).$$

It is seen that

$$\text{comp}\ \Lambda_A(X_0) = \mathscr{B} = \text{int}\ \Lambda_B(X_0),$$
$$\text{comp}\ \Lambda_B(X_0) = \mathscr{A} = \text{int}\ \Lambda_A(X_0).$$

By exactly the same approach, Case (2) and Case (3) imply that

$$\text{comp}\ \Lambda_B(X_0) = \mathscr{B} = \text{int}\ \Lambda_A(X_0),$$
$$\text{comp}\ \Lambda_A(X_0) = \mathscr{A} = \text{int}\ \Lambda_B(X_0),$$

while Case (2) and Case (4) imply that $\Lambda_A(X_0) = \Lambda_\Sigma(X_0)$, another impossibility, because int $\Lambda_\Sigma(X_0) = \phi$.

4. Penetration of Isovalued Surfaces

Let $S = \{X(t) \mid X(0) = X_0, t > 0\}$ be a smooth path with positive length starting at $X_0 \in \Sigma(a)$, $\Sigma(a)$ an isovalued surface in a given regular region $\mathscr{R}$. By S penetrating into $A/\Sigma(a)$ or $B/\Sigma(a)$ we mean that there exists a positive ϵ such that S lies completely in the interior of $A/\Sigma(a)$ [or $B/\Sigma(a)$] for $0 < t < \epsilon$. If a player under the restriction of his vectogram can select a path penetrating into $A/\Sigma(a)$ or $B/\Sigma(a)$, we say that he can *make X penetrate into* $A/\Sigma(a)$ *or* $B/\Sigma(a)$.

Lemma 4.1. The following statements hold:

(i) If $\dot{X}(0^+) \in \text{int}\ \Lambda_A(X_0)$ [or int $\Lambda_B(X_0)$], then S must penetrate into $A/\Sigma(a)$ [or $B/\Sigma(a)$].

(ii) If S penetrates into $A/\Sigma(a)$ [or $B/\Sigma(a)$], then $\dot{X}(0^+) \in \Lambda_A(X_0)$ [or $\Lambda_B(X_0)$].

Proof. (i) Suppose that the assertion were false. Then, we could find a point sequence $\{X_n\}$ on S such that $\{X_n\} \subset \Sigma(a) \cup B/\Sigma(a)$ and $\{X_n\}$ converges to X_0 in the direction of $\dot{X}(0^+)$. But this implies that $\dot{X}(0^+) \in \Lambda_B(X_0)$. In view of (13)–(14) of Lemma 3.1, $\dot{X}(0^+) \notin \text{int } \Lambda_A(X_0)$, resulting in a contradiction.

(ii) Suppose that the assertion were false. Then, by the Lemma 3.1, $\dot{X}(0^+) \in \text{int } \Lambda_B(X_0)$. By (i), S would penetrate into $B/\Sigma(a)$ rather than $A/\Sigma(a)$, resulting in a contradiction.

Corollary 4.1. A player can make X penetrate into $A/\Sigma(a)$ or $B/\Sigma(a)$ if there exists a velocity vector of his vectogram lying in the interior of $\Lambda_A(X_0)$ or $\Lambda_B(X_0)$.

Proof. It follows immediately from the Lemma 4.1.

Corollary 4.2. Let (μ^*, ν^*) be an optimal pair of strategies in a regular region $\mathscr{R}$. Then, throughout $\mathscr{R}$, $f(X, \mu^*(X), \nu^*(X)) \in \Lambda_\Sigma(X)$.

Proof. Since, under the optimal play, neither player can make X penetrate into $A/\Sigma(V(X))$ or $B/\Sigma(V(X))$, by Corollary 4.1 and Lemma 3.1, we have

$$\begin{aligned} f[X, \mu^*(X), \nu^*(X)] &\in \text{comp}(\text{int } \Lambda_A(X)) \cap \text{comp}(\text{int } \Lambda_B(X)) \\ &= \Lambda_B(X) \cap \Lambda_A(X) = \Lambda_\Sigma(X). \end{aligned}$$

5. First Necessary Condition for Optimal Strategies and Isaacs' Main Equation

Theorem 5.1. Suppose that μ^* and ν^* are optimal strategies in a regular region $\mathscr{R}$. Then, at each point X of $\mathscr{R}$, the following conditions hold:

$$f(X, \mu^*(X), \Psi) \subset \Lambda_B(X), \tag{15}$$

$$f(X, \Phi, \nu^*(X)) \subset \Lambda_A(X). \tag{16}$$

Proof. Suppose that (15) does not hold at a point X_0 of $\mathscr{R}$. Let $\mu^*(X_0) = \phi^*$ and $\nu^*(X_0) = \psi^*$. Then, there exists a vector $U \neq f(X_0, \phi^*, \psi^*)$ of $f(X_0, \phi^*, \Psi)$ such that $U \notin \Lambda_B(X)$. By Lemma

3.1, $U \in \text{int}\, \Lambda_A(X_0)$. In view of Corollary 4.1, the maximizing player can select U to make X penetrate into $A/\Sigma(V(X_0))$ to increase his payoff. This contradicts the assumption that ψ^* is the optimal control of the maximizing player at X_0. Hence, (15) must hold. A similar proof holds for (16).

Corollary 5.1. (*Isaacs' Main Equation*). Suppose that $V(X)$ has continuous derivatives $\nabla V(X) = (V_{x_0}, V_{x_1}, \ldots, V_{x_n})$ in a subregion $\mathscr{R}^0$ of $\mathscr{R}$. Let $X \in \mathscr{R}^0$ and $f(X, \mu^*(X), \nu^*(X)) = f(X, \phi^*, \psi^*)$. Then,

$$\nabla V(X) \cdot f(X, \phi^*, \psi^*) = 0, \tag{17}$$

$$\min_{\phi \in \Phi} \nabla V(X) \cdot f(X, \phi, \psi^*) = 0, \tag{18}$$

$$\max_{\psi \in \Psi} \nabla V(X) \cdot f(X, \phi^*, \psi) = 0, \tag{19}$$

$$\min_{\phi \in \Phi} \max_{\psi \in \Psi} \nabla V(X) \cdot f(X, \phi, \psi) = \max_{\psi \in \Psi} \min_{\phi \in \Phi} \nabla V(X) \cdot f(X, \phi, \psi) = 0. \tag{20}$$

Proof. Note, if $|\nabla V| = 0$, then (17)–(20) hold trivially. Hence, we may assume that $|\nabla V| \neq 0$. In $\mathscr{R}^0$, as ∇V is continuous and normal to $\Sigma(V(X))$, the isovalued surface passing through X, we have

$$\Lambda_\Sigma(X) = \{U \mid \nabla V(X) \cdot U = 0\}.$$

By (13)–(14) of Lemma 3.1 and ∇V pointing into the increasing direction of V, it is sccn that

$$\Lambda_A(X) = \{U \mid \nabla V(X) \cdot U \geqslant 0\}, \tag{21}$$

$$\Lambda_B(X) = \{U \mid \nabla V(X) \cdot U \leqslant 0\}. \tag{22}$$

The assertion (17) follows immediately from Corollary 4.2. The assertions (18) and (19) then follow immediately from the Theorem 5.1 and assertion (17). In order to establish (20), observe that, as

$$f(X, \phi^0, \psi^*) \in \Lambda_A(X) \qquad \text{for any } \phi^0 \in \Phi,$$

we have

$$\max_{\psi \in \Psi} \nabla V(X) \cdot f(X, \phi^0, \psi) \geqslant \nabla V(X) \cdot f(X, \phi^0, \psi^*) \geqslant 0.$$

Because of (19), we have

$$\min_{\phi \in \Phi} \max_{\psi \in \Psi} \nabla V(X) \cdot f(X, \phi, \psi) = 0.$$

The equality

$$\max_{\psi \in \Psi} \min_{\phi \in \Phi} \nabla V(X) \cdot f(X, \phi, \psi) = 0$$

can be established similarly.

Remark 5.1. In the proof of Corollary 5.1, we did not use Isaacs' minimax assumption.

6. Semipermeable Direction

Definition 6.1. Given $X \in \mathscr{E}$, $\phi^0 \in \Phi$, $\psi^0 \in \Psi$, the vector $f(X, \phi^0, \psi^0)$ represents a semipermeable direction iff $f(X, \phi^0, \psi^0) \neq 0$ and there exists a homogeneous nonzero linear form F over E^{n+1} such that

$$F[f(X, \phi^0, \psi^0)] = 0, \tag{23}$$

$$F[f(X, \phi^0, \psi)] \leqslant 0 \qquad \text{for all } \psi \in \Psi, \tag{24}$$

$$F[f(X, \phi, \psi^0)] \geqslant 0 \qquad \text{for all } \phi \in \Phi. \tag{25}$$

If $f(X, \phi^0, \psi^0)$ represents a semipermeable direction, we call (ϕ^0, ψ^0) a pair of *semipermeable controls* at X.

Theorem 6.1. The vector $f(X, \phi^0, \psi^0)$ represents a semipermeable direction iff it is nonzero and the vectograms $f(X, \phi^0, \Psi)$ and $f(X, \Phi, \psi^0)$ are completely contained in the two different closed halfspaces separated by an n-dimensional hyperplane containing $f(X, \phi^0, \psi^0)$.

Proof. *Sufficiency.* Let λ be the unit normal vector of the hyperplane such that $\lambda \cdot f(X, \phi, \psi^0) \geqslant 0$ for all $\phi \in \Phi$. This is possible by our assumptions. Put $F(U) = \lambda \cdot U$ for $U \in E^{n+1}$. Then, F is a homogeneous nonzero linear form of $f(X, \phi, \psi)$. Furthermore, F satisfies (23)–(25).

Necessity. Since $f(X, \phi^0, \psi^0)$ represents a semipermeable direction, there exists a nonzero homogeneous form $F(U)$ such that (23)–(25) are satisfied. Observe that $\{U \mid F(U) = 0\}$ is an n-dimensional hyperplane containing $f(X, \phi^0, \psi^0)$; $\{U \mid F(U) \geqslant 0\}$ and $\{U \mid F(U) \leqslant 0\}$ are the two closed halfspaces containing $f(X, \Phi, \psi^0)$ and $f(X, \phi^0, \Psi)$ respectively.

Given a set S, we shall denote its *convex hull* by $\mathscr{H}(S)$.

Theorem 6.2. Suppose that $f(X, \phi^0, \psi^0)$ represents a semipermeable direction. Then, $f(X, \phi^0, \psi^0) \in \partial\mathscr{H}(f(X, \phi^0, \Psi)) \cap \partial\mathscr{H}(f(X, \Phi, \psi^0))$.

Proof. Observe that (23) and (24) imply that F, with respect to $f(X, \phi^0, \Psi)$, takes its maximum value at $f(X, \phi^0, \psi^0)$. As F is a nonzero linear form, the maximal point must be on the boundary of $\mathscr{H}[f(X,\phi^0,\Psi)]$. That is,

$$f(X, \phi^0, \psi^0) \in \partial\mathscr{H}[f(X, \phi^0, \Psi)].$$

On the other hand, (23) and (25) imply that F, with respect to $f(X, \Phi, \psi^0)$, has its minimal value at $f(X, \phi^0, \psi^0)$. Since F is a nonzero linear form, the minimal point must be on the boundary of $\mathscr{H}[f(X, \Phi, \psi^0)]$. That is, $f(X, \phi^0, \psi^0) \in \partial\mathscr{H}(f(X, \Phi, \psi^0))$. Hence,

$$f(X, \phi^0, \psi^0) \in \partial\mathscr{H}(f(X, \Phi, \psi^0)) \cap \partial\mathscr{H}(f(X, \phi^0, \Psi)).$$

Remark 6.1. Theorem 6.2 can be used to identify a velocity vector to represent a semipermeable direction. For instance, if f is linear with respect to ϕ and ψ, then, throughout the playing space $\mathscr{E}$, except in degenerate cases, (ϕ^0, ψ^0) is a pair of semipermeable controls only if (ϕ^0, ψ^0) are extreme points of Φ and Ψ, respectively.

7. Second Necessary Condition for Optimal Strategies and Necessary Condition for a Regular Region to Exist

Let $\mathscr{R}$ be a regular region with the decomposition

$$D = \{\mathscr{R}_i \mid i = 1, 2, \ldots, m\}.$$

Then, at each point X_0 of each $\mathscr{R}_i$, there passes an isovalued surface Σ such that the surface is smooth at X_0 with a continuous normal vector. We shall let $\lambda(X_0)$ be the unit normal vector at X_0 to the surface Σ such that $\lambda(X_0)$ points into A/Σ. Note that $\lambda(X_0)$ essentially is an adjoint variable of the game (see any of Refs. 3–7). Following these definitions, and since Σ is smooth at X_0, we have

$$\Lambda_\Sigma(X_0) = \{U \mid \lambda(X_0) \cdot U = 0\}, \tag{26}$$

$$\Lambda_A(X_0) = \{U \mid \lambda(X_0) \cdot U \geqslant 0\}, \tag{27}$$

$$\Lambda_B(X_0) = \{U \mid \lambda(X_0) \cdot U \leqslant 0\}. \tag{28}$$

Now, suppose that $X_0 \in \mathscr{R}$, and

$$X_0 \in \bigcap_{k=1}^{\gamma} \partial\mathscr{R}_{i_k}, \qquad 1 \leqslant \gamma \leqslant m, \qquad \mathscr{R}_{i_k} \in D.$$

Then, the isovalued surface Σ may not be smooth, and $\lambda(X)$ may not be defined at X_0. In such a case, we let

$$\mathscr{Q} = \{X(t) \mid t_1 < t < t_2, X(t_0) = X_0, t_1 < t_0 < t_2\}$$

be the optimal path passing through X_0. Since $\mathscr{Q}$ can intersect each $\partial\mathscr{R}_i$ at most at a finite number of points, there exist t_1' and t_2',

$$t_1 < t_1' < t_0 < t_2' < t_2,$$

such that, for some $1 \leqslant k^1, k^2 \leqslant \gamma$,

$$X(t) \in \mathscr{R}_{i_{k^1}} \qquad \text{for } t_1' < t < t_0,$$

$$X(t) \in \mathscr{R}_{i_{k^2}} \qquad \text{for } t_0 < t < t_2'.$$

We define

$$\lambda^{\pm}(X_0) = \lim_{\substack{\epsilon \to 0 \\ \epsilon > 0}} \lambda(X(t_0 \pm \epsilon)). \tag{29}$$

The existence of $\lambda \pm (X_0)$ is guaranteed by Assumption 3.4 of a regular region. We also define the following limiting cones:

$$\Lambda_\Sigma^{\pm}(X_0) = \{U \mid \lambda^{\pm}(X_0) \cdot U = 0\}, \tag{30}$$

$$\Lambda_A^{\pm}(X_0) = \{U \mid \lambda^{\pm}(X_0) \cdot U \geqslant 0\}, \tag{31}$$

$$\Lambda_B^{\pm}(X_0) = \{U \mid \lambda^{\pm}(X_0) \cdot U \leqslant 0\}. \tag{32}$$

We shall extend the above definitions of the limiting normal vector and cones to every point of $\mathscr{R}$ in an obvious way. We first state the following result.

Lemma 7.1. Let (μ^*, ν^*) be a pair of optimal strategies. Then, at each point of $\mathscr{R}$,

$$f(X, \mu^*(X), \Psi) \subset \Lambda_B^+(X), \qquad f(X, \Phi, \nu^*(X)) \subset \Lambda_A^+(X).$$

Proof. First, suppose that $X \in \mathscr{R}_i$ for some $\mathscr{R}_i \in D$. Then,

$$\lambda^+(X) = \lambda(X) = \lambda^-(X), \qquad \Lambda_A^+(X) = \Lambda_A(X), \qquad \Lambda_B^+(X) = \Lambda_B(X).$$

Our assertion then follows immediately from Theorem 5.1.

Next, suppose that

$$X_0 \in \bigcap_{k=1}^{\gamma} \partial\mathscr{R}_{i_k}, \qquad 1 \leqslant \gamma \leqslant m, \qquad \mathscr{R}_{i_k} \in D.$$

Let

$$\mathscr{Q}^+ = \{X^*(t) \mid 0 \leqslant t < \epsilon, X(0) = X_0\}$$

be a part of the optimal path starting at X_0. By Assumption 3.4, when ϵ is sufficiently small, we have $(\mathscr{Q}^+ - X_0) \in \mathscr{R}_{i_\alpha}$ for some $1 \leqslant \alpha \leqslant \gamma$.

In order to show that

$$f(X_0, \mu^*(X_0), \Psi) \subset \Lambda_B^+(X_0), \qquad f(X_0, \Phi, \nu^*(X_0)) \subset \Lambda_A^+(X_0),$$

suppose that $f(X_0, \mu^*(X_0), \Psi) \not\subset \Lambda_B^+(X_0)$. Then, there exist $\psi^0 \in \Psi$ such that

$$f(X_0, \mu^*(X_0), \psi^0) \in \text{int } \Lambda_A^+(X_0).$$

That is,

$$\lambda^+(X_0) \cdot f(X_0, \mu^*(X_0), \psi^0) > 0. \tag{33}$$

Since $\mu^*(X_0)$ is right-hand continuous (see the convention of Section 1), so is $f(X_0, \mu^*(X_0), \psi^0)$ (ψ^0 is fixed). That is,

$$\lim_{\substack{t\to 0\\ t>0}} f(X^*(t), \mu^*(X^*(t)), \psi^0) = f(X_0, \mu^*(X_0), \psi^0). \tag{34}$$

In view of (29), (33), (34), there is a t', $0 < t' < \epsilon$, such that

$$\lambda(X^*(t')) \cdot f[X^*(t'), \mu^*(X(t')), \psi^0] > 0.$$

That is,

$$f[X^*(t'), \mu^*(X^*(t')), \psi^0] \notin \Lambda_B(X^*(t')) \qquad \text{even if } X^*(t') \in \mathscr{R}_{i_\alpha}.$$

This contradicts the result established in the previous case. So,

$$f[X_0, \mu^*(X_0), \Psi] \subset \Lambda_B^+(X_0).$$

Similarly, one can show that

$$f[X_0, \Phi, \nu^*(X_0)] \subset \Lambda_A^+(X_0).$$

Now, recall that at each $X \in \mathscr{R}$, $f(X, \mu^*(X), \nu^*(X))$ is a time-forward derivative. So, we have that $f(X, \mu^*(X), \nu^*(X)) \in \Lambda_\Sigma^+(X)$. This observation and Lemma 7.1 yield the following second necessary condition of optimal strategies.

Theorem 7.1. Let (μ^*, ν^*) be a pair of optimal strategies in a regular region $\mathscr{R}$. Then, throughout $\mathscr{R}$, $f(X, \mu^*(X), \nu^*(X))$ represents a semipermeable direction with respect to $F(U) = \lambda^+(X) \cdot U$.

Remark 7.1. Theorem 7.1 says that, in order for (μ^*, ν^*) to be a pair of optimal (saddle-point) strategies in $\mathscr{R}$, at each point of $\mathscr{R}$, $(\mu^*(X), \nu^*(X))$ must form a pair of saddle-point controls with respect to the admissible control sets Φ and Ψ and a payoff of $\lambda \cdot f(X, \phi, \psi)$ with $\lambda \neq 0$.

Following this theorem, we have a necessary condition for a region to be regular.

Corollary 7.1. Let $\mathscr{R}$ be a regular region. Then, there must exist at least a pair of strategies (μ^0, ν^0) such that, throughout $\mathscr{R}$, $f(X, \mu^0(X), \nu^0(X))$ represents a semipermeable direction.

8. Semipermeable Surfaces and Their Smooth Extension

Definition 8.1. Given an initial state $X^0 \in \mathscr{R}$ ($\mathscr{R}$ an arbitrary region of $\mathscr{E}$) and an admissible strategy μ^0 (or ν^0), the attainable states $A(X^0, \mu^0)$ [or $A(X^0, \nu^0)$] in $\mathscr{R}$ resulting from μ^0 (or ν^0) will be defined by

$$A(X^0, \mu^0) = \left\{ X \in \mathscr{R} \,\middle|\, \begin{array}{l} \exists \text{ an admissible } \nu \text{ such that } \dot{X} = f[X, \mu^0(X), \nu(X)] \text{ will} \\ \text{transfer } X^0 \text{ to } X \end{array} \right\},$$

$$A(X^0, \nu^0) = \left\{ X \in \mathscr{R} \,\middle|\, \begin{array}{l} \exists \text{ an admissible } \mu \text{ such that } \dot{X} = f[X, \mu(X), \nu^0(X)] \text{ will} \\ \text{transfer } X^0 \text{ to } X \end{array} \right\}.$$

Definition 8.2. Let $\mathscr{R}$ be a region of $\mathscr{E}$. The surface $\Sigma \subset \mathscr{R}$ is a *semipermeable surface* in $\mathscr{R}$ if Σ nicely separates a neighborhood N of Σ in $\mathscr{R}$ into two parts N_A and N_B and there exists a pair of admissible strategies (μ^0, ν^0) over $\mathscr{R}$ such that

$$A(X^0, \mu^0) \cap N \subset N_A \cup \Sigma, \qquad A(X^0, \nu^0) \cap N \subset N_B \cup \Sigma$$

for all $X^0 \in \Sigma$.

Remark 8.1. The definition of a semipermeable surface in a region $\mathscr{R}$ is independent of the selection of a neighborhood of Σ. One can show that in $\mathscr{R}$, any neighborhood of Σ separated by Σ has the properties required in the definition.

Remark 8.2. If Σ is a semipermeable surface in $\mathscr{R}$ with strategies (μ^0, ν^0), then each path starting at a point of Σ and satisfying $\dot{X} = f(X, \mu^0(X), \nu^0(X))$ will stay on Σ as long as the path is in a neighborhood and separated by Σ.

Remark 8.3. Let (μ^0, ν^0) be defined over a region $\mathscr{R}$, and let M be an $(n-1)$-dimensional manifold in $\mathscr{R}$. Suppose that

$$\Sigma(t_1, t_2) = \left\{ X(t) \,\middle|\, \begin{array}{c} \dot{X} = f(X, \mu^0(X), \nu^0(X)), \quad X(t_0) = X_0 \in M, \\ t_1 < t, \quad t_0 < t_2 \end{array} \right\}$$

is a semipermeable surface in $\mathscr{R}$. Then,

$$\Sigma(t', t'') = \{X(t) \mid X(t) \in \Sigma(t_1, t_2), t_1 \leqslant t' < t'' \leqslant t_2\}$$

is also a semipermeable surface in $\mathscr{R}$.

Theorem 8.1.[12] Each isovalued surface Σ of a regular region $\mathscr{R}$ is semipermeable.

Proof. From the assumption on $\mathscr{R}$, Σ nicely separates $\mathscr{R}$. Let μ^* and ν^* be a pair of optimal strategies in $\mathscr{R}$. Given $X^0 \in \Sigma$, it suffices to show that

$$A(X^0, \mu^*) \subset (B/\Sigma) \cup \Sigma, \tag{35}$$

$$A(X^0, \nu^*) \subset (A/\Sigma) \cup \Sigma. \tag{36}$$

Suppose that (35) is false. Then, there is an admissible strategy ν^0 such that the path resulting from μ^* and ν^0 and starting at X^0 will reach some point, say X_A of A/Σ. As $V(X^0) < V(X_A)$, ν^* cannot be the optimal strategy, resulting in a contradiction. So,

$$A(X^0, v^*) \subset (B/\Sigma) \cup \Sigma.$$

Similarly, one can show that

$$A(X^0, \mu^*) \subset (A/\Sigma) \cup \Sigma.$$

Some relations between semipermeable directions and semipermeable surfaces can be stated as follows.

Theorem 8.2. Let (μ^0, ν^0) be admissible, and continuous in a region $\mathscr{R}$ of $\mathscr{E}$, and let $M_0 \subset \mathscr{R}$ be an $(n-1)$-dimensional manifold. Suppose that

$$\Sigma = \{X(t) \mid \dot{X} = f[X, \mu^0(X), \nu^0(X)], X(t_0) = X_0 \in M_0, t_1 < t, t_0 < t_2\}$$

[12] This result has been shown by Isaacs under a stronger assumption. Leitmann and his associates also have derived this result (see Refs. 1, 5–7).

is a smooth n-dimensional manifold, which nicely separates $\mathscr{R}$ into two parts $\mathscr{R}_A$ and $\mathscr{R}_B$. Then, (i) if Σ is a semipermeable surface with strategies (μ^0, ν^0), at each point X of Σ, $f(X, \mu^0(X), \nu^0(X))$ represents a semipermeable direction; and (ii) if $n(X)$ is the unit normal vector pointing into $\mathscr{R}_A$ or $\mathscr{R}_B$ at a point X of Σ and if at each point X

$$n(X) \cdot f(X, \mu^0(X), \psi) \leqslant 0 \qquad \text{[equality holds only if } \psi = \nu^0(X)],$$

$$n(X) \cdot f(X, \phi, \nu^0(X)) \geqslant 0 \qquad \text{[equality holds only if } \phi = \mu^0(X)],$$

then Σ is a semipermeable surface with strategies (μ^0, ν^0).

Proof. Let $\lambda(X)$ be the unit normal vector to Σ pointing into $\mathscr{R}_A$ at $X \in \Sigma$. From the assumption, we find that

$$\lambda(X) \cdot f[X, \mu^0(X), \nu^0(X)] = 0, \qquad \Lambda(\mathscr{R}_A, X) = \{U \mid \lambda(X) \cdot U \geqslant 0\},$$

$$\Lambda(\mathscr{R}_B, X) = \{U \mid \lambda(X) \cdot U \leqslant 0\},$$

where $\Lambda(\mathscr{R}_A, X)$ [or $\Lambda(\mathscr{R}_B, X)$] is the tangent cone of $\mathscr{R}_A$ (or $\mathscr{R}_B$) at X.

We first prove Assertion (i). Since Σ is a semipermeable surface with strategies (μ^0, ν^0), without loss of generality we can assume that, for all $X \in \Sigma$,

$$A(X, \mu^0) \subset \Sigma \cup \mathscr{R}_A, \tag{37}$$

$$A(X, \nu^0) \subset \Sigma \cup \mathscr{R}_B. \tag{38}$$

By exactly the same approach as for Lemma 4.1, (37) implies that, for all $X \in \Sigma$,

$$f[X, \mu^0(X), \psi] \notin \operatorname{int} \Lambda(\mathscr{R}_B, X) \qquad \text{for all } \psi \in \Psi.$$

That is,

$$f[X, \mu^0(X), \Psi] \subset \Lambda(\mathscr{R}_A, X).$$

Similarly,

$$f[X, \Phi, \nu^0(X)] \subset \Lambda(\mathscr{R}_B, X).$$

In view of the definition of $\Lambda(\mathscr{R}_A, X)$ and $\Lambda(\mathscr{R}_B, X)$, we are sure that $f[X, \mu^0(X), \nu^0(X)]$ represents a semipermeable direction throughout Σ.

Next, we prove Assertion (ii). Suppose that Σ is not a semipermeable surface with strategies (μ^0, ν^0). Then, there is a point X such that either

$$A(X, \mu^0) \cap A(X, \nu^0) \cap \mathscr{R}_A \neq \phi$$

or

$$A(X, \mu^0) \cap A(X, \nu^0) \cap \mathscr{R}_B \neq \phi.$$

By the same approach as in Section 4, if $A(X, \mu^0) \cap A(X, \nu^0) \cap \mathscr{R}_A \neq \phi$, there exist ϕ' and ψ' such that

$$f[X, \mu^0(X), \psi'] \in \Lambda(\mathscr{R}_A, X), \qquad f[X, \phi', \nu^0(X)] \in \Lambda(\mathscr{R}_A, X).$$

That is,

$$\lambda(X) \cdot f[X, \mu^0(X), \psi'] \geqslant 0, \qquad \lambda(X) \cdot f[X, \phi', \nu^0(X)] \geqslant 0.$$

However, since $\phi' \neq \mu^0(X)$ and $\psi' \neq \nu^0(X)$ [otherwise the resulting paths, starting at X, and satisfying, respectively, $\dot{X} = f(X, \mu^0(X), \psi')$ and $\dot{X} = f(X, \phi', \nu^0(X))$ cannot enter $\mathscr{R}_A$], and since $n(X) = \pm \lambda(X)$, we cannot have the above inequalities simultaneously [from the assumption in (ii)]. This leads to a contradiction.

Remark 8.4. Assertion (ii) is a partial converse of Assertion (i). Even in E^2, under a certain degenerate case, it is not necessarily true that a path resulting from a vector, which represents a semipermeable direction, is a semipermeable surface. In proving Theorem 11.1 of Ref. 10, the author used this unwarranted assumption. Although the main results still hold, one has to put a stronger assumption to rule out the degenerate cases.

Definition 8.3. Let (μ^0, ν^0) be admissible and continuous in a region $\mathscr{R}_0$, and let M_0 be an $(n-1)$-dimensional manifold in $\mathscr{R}_0$. Suppose that

$$\Sigma(t_1, t_2) = \left\{ X(t) \,\middle|\, \begin{array}{c} \dot{X} = f(X, \mu^0, \nu^0), \quad X(t_0) = X_0 \in M_0, \\ t_1 < t, \quad t_0 < t_2 \end{array} \right\} \tag{39}$$

is a smooth semipermeable surface in $\mathscr{R}_0$ with strategies (μ^0, ν^0) and

$$\{\lim_{\substack{t \to t_2 \\ t < t_2}} X(t) \mid X(t) \in \Sigma(t_1, t_2)\} = M_1.$$

We say that the semipermeable surface $\Sigma(t_1, t_2)$ has a *smooth extension* at M_1 if there exists a pair of admissible strategies (μ', ν') which are continuous over $\mathscr{R}_1 \supset \mathscr{R}_0$, $\mathscr{R}_1 \neq \mathscr{R}_0$, such that

$$\mu'(X) = \mu^0(X) \qquad \text{if} \qquad X \in \mathscr{R}_0, \tag{40}$$

$$\nu'(X) = \nu^0(X) \qquad \text{if} \qquad X \in \mathscr{R}_0, \tag{41}$$

$$\Sigma(t_1, t_3) = \left\{ X(t) \,\middle|\, \begin{array}{c} \dot{X} = f[X, \mu(X), \nu(X)], \quad X(t_0) = X_0 \in M_0, \\ t_1 < t, \quad t_2 < t_3 \end{array} \right\} \tag{42}$$

is a smooth semipermeable surface in $\mathscr{R}_1$ such that

$$M_1 \cup \Sigma(t_1, t_2) \subset \Sigma(t_1, t_3), \tag{43}$$

$$\text{each } X(t) \in \Sigma(t_1, t_3) \text{ with } t > t_2 \text{ is not contained in } \Sigma(t_1, t_2) \cup M_1. \tag{44}$$

Remark 8.5. Suppose that (μ^0, ν^0) are admissible and continuous in a region $\mathscr{R}$ and

$$\Sigma(t_1, t_2) = \left\{ X(t) \,\middle|\, \begin{array}{l} \dot{X} = f(X, \mu^0, \nu^0), \quad X(t_0) = X_0 \in M_0, \\ \qquad\qquad t_1 < t, \quad t_0 < t_2 \end{array} \right\}$$

is a smooth semipermeable surface in $\mathscr{R}$. Let t' be a point in the open interval (t_1, t_2). Let

$$M' = \lim_{t \to t'} \{X(t) \mid X(t) \in \Sigma(t_1, t_2)\},$$

$$\Sigma(t_1, t') = \{X(t) \mid X(t) \in \Sigma(t_1, t_2), t_1 < t < t'\}.$$

Then, obviously, $\Sigma(t_1, t')$ has a smooth semipermeable extension at M'.

Remark 8.6. Suppose that (μ^0, ν^0) are admissible and continuous over a closed subregion $\mathscr{R}^0$ of $\mathscr{E}$, $\mathscr{E} - \mathscr{R}^0 \neq \phi$. There exists a pair of strategies (μ', ν') which is admissible[13] and continuous in $\mathscr{R}' \supset \mathscr{R}^0$ such that, for $X \in \mathscr{R}^0$, $\mu'(X) = \mu^0(X)$ and $\nu'(X) = \nu^0(X)$. From this observation and Remark 8.5, we may treat Theorem 8.2 as a necessary and sufficient condition for a semipermeable surface to have a smooth extension at a particular manifold.

9. Transition Surfaces[14]

Recall that, given a regular region $\mathscr{R}$, there is a decomposition $D = \{\mathscr{R}_i \mid i = 1, 2,..., m\}$ such that the optimal strategies (μ^*, ν^*) are continuous at the interior of each $\mathscr{R}_i$. However, the optimal strategies may be discontinuous at the boundary of $\mathscr{R}_i$. The boundaries of $\mathscr{R}_i$ are essentially singular surfaces.

It is known that, unless we can find the boundaries of $\mathscr{R}_i$ or the decompositon of $\mathscr{R}$, a complete solution to nontrivial differential games is impossible. We will study the behavior of isovalued surfaces near the boundary of $\mathscr{R}_i$ so that we can find a way to decompose a regular region. In order to simplify our presentation, we shall make the following assumptions on a regular region $\mathscr{R}$ and leave the extensions to the reader.

Assumption 9.1. The pair of optimal strategies (μ^*, ν^*) are unique in $\mathscr{R}$ and piecewise continuous with respect to the decomposition $D = \{\mathscr{R}_1, \mathscr{R}_2\}$ of $\mathscr{R}$. At the common boundary $M = \bar{\mathscr{R}}_1 \cap \bar{\mathscr{R}}_2$ of $\mathscr{R}_1$ and

[13] See Ref. 9, p. 149, Tietze theorem.

[14] For other types of singular surfaces, see Refs. 1 and 6.

$\mathscr{R}_2$, at least one of (μ^*, ν^*) has a jump discontinuity along the optimal path.

Assumption 9.2. Let Σ^* be an arbitrary isovalued surface in $\mathscr{R}$ extending into both $\mathscr{R}_1$ and $\mathscr{R}_2$. Then, there exists a neighborhood $\mathscr{N}$ of M_0, $M_0 = \Sigma^* \cap M$, such that

$$\Sigma^* \cap \mathscr{N} = \left\{ X(t) \,\middle|\, \begin{matrix} \dot{X} = f[X, \mu^*(X), \nu^*(X)], \quad X(t_0) = X_0 \in M_0\,, \\ t_1 < t, \quad t_0 < t_2 \end{matrix} \right\},$$

$$X(t) \in \mathscr{R}_1 \quad \text{if} \quad t_1 < t < t_0\,,$$
$$X(t) \in \mathscr{R}_2 \quad \text{if} \quad t_0 < t < t_2\,.$$

Definition 9.1. The nonempty common boundary M of $\mathscr{R}_1$ and $\mathscr{R}_2$ which satisfies Assumptions 9.1 and 9.2 is called a *transition surface.* When only one optimal strategy has discontinuity at M, M is called a *single transition surface*; otherwise, it is called a *double transition surface.* The significance of this distinction will be explored in the next section.

Now, suppose that M is a transition surface in $\mathscr{R}$. Then, for every isovalued surface Σ^* in $\mathscr{R}$ which extends into both $\mathscr{R}_1$ and $\mathscr{R}_2$ and intersects M at $M_0 \neq \phi$, there exists a neighborhood $\mathscr{N}$ of M_0 such that the parametric expression in Assumption 9.2 holds for $\Sigma^* \cap \mathscr{N}$. Define

$$\Sigma^*(t_1\,, t_0) = \{X(t) \in \Sigma^* \cap \mathscr{N} \mid t_1 < t < t_0\},$$
$$\Sigma^*(t_0\,, t_2) = \{X(t) \in \Sigma^* \cap \mathscr{N} \mid t_0 < t < t_2\}.$$

It is seen that

$$\Sigma^* \cap \mathscr{N} = \Sigma^*(t_1\,, t_0) \cup M_0 \cup \Sigma^*(t_0\,, t_2).$$

We show the following result.

Theorem 9.1. An isovalued surface $\Sigma^*(t_1\,, t_0)$ *cannot* have a smooth semipermeable extension at $M_0 = \Sigma^* \cap M$.

Proof. (see Fig. 1). Suppose that $\Sigma^*(t_1\,, t_0)$ has a smooth semipermeable extension. Then, there exist (μ^0, ν^0) which are defined and continuous in a region $\mathscr{R}^0 \supset [\mathscr{N} \cap (\mathscr{R}_1 \cup M)]$ such that [15]

$$(\mu^0(X), \nu^0(X)) = (\mu^*(X), \nu^*(X)) \quad \text{if} \quad X \in \mathscr{N} \cap \mathscr{R}_1\,, \tag{45}$$

$$\Sigma^0 = \left\{ X(t) \,\middle|\, \begin{matrix} \dot{X} = f[X, \mu^0(X), \nu^0(X)], \quad X(t_0) = X_0 \in M_0\,, \\ t_1 < t, \quad t_0 < t_3 \end{matrix} \right\} \tag{46}$$

[15] The parametric representation is possible from the definition of smooth semipermeable extension and Assumptions 9.1. and 9.2 by simply reassigning the initial points on the interval $(t_1\,, t_3)$.

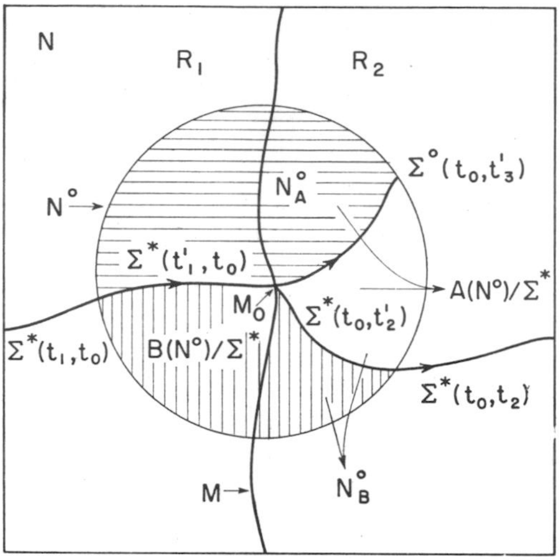

Fig. 1

is a smooth semipermeable surface in $\mathscr{R}^0$,

$$X(t) \notin \Sigma(t_1, t_0) \cup M_0 \qquad \text{if} \qquad t_0 < t < t_3. \tag{47}$$

Let $\mathscr{N}^0$ be a neighborhood of M_0 so that $\mathscr{N}^0 \subset \mathscr{N} \cap \mathscr{R}^0$ and $\mathscr{N}^0$ is separated by both Σ^* and Σ^0. Let

$$A(\mathscr{N}^0)/\Sigma^* = \mathscr{N}^0 \cap A/\Sigma^*, \qquad B(\mathscr{N}^0)/\Sigma^* = \mathscr{N}^0 \cap B/\Sigma^*,$$

$$\Sigma^*(t_1', t_0) = \mathscr{N}^0 \cap \Sigma^*(t_1, t_0), \qquad \Sigma^*(t_0, t_2') = \mathscr{N}^0 \cap \Sigma^*(t_0, t_2),$$

and denote the two parts of $\mathscr{N}^0$ separated by Σ^0 by $\mathscr{N}_A{}^0$ and $\mathscr{N}_B{}^0$. Let

$$\Sigma^0(t_0, t_3') = \{X(t) \in \Sigma^0 \cap \mathscr{N}^0 \mid t_0 < t < t_3' \leqslant t_3\}.$$

Observe that $\Sigma^0(t_0, t_3')$ cannot have an intersection with Σ^* in $\mathscr{N}^0$. To see the point, suppose that the assertion were false. Let $\Sigma^0(t_0, t_3')$ intersect Σ^* at some point $\tilde{X}$ (see Fig. 2). Let $X_0 \in M_0$ be the point being transfered to $\tilde{X}$ by $f(X, \mu^0(X), \nu^0(X))$. Since (μ^*, ν^*) have discontinuity at M_0 and (μ^0, ν^0) are continuous at M_0, we have

$$(\mu^0(X_0), \nu^0(X_0)) \neq (\mu^*(X_0), \nu^*(X_0)).$$

Since we have a terminal payoff, the above supposition results in a contradiction to the assumption that the pair of optimal strategies is unique and that $\Sigma^0(t_0, t_3')$ and Σ^* are semipermeable. Thus, $\Sigma^0(t_0, t_3')$

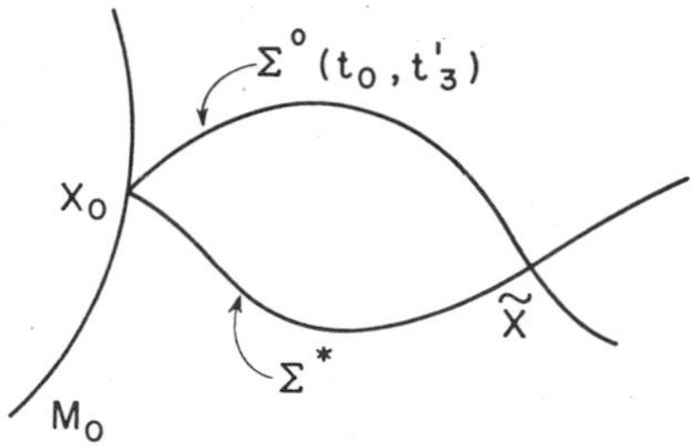

Fig. 2

is completely contained either in $A(\mathcal{N}^0)/\Sigma^*$ or in $B(\mathcal{N}^0)/\Sigma^*$, and exactly one of $\{\mathcal{N}_A^0, \mathcal{N}_B^0\}$ will be contained in one of $\{A(\mathcal{N}^0)/\Sigma^*, B(\mathcal{N}^0)/\Sigma^*\}$.

Without loss of generality, assume that $\mathcal{N}_A^0$ is contained in either $A(\mathcal{N}^0)/\Sigma^*$ or $B(\mathcal{N}^0)/\Sigma^*$. First, suppose that $\mathcal{N}_A^0 \subset A(\mathcal{N}^0)/\Sigma^*$ (see Fig. 1), so that $[\mathcal{N}_A^0 \cup \Sigma^0(t_0, t_3')] \subset A(\mathcal{N}^0)/\Sigma^*$. Let $X' \in \Sigma^*(t_1', t_0)$, which is a portion of an isovalued surface. In view of the derivation of Theorem 8.1, we know that

$$\mathcal{N}^0 \cap A(X', \nu^*) \subset (A/\Sigma^* \cup \Sigma^*) \cap \mathcal{N}^0.$$

Since (μ^*, ν^*) is the unique pair of optimal strategies, except for the path starting X' and satisfying $\dot{X} = f[X, \mu^*(X), \nu^*(X)]$, all the points $\mathcal{N}^0 \cap A(X', \nu^*)$ are contained in $A/\Sigma^* \cap \mathcal{N}^0 = A(\mathcal{N}^0)/\Sigma^*$. This observation and the definitions $\mathcal{N}_A^0$ and $A(\mathcal{N}^0)/\Sigma^*$ yield that at least one point of $A(X', \nu^*)$ is contained in $\mathcal{N}_A^0$. Since

$$\mathcal{N}^0 \cap \Sigma^0 = [\Sigma^*(t_1', t_0) \cup M_0 \cup \Sigma^0(t_0, t_3')] \cap \mathcal{N}^0$$

is a semipermeable surface with strategies (μ^0, ν^0), we have

$$A(X, \nu^0) \subset [\mathcal{N}_A^0 \cup \Sigma^0(t_0, t_3')] \subset A(\mathcal{N}^0)/\Sigma^*$$

for all $X \in \Sigma^0(t_0, t_3')$. Now, let the ψ-player use strategies ν^0 instead of ν^* at $[\Sigma^*(t_1', t_0) \cup M_0 \cup \Sigma^0(t_0, t_3')]$; it is seen that he can eventually make X penetrate into $A(\mathcal{N}^0)/\Sigma^*$ to improve his payoff, because $A(X, \nu^0) \subset A(\mathcal{N}^0)/\Sigma^*$. This contradicts the assumption that (μ^*, ν^*) are the optimal strategies.

Next, suppose that $\mathcal{N}_A^0 \subset B(\mathcal{N}^0)/\Sigma^*$, so that

$$\mathcal{N}_A^0 \cup \Sigma^0(t_0, t_3') \subset B(\mathcal{N}^0)/\Sigma^*.$$

Similar to the first case, one can show that

$$A(X, \mu^0) \subset [\mathcal{N}_A^0 \cup \Sigma^0(t_0, t_3')] \subset B(\mathcal{N}^0)/\Sigma^* \qquad \text{for } X \in \Sigma^0(t_0, t_3').$$

If instead of μ^*, the ψ-player selects μ^0 at $[\Sigma^*(t_1', t_0) \cup M_0 \cup \Sigma^0(t_0, t_3')]$, it is seen that he can eventually make X penetrate into $B(\mathcal{N}^0)/\Sigma^*$ to improve his payoff, because $A(X, \mu^0) \subset B(\mathcal{N}^0)/\Sigma^*$. Again, this contradicts the assumption that (μ^*, ν^*) are the optimal strategies in $\mathscr{R}$.

Since each isovalued surface is a piecewise smooth semipermeable surface, in view of the theorem, we have the following corollary.

Corollary 9.1. Let M be the transition surface in a regular region which satisfies Assumptions 9.1 and 9.2. Then, it is necessary that there exists (a) a neighborhood $\mathcal{N}$ of M, which is also nicely separated by M into $\mathcal{N}_1$ and $\mathcal{N}_2$ and (b) at least a pair of strategies (μ^0, ν^0) which are continuous in $\mathcal{N}_1$ such that (i) each path which starts in $\mathcal{N}_1$ and satisfies $\dot{X} = f[X, \mu^0(X), \nu^0(X)]$ will lie in $\mathcal{N}$ and will either get into $\mathcal{N}_2$ or have a limit point at M, and (ii) in $\mathcal{N}_1$, there exists a family of surfaces consisting of the paths described in (i) such that each surface forms a smooth semipermeable surface in $\mathcal{N}_1$, but it *cannot* have a smooth semipermeable extension at $\Sigma \cap M$.

Remark 9.1. Corollary 9.1, together with Theorem 8.2, may be used to determine candidates for transition surfaces before we actually solve a differential game. When our problem has a more specific structure, more restrictive and exact necessary conditions may be derived (Refs. 10–11). In fact, we are going to apply those results to locate transition surfaces for linear differential games where $f(X, \phi, \psi)$ is linear in ϕ and ψ.

10. Smooth and Jump Condition

It is known that the adjoint variable $\lambda(X)$ may not be continuous even in a regular region, no matter whether $f(X, \phi, \psi)$ is linear or nonlinear in (ϕ, ψ) (see Refs. 10–12). Following the definition of Section 7, we shall investigate the behavior of $\lambda^{\pm}(X)$ at transition surfaces.

Theorem 10.1. Let M be a smooth n-dimensional manifold in a regular region $\mathscr{R}$. Suppose that M is a transition surface in $\mathscr{R}$. Then, the following results hold[16]:

(i) $\lambda^+(X_0)$, $\lambda^-(X_0)$, and the normal vector $\lambda_M(X_0)$ of M at X_0 are

[16] By a different approach, Berkovitz has derived (ii) and a different form of (i) (see Ref. 13). Result (i) has been derived and called a jump condition by Leitmann (see Refs. 5–7). We shall call (ii) a smooth condition.

linearly dependent. That is, there exist C_1, C_2, C_3, not all zeros, such that

$$C_1\lambda^+(X_0) + C_2\lambda^-(X_0) + C_3\lambda_M(X_0) = 0.$$

(ii) Suppose that $\mathcal{Q}$ is an optimal path which does not tangentially enter nor tangentially leave M when it passes through the point $X_0 \in M$. Then, $\lambda^+(X_0) = \lambda^-(X_0)$ if, along $\mathcal{Q}$, there is only one optimal strategy that has a discontinuity at X_0.

Proof. (i) Note, from Section 7, that $\Lambda_\Sigma^+(X_0)$ and $\Lambda_\Sigma^-(X_0)$ are the two limiting tangent planes of Σ along $\mathcal{Q}$ at the point X_0 (see Fig. 3). Let $T_{M\cap\Sigma}(X_0)$ be the tangent plane of $M \cap \Sigma$ at X_0. Note that $\lambda^+(X_0)$, $\lambda^-(X_0)$, and $\lambda_M(X_0)$ are three normal vectors at X_0 to $T_{M\cap\Sigma}(X_0)$. Since $T_{M\cap\Sigma}(X_0)$ has dimensionality $n - 1$, and since all the vectors in question have $n + 1$ components, $\lambda^+(X_0)$, $\lambda^-(X_0)$, and $\lambda_\Sigma(X_0)$ must be linearly dependent.

(ii) Without loss of generality, we may assume that, along an optimal path $\mathcal{Q}$, μ^* is continuous at X_0 and ν^* is discontinuous. Now, suppose that $\lambda^+(X_0) \neq \lambda^-(X_0)$. Let

$$f^\pm(X_0) = \lim_{\substack{\epsilon\to 0\\ \epsilon>0}} f[X(t_0 \pm \epsilon), \mu^*(X(t_0 \pm \epsilon)), \nu^*(X(t_0 \pm \epsilon))] \neq 0,$$

where $X(t)$ is a parametric representation of $\mathcal{Q}$, as stated in Section 7. The

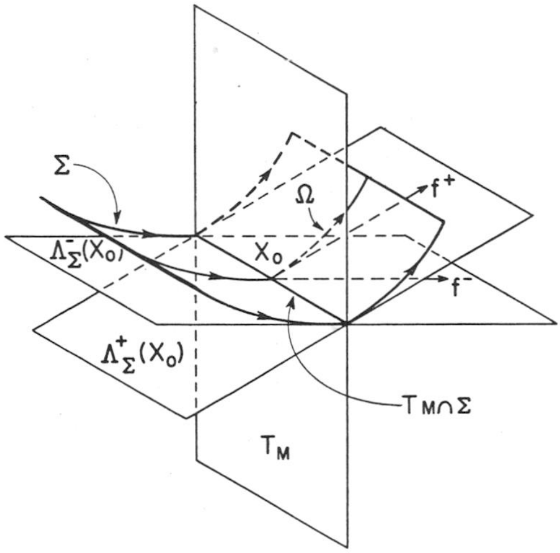

Fig. 3

nonzero property comes from Assumption 3.4. Since $X(t)$ and $\mu^*(X(t))$ are continuous at t_0, we have

$$f^{\pm}(X_0) = \lim_{\substack{\epsilon\to 0\\ \epsilon>0}} f[X_0, \mu^*(X_0), \nu^*(X(t_0 \pm \epsilon))]. \tag{48}$$

By Lemma 7.1, we have

$$f^{\pm}(X_0) \in \Lambda_B{}^+(X_0). \tag{49}$$

Since $\lambda^+(X_0) \neq \lambda^-(X_0)$, and since $\mathscr{Q}$ does not enter or leave M tangentially, the two limiting planes $\Lambda_\Sigma{}^{\pm}(X_0)$ of Σ cannot have an intersection other than in $T_{M\cap\Sigma}(X_0)$. Otherwise, $\Lambda_\Sigma{}^+(X_0) \cap \Lambda_\Sigma{}^-(X_0)$ has dimensionality n, or $\Lambda_\Sigma{}^+(X_0) = \Lambda_\Sigma{}^-(X_0)$, which contradicts the assumption $\lambda^+(X_0) \neq \lambda^-(X_0)$ and nontangentiality. Observe that the tangent plane of M at X_0 separates E^{n+1} into two open halfspaces. Since M is a transition surface, f^+ and f^- must be in the same halfspace. Since $f^+(X_0) \in \Lambda_\Sigma{}^+(X_0)$ and $f^-(X_0) \in \Lambda_\Sigma{}^-(X_0)$, (49) can be strengthened into

$$f^-(X_0) \in \text{int } \Lambda_B{}^+(X_0).$$

So, $f^+(X_0)$ lies on the top of $\Lambda_\Sigma{}^-(X_0)$. That is, $f^+(X_0) \cdot \lambda^-(X_0) > 0$. Consequently, we have

$$f^+(X_0) \in \text{int } \Lambda_A{}^-(X_0). \tag{50}$$

Recall, from right-hand continuity, that

$$\lim_{\substack{\epsilon\to 0\\ \epsilon>0}} \nu^*(X(t_0 + \epsilon)) = \nu^*(X_0). \tag{51}$$

Combining (48), (50), and (51), we have

$$f[X_0, \mu^*(X_0), \nu^*(X_0)] \in \text{int } \Lambda_A{}^-(X_0).$$

Let $\psi^* = \nu^*(X_0)$. By the definition of limiting tangent cone $\Lambda_A{}^-(X_0)$ and the continuity of $f(X(t), \mu^*(X(t)), \psi^*)$ (the control ψ is fixed at ψ^*), there exists t', $t_1 < t' < t_0$, such that

$$f[X(t'), \mu^*(X(t')), \psi^*] \in \text{int } \Lambda_A(X(t')).$$

But this contradicts Theorem 5.1.

References

1. Isaacs, R., *Differential Games*, John Wiley and Sons, New York, 1965.
2. Hestenes, M. R., *Calculus of Variations and Optimal Control Theory*, John Wiley and Sons, New York, 1966.
3. Leitmann, G., *An Introduction to Optimal Control*, McGraw-Hill Book Company, New York, 1966.
4. Blaquiere, A., and Leitmann, G., *On the Geometry of Optimal Processes*, Chapter 7, Topics in Optimization, Edited by G. Leitmann, Academic Press, New York, 1967.
5. Leitmann, G., *Two-Person Zero-Sum Games*, Paper presented at the 36th National Meeting of the Operations Research Society of America, Miami, Florida, 1969.
6. Leitmann, G., and Mon, G., *On a Class of Differential Games*, Paper presented at the Colloquium in Advanced Problems and Methods for Space Flight Optimization, Liege, Belgium, 1967.
7. Blaquiere, A., Gerard, F., and Leitmann, G., *Quantitative and Qualitative Games*, Academic Press, New York, 1969.
8. Stalford, H., and Leitmann, G., *Sufficient Condition for Optimality in Two-Person Zero-Sum Differential Games with State and Strategy Constraints*, University of California at Berkeley, Report No. ORC-70-2, 1970.
9. Dugundji, J., *Topology*, Allyn and Bacon, New York, 1966.
10. Yu, P. L., *Qualification of Optimal Strategies and Singular Surfaces of a Class of Differential Games*, Journal of Optimization Theory and Applications, Vol. 6, No. 6, 1970.
11. Yu, P. L., *Semipermeable Directions and Transition Surface of a Class of Differential Games*, Journal of Optimization Theory and Applications, Vol. 6, No. 6, 1970.
12. Yu, P. L., *On Singular Surfaces of Linear Differential Games*, University of Rochester, Center for System Science, Report No. CSS-70-01.
13. Berkovitz, L. D., *A Variational Approach to Differential Games*, Advances in Game Theory, Annals of Mathematics, Study No. 52, Princeton University Press, Princeton, New Jersey, 1964.
14. Bradley, J., and Yu, P. L., *Semipermeable Directions in Linear Differential Games*, Journal of Mathematical Analysis and Applications, Vol. 48, No. 3, 1974.
15. Bradley, J., and Yu, P. L., *Some Basic Properties of the Payoffs Defined by Closed-Loop Strategies*, Vol. 7, No. 2, 1976.
16. Bradley, J., and Yu, P. L., *A Concept of Optimal Strategies in Differential Games*, Vol. 7, No. 2, 1976.

XIII

On a Nonsingular Singular Surface of a Differential Game[1]

D. J. Wilson

Abstract. A differential game in which nonterminating play is assigned a fixed value is examined. The solution exhibits a new kind of singular surface which an optimal path takes an infinite time to cross. By playing nonoptimally in a sufficiently small neighborhood of this surface, the player who prefers the optimal (terminating) value to the value of nonterminating play can force the state across this surface and obtain a value arbitrarily close to the optimal.

1. Introduction

The phenomenon of bang-bang-bang surfaces in differential games has been described by Isaacs (Ref. 1), who gave the following example by way of illustration. The game takes place in the playing region $\mathscr{E} = \{(x, y) \in R^2; y \geqslant -1\}$ and finishes when the state reaches the terminal surface $\mathscr{C} = \{(x, y) \in R^2; y = -1\}$. The kinetic equations governing the motion of the state are

$$\dot{x} = \varphi(-\tfrac{1}{2}\sqrt{3}y) + \sqrt{3}(2 - \tfrac{1}{2}y) + 2\sqrt{3}\cos\psi,$$
$$\dot{y} = \varphi(3 + \tfrac{1}{2}y) + (\tfrac{1}{2}y - 1) + 2\sqrt{3}\sin\psi.$$

A player P, controlling $\varphi \in [-1, 1]$, tries to minimize and a player E, controlling ψ(arbitrary), tries to maximize the payoff H, which is terminal and is given by $H(x, y) = -x$.

Isaacs proposed a solution to this differential game under the assumption that both players suffer a drastic penalty if the game does not terminate (Ref. 1). However, such an assumption makes the game non-

[1] This work was carried out with the support of a CSIRO postgraduate studentship.

zero sum and introduces the conceptual difficulties usually attendant on such games. We shall solve the game under the assumption that the nonterminating play has a certain value V_∞ for E and that it has value $-V_\infty$ for P (we shall allow $V_\infty = +\infty$ or $V_\infty = -\infty$). The solution obtained exhibits an interesting new kind of *singular* surface.

2. Nonzero-Sum Game

The optimal paths of the solution given by Isaacs, under the assumption that nontermination is anathema to both players, are sketched in Fig. 1. The optimal play has two stages.

Primary stage. For $y \in [-1, 0]$ (Region I of Fig. 1), the optimal paths are straight lines of slope $-1/\sqrt{3}$. The optimal values $\bar{\varphi}$, $\bar{\psi}$ of φ, ψ are respectively given by

$$\bar{\varphi} = 1, \qquad \bar{\psi} = -2\pi/3$$

Secondary stage. For $y \geqslant 0$ (Regions II, III, IV of Fig. 1), the optimal paths are vertical straight lines, and the optimal strategies of P and E are given by

$$\bar{\varphi} = -1, \qquad \bar{\psi} = -\pi.$$

However, the wisdom of this strategy may be questioned when one considers the following statements, which are not difficult to verify: (i) P can force termination against any defense by E in the whole of $\mathscr{E}$;

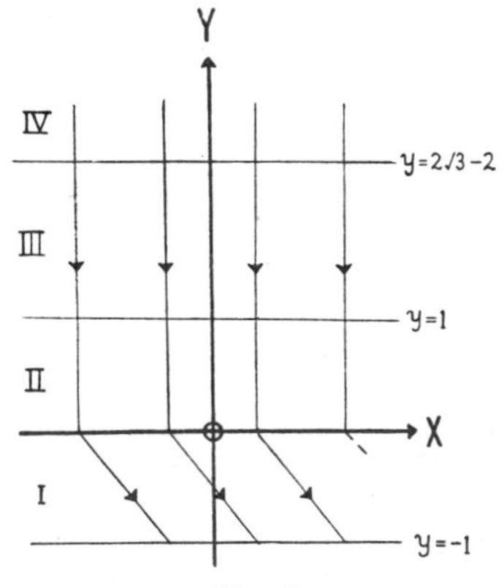

Fig. 1

(ii) P can prevent termination against any defense by E in Region IV (that is, when $y \geqslant 2\sqrt{3} - 2$); (iii) P can prevent termination against E's primary optimal strategy in Regions III and IV (that is, for $y \geqslant 1$); (iv) P can prevent termination against E's secondary optimal strategy in the whole of $\mathscr{E}$; (v) E can force termination against any defense by P in Regions I, II, III; and (vi) E can prevent termination against P's primary optimal strategy in the whole of $\mathscr{E}$.

In view of (vi), it may well be asked whether E can force P to abandon his primary optimal strategy by using a strategy which will prevent termination. In Region I, where P's primary optimal strategy is considered *truly optimal*, it seems plausible that E may be able to blackmail P into playing nonoptimally and thereby incurring an increase in payoff. However, in Region I, P can counter such perverse tactics by E by playing optimally for a sufficient length of time and then forcing termination at his leisure. If the state ever reaches the upper boundary of Region I, P can force it back into the interior by playing his secondary optimal strategy for a short time before reverting to his primary optimal strategy. With this strategy, P ensures that the payoff which he can attain by forcing termination decreases steadily as the game progresses; and, if E at any time sees the error of his ways and reverts to his primary optimal strategy, then the payoff which he attains is much worse than if he had played optimally from the start. A typical play in which E pursues this refractory strategy is represented in Fig. 2. In this case, P has decided to settle for a payoff of $-p$.

For play starting in Region II, however, the question is not so easily settled, because the primary optimal paths extend into Region II and

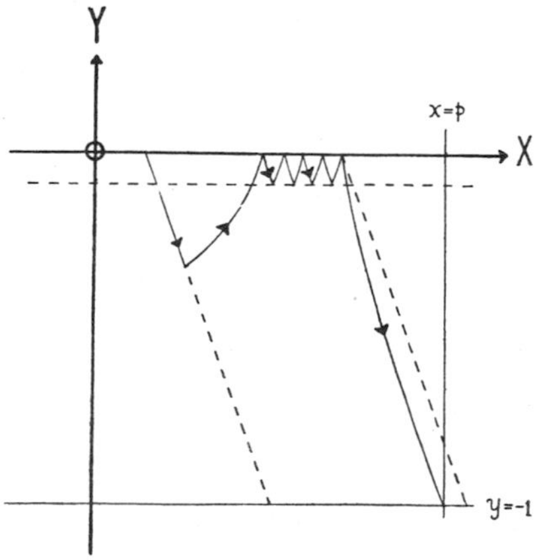

Fig. 2

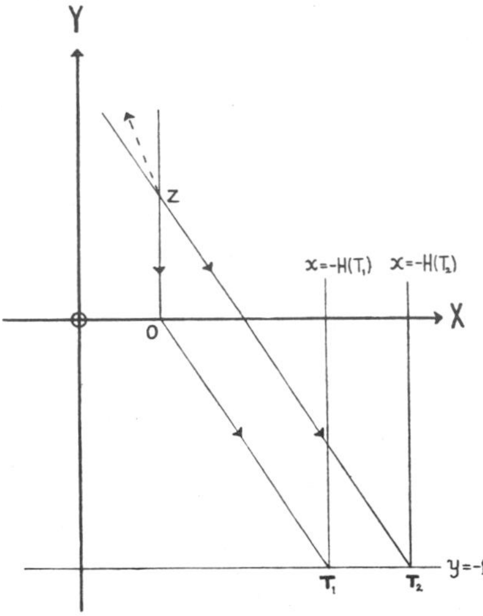

Fig. 3

constitute valid solutions of the kinetic and differential equations governing the value. For example, consider a play of the game which begins in the state Z indicated in Fig. 3. In such a play, P can keep the state on the right of the primary optimal path ZT_2 by using his primary optimal strategy. This ensures that, if and when the game terminates, the payoff does not exceed $H(T_2)$. On the other hand, E can keep the state on the left of the secondary–primary optimal path ZOT_1 by using his secondary optimal strategy in Region II and his primary optimal strategy in Region I. This ensures that, if and when the game finishes, the payoff is not less than $H(T_1)$. However, since $H(T_1) > H(T_2)$, these terminating conditions are contradictory. In fact, if both P and E persist in playing these strategies, the state moves along the dotted path of Fig. 3, the game never terminates, and both players sustain a drastic penalty. It seems clear that both players should be prepared to tolerate a less favorable payoff in order that the game may terminate. How much should each player concede? And for how long should one player persist with a given strategy when the other remains obtuse in the face of this reasoning and refuses to cooperate?

Similar problems arise when one considers optimal play in Region IV. Here, E is completely at P's mercy, for P can prevent termination of the game against any strategy by E and demand an exorbitant price (supposing that communication between the players is possible) for allowing the hapless E to terminate the game. Should E submit to these demands or should he resign himself to a fate worse than death, taking consolation in the fact that the blackguard P must either relent or also share in his suffering?

We offer no answer to these questions, which are typical of those which occur and remain unanswered in the theory of non zero-sum games. By way of avoiding the issue, we shall instead solve the following zero-sum game.

3. Zero-Sum Game

As mentioned in Section 1, we alter the game by letting the payoff be $H(x, y)$ if the game terminates at $(x, y) \in \mathscr{C}$ and V_∞ if it does not terminate. In solving this game, we shall use the notation of Ref. 2 throughout, except that we shall use normal *forward* time exclusively, instead of the reverse time employed in Ref. 2.

The ME_1 is (Ref. 2, p. 68)

$$0 = \min_\varphi \max_\psi \{\varphi[(3 + \tfrac{1}{2}y) V_y - \tfrac{1}{2}\sqrt{3}yV_x] + 2\sqrt{3}(V_x \cos\psi + V_y \sin\psi) + \sqrt{3}(2 - \tfrac{1}{2}y) V_x + (\tfrac{1}{2}y - 1) V_y\}.$$

The optimal strategies are thus given by

$$\cos\bar{\psi} = V_x/\rho, \qquad \sin\bar{\psi} = V_y/\rho, \qquad \bar{\varphi} = -\operatorname{sign} A,$$

where

$$\rho = \sqrt{(V_x^2 + V_y^2)}, \qquad A = (3 + \tfrac{1}{2}y) V_y - \tfrac{1}{2}\sqrt{3}yV_x.$$

The ME_2 is (Ref. 2, p. 69)

$$0 = -|A| + 2\sqrt{3}\rho + \sqrt{3}(2 - \tfrac{1}{2}y) V_x + (\tfrac{1}{2}y - 1) V_y,$$

and the differential equations for V_x, V_y along the optimal paths are

$$\dot{V}_x = 0, \tag{1}$$

$$\dot{V}_y = V_x(\tfrac{1}{2}\sqrt{3}\varphi + \tfrac{1}{2}\sqrt{3}) - V_y(\tfrac{1}{2}\varphi + \tfrac{1}{2}). \tag{2}$$

The terminal surface $\mathscr{C}$ can be parameterized by the function

$$h_1(s) = s, \qquad h_2(s) = -1.$$

The value V on the terminal surface is given by

$$V(s) = H(s, -1) = -s.$$

Thus,

$$V'(s) = -1 = V_x h_1'(s) + V_y h_2'(s) = V_x . \tag{3}$$

Thus, if all points in a neighborhood of a given point (x, y) lie on a unique optimal path, then it follows from (1) and (3) that $V_x(x, y) = -1$·

Substitution of -1 for V_x in the ME_2 yields

$$0 = -\left|(3 + \tfrac{1}{2}y) V_y + \tfrac{1}{2}\sqrt{3}y\right| + 2\sqrt{3}\sqrt{(V_y^2 + 1)} - \sqrt{3}(2 - \tfrac{1}{2}y) + (\tfrac{1}{2}y - 1) V_y . \tag{4}$$

We now consider two cases.

(a) If $A \geqslant 0$, then (4) reduces to

$$2V_y + \sqrt{3} = \sqrt{3}\sqrt{(V_y^2 + 1)}. \tag{5}$$

Squaring (5) yields

$$V_y(V_y + 4\sqrt{3}) = 0,$$

that is,

$$V_y = 0 \qquad \text{or} \qquad V_y = -4\sqrt{3}.$$

The second answer is impossible for any value of y, since it does not satisfy (5). The first answer ($V_y = 0$) satisfies (5) but is only possible when it renders $A \geqslant 0$; that is, for $y \geqslant 0$.

(b) If $A \leqslant 0$, then (4) reduces to

$$(y + 2) V_y + \sqrt{3}(y - 2) = -2\sqrt{3}\sqrt{(V_y^2 + 1)}. \tag{6}$$

Squaring (6) yields

$$[V_y(y^2 + 4y - 8) + \sqrt{3}y(y - 4)](V_y + \sqrt{3}) = 0,$$

that is,

$$V_y = -\sqrt{3} \qquad \text{or} \qquad V_y = -\sqrt{3}\, y(y - 4)/(y^2 + 4y - 8).$$

These solutions will only be valid when they render $A \leqslant 0$ and the left-hand side of (6) negative. Thus, $V_y = -\sqrt{3}$ is a valid solution in

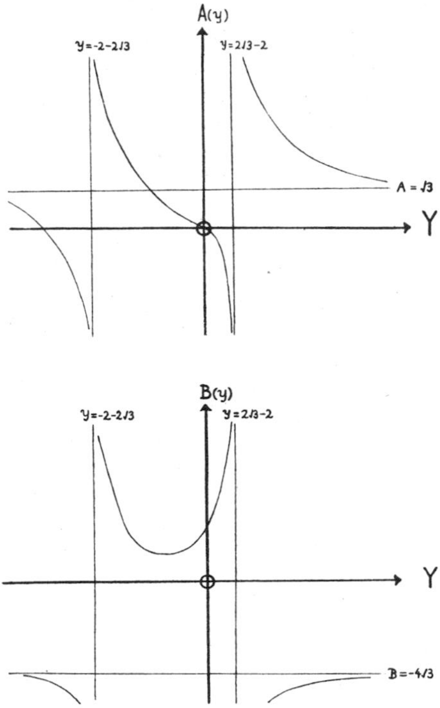

Fig. 4

the whole of $\mathscr{E}$, and $V_y = -\sqrt{3}\ y(y-4)/(y^2+4y-8)$ is a valid solution whenever

$$A(y) = -\sqrt{3}(3+\tfrac{1}{2}y)\,y(y-4)/(y^2+4y-8) + \tfrac{1}{2}\sqrt{3}y \leqslant 0, \qquad (7)$$

$$B(y) = \sqrt{3}(2-y) + \sqrt{3}\,y(y-4)(y+2)/(y^2+4y-8) > 0. \qquad (8)$$

The sketch of the graphs of A and B in Fig. 4 indicates that conditions (7) and (8) are satisfied for $y \in [0, 2\sqrt{3}-2]$.

Thus, there are three distinct possibilities for the value of V_y; namely:

(i) $V_y = -\sqrt{3}$ in the whole of $\mathscr{E}$,
(ii) $V_y = 0$ for $y \geqslant 0$,
(iii) $V_y = -\sqrt{3}\ y(y-4)/(y^2+4y-8)$ for $y \in [0, 2\sqrt{3}-2)$.

The first solution above gives the primary optimal paths; the second, which has also been noted by Ciletti in Ref. 3, gives the secondary optimal paths; and the third gives a distinct family of optimal paths in part of Region III of Fig. 1. The values of V_y given by (ii) and (iii) coincide on the surface $y = 0$; apparently for this reason, possibility (iii) escaped observation in Refs. 1 and 3.

4. Solution of the Zero-Sum Game

We now use the possible values of V_y given by (i), (ii), (iii) of the last section to integrate the path equations and obtain the solution of the game. In Region I, only one value of V_y is possible [that given by (i)], and the path equations have a unique solution. In Regions II and III, all three possible values of V_y may be used to integrate the path equations, and therefore three families of candidates for optimal paths are obtained. In each case, one of these families may be rejected as irrelevant to the game under consideration [the family given by (iii) in the case of Region II, and the family given by (i) in the case of Region III]. The two regions can then be divided into disjoint portions, in each of which the members of one of the remaining families may be regarded as optimal. In Region IV, two values of V_y are possible, but only one of these [that given by (ii)] yields optimal paths. In certain parts of Regions II, III, IV, it is found that all paths are optimal and lead to a value of V_∞.

We let the time of termination be $t = 0$ and integrate back from the terminal surface. On the terminal surface $\mathscr{C}$, the only possible value for V_y is given by (i) of the last paragraph; that is, $V_y = -\sqrt{3}$. Integration of the path equations now gives

$$V_x \equiv -1, \qquad V_y \equiv -\sqrt{3},$$

and

$$x(t) = s - 2\sqrt{3} + 2\sqrt{3}\exp(t), y(t) = 1 - 2\exp(t) = ([s - x(t)]/\sqrt{3}) - 1,$$

where $s = x(0)$ is the terminal abcissa of the optimal path. The optimal strategies are determined by

$$\cos\bar{\psi} = -\tfrac{1}{2}, \qquad \sin\bar{\psi} = -\tfrac{1}{2}\sqrt{3}, \qquad \bar{\varphi} = 1.$$

These are the primary optimal paths, which are lines of slope $-\sqrt{\tfrac{1}{3}}$ extending into $\mathscr{E}$ up to the line $y = 1$ which they reach at $t = -\infty$.

At $y = 0$, solutions (ii) and (iii) of Section 3 both become possible. We therefore let $y = 0$ be the new terminal surface and consider first the case where $V_y = 0$.

In this case, integration of the path equations gives

$$V_x \equiv -1, \qquad V_y \equiv 0,$$

and

$$x \equiv s - \sqrt{3}, \qquad y(t) = -4(t - \beta),$$

where $\beta = \log \frac{1}{2}$ is the time when the primary path reaches the line $y = 0$. The optimal strategies are determined by

$$\cos \bar{\psi} = -1, \qquad \sin \bar{\psi} = 0, \qquad \bar{\varphi} = -1.$$

These vertical straight lines are the secondary optimal paths.

In the second case, where $V_y = -\sqrt{3}\, y(y - 4)/(y^2 + 4y - 8)$, the initial value of V_y on $y = 0$ is again 0. However, this case is distinguished from the first by the fact that A must become negative above the line $y = 0$, thus giving $\bar{\varphi} = 1$. The paths then obtained by integrating the path equations move into Region II as *t increases* and approach the line $y = 1$ asymptotically as $t \to \infty$. We omit the expressions for these paths, as they are quite lengthy and they appear to have no relevance to the game under consideration. These paths are sketched in Fig. 5.

In Regions III and IV, use of the value $V = -\sqrt{3}$ gives paths of a similar kind which *start* from the line $y = 1$ at $t = -\infty$ and move upwards into Region III as t increases. These paths are straight lines with the same slope as the primary optimal paths and also seem to have no significance in the game which we are considering.

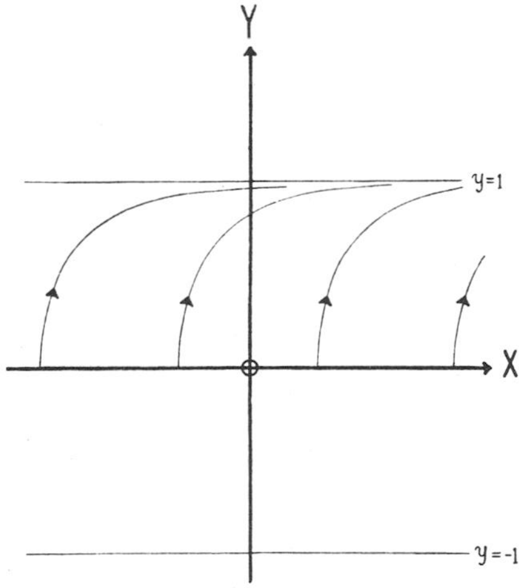

Fig. 5

If a play of the game starts at a point of Region II from which the primary optimal path leads to a payoff $v \geqslant V_\infty$, then playing his primary optimal strategy ensures E that the play terminates with a payoff no smaller than v. On the other hand, by playing *his* primary optimal strategy, P ensures that, if the game terminates, the payoff will be no larger than v; and, if the game does not terminate, then the payoff is V_∞, which is also not larger than v. Thus, the value of the game under these circumstances is v, and the primary optimal paths are truly optimal. If the game starts at a point of Regions II, III, IV from which the secondary–primary optimal path leads to a payoff $u \leqslant V_\infty$, then, by playing his secondary–primary optimal strategy (or an appropriate deviation therefrom, as indicated in Fig. 2, whenever E plays non-optimally), P ensures that play terminates with a payoff no larger than u. On the other hand, by playing *his* secondary–primary optimal strategy, E ensures that the payoff is no smaller than u, whether termination occurs or not. In this case, therefore, the secondary–primary optimal paths are truly optimal, and the value of the game is u. The truly optimal paths for Regions I and II are sketched in Fig. 6.

In the outstanding portion of Region II which lies between the primary optimal path with value V_∞ and the secondary optimal path with value V_∞ (that is, the cross-hatched portion of Fig. 6), P can prevent any termination with a payoff larger than V_∞, and E can prevent any termination with a payoff smaller than V_∞. Throughout this area, the value is therefore V_∞; and in its interior, the strategies of the players

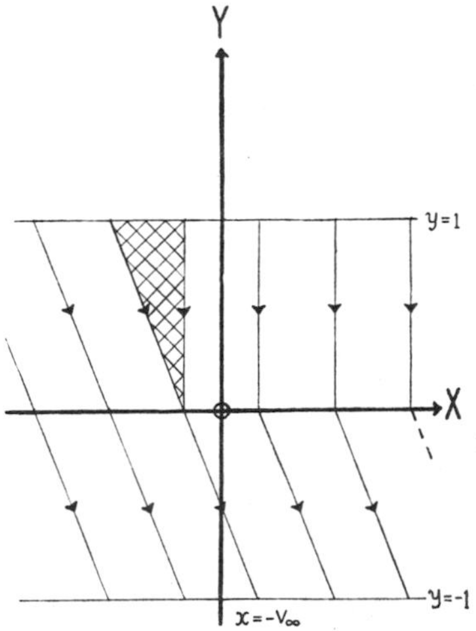

Fig. 6

are immaterial, all paths being optimal. Similar reasoning applies to points of Region IV, from which the secondary–primary optimal paths lead to a payoff greater than V_∞. Consequently, in the portion of Region IV to the left of the secondary optimal path with value V_∞, all paths are optimal, and the value is V_∞.

The only area of the playing space $\mathscr{E}$ in which the optimal strategies now remain to be determined is the area of Region III lying to the left of the secondary optimal path with value V_∞. It seems clear that, since E can force termination from a starting point of Region III, then, if this point is far enough to the left, it should lie on an optimal path leading to a value greater than V_∞. However, the only possible way to construct such optimal paths is by using the value

$$V_y = -\sqrt{3}\, y(y-4)/(y^2+4y-8);$$

and, when one tries to construct optimal paths backward from the line $y = 1$ by using this value of V_y, the path equations make x and y identically constant (that is, the *optimal paths* consist of single points on the line $y = 1$). To overcome this difficulty, we construct paths backward *and forward* from a line $y = p$, with $1 < p < 2\sqrt{3} - 2$ and

$$V_x = -1, \qquad V_y = -\sqrt{3}\, p(p-4)/(p^2+4p-8)$$

as starting values on this line. We set

$$\epsilon = \sqrt{3}\, p(p-4)/(p^2+4p-8) - \sqrt{3},$$

so that $\epsilon > 0$, and

$$p = (2\sqrt{3}/\epsilon)\sqrt{[(\epsilon+\sqrt{3})^2+1]} - 2 - 4\sqrt{3}/\epsilon.$$

We also adopt a new time variable, letting $t = 0$ when the paths cross the line $y = p$.

Integration of the path equations then gives

$$V_x \equiv -1, \qquad V_y = -\epsilon \exp(-t) - \sqrt{3},$$

and

$$y = (2\sqrt{3}/\epsilon)\exp(t)\{\sqrt{([\epsilon\exp(-t)+\sqrt{3}]^2+1)} - 2\} - 2,$$

$$x = 4\sqrt{3}t + (12/\epsilon)\exp(t) + 6\,\operatorname{arcsinh}[\epsilon\exp(-t)+\sqrt{3}]$$
$$-3\sqrt{\{[(4/\epsilon)\exp(t)+\sqrt{3}]^2+1\}} - 4\sqrt{3}\,\operatorname{arcsinh}[(4/\epsilon)\exp(t)+\sqrt{3}] + k,$$

where k is a constant. The optimal strategies are determined by

$$\cos \bar{\psi} = (y^2 + 4y - 8)/2\sqrt{(y^4 - 4y^3 + 12y^2 - 16y + 16)},$$
$$\sin \bar{\psi} = \sqrt{3}y(y - 4)/2\sqrt{(y^4 - 4y^3 + 12y^2 - 16y + 16)},$$
$$\bar{\varphi} = 1.$$

From the above expressions we find that

$$\lim_{t\to\infty} y = 1,$$

$$\lim_{t\to\infty} x = k - 3\sqrt{3} + 6\log(2 + \sqrt{3}) + 4\sqrt{3}\log(\epsilon/8),$$

$$\lim_{t\to-\infty} y = 2\sqrt{3} - 2,$$

and $x \to -\infty$ as $t \to -\infty$. Also, by substituting the expressions for y and V_y in the Kinetic equations, we obtain

$$\lim_{t\to\infty} (\dot{y}/\dot{x}) = -1/\sqrt{3}.$$

Thus, these paths join smoothly onto the primary optimal paths along the line $y = 1$ and extend upwards into Region III, approaching the line $y = 2\sqrt{3} - 2$ asymptotically. They are sketched in Fig. 7.

A path of the type just described will be truly optimal in Region III if it joins onto a primary optimal path leading to a value greater than V_∞.

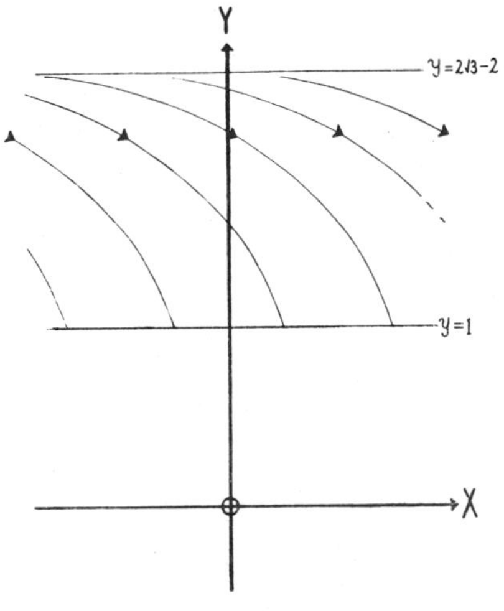

Fig. 7

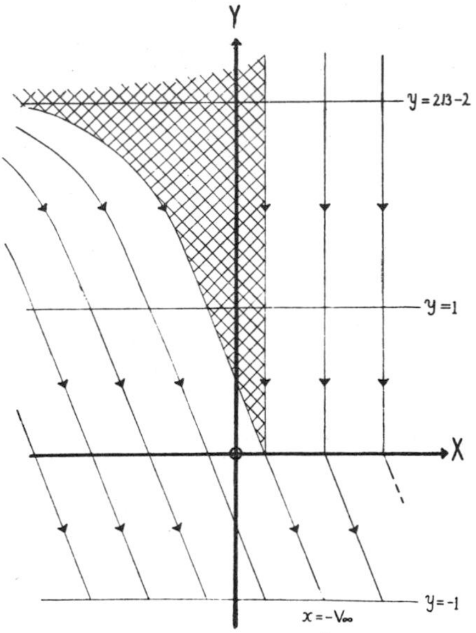

Fig. 8

We have already noted that the secondary optimal paths are truly optimal in Region III when they lead to a value less than V_∞. In the outstanding portion of Region III, all paths are optimal and lead to a value of V_∞.

The solution of the zero-sum game has now been completely described and is represented in Fig. 8. In the cross-hatched region, all paths are optimal and lead to a value of V_∞. The optimal paths to the left of the diagram take an infinitely long time to cross the line $y = 1$. For this reason, E has no optimal strategy in the left-hand part of Region III, but only an ϵ-optimal strategy for every positive ϵ. E can play his *optimal* strategy until the state is close enough to the line $y = 1$ for him to force it across this line with a loss in payoff of at most ϵ. He can then force the state across the line (for example, by playing a strategy $\bar{\psi}$, with $\sin \bar{\psi} = -1$) and switch to his primary optimal strategy. With this strategy, E ensures that the payoff will lie within ϵ of the value.

5. Conclusion and Comments

In this paper, we have determined the solution of a differential game in which nonterminating play has been assigned a payoff V_∞. In the case when $V_\infty = +\infty$, this solution is the same as that given by Isaacs in Ref. 1. This occurs when P (the minimizing player) prefers any terminal value to a play which does not terminate, and E (the maximizing player) prefers nonterminating play to any payoff obtained by termination.

When V_∞ is finite, the left-hand portion of the line $y = 1$ constitutes a phenomenon which we believe has not previously been encountered in differential games. The value is continuous and has continuous partial derivatives across this line, and the optimal paths and strategies are smooth. Indeed, the surface hardly merits the title *singular* (hence, the title of this paper), its only claim to singularity being the very large (infinite, in fact) amount of time required for the state to pass through it when both players use their *optimal* strategies. When $V_\infty = -\infty$, the whole of the line $y = 1$ exhibits this peculiar behavior.

6. Addendum

I would like to thank the reviewer for several useful comments concerning the clarity of this paper. He has also suggested the rather neat title *temporal barrier* for the surface exhibited herein, and raised an important point which deserves some discussion.

The posited optimality of the solution given here rests on the method by which it was obtained, namely, by eliminating solutions of the necessary condition for optimality until only one reasonable solution is left. Whether or not this remaining solution is indeed *optimal* still remains to be proved rigorously. Because of the number and unusual nature of the solutions to the necessary conditions, the applicability of standard sufficiency theorems to this example has been queried. In this particular case, we feel that the best approach for proving optimality is by piecewise application of Isaacs' verification theorem (Ref. 2, p. 72) to modified K-strategies based on the tactics $\bar{\phi}, \bar{\psi}$. The K-strategies must be modified in two ways. In the neighborhood of the left-hand portion of the line $y = 1$, E's K-strategy is not given by the tactic $\bar{\psi}$ but by the condition that penetration must be effected. The second modification occurs when E plays nonoptimally, as indicated in Fig. 2. In this case, P must terminate the game at some indefinite time of his own choosing, and this requires that he depart from the *optimal* tactic $\bar{\phi}$ when selecting his K-strategy. By this means, we believe that a proof of the ϵ-optimality of these K-strategies, although tedious, should not be difficult to construct.

References

1. Isaacs, R. *Differential Games: Their Scope, Nature, and Future*, Journal of Optimization Theory and Applications, Vol. 3, No. 5, 1969.
2. Isaacs, R., *Differential Games*, John Wiley and Sons, New York, 1965.
3. Ciletti, M. D., *On the Contradiction of Bang-Bang-Bang Surface in Differential Games*, Journal of Optimization Theory and Applications, Vol. 5, No. 3, 1970.

XIV

Equilibrium Existence Results for Simple Dynamic Games[1]

P. R. Kleindorfer and M. R. Sertel

Abstract. The investigation of equilibrium existence in a salient class of multistage games is simplified and generalized by reducing them to singlestage games which correspond also to competitive economies. Thus, economic theory provides much of the motivation as well as method for the study.

Working in locally convex topological vector spaces, a fixed-point theorem of Fan is applied to show the existence of the Nash equilibria studied. En route, but also as a matter of interest in itself, certain topological foundations of the equilibrium analysis used are laid out.

A particular feature of generality of the dynamic games studied is that the feasible control regions of individual players are allowed to depend on the past state-history and control-history. Another such feature is that the next-state map is allowed certain nonlinearities.

[1] This paper is based on the more complete paper (Ref. 11) which was presented at the IEEE Conference on Decision and Control in New Orleans, 1972. The authors believe that the more complete version (Ref. 11) also makes the connection between dynamic games and economies rather more transparent than does the present version and that, from the mathematical economic theorist's viewpoint, Ref. 11 is probably to be considered as a more proper exposition.

1. Introduction

The purpose of this paper is to bring some simplicity and generality to the investigation of equilibrium existence in certain *simple* dynamic games. The way we deal with these objects is, essentially, by reducing them to singlestage games, these latter being, from the viewpoint of equilibrium existence results, at once more general and less cumbersome. In Section 2, we define notions related to the simple dynamic games of interest. Then Section 3 provides some mathematical equipment for the task of Section 4. In Section 4, first we show that the simple dynamic games in question have nonempty and compact sets of equilibria. We then illustrate by example that a general class of discrete-time, deterministic games with convex performance criteria is covered by the equilibrium existence results just described. This class includes dynamic games in which certain nonlinearities in the next-state map are allowed and for which controls are restricted to compact regions, these regions themselves varying as a function of state.

Historically, the study of the existence of competitive equilibrium in games and economies exhibits quite a long-standing extensive literature. A crucial turning point in that literature is afforded by the Arrow and Debreu (Ref. 1) study, benefiting from Debreu's (Ref. 2) earlier investigation and using a FPT (fixed-point theorem) of Eilenberg and Montgomery (Ref. 3). This work is generalized in Prakash and Sertel (Refs. 4, 14, 15, 16, 17). Our present results follow in the spirit of these in their reliance on fixed-point theory to establish the existence of equilibria. Alternative approaches to the equilibrium existence question for dynamic games can be found in Kuhn and Szegö (Ref. 5) and through the survey paper by Ho (Ref. 6).

We work in locally convex spaces. Such spaces include normed spaces and the conjugate space of the Banach space of all real-valued continuous functions on a given compact space. This conjugate space, in turn, is the natural habitat of probability measures under the weak (that is, w^*) topology (see Parthasarathy, Ref. 7). Our working in locally convex spaces is motivated by the hope and conjecture that equilibrium existence results for stochastic dynamic games may also be obtained by the approach pursued here. The reader who so wishes may, of course, think in Euclidean terms throughout.

Notation and Conventions. For the set of natural numbers, we denote $\mathcal{N} = \{1, 2,...\}$. For fixed $n \in \mathcal{N}$, we denote $N = \{1, 2,..., n\}$, $N_0 = \{0\} \cup N$. Given any set Z and any $m \in \mathcal{N}$, we consider the family $\{{}^iZ = Z \mid i = 1,.., m\}$ and denote the Cartesian product $\prod_{i=1}^{m} {}^iZ =$

${}^1Z \times \cdots \times {}^mZ$ by ${}_mZ$, generic elements being in lower case: ${}^mz \in {}^mZ$ and ${}_mz \in {}_mZ$. When a given ${}_mz = ({}^1z, \ldots, {}^mz)$ is understood, we denote projections and components according to $\pi_{({}_kZ)}({}_mz) = {}_kz$ and $\pi_{({}^kZ)}({}_mz) = {}^kz$, $k = 1, \ldots, m$.

Every Cartesian product of topological spaces is understood to carry the product topology.

We abbreviate *locally convex Hausdorff topological vector space* to LCHTVS.

Let Z be a set. Then, $[Z]$ denotes the set of nonempty subsets of Z. When Z is equipped with a topology, $\mathscr{C}(Z)$ denotes the set of nonempty closed subsets of Z and $\mathscr{K}(Z)$ denotes the set of nonempty compact subsets of Z. When Z lies in a real vector space, $\mathscr{Q}(Z)$ denotes the set of nonempty convex subsets of Z. We abbreviate $\mathscr{C}(Z) \cap \mathscr{Q}(Z)$ to $\mathscr{C}\mathscr{Q}(Z)$ and $\mathscr{K}(Z) \cap \mathscr{Q}(Z)$ to $\mathscr{K}\mathscr{Q}(Z)$.

When discussing the continuity properties of set-valued maps, we follow Michael (Ref. 8). In particular, for P and Q, topological spaces, let $F: P \to [Q]$ be a map. We say that F is *lower semicontinuous* (resp., *upper semicontinuous*) iff the set $\{p \in P \mid F(p) \cap V \neq \varnothing\}$ is open (resp., closed) in P whenever $V \subset \mathscr{Q}$ is open (resp., closed). We say that F is *continuous* iff it is both lsc and usc. Here, the abbreviations lsc and usc stand for lower semicontinuous and upper semicontinuous, respectively.

Finally, R denotes the set of reals; whenever considered as a topological space, R carries the usual topology.

2. Simple Dynamic Games

The purpose of this section is to define simple dynamic games and their equilibria. We incorporate in the definition of simple dynamic games the major convexity, continuity, and linearity assumptions used in demonstrating existence of their equilibria.

Definition 2.1. A *simple dynamic game* (s.d.g.) is an ordered ninetuple

$$S = \langle n, W, Y, \underline{Y}, \delta, \Omega, U, T, A \rangle,$$

where the elements of S are defined as follows.

(i) $n \in \mathscr{N}$ is called the *planning horizon*.

(ii) $W = \{X_\alpha \neq \varnothing \mid \alpha \in A\}$ is a family of nonempty sets X_α, each of which is assumed compact and convex in some LCHTVS, and from which we define

$$X = \prod_A X_\alpha \qquad \text{and} \qquad X^\alpha = \prod_{A\setminus\{\alpha\}} X_\beta, \qquad (\alpha \in A).$$

The spaces X, ${}_nX$, ${}_nX_\alpha$, and ${}_nX^\alpha$, called the *control space, plan space, α-plan space,* and *α-exclusive plan space*, respectively, are consequently all compact and convex in some LCHTVS.

(iii) $Y \neq \varnothing$, is a closed and convex subset, called the *state space*, in some LCHTVS.

(iv) $\underline{Y}$ stands for the *set of feasible initial states*, and we assume that $\underline{Y} \in \mathscr{K}\mathscr{Q}(Y)$.

(v) $\delta\colon X \times Y \to Y$ is a function, called the ***next-state map***. δ is assumed continuous and, for each $x^\alpha \in X^\alpha$ ($\alpha \in A$), linear[2] on $X_\alpha \times \{x^\alpha\} \times \underline{Y}$. From δ, we inductively define the derived functions

$$^k\delta\colon {}_nX \times \underline{Y} \to Y \qquad \text{and} \qquad {}_k\delta\colon {}_nX \times \underline{Y} \to \underline{Y} \times {}_kY$$

as follows:

$$^k\delta({}_nx, \underline{y}) = \delta(^kx, {}^{k-1}\delta({}_nx, \underline{y})),$$

$${}_k\delta({}_nx, \underline{y}) = ({}^0\delta({}_nx, \underline{y}), \ldots, {}^k\delta({}_nx, \underline{y})),$$

where

$$^0\delta\colon {}_nX \times \underline{Y} \to Y \qquad \text{is defined by} \quad {}^0\delta({}_nx, \underline{y}) = \underline{y},$$

and where

$${}_nx \in {}_nX, \qquad \underline{y} \in \underline{Y}, \qquad k \in N.$$

(vi) $\Omega = \{\omega_\alpha : \underline{Y} \to \mathscr{C}\mathscr{Q}(X_\alpha) \mid \alpha \in A\}$ is a family of continuous maps.

(vii) $U = \{u_\alpha : {}_nX \times \underline{Y} \times {}_nY \to R \mid \alpha \in A\}$ is a family of continuous real-valued functions. For each $\alpha \in A$, the so-called *utility function* u_α of α is assumed quasiconcave on ${}_nX_\alpha \times \{{}_nx^\alpha, \underline{y}\} \times {}_nY$ for each $({}_nx^\alpha, \underline{y}) \in {}_nX^\alpha \times \underline{Y}$.

(viii) $T = \{t_\alpha : X \times Y \to \mathscr{C}(X_\alpha) \mid \alpha \in A\}$ is a family of continuous maps, whence, for each $\alpha \in A$, we inductively define the derived maps $^kt_\alpha : {}_nX \times \underline{Y} \to \mathscr{C}(X_\alpha)$ and ${}_kt_\alpha : {}_nX \times \underline{Y} \to \mathscr{C}({}_kX_\alpha)$ as follows[3]:

$$^kt_\alpha({}_nx, \underline{y}) = t_\alpha(^kx, {}^{k-1}\delta({}_nx, \underline{y})),$$

$${}_kt_\alpha({}_nx, \underline{y}) = \prod_{j=0}^{k-1} {}^jt_\alpha({}_nx, \underline{y}),$$

---

[2] For example, if $A = \{1, \ldots, m\}$ and Y and X_α, $\alpha \in A$, are subsets of R, then $\delta(x, y) = y + x_1 x_2 \cdots x_m$ satisfies the required linearity and continuity assumptions.

[3] The reader may find it convenient to assume on first reading that $t_\alpha(x^\alpha, y) = X_\alpha$ for all $(x^\alpha, y) \in X^\alpha \times Y$. This assumption implies, in particular, that ${}_kt_\alpha({}_nx, \underline{y}) = {}_kX_\alpha$ for all $({}_nx, \underline{y}) \in {}_nX \times \underline{Y}$, which, as will become clear below, corresponds to the case where the feasible control region is fixed.

where

$$^0t_\alpha : {}_nX \times \underset{\sim}{Y} \to \mathscr{C}(X_\alpha) \qquad \text{is defined by} \quad {}^0t_\alpha({}_nx, \underset{\sim}{y}) = \omega_\alpha(\underset{\sim}{y}),$$

and where

$${}_nx \in {}_nX, \qquad \underset{\sim}{y} \in \underset{\sim}{Y}, \qquad k \in N.$$

(ix) A, called the *personnel*, is a self-indexed, nonempty family of maps

$$\alpha: {}_nX^\alpha \times \underset{\sim}{Y} \to [{}_nX_\alpha],$$

each of which is called a *player* and such that, for any $({}_nx^\alpha, \underset{\sim}{y}) \in {}_nX^\alpha \times \underset{\sim}{Y}$,

$$\alpha({}_nx^\alpha, \underset{\sim}{y}) = \{{}_nx_\alpha \in {}_n\hat{t}_\alpha({}_nx^\alpha, \underset{\sim}{y}) \mid u_\alpha({}_nx_\alpha, {}_nx^\alpha, {}_n\delta({}_nx_\alpha, {}_nx^\alpha, \underset{\sim}{y})) = \sup_{{}_n\hat{t}_\alpha({}_nx^\alpha, \underset{\sim}{y})} u_\alpha(\cdot, {}_nx^\alpha, {}_n\delta(\cdot, {}_nx^\alpha, \underset{\sim}{y}))\},$$

where

$${}_n\hat{t}_\alpha : {}_nX^\alpha \times \underset{\sim}{Y} \to [{}_nX_\alpha] \qquad \text{is defined for each } \alpha \in A \text{ and each}$$

$$({}_nx_\alpha, \underset{\sim}{y}) \in {}_nX^\alpha \times \underset{\sim}{Y} \qquad \text{by} \quad {}_n\hat{t}_\alpha({}_nx^\alpha, \underset{\sim}{y}) = \{{}_nx_\alpha{}' \in {}_nX_\alpha \mid {}_nx_\alpha{}' \in {}_nt_\alpha({}_nx_\alpha{}', {}_nx^\alpha, \underset{\sim}{y})\}.$$

◁

The key to understanding the above formulation of dynamic games is Definition 2.1(ix), which we now elucidate briefly. Given any initial state $\underset{\sim}{y}$ and any α-exclusive plan ${}_nx^\alpha$, each player α is assumed to develop an α-plan obeying Definition 2.1(ix). In doing so, player α takes the given $({}_nx^\alpha, \underset{\sim}{y})$ and computes the set $\alpha({}_nx^\alpha, \underset{\sim}{y})$ of solutions to the following optimal control problem:

$$\max_{{}_nX_\alpha \times \{{}_nx^\alpha, \underset{\sim}{y}\} \times {}_nY} \quad u_\alpha({}_nx_\alpha, {}_nx^\alpha, \underset{\sim}{y}, {}_ny),$$

subject to

$${}^ky = \delta({}^kx_\alpha, {}^kx^\alpha, {}^{k-1}y), \qquad k \in N, \ {}^0y = \underset{\sim}{y},$$

$${}^{k+1}x_\alpha \in {}^kt_\alpha({}_nx_\alpha, {}_nx^\alpha, \underset{\sim}{y}), \qquad k+1 \in N,$$

where ${}_nx = ({}_nx_\alpha, {}_nx^\alpha) = ({}^1x, \ldots, {}^nx)$ and ${}_ny = ({}^1y, \ldots, {}^ny)$ are the sequences of, respectively, controls and states over the planning horizon, left superscript being the *time* index. Note that, by Definition 2.1(viii),

$${}^1x_\alpha \in {}^0t_\alpha({}_nx, \underset{\sim}{y}) = \omega_\alpha(\underset{\sim}{y}),$$

the feasible control region for ${}^1x_\alpha$, while, for $k = 1, \ldots, n-1$, we have

$${}^{k+1}x_\alpha \in t_\alpha({}^kx_\alpha, {}^kx^\alpha, {}^{k-1}y),$$

so that the feasible control region for ${}^k x_\alpha$ is determined by the initial state and the controls preceding the kth.

We postpone until the next section a discussion of the conditions under which a s.d.g. is *well defined*, in the sense that, with an eye on (ix), $\alpha({}_n x^\alpha, \underline{y}) \neq \varnothing$ for all $({}_n x^\alpha, \underline{y}) \in {}_n X^\alpha \times \underline{Y}$, $\alpha \in A$.

Remark and Notation. Let S be a s.d.g. specified as in Definition 2.1, and suppose that $\underline{Y} \supset \underline{Y}' \neq \varnothing$. Then, clearly,

$$S' = \langle n, W, Y, \underline{Y}', \delta, \Omega, U, T, A \rangle$$

is also a s.d.g. When S and S' are related in this fashion, we say that S' is a *restriction* of S. When $\underline{Y}' = \{\underline{y}\} \subset \underline{Y}$, S' will be called the restriction of S to the initial state $\underline{y}$. When S is understood, we denote S' in this case by $S(\underline{y})$.

Definition 2.2. The *evolution* of a simple dynamic game

$$S = \langle n, W, Y, \underline{Y}, \delta, \Omega, U, T, A \rangle$$

is a map

$$E: {}_n X \times \underline{Y} \to [{}_n X \times \underline{Y}]$$

defined, at each $({}_n x, \underline{y}) \in {}_n X \times \underline{Y}$, by[4]

$$E({}_n x, \underline{y}) = \Big(\prod_A \alpha({}_n x^\alpha, \underline{y})\Big) \times \{\underline{y}\}.$$

A point $({}_n x, \underline{y})$ will be called an *equilibrium* of S iff $({}_n x, \underline{y}) \in E({}_n x, \underline{y})$. The set of equilibria of s is called the *contractual set* of S. $\lhd$

It may be verified that the contractual set of S consists of precisely those points $({}_n \hat{x}, \hat{\underline{y}}) \in {}_n X \times \underline{Y}$ which, for each $\alpha \in A$, satisfy

$$\begin{aligned} u_\alpha({}_n\hat{x}_\alpha, {}_n\hat{x}^\alpha, {}_n\delta({}_n\hat{x}_\alpha, {}_n\hat{x}^\alpha, \hat{\underline{y}})) &= \sup_{{}_n x_\alpha' \in {}_n t_\alpha({}_n x_\alpha', {}_n\hat{x}^\alpha, \hat{\underline{y}})} u_\alpha({}_n x_\alpha', {}_n\hat{x}^\alpha, {}_n\delta({}_n x_\alpha', {}_n\hat{x}^\alpha, \hat{\underline{y}})) \\ &= \sup_{{}_n X_\alpha \times \{{}_n\hat{x}^\alpha, \hat{\underline{y}}\} \times {}_n Y} u_\alpha({}_n x_\alpha, {}_n\hat{x}^\alpha, \hat{\underline{y}}, {}_n y), \end{aligned} \tag{1}$$

subject to

$$\begin{aligned} {}^k y &= \delta({}^k x_\alpha, {}^k\hat{x}^\alpha, {}^{k-1} y), & k \in N, {}^0 y = \hat{\underline{y}}, \\ {}^{k+1} x_\alpha &\in {}^k t_\alpha({}_n x_\alpha, {}_n\hat{x}^\alpha, \hat{\underline{y}}), & k + 1 \in N. \end{aligned}$$

---

[4] Note that E is actually a map into $\prod_A [{}_n X_\alpha] \times [\underline{Y}] \subset [{}_n X \times \underline{Y}]$.

In order to focus more clearly on the equilibrium existence question, we now show the relationship of simple dynamic games to singlestage games (see Footnote 1). To do so, take a new object (would-be player) $a \notin A$, and define $A_+ = A \cup \{a\}$. Also, define the sets

$$Z_a = \underline{Y}, \quad Z_\alpha = {}_nX_\alpha, \quad \alpha \in A; \quad Z = \prod_{A_+} Z_\alpha, \quad Z^\alpha = \prod_{A_+\setminus\{\alpha\}} Z_\beta, \quad \alpha \in A_+, \tag{2}$$

and the functions

$$\{v_\alpha : Z \to R \mid \alpha \in A_+\} \qquad \text{and} \qquad {}_nt_a : Z \to \mathscr{C}(Z_a)$$

by

$$\begin{aligned} v_\alpha(z) &= u_\alpha({}_nx, {}_n\delta({}_nx, \underline{y})), \qquad \alpha \in A, \\ v_a(z) &= \sqrt{2}, \\ {}_nt_a(z) &= Z_a, \end{aligned} \tag{3}$$

where $z = ({}_nx, \underline{y}) \in Z$.

Then, from (1), the equilibrium existence question for simple dynamic games reduces to determining conditions on v_α, ${}_nt_\alpha$, and Z such that there exists $\hat{z} \in Z$ satisfying

$$v_\alpha(\hat{z}_\alpha, \hat{z}^\alpha) = \sup_{z_\alpha \in {}_nt_\alpha(z_\alpha, \hat{z}^\alpha)} v_\alpha(z_\alpha, \hat{z}^\alpha), \qquad \alpha \in A_+. \tag{4}$$

When for all $z \in Z$, ${}_nt_\alpha(z) = Z_\alpha$ holds for each $\alpha \in A_+$, this is precisely the usual equilibrium existence question posed for singlestage games (see Debreu, Ref. 2). To deal with the equilibrium existence question in the present more general case, we need to use the results summarized in the next section.

3. Topological Foundations

This section provides some key topological results for equilibrium analysis in game theory. We immediately apply these results to the singlestage equilibrium existence question posed at the end of the preceding section.

We first recall the following closed-graph theorem which may be derived in a straightforward fashion[5] from Lemma 2 of Fan (Ref. 9).

Proposition 3.1. Given compact Hausdorff spaces P and Q and

[5] See Ref. 14, Propositions 3.2 and 3.3, for details.

two maps $f\colon P \to Q$, $F\colon P \to \mathscr{C}(Q)$, denote the graphs of f and F by g and G, respectively:

$$g = \{(p, q) \mid p \in P, \quad q = f(p)\},$$

$$G = \{(p, q) \mid p \in P, \quad q \in F(p)\}.$$

It is iff g (resp., G) $\subset P \times Q$ is closed that f is continuous (resp., F is usc). $\lhd$

Various special forms of the following proposition have been the focus of much recent activity in the mathematical programming literature (see, e.g., Hogan, Ref. 10). For a proof of the proposition in its following general form, see Sertel, Ref. 4, Corollary 3.1.4.3.

Proposition 3.2. Let P and Q be as in Proposition 3.1. Let $u\colon P \times Q \to R$ be a continuous real-valued function, and let $s\colon Q \to \mathscr{C}(P)$ be a continuous point-to-set map. Defining $v\colon Q \to R$ by

$$v(q) = \sup_{p \in s(q)} u(p, q) \qquad (q \in Q),$$

v is then continuous. $\rhd$

With the aid of the above two propositions, we can now prove the following important fact.

Theorem 3.1. Let $u\colon P \times Q \to R$ be a continuous function on a nonempty compact Hausdorff space $P \times Q$, let $s\colon Q \to \mathscr{C}(P)$ be continuous, and define a map α on Q by

$$\alpha(q) = \{p \in s(q) \mid u(p, q) \geqslant \sup_{s(q)} u(\cdot, q)\} \qquad (q \in Q).$$

Then, α maps Q upper semicontinuously into $\mathscr{C}(P)$.

Proof. Given any $q \in Q$, $s(q)$ is compact, being closed in the compact P; hence, $s(q) \times \{q\}$ is compact, so that the continuous u attains a supremum on it. This shows that $\alpha(q) \neq \varnothing$ for each $q \in Q$. To establish the upper semicontinuity of α, we will use repeatedly the closed-graph theorem (Proposition 3.1) and show that the graph

$$\Gamma_\alpha = \{(q, p) \mid q \in Q, p \in \alpha(q)\}$$

of α is closed. From the closedness of Γ_α in the compact $Q \times P$, it will follow that Γ_α is compact, so that, for each $q \in Q$,

$$\alpha(q) = \pi_P((\{q\} \times P) \cap \Gamma_\alpha)$$

is compact, hence, closed in P.

We now show that $\Gamma_\alpha \subset Q \times P$ is closed. As s is continuous, it is upper semicontinuous, so that, by Proposition 3.1, its graph

$$\Gamma_s = \{(q, p) \mid q \in Q, p \in s(q)\} \subset Q \times P$$

is closed. As u is continuous on the compact $P \times Q$, $u(P \times Q) \subset R$ is compact and, of course, Hausdorff. By Proposition 3.2, the function v: $Q \to R$ defined, at each $q \in Q$, by

$$v(q) = \sup_{s(q) \times \{q\}} u$$

is continuous, so that its graph

$$\Gamma_v = \{(q, r) \mid q \in Q, r = v(q)\} \subset Q \times v(Q) \subset Q \times u(P \times Q)$$

is closed. Hence, $\Gamma_v \times P$ is closed. By continuity of u, so is the graph

$$\Gamma_u = \{(p, q, r) \mid (p, q) \in P \times Q, r = u(p, q)\} \subset P \times Q \times u(P \times Q)$$

of u closed. Thus, $(\Gamma_v \times P) \cap \Gamma_u$ is closed, hence compact. Therefore, its projection G into $P \times Q$ is compact, hence closed. Now, Γ_α is nothing but $G \cap \Gamma_s$, which, clearly, is closed. This completes the proof.

Corollary 3.1. In the last theorem, assume also that P is convex in a real vector space, that u is quasiconcave on $P \times \{q\}$ for each $q \in Q$, and that $s(Q) \subset \mathscr{C}\mathscr{Q}(P)$. Then, α maps Q upper semicontinuously into $\mathscr{C}\mathscr{Q}(P)$.

Proof. Let $q \in Q$. All we need to show is that $\alpha(q)$ is convex. Now, $\alpha(q)$ is nothing but the intersection of $s(q)$ with $\{p \in P \mid u(p, q) \geqslant v(q)\}$, the former of which is convex by assumption, and the latter of which is convex by quasiconcavity of u on $P \times \{q\}$. This shows that $\alpha(q)$ is convex, completing the proof. $\lhd$

This section culminates with the following proposition, which serves as the foundation on which the proof of our main result in the next section rests.

Proposition 3.3. Let S be a s.d.g., as in Definition 2.1, such that, using the notation of (3) and (4) in Section 2,

(i) T is as in Definition 2.1(viii) and, for each $\alpha \in A$, the section ${}_nT_\alpha(z^\alpha) = \{(z_\alpha, z_\alpha') \in Z_\alpha \times Z_\alpha \mid z_\alpha' \in {}_nt_\alpha(z_\alpha, z^\alpha)\}$ is convex for each $z^\alpha \in Z^\alpha$. Then,

(a) for each $\alpha \in A$, the graph

$$ {}_nT_\alpha = \{(z_\alpha, z^\alpha, z_\alpha') | (z_\alpha, z^\alpha) \in Z, z_\alpha' \in {}_nt_\alpha(z_\alpha, z^\alpha)\} \subset Z \times Z_\alpha $$

of ${}_nt_\alpha$ is closed;

(b) for each $\alpha \in A$, $v_\alpha : Z \to R$, as given in (3)–(4), is continuous and, for each $z^\alpha \in Z^\alpha$, quasiconcave on $Z_\alpha \times \{z^\alpha\}$;

(c) for each $\alpha \in A$, ${}_n\hat{t}_\alpha$ maps Z^α upper semicontinuously into $\mathscr{C}\mathscr{Q}(Z_\alpha)$;

(d) for each $\alpha \in A$, α maps Z^α into $\mathscr{C}\mathscr{Q}(Z_\alpha)$; and

(e) S is well defined.

If, furthermore,

(ii) for each $\alpha \in A$, ${}_n\hat{t}_\alpha$ is lower semicontinuous, then also

(f) for each $\alpha \in A$, α is upper semicontinuous; and

(g) the contractual set C of S is nonempty and compact.

Proof. (a) By Proposition 3.1, ${}_nT_\alpha$ will be closed iff ${}_nt_\alpha$ is usc. But ${}_nt_\alpha$ will be usc iff each of its components ${}^kt_\alpha$ is so, $k + 1 \in N$. In fact, since ${}^kt_\alpha$ is composed [see Definition 2.1(viii)] of t_α [continuous by Definition 2.1(viii)], projection $\pi({}^kX)$, and ${}^{k-1}\delta$ [continuous, by Definition 2.1(v)], ${}^kt_\alpha$ is continuous, hence upper semicontinuous, $k \in N$. Also, ${}^0t_\alpha$ is continuous by Definition 2.1 (viii) and (vi). Thus, ${}_nt_\alpha$ is upper semicontinuous and ${}_nT_\alpha$ is closed, proving (a).

(b) The continuity of v_α follows from that of u_α [Definition 2.1(vii)] and of ${}_n\delta$, where the continuity of ${}_n\delta$ follows from that of δ [Definition 2.1(v)]. The asserted quasiconcavity follows from the quasiconcavity of u_α [Definition 2.1(vii)] and from the linearity of ${}_n\delta$ on ${}_nX_\alpha \times \{{}_nx^\alpha\} \times \underline{Y}$ for each ${}_nx \in {}_nX^\alpha$, the latter of which in turn follows from the linearity properties of δ [Definition 2.1(v)]. Thus, (b) is proved.

(c) Let $\alpha \in A$ and $z^\alpha \in Z^\alpha$. The graph ${}_nT_\alpha(z^\alpha)$ of the restriction ${}_nt_\alpha(., z^\alpha)$ of ${}_nt_\alpha$ to $Z_\alpha \times \{z^\alpha\}$ is convex by assumption, and it is closed since ${}_nT_\alpha$ is closed by (a). As Z_α is compact and convex, ${}_nt_\alpha(., z^\alpha)$ is therefore into $\mathscr{C}\mathscr{Q}(Z_\alpha)$; by Proposition 3.1, it is also upper semicontinuous. Hence, by Fan's FPT (Ref. 9, Theorem 1), its graph ${}_nT_\alpha(z^\alpha)$ intersects the diagonal Δ_α of $Z_\alpha \times Z_\alpha$. Furthermore, this intersection is that of two closed and convex sets, and so is closed and convex. Being closed in the compact $Z_\alpha \times Z_\alpha$, it is compact. Moreover, its projection into Z_α is nothing but ${}_n\hat{t}_\alpha(z^\alpha)$, so that ${}_n\hat{t}_\alpha(z^\alpha) \in \mathscr{C}\mathscr{Q}(Z_\alpha)$ and ${}_n\hat{t}_\alpha$ is into $\mathscr{C}\mathscr{Q}(Z_\alpha)$. Now, the graph of ${}_n\hat{t}_\alpha$ is simply

$$\pi_{(Z^\alpha \times Z_\alpha)}({}_nT_\alpha \cap \{(z_\alpha, y^\alpha, y_\alpha) \in Z_\alpha \times Z^\alpha \times Z_\alpha \mid x_\alpha = y_\alpha\}),$$

which is compact, hence closed. Thus, by Proposition 3.1, ${}_n\hat{t}_\alpha$ is also upper semicontinuous. This completes the proof of (c).

(d) From (c), we see that ${}_n\hat{t}_\alpha(z^\alpha)$ is closed, hence compact, so that the restriction of the continuous [see (b)] v_α to the compact $Z_\alpha \times \{z^\alpha\}$ attains a supremum on a nonempty, closed subset of ${}_n\hat{t}_\alpha(z^\alpha)$. Clearly, this subset is nothing but $\alpha(z^\alpha)$, and from Corollary 3.1 it is convex since, by (b), v_α is quasiconcave on $Z_\alpha \times \{z^\alpha\}$ and, by (c), ${}_n\hat{t}_\alpha(z^\alpha)$ is convex. This proves (d).

(e) As $[Z_\alpha] \supset \mathscr{C}\mathscr{Q}(Z_\alpha)$, from (d) we see that $\alpha(Z^\alpha) \subset [Z_\alpha]$, i.e., that S is a well-defined s.d.g., proving (e).

For (f) and (g), now assume (ii).

(f) Being upper semicontinuous by (c) and lower semicontinuous by (ii), ${}_n\hat{t}_\alpha$ is now continuous. Applying Theorem 3.1, (f) is proved.

(g) From (d) and (f), for each $\alpha \in A$, α maps Z^α upper semicontinuously into $\mathscr{C}\mathscr{Q}(Z_\alpha)$. Thus, the evolution E of S maps Z upper semicontinuously into $\mathscr{C}\mathscr{Q}(Z)$, so that, by Fan's FPT (Ref. 9), E has a fixed point. In other words, S has an equilibrium and $C \neq \varnothing$. Actually, since E is upper semicontinuous, by Proposition 3.1 its graph

$$\Gamma_E = \{(z, z') \in Z \times Z \mid z' \in E(z)\}$$

is closed, hence compact in the compact space $Z \times Z$. As $C = \pi_Z(\Gamma_E \cap \Delta)$, where Δ is the diagonal of $Z \times Z$, C is also compact. This completes the proof.

4. Equilibrium Existence Results for Dynamic Games

This section presents our major results. Examples are provided at the end to illustrate applications to s.d.g.'s with certain special characteristics of interest.

Theorem 4.1. Let $S = \langle n, W, Y, \underline{Y}, \delta, \Omega, U, T, A\rangle$ be specified as in Definition 2.1, where

(i) for each $\alpha \in A$, $t_\alpha : X \times Y \to \mathscr{C}(X_\alpha)$ is such that the section

$${}^k\Gamma_\alpha(x^\alpha) = \{(x_\alpha, x^\alpha, y, x_\alpha') \in X_\alpha \times \{x^\alpha\} \times {}^k\bar{D} \times X_\alpha \mid x_\alpha' \in t_\alpha(x_\alpha, x^\alpha, y)\}$$

is convex for each $x^\alpha \in X^\alpha$ and each $k \in \{1,..., n-1\}$, where ${}^k\bar{D} \subset Y$ is the closed convex hull of the set

$${}^kD = {}^{k-1}\delta({}_nX \times \underline{Y}).$$

Then,

(a) S is a simple dynamic game;

(b) for each $\alpha \in A$, α maps ${}_nX^\alpha \times \underline{Y}$ upper semicontinuously into $\mathscr{C}\mathscr{Q}({}_nX_\alpha)$; and

(c) the contractual set of S [in fact, of every restriction of S to a $\underline{Y}' \in \mathscr{K}\mathscr{Q}(\underline{Y})$] is nonempty and compact.

Proof. (a) and (b). By Proposition 3.3, S will be a well-defined s.d.g. and each $\alpha \in A$ will have the asserted continuity and convexity properties if S fulfills the hypotheses (i) and (ii) of that proposition. The convexity of the sections

$${}_nT_\alpha({}_nx^\alpha, \underline{y}) = {}_nT_\alpha(z^\alpha)$$

is a straightforward matter of computation, using Definition 2.1(viii) of ${}_nt_\alpha$ and the linearity and convexity assumptions made in Definition 2.1(v) and (vi) and Theorem 4.1(i). We therefore verify only (ii) of Proposition 3.3.

Fix $\alpha \in A$. To establish the lower semicontinuity of ${}_n\hat{t}_\alpha$, let

$$V = \prod_{k \in N} {}^kV \subset {}_nX_\alpha$$

be any basic open set in the product topology of ${}_nX_\alpha$. It suffices to show that

$$U = \{({}_nx^\alpha, \underline{y}) \mid V \cap {}_n\dot{t}_\alpha({}_nx^\alpha, \underline{y}) \neq \varnothing\} \subset {}_nX^\alpha \times \underline{Y}$$

is open. This being trivially so when $V = \varnothing$, we assume henceforth that $V \neq \varnothing$. It is useful to define $Q = {}_nX^\alpha \times \underline{Y}$ and, using the Definition 2.1(ix) of ${}_n\hat{t}_\alpha$, to rewrite U as

$$U = \{q \in Q \mid \exists\ {}_nx_\alpha \in V$$

with

$${}^kx_\alpha \in {}^{k-1}t_\alpha({}_nx_\alpha, q) \text{ for each } k \in N\}.$$

We now proceed to show that U contains a nbd of each of its points. Toward that, suppose that $\hat{q} \in U$. We will construct a set $\hat{U}$ and complete our proof by showing that $\hat{U}$ is a nbd of $\hat{q}$ contained in U. Fix ${}_n\bar{x}_\alpha \in {}_nX_\alpha$, and define ${}^0P = \{{}_n\bar{x}_\alpha\}$ and, for each $k \in N$, ${}^kP = X_\alpha$, writing[6]

$${}^0P \times {}^1P \times \cdots \times {}^kP = {}_kP,$$

with generic elements ${}^kp \in {}^kP$ and ${}_kp \in {}_kP$ $(k \in N_0)$. For each $k+1 \in N$, define the function

$${}^k\chi\colon {}_nX_\alpha \times {}_kX_\alpha \to {}_nX_\alpha$$

by

$${}^k\chi({}_nx_\alpha{}', {}_kx_\alpha) = ({}_kx_\alpha, {}^{k+1}x_\alpha{}',\dots, {}^nx_\alpha{}'), \qquad ({}_nx_\alpha, {}_kx_\alpha) \in {}_nX_\alpha \times {}_kX_\alpha,$$

and the map

$${}^k\tau\colon {}_kP \times Q \to [{}^{k+1}P]$$

by

$${}^k\tau({}_kp, q) = {}^kt_\alpha({}^k\chi({}_kp), q), \qquad ({}_kp, q) \in {}_kP \times Q.$$

From Definition 2.1(viii), we see that, for any $k + 1 \in N$ and for any $({}_nx_\alpha, q) \in {}_nX_\alpha \times Q$, by writing

$${}_kp = ({}_n\bar{x}_\alpha, {}_kx_\alpha),$$

one obtains

$${}^kt_\alpha({}_nx_\alpha, q) = {}^k\tau({}_kp, q).$$

From this, we are able to rewrite U as

$$U = \{q \in Q \mid \exists\, ({}^1p,\dots, {}^np) \in V \quad \text{with} \quad {}^kp \in {}^{k-1}\tau({}_{k-1}p, q) \quad \text{for each} \quad k \in N\}.$$

Since $\hat{q} \in U$, there exists $({}^1\hat{p},\dots, {}^n\hat{p}) \in V$ with

$${}^k\hat{p} \in {}^{k-1}\tau({}_{k-1}\hat{p}, \hat{q}) \qquad \text{for} \quad k \in N.$$

Also, for each $k + 1 \in N$, ${}^k\tau$ is continuous, since ${}^kt_\alpha$ is so [as seen in the proof of Proposition 3.3(a), above] and since ${}^k\chi$ is obviously so. Thus, for each $k + 1 \in N$, ${}^k\tau$ is certainly lower semicontinuous.

To construct $\hat{U}$, set

$${}^kV_n = {}^kV \ (k \in N),$$

and, for each $k + 1 \in N$ (following the order $k = n - 1, n - 2,\dots, 0$),

[6] The reader will, please, excuse our momentary departure, in defining ${}_kP$, from our usual notational convention as announced in Section 1.

inductively define open sets ${}^kW$, ${}^jV_k$ $(j \in \{1,\dots, k\})$, and ${}^kU$, obeying

$$ {}^kW = \left\{({}^1p,\dots, {}^kp, q) \mid \left(\bigcap_{j=k+1}^{n} {}^{k+1}V_j\right) \cap {}^k\tau({}_kp, q) \neq \varnothing\right\}, $$

with

$$ ({}^1\hat{p},\dots, {}^k\hat{p}, \hat{q}) \in \left(\bigcap_{j=1}^{n} {}^jV_k\right) \times {}^kU \subset {}^kW, \qquad {}^jV_k \subset {}^kP, \qquad {}^kU \subset Q. $$

To check that this is possible, we use the following facts inductively (in the order $k = n - 1, n - 2,\dots, 0$):

(i) ${}^k\tau$ is lower semicontinuous and $\bigcap_{j=k+1}^{n} {}^{k+1}V_j$ is open, so ${}^kW$ is open;

(ii) ${}^{k+1}\hat{p} \in \bigcap_{j=k+1}^{n} {}^{k+1}V_j$, and ${}^{k+1}\hat{p} \in {}^k\tau({}_k\hat{p}, \hat{q})$, so $({}^1\hat{p},\dots, {}^k\hat{p}, \hat{q}) \in {}^kW$;

(iii) using (i) and (ii),

$$ \left(\prod_{j=1}^{k} {}^jV_k\right) \times {}^kU \subset {}^kW $$

is chosen as a basic open set (in the product topology of ${}^1P \times \cdots \times {}^kP \times Q$) owning the point $({}^1\hat{p},\dots, {}^k\hat{p}, \hat{q})$.

Set $\hat{U} = \bigcap_{k=0}^{n-1} {}^kU$. From its construction, it is clear that $\hat{U}$ is open and that $\hat{q} \in \hat{U}$. It remains to show only that $\hat{U} \subset U$. For this, we take any $\tilde{q} \in \hat{U}$ and show that $\tilde{q} \in U$ by inductively constructing a point $({}^1\tilde{p},\dots, {}^n\tilde{p}) \in V$ such that

$$ {}^{k+1}\tilde{p} \in {}^k\tau({}_k\tilde{p}, \tilde{q}) $$

for every $k + 1 \in N$. Since

$$ \tilde{q} \in \hat{U} \subset {}^0U \subset {}^0W, $$

we have that

$$ \left(\bigcap_{j=1}^{n} {}^1V_j\right) \cap {}^0\tau({}^0p, \tilde{q}) \neq \varnothing . $$

Choose ${}^1\tilde{p}$ as any point of this intersection. Now assume that, for $\tilde{k} + 1 \in N$,

$$ {}^k\tilde{p} \in \left(\bigcap_{j=k}^{n} {}^kV_j\right) \cap {}^{k-1}\tau({}_{k-1}\tilde{p}, \tilde{q}) $$

is chosen for every $k \in \{1,..., \tilde{k}\}$. Then,

$$^{k}\tilde{p} \in \bigcap_{j=k}^{n} {}^{k}V_j \ (k \in \{1,..., \tilde{k}\}) \qquad \text{and} \qquad \tilde{q} \in \hat{U} \subset {}^{\tilde{k}}U$$

being clear, it follows that

$$({}^{1}\tilde{p},..., {}^{\tilde{k}}\tilde{p}, \tilde{q}) \in \left(\prod_{j=1}^{\tilde{k}} {}^{j}V_{\tilde{k}}\right) \times {}^{\tilde{k}}U \subset {}^{\tilde{k}}W.$$

Now, ${}^{\tilde{k}}W \neq \varnothing$, so that $(\bigcap_{j=\tilde{k}+1}^{n} {}^{\tilde{k}+1}V_j) \cap {}^{\tilde{k}}\tau({}_{\tilde{k}}\tilde{p}, \tilde{q})$ is nonempty and affords an element ${}^{\tilde{k}+1}\tilde{p}$. This shows that the desired $({}^{1}\tilde{p},..., {}^{n}\tilde{p})$ exists, so that $\hat{U} \subset U$, establishing that S satisfies Proposition 3.3(ii), and thus completing the proof of (a) and (b).

(c) That S has a nonempty and compact contractual set is now a direct consequence of Proposition 3.3(g). That any restriction of S to a nonempty compact and convex set $\underline{Y}' \subset \underline{Y}$ has a nonempty and compact contractual set follows (see Remark and Notation in Section 2).

Remark 4.1. If, in Theorem 3.1 one takes the restriction of S to the initial state $\underline{y} \in \underline{Y}$, then Theorem 3.1(c) implies that $S(\underline{y})$ has an equilibrium for all $\underline{y} \in \underline{Y}$, that is, for each $\underline{y} \in \underline{Y}$, there exists ${}_{n}\mathscr{X} \in {}_{n}\hat{X}$ satisfying (1) with $\hat{y} = \underline{y}$. Thus, under the hypotheses of the last theorem, there is an equilibrium plan for each feasible initial state. Theorem 4.1 also shows that there exist well-defined simple dynamic games, so that Definition 2.1 is not a self-contradiction. Of course, this well-definedness holds under much weaker conditions than those assumed in Definition 2.1 and Theorem 4.1. For example, it can be shown, following the proofs of Proposition 3.3(a), Proposition 3.3(c), and Theorem 4.1, that ${}_{n}\hat{t}_{\alpha}$ will be a continuous map of ${}_{n}X^{\alpha} \times Y$ into $\mathscr{C}({}_{n}X_{\alpha})$ if ω_{α}, t_{α}, and δ are only assumed continuous without any additional convexity or linearity assumptions. Given this, and recalling Definition 2.1(ix), S will be well-defined even if we discard all convexity (quasiconcavity and linearity) assumptions in Definition 2.1.

Remark 4.2. In the simple dynamic games that we have considered so far, the feasible control region to which the α-control ${}^{k+1}x_{\alpha}$ is constrained to belong depends on the previous control ${}^{k}x$ and state ${}^{k-1}y$. See the discussion immediately preceding the Remark and Notation in Section 2. The reader will note, however, that the case where the control region in question also depends on ${}^{k}y$ falls naturally under the case that we consider generally. Below, we illustrate this by considering a dependence of these regions on ${}^{k}y$ and, while this dependence is on ${}^{k}y$ alone, the reader will easily be able to extend the simple idea involved to the general case where the dependence on ${}^{k}x$ and ${}^{k-1}y$ is not through ${}^{k}y$ alone.

Example 4.1. Consider a s.d.g. S in which, for each $\alpha \in A$, $\bar{\omega}_\alpha : Y \to \mathscr{C}\mathscr{Q}(X_\alpha)$ is an extension of ω_α, t_α is given by

$$t_\alpha = \bar{\omega}_\alpha \circ \delta,$$

and $\bar{\omega}_\alpha$ is continuous with the graphs

$${}^{k}\Omega_\alpha = \{(x_\alpha, y) \in X_\alpha \times {}^{k}\bar{D} \mid x_\alpha \in \bar{\omega}_\alpha(y)\}$$

convex for each $k \in N$, where ${}^{k}\bar{D}$ is as in Theorem 4.1(i).

In this case, the fashion in which the α-controls ${}^{k+1}x_\alpha$ are constrained to their respective control regions is, thus, as follows:

$${}^{k+1}x_\alpha \in \bar{\omega}_\alpha({}^{k}y), \qquad k + 1 \in N, \alpha \in A.$$

Moreover, it is straightforward to check that S satisfies Theorem 4.1(i) so that the conclusions of Theorem 4.1 are valid.

The following example illustrates the application of Theorem 4.1 to a salient class of deterministic dynamic games.

Example 4.2. Let A be a finite set, and let

$$\{m_\alpha \in \mathscr{N} \mid \alpha \in A\}, \qquad \{l_\alpha \in \mathscr{N} \mid \alpha \in A\}$$

and $m \in \mathscr{N}$ be given. We specify the elements of a s.d.g.

$$S = \langle n, W, Y, \underline{Y}, \delta, \bar{\Omega}, U, T, A\rangle$$

as follows:

(i) $X_\alpha \in \mathscr{K}\mathscr{Q}({}_{m_\alpha}R)$ $(\alpha \in A)$;

(ii) $Y = {}_{m}R$;

(iii) $\underline{Y} \in \mathscr{K}\mathscr{Q}(Y)$;

(iv) δ satisfies Definition 2.1 (v);

(v) $\bar{\omega}_\alpha : Y \to \mathscr{C}\mathscr{Q}(X_\alpha)$ is defined by

$$\bar{\omega}_\alpha(y) = \{x_\alpha \in X_\alpha \mid \psi_\alpha(x_\alpha, y) \geqslant 0\},$$

where $\psi_\alpha : X_\alpha \times Y \to {}_{l_\alpha}R$ is a given continuous function, quasiconcave in x_α, $\alpha \in A$;

(vi) $t_\alpha : X \times Y \to [X_\alpha]$ is defined as in Example 4.1, so that

$$t_\alpha(x, y) = (\bar{\omega}_\alpha(\delta(x, y)), \qquad (x, y) \in X \times Y \quad (\alpha \in A);$$

(vii) $u_\alpha : {}_nX \times \underline{Y} \times {}_nY \to R$ is defined for each $\alpha \in A$ and

$$({}_nx, \underline{y}, {}_ny) \in {}_nX \times \underline{Y} \times {}_nY$$

by

$$u({}_nx, \underline{y}, {}_ny) = {}^nf_\alpha({}^ny) + \sum_{k=0}^{n-1} {}^kf_\alpha({}^{k+1}x, {}^ky),$$

where ${}^0y \in \underline{Y}$, and where

$$\{{}^kf_\alpha : X \times Y \to R \mid k+1 \in N, \alpha \in A\} \qquad \text{and} \qquad \{{}^nf_\alpha : Y \to R \mid \alpha \in A\}$$

are each a family of continuous and concave functions;

(viii) α satisfies Definition 2.1(ix), $\alpha \in A$.

Suppose, in addition, that the following condition holds:

(ix) For all $\alpha \in A$, $\bar{\omega}_\alpha$ satisfies Example 4.1.

Note that these conditions on $\bar{\omega}_\alpha$ will be satisfied if ψ_α is linear or if ψ_α satisfies: (i) $\psi_\alpha : X_\alpha \times Y \to {}_{l_\alpha}R$ is continuous; (ii) for each $y \in Y$, ψ_α is concave on $X_\alpha \times \{y\}$; (iii) for all $y \in Y$, there exists $x_\alpha \in X_\alpha$ with $\psi_\alpha(x_\alpha, y) > 0$; and (iv) ψ_α is concave on $X_\alpha \times {}^k\bar{D}$, $k \in N$. See, e.g., Hogan, Ref. 10, Theorems 10 and 12.

It is easily verified that S, as specified above, satisfies the requirements of Definition 2.1 and Theorem 4.1. Therefore, by Theorem 4.1 and recalling (1) and Remark 4.1, for each $\hat{\underline{y}} \in \underline{Y}$, there exists an ${}_n\hat{x} \in {}_nX$ satisfying, for all $\alpha \in A$,

$$u_\alpha({}_n\hat{x}, \hat{\underline{y}}, {}_n\delta({}_n\hat{x}, \hat{\underline{y}})) = \max_{{}_nX_\alpha \times \{({}_n\hat{x}^\alpha, \hat{\underline{y}})\} \times {}_nY} \left({}^nf_\alpha({}^ny) + \sum_{k=0}^{n-1} {}^kf_\alpha({}^{k+1}x_\alpha, {}^{k+1}\hat{x}^\alpha, {}^ky)\right)$$

subject to

$$ {}^ky = \delta({}^kx_\alpha, {}^k\hat{x}^\alpha, {}^{k-1}y), \qquad {}^0y = \hat{\underline{y}},$$

$$\psi_\alpha({}^kx_\alpha, {}^{k-1}y) \geq 0, \qquad {}^kx_\alpha \in X_\alpha,$$

with $k \in N$.

5. Concluding Remarks

This paper has shown (Theorem 4.1) the existence of equilibria for certain *simple* dynamic games. The applicability of these results to deterministic dynamic games has been illustrated by Example 4.2. A central theme of this research, explored more fully in Kleindorfer and Sertel (Ref. 11), has been the relevance of static games to the study of equilibrium existence in dynamic games.

We suggest several directions in which to extend our results. Foremost among these is, no doubt, dealing explicitly with informational imperfections on the part of players as well as with randomness. To do so will certainly require a more detailed examination of closed-loop and open-loop phenomena (see Starr and Ho, Refs. 12 and 13). As can be seen from (1) of Section 2, the equilibria whose existence is established here correspond to open-loop equilibria of dynamic games as they are usually defined (see, e.g., Ref. 6). However, the existence of closed-loop equilibria can also be studied by similar methods. Essentially, one starts with a simple dynamic game S and derives from it a dynamic game S' by replacing the α-control spaces ${}^kX_\alpha$ of S by the function space of strategy maps (or control laws) mapping the observed state and control history (until time k) into ${}^kX_\alpha$. The nontrivial issue to be resolved is the determination of conditions under which the dynamic game S' resulting from this transformation will inherit from S the convexity, linearity, continuity, and compactness properties that play such an important role in the approach we have used in this paper.

References

1. Arrow, K. J., and Debreu, G., *Existence of an Equilibrium for a Competitive Economy*, Econometrica, Vol. 22, No. 3, 1954.
2. Debreu, G., *A Social Equilibrium Existence Theorem*, Proceedings of the National Academy of Sciences, Vol. 38, No. 10, 1952.
3. Eilenberg, S., and Montgomery, D., *Fixed Point Theorems for Multi-Valued Transformations*, American Journal of Mathematics, Vol. 68, No. 2, 1946.
4. Sertel, M. R., *Elements of Equilibrium Methods for Social Analysis*, Massachusetts Institute of Technology, Ph.D. Thesis, 1971.
5. Kuhn, A., and Szegö, G., Editors, *The Theory of Differential Games and Related Topics*, North Holland Publishing Company, Amsterdam, Holland, 1971.
6. Ho, Y. C., *Differential Games, Dynamic Optimization, and Generalized Control Theory*, Journal of Optimization Theory and Applications, Vol. 6, No. 3, 1970.
7. Parthasarathy, K. P., *Probability Measures on Metric Spaces*, Academic Press, New York, New York, 1967.

8. MICHAEL, E., *Topologies on Spaces of Subsets*, Transactions of the American Mathematical Society, Vol. 71, No. 4, 1951.
9. FAN, K., *Fixed-Point and Minimax Theorems in Locally Convex Topological Linear Spaces*, Proceedings of the National Academy of Sciences, Vol. 38, No. 2, 1952.
10. HOGAN, W. F., *Point to Set Maps in Mathematical Programming*, SIAM Review, Vol. 5, No. 3, 1973.
11. KLEINDORFER, P., and SERTEL, M. R., *Equilibrium Existence Results for Simple Economies and Dynamic Games*, International Institute of Management, West Berlin, Germany, Preprint No. I/72-22, 1972.
12. STARR, A. W., and HO, Y. C., *Nonzero-Sum Differential Games*, Journal of Optimization Theory and Applications, Vol. 3, No. 3, 1969.
13. STARR, A. W., and HO, Y. C., *Further Properties of Nonzero-Sum Differential Games*, Journal of Optimization Theory and Applications, Vol. 4, No. 4, 1969.
14. PRAKASH, P., *Foundations of Systems: For Decision, Planning, Control and Social/Economic Analysis*, Massachusetts Institute of Technology, Ph.D. Thesis, 1971.
15. PRAKASH, P., and SERTEL, M. R., *Fixed Point and Minimax Theorems in Semi-Linear Spaces*, Massachusetts Institute of Technology, Sloan School of Management, Working Paper, No. 484-70, 1970.
16. PRAKASH, P., and SERTEL, M. R., *Existence of Non-cooperative Equilibria in Social Systems*, Discussion Paper No. 92 (revised November, 1974), The Center for Mathematical Studies in Economics and Management Science, Northwestern University, Presented in August, 1971 to the NSF Conference on Control Theory of Partial Differential Equations, University of Maryland.
17. PRAKASH, P., and SERTEL, M. R., *Topological Semivector Spaces: Convexity and Fixed Point Theory*, Semigroup Forum, Vol. 9, No. 2, 1974.

XV

A Class of Trilinear Differential Games[1]

S. CLEMHOUT AND H. Y. WAN, JR.

Abstract. This paper characterizes a class of N-person, general sum differential games for which the optimal strategies only depend upon remaining playing time. Such strategies can be easily characterized and determined, and the optimal play can be easily analyzed.

1. Introduction

Much has been established both for the necessary conditions and the sufficient conditions pertaining to the optimal play in n-person, general sum differential games (Refs. 1–4). The major difficulty blocking the application of such game models is that equilibrium strategies are extremely hard to determine or characterize. Specifically, the adjoint system arising from such games usually involves partial derivatives of unknown optimal strategies. Beside the linear–quadratic games and some special examples (Ref. 1), the literature contained no other known class of solvable n-person general sum games until recently (see Refs. 5–6). Here, we present another class of solvable games where the optimal strategies depend only on time. An economic application is reported separately (Ref. 7).

[1] We acknowledge the helpful comments of G. Leitmann and an anonymous referee.

2. Problem Statement

Consider a dynamic system

$$\dot{x} = \left[\sum_{j=1}^{N} (c_j(u_j) >< d^j) + E(t) \right] x + b(t), \tag{1}$$

where $x \in R^M$ is the state vector; d^j is an M-dimensional constant vector, $j = 1,\ldots, N$; $b(t)$ is an M-dimensional vector which may depend on time t, $b(\cdot)$ Lebesgue measurable and bounded; $c_j(\cdot)$ is a twice-differentiable, M-dimensional vector function of a vector u_j in R^{S_j} for some S_j, $j = 1,\ldots, N$; $E(t)$ is an $M \times M$ matrix which may depend on time, $E(\cdot)$ Lebesgue measurable and bounded; $c_j >< d^j$ is an $M \times M$ dyad matrix, that is, the matrix product of an M-dimensional column vector c_j postmultiplied by an M-dimensional row vector d^j.

The payoff for player i is

$$J_i = \int_{t_0}^{t_f} L_i \, dt, \qquad i = 1,\ldots, N, \tag{2}$$

where t_0 and t_f are fixed beginning and ending playing time and

$$L_i = \left(\sum_{j=1}^{N} c_{ij}^0(u_j) \, d^j + e_i^0(t) \right) \cdot x + b_i^0(t), \tag{3}$$

where $c_{ij}^0(\cdot)$ is a twice-differentiable scalar function of u_j; e_i^0 is an M-dimensional vector which may depend on t, $e_i^0(\cdot)$ Lebesgue measurable and bounded; and b_i^0 as a scalar which may be a function of time, $b_i^0(\cdot)$ Lebesgue measurable and bounded.

The control vector for player i is

$$u_i \in U^i \subset R^{S_i}, \qquad U^i \text{ compact and convex.} \tag{4}$$

The *playing space* is

$$S = \{x \mid Dx > 0\} \times (-\infty, t_f), \tag{5}$$

where

$$D = \begin{bmatrix} d^1 \\ \cdots\cdots \\ \vdots \\ \cdots\cdots \\ d^N \end{bmatrix}. \tag{6}$$

The *playable strategy class* is P, consisting of a class of

$$\sigma(\cdot) \triangleq (\sigma^1(\cdot),..., \sigma^N(\cdot)): S \to \prod_{j=1}^{N} U^j, \tag{7}$$

such that the following conditions are satisfied:

(i) $\sigma(\cdot)$ is Borel measurable;

(ii) for any $s = (x_0, \tau) \in S$, there exists at least one pair of $(v(\cdot), \phi(\cdot))$, where

(α) $v(\cdot) \triangleq (v^1(\cdot),..., v^N(\cdot))$: $[t_0, t_f] \to \prod_{i=1}^{N} U^i$ is Lebesgue measurable and bounded,

(β) $\phi(\cdot)$: $[t_0, t_f] \to \{x \mid Dx > 0\}$ is absolutely continuous,

(γ) $v(t) = \sigma(\phi(t), t)$ for all $t \in [t_0, t_f]$,

(δ) $\dot{\phi}(t) = \{\sum_{j=1}^{N} (c_j[v_j(t)] >< d^j) + E(t)\} \phi(t) + b(t), \phi(\tau) = x_0$;

(iii) if $\bar{\sigma}$ and $\bar{\bar{\sigma}}$ are both in P, then, for $t_1 \in (-\infty, t_f)$, a mapping σ is also in P if

$$\sigma(x, t) = \begin{cases} \bar{\sigma}(x, t) \leqslant t_1, \\ \bar{\bar{\sigma}}(x, t) > t_1. \end{cases}$$

A *play* is an ordered triplet $(s, \sigma(\cdot), \phi(\cdot))$ such that $s \in S$, $\sigma(\cdot) \in P$, and there exists $(v(\cdot), \phi(\cdot))$ generated by σ and the initial condition (x_0, τ) as in the definition of P above. Obviously, the payoff function depends upon the play, i.e.,

$$J_i = J_i(s, \sigma(\cdot), \phi(\cdot)), \qquad i = 1,..., N. \tag{8}$$

A strategy $\sigma^*(\cdot) \in P$ is a Nash equilibrium iff the following condition is satisfied: let $(s, \sigma^*(\cdot), \phi^*(\cdot))$ be any one play associated with s and $\sigma^*(\cdot)$, $(s, {}^i\sigma(\cdot), \phi^i(\cdot))$ be any one play associated with s and ${}^i\sigma(\cdot) = (\sigma^{1*}(\cdot),..., \sigma^{i-1*}(\cdot), \sigma^i(\cdot), \sigma^{i+1*}(\cdot),..., \sigma^{N*}(\cdot)) \in P$ [${}^i\sigma(\cdot)$ may be the same as $\sigma^*(\cdot)$], for any $i = 1,..., N$; then,

$$J_i(s, \sigma^*(\cdot), \phi^*(\cdot)) \geqslant J_i(s, {}^i\sigma(\cdot), \phi^i(\cdot)) \qquad \text{for all} \quad s \in S. \tag{9}$$

3. Characterization of a Nash Equilibrium

Consider now the following system:

$$\bar{\sigma}_i(x, t) = \bar{v}_i(t) \qquad \text{for all } (x, t) \in S$$

such that

$$c_{ii}^0(\bar{v}_i(t)) + \overline{p_i}(t) \cdot c_i(\bar{v}_i(t)) = \max_{u_i \in U^i} c_{ii}^0(u_i) + \overline{p_i}(t) \cdot c_i(u_i), \tag{10}$$

$$\dot{\overline{p_i'}} = -\left[\sum_{j=1}^{N} (c_{ij}^0(\bar{v}_j(t))\, d^j) + e_i{}^0 + \overline{p_i'}\left(\sum_{j=1}^{N} c_j(\bar{v}_j(t)) >< d^j + E\right)\right], \quad \overline{p_i}(t_f) = 0,$$

$$i = 1,\ldots, N, \tag{11}$$

$$\bar{x}(t) = x_0 + \int_0^t \left\{\left(\sum_{j=1}^{N} (c_j(\bar{v}_j(\tau)) >< d^j) + E(\tau)\right) \bar{x}(\tau) + b(\tau)\right\} d\tau. \tag{12}$$

One now has the following proposition.

Proposition 3.1. If (i) the maximization problem in (10) has a unique solution for all i, $1 \leqslant i \leqslant N$, and all $t \in [t_0, t_f]$, and if (ii) $D\bar{x}(t) > 0$ for $t \in [t_0, t_f]$, then $(\bar{\sigma}_1(\cdot),\ldots, \bar{\sigma}_N(\cdot)) = \bar{\sigma}(\cdot)$ is a Nash equilibrium.

Remark 3.1. If either $[c_{ii}^0(\cdot) + \bar{p}_i(t) \cdot c_i(\cdot)]$ is a nonvanishing concave function while U^i is strictly convex or $[c_{ii}^0(\cdot) + \bar{p}_i(t) \cdot c_i(\cdot)]$ is nonvanishing and strictly concave while U^i is convex, then the existence of the maximum implies the uniqueness of $\bar{v}_i(t)$.

Remark 3.2. There may exist other Nash equilibria which we cannot yet determine in full.

4. Verification

According to the Leitmann–Stalford verification theorem (Ref. 4), the strategy N-tuple $\bar{\sigma}(\cdot)$ is an optimal play if three conditions are satisfied: (a) it is playable, (b) there exists $p_i(t)$, $i = 1,\ldots, N$, such that $p_i(t_f) \cdot [\bar{x}(t_f) - x] = 0$ for all x such that $Dx > 0$, and (c) for the same $p_i(t)$, the following condition holds for all $t \in (-\infty, t_f)$, for all x such that $Dx > 0$, and for all $u_i \in U^i$,

$$H^i\{t, \bar{x}(t), \bar{v}(t), p_i\}$$

$$- H^i\{t, x, \bar{v}_1(t),\ldots, \bar{v}_{i-1}(t), u_i, \bar{v}_{i+1}(t),\ldots, \bar{v}_N(t), p_i\} + p_i \cdot (\bar{x} - x) \geqslant 0, \tag{13}$$

where

$$H^i = L_i + p_i \cdot f = (1\ p_i) \left[\begin{array}{c|c} b_i^0 & \sum_1^N c_{ij}^0 d^j + e_i^0 \\ \hline b & \sum_1^N (c^j >< d^j) + E \end{array}\right] \left[\frac{1}{x}\right].$$

Condition (a) is implied by requirements (i) and (ii) in Proposition 3.1. Condition (b) is satisfied from (11) with $p_i(t) = \bar{p}_i(t)$. All that is left to be verified is condition (c), that is, Ineq. (13) with $p_i(t) = \bar{p}_i(t)$. Using (1), (3), (10), (11), and (12), one can rewrite (13) as

$$\begin{aligned}
&\left\{\left[\sum_{j=1}^N c_{ij}^0(\bar{v}_j(t))\, d^j + e_i^0\right] \cdot \bar{x} + b_i^0\right\} + \bar{p}_i \cdot \left\{\left[\sum_{j=1}^N (c_j(\bar{v}_j(t)) >< d^j) + E\right] \bar{x}(t) + b\right\} \\
&\quad - \left\{\left[\sum_{i \neq j=1}^N (c_{ij}^0(\bar{v}_j(t))\, d^j) + c_{ii}^0(u_i)\, d^i + e_i^0\right] \cdot x + b_i^0\right\} \\
&\quad - \bar{p}_i' \left\{\left[\sum_{i \neq j=1}^N (c_j(\bar{v}_j(t)) >< d^j) + c_i(u_i) >< d^i + E\right] x + b\right\} \\
&\quad - \left[\sum_{j=1}^N (c_{ij}^0(\bar{v}_j(t))\, d^j) + e_i^0 + \bar{p}_i' \left(\sum_{j=1}^N c_j(\bar{v}_j(t)) >< d^j + E\right)\right] \cdot (\bar{x} - x) \\
&= \left\{\sum_{j \neq i} ([c_{ij}^0(\bar{v}_j(t))\, d^j + e_i^0] + \bar{p}_i' c_j(\bar{v}_j(t)) >< d^j) + E\right\} [(\bar{x} - x) - (\bar{x} - x)] \\
&\quad + (\{[c_{ii}^0(\bar{v}_i(t))\, d^i + e_i^0 + \bar{p}_i' c_i(\bar{v}_i(t)) >< d^i] \cdot \bar{x} \\
&\quad - [c_{ii}^0(u_i)\, d^i + e_i^0 + \bar{p}_i' c_i(u_i) >< d^i] \cdot x\} \\
&\quad - [c_{ii}^0(\bar{v}_i(t))\, d^i + e_i^0 + \bar{p}_i' c_i(\bar{v}_i(t)) >< d^i] \cdot (\bar{x} - x)) \\
&= 0 + (\{[c_{ii}^0(\bar{v}_i(t)) + \bar{p}_i \cdot c_i(\bar{v}_i(t))] - [c_{ii}^0(u_i) + \bar{p}_i \cdot c_i(u_i)]\}(d^i \cdot x)).
\end{aligned} \tag{14}$$

By requirement (ii), $d^i \cdot x > 0$. From (10), the term in { } in the last member of (14) is nonnegative. Consequently, Ineq. (13) holds. This completes the verification.

Remark 4.1. Since H^i is linear in p_i, x, and (c_{ij}^0, c_j), $j = 1,..., N$, the game is named *trilinear*.

Remark 4.2. The crucial feature in this game structure is that, for all $j = 1,\ldots, N$, the same vector d^j appears in (1) as well as (3). This allows the factoring out of $d^j \cdot x$ in (14). A slight generalization appears possible by allowing $d^j(\cdot)$ to be a Lebesgue measurable function of time. The definition of the playable space also has to be changed accordingly.

Remark 4.3. For N-person, general sum differential games, the adjoint system involves terms containing gradients of unknown strategies (e.g., Ref. 3). Since, in this type of games, optimal strategies depend upon time only, the abovementioned terms will vanish.

References

1. CASE, J. H., *Toward a Theory of Many Player Differential Games*, SIAM Journal on Control, Vol. 7, No. 2, 1969.
2. STARR, A. W., and HO, Y. C., *Non-zero Sum Differential Games*, Journal of Optimization Theory and Applications, Vol. 3, No. 3, 1969.
3. LEITMANN, G., *Differential Games*, Differential Games: Theory and Applications, Edited by M. D. Ciletti and A. W. Starr, American Society of Mechanical Engineering, New York, New York, 1970.
4. LEITMANN, G., and STALFORD, H., *Sufficiency for Optimal Strategies in Nash Equilibrium Games*, Techniques of Optimization, Edited by A. V. Balakrishnan, Academic Press, New York, New York, 1972.
5. LEITMANN, G., and LIU, P. T., *A Differential Game Model of Labor-Management Negotiations During a Strike*, Journal of Optimization Theory and Applications, Vol. 13, No. 4, 1974.
6. CLEMHOUT, S., LEITMANN, G., and WAN, H. Y., JR., *A Model of Bargaining Under Strike—The Differential Game View*, Cornell Working Paper in Economics, No. 66, 1973.
7. CLEMHOUT, S., LEITMANN, G., and WAN, H. Y., JR., *A Differential Game of Oligopoly*, Cybernetics, Vol. 3, No. 1, 1973.

XVI

Introduction to Chapters on Two-Person Decision Problems under Uncertainty

Y. C. Ho and F. K. Sun

The following four chapters represent reprints of two papers and one technical note which appeared in the Journal of Optimization Theory and Applications and one related, but previously unpublished, technical note. All deal with the subject of information in two or more person decision problems under uncertainty. Chapter XVII explores the question of the value of information in simple zero-sum problems involving two teams. More specifically, we illustrate that the value of communication among team members may be negative when such communication cannot be kept secret from the opposing team. The next three chapters are to be taken together as a group treating the problem of linear-quadratic-Gaussian stochastic differential games under a special class of information structure. This class is characterized by the requirement that either the two players have identical information or one of the players always knows more than the other player. Under this structure, explicit strategies and value can be computed and implemented. In addition to these results, we discuss their relationships to Kalman–Bucy filtering and stochastic control and to the value of the public information concept introduced in Chapter XVII.

XVII

Value of Information in Two-Team Zero-Sum Problems[1]

Y. C. Ho and F. K. Sun

Abstract. The situation in which two groups of people have conflicts of interest is considered as a two-team zero-sum game problem. Two special cases of this problem are solved to illustrate that communication among members of a team may not be worthwhile and extra information need not always be desired by decision makers. In the appendix, it is shown that the optimal saddle-point solution exists and is still affine for the general problem with quadratic Gaussian performance index.

1. Introduction

The value of information is often an important design consideration in statistical decision problems or decentralized statistical decision problems (team theory). It is always taken as obvious that this value must be nonnegative, in the sense that extra information cannot hurt, since one always has the option of disregarding the information. It is not generally realized that this fact is no longer true in game-theoretic consideration (Ref. 1). This is not too surprising since the question of *information* is completely suppressed in the strategic or characteristic

[1] The research reported in this paper was made possible through support extended to the Division of Engineering and Applied Physics, Harvard University, by the U.S. Office of Naval Research under the Joint Services Electronics Program, Contract No. N00014-67-A-0298-0006, and by the National Science Foundation, Grant No. GK-31511.

function form of a game. In this paper, we shall address such problems. The simplest problem is to consider the case of zero-sum games. In particular, we visualize the situation of two opposing teams of decision makers where communication among decision makers cannot be kept secret. Such a situation is very frequently encountered in market competition of business firms, in military maneuvers, or more often in parlor games such as bridge.

It should be emphasized that we have very little knowledge and results in these emerging problems. This paper only makes a beginning by solving a class of relatively simple examples to illustrate some insights.

2. Two Special Cases of the Proposed Problem

Statement of First Problem. Consider three members A_1, B_1, and B_2 in a zero-sum game, where A_1 is the opponent of B_1 and B_2, B_1 and B_2 form a team against A_1. Let μ_1, s_1, and s_2 be the observed state of A_1, B_1, and B_2, respectively. Assume that $x = (\mu_1, s_1, s_2)^t$ is jointly normal with zero mean and covariance matrix $\Sigma = (\Sigma_{ij})$. Furthermore, consider four different information structures due to different types of transmission of information as follows.

(a) No communication exists between the members of the team:

$$I_a \begin{cases} y_1 = \eta_1(x) = [100]x = \mu_1, & \text{information of } A_1, \\ z_1 = \delta_1(x) = [010]x = s_1, & \text{information of } B_1, \\ z_2 = \delta_2(x) = [001]x = s_2, & \text{information of } B_2. \end{cases}$$

(b) B_1 sends all information he knows to B_2. Due to the revelation of information in the process of transmission, A_1 also receives the same amount of information:

$$I_b \begin{cases} y_1 = \eta_1(x) = \begin{bmatrix}100\\010\end{bmatrix}x, \\ z_1 = \delta_1(x) = [010]x, \\ z_2 = \delta_2(x) = \begin{bmatrix}010\\001\end{bmatrix}x. \end{cases}$$

(c) Similar to (b), except B_2 sends information to B_1:

$$I_c \begin{cases} y_1 = \eta_1(x) = \begin{bmatrix}100\\001\end{bmatrix}x, \\ z_1 = \delta_1(x) = \begin{bmatrix}010\\001\end{bmatrix}x, \\ z_2 = \delta_2(x) = [001]x. \end{cases}$$

(d) B_1 sends all information he knows to B_2, and so does B_2 to B_1. Due to the revelation of information in the process of transmission, A_1 receives all these communications:

$$I_d \begin{cases} y_1 = \eta_1(x) = \begin{bmatrix} 100 \\ 010 \\ 001 \end{bmatrix} x, \\ z_1 = \delta_1(x) = \begin{bmatrix} 010 \\ 001 \end{bmatrix} x, \\ z_2 = \delta_2(x) = \begin{bmatrix} 010 \\ 001 \end{bmatrix} x. \end{cases}$$

Let the performance index be

$$J_{(\cdot)} = E\left\{[a_1 b_1 b_2] \begin{bmatrix} -1 & r_{11} & r_{12} \\ r_{11} & 1 & q_{12} \\ r_{12} & q_{12} & 1 \end{bmatrix} \begin{bmatrix} a_1 \\ b_1 \\ b_2 \end{bmatrix} + 2[a_1 b_1 b_2] \begin{bmatrix} 1 & 0 & 0 \\ 0 & -1 & 0 \\ 0 & 0 & -1 \end{bmatrix} x \right\},$$

where the subscript of J denotes the performance index with respect to the different cases of information structure and a_1, b_1, and b_2 are control variables of A_1, B_1, and B_2, respectively. We assume that $1 - q_{12}^2 > 0$, so that J is convex in b_1 and b_2. Then, the problem is to find the optimal control law

$$a_1 = \alpha_1^*(y_1) \in \mathscr{A}_{\eta_1}, \qquad b_1 = \beta_1^*(z_1) \in \mathscr{B}_{\delta_1}, \qquad b_2 = \beta_2^*(z_2) \in \mathscr{B}_{\delta_2},$$

such that

$$J_{(\cdot)}(\alpha_1, (\beta_1^*, \beta_2^*)) \leqslant J_{(\cdot)}(\alpha_1^*, (\beta_1^*, \beta_2^*)) \leqslant J_{(\cdot)}(\alpha_1^*, (\beta_1, \beta_2))$$

for all

$$(\alpha_1, \beta_1, \beta_2) \in \mathscr{A}_{\eta_1} \times \mathscr{B}_{\delta_1} \times \mathscr{B}_{\delta_2},$$

where $\mathscr{A}_{\eta_1}$, $\mathscr{B}_{\delta_1}$, and $\mathscr{B}_{\delta_2}$ are the admissible class of decision rules (strategies) for a_1, b_1, and b_2, usually taken to be the class of all μ_1-, s_1-, s_2-measurable functions from $x \mapsto a_1$, b_1, b_2, respectively. We wish to compare $J_{(a)}$, $J_{(b)}$, $J_{(c)}$, and $J_{(d)}$ induced by four different types of information structure.

Solution of the First Problem. In the appendix, we will show that, under certain conditions which the above case specifies, the optimal saddle-point strategies of the two teams are linear in their information variables. Furthermore, the expected value of J takes on a very simple form. Thus, we can explicitly compare $J_{(\cdot)}$ for the various cases under consideration.

Case (1). $r_{11} = r_{12} = q_{12} = 0$, there is no coupling among A_1, B_1, and B_2. It simply yields that

$$J_a^* = J_b^* = J_c^* = J_d^* = \Sigma_{11} - \Sigma_{22} - \Sigma_{33}.$$

The asterik denotes J evaluated at optimal strategy. This result is intuitively clear. Since no coupling exists, none of the players can take the advantage of knowing more. However, if the transmission of information has nonzero associated cost, B_1 and B_2 should prefer I_a, that is, no communication.

Case (2). $r_{11} = r_{12} = 0$, no coupling exists between A_1 and B_1, neither between A_1 and B_2, A_1 cannot take advantage of knowing more in this case. The problem is simply reduced to a pure team decision problem. B_1 and B_2 should make all their effort to know each other. Mathematically, we have that

$$J_d^* \leqslant J_b^* \leqslant J_a^*, \qquad J_d^* \leqslant J_c^* \leqslant J_d^*, \qquad J_b^* \leqslant J_c^*$$

only if $\Sigma_{33} \geqslant \Sigma_{22}$; that is, if only one-way transmission is permitted, one with smaller variance of information should transmit his information to his partner. Furthermore, the desirability of communication is decreased as the correlation of their information is increased.

Case (3). $r_{11} = 0$, no coupling exists between A_1 and B_1. For simplicity, we assume that μ_1, s_1 and s_2 are statistically independent. Then,

$$J_a^* = \Sigma_{11} - \Sigma_{22} - \Sigma_{33}, \quad J_b^* = \Sigma_{11} - [(1 + r_{12}^2)/(1 + r_{12}^2 - q_{12}^2)]\, \Sigma_{22} - \Sigma_{33}.$$

Since

$$1 \leqslant 1/[1 - q_{12}^2/(1 + r_{12}^2)] \leqslant 1/(1 - q_{12}^2),$$

we have that $J_b^* \leqslant J_a^*$ and J_b^* equals J_a^* as r_{12}^2 goes to infinity, i.e., it always pays for B_1 to send information to B_2. But, due to the existence of r_{12}, A_1 can take the advantage of knowing B_2. Therefore, it is not as worthwhile for B_1 to send information to B_2 as in Case (2). Furthermore,

$$J_c^* = \Sigma_{11} - \Sigma_{22} - [1/(1 + r_{12}^2 - q_{12}^2)]\, \Sigma_{33}.$$

Hence, if r_{12} is large enough such that

$$1/(1 + r_{12}^2 - q_{12}^2) < 1,$$

then it may not even be worthwhile for B_2 to send information to B_1 since A_1 would then know what B_2 knows completely. Finally, we have

$$J_d^* = \Sigma_{11} - \left\{\frac{1}{[1 - r_{12}^2/(1 + r_{12}^2)]}\right\} \Sigma_{22} - [1/(1 + r_{12}^2 - q_{12}^2)]\, \Sigma_{33}\,.$$

From this, we have $J_d^* \leqslant J_c^*$, but $J_d^* \leqslant J_b^*$ only if

$$1/(1 + r_{12}^2 - q_{12}^2) \geqslant 1.$$

From the above discussion, it is evident that decision makers in this environment should choose the communication system carefully. Otherwise, it may end up to be a case worse than no communication.

Statement of the Second Problem. Consider a stochastic zero-sum game with two decision makers A and B and a scalar quadratic performance index

$$J = E\{\mathscr{J}(x, a, b)\} = E\{(x + a + b)^2\, c - pa^2 + qb^2\},$$

where x is $N(0, \sigma_x^2)$, $p > c > 0$, $q > 0$, and a, b are scalar control variables of A and B, respectively. Assume that the information obtained by both players is only through a broadcasting channel, that is, both players receive the same information. We call such information *public information*. Let $y = \eta(x)$ and $z = \eta(x)$ be such public information, and $\eta(x)$ and x are jointly normal with means $\binom{\bar{\eta}}{0}$ and covariance matrix

$$\begin{bmatrix} \sigma_\eta^2 & \sigma_{\eta x} \\ \sigma_{\eta x} & \sigma_x^2 \end{bmatrix}.$$

Then, we say that $(\alpha^*, \beta^*) \in \mathscr{A}_\eta \times \mathscr{B}_\eta$ is a pair of optimal strategies iff

$$J(\alpha, \beta^*) \leqslant J(\alpha^*, \beta^*) \leqslant J(\alpha^*, \beta)$$

for all $\alpha \in \mathscr{A}_\eta$ and $\beta \in \mathscr{B}_\eta$, that is, in the sense of saddle point.

By simple calculation, we obtain

$$\begin{bmatrix} \alpha^* \\ \beta^* \end{bmatrix} = \begin{bmatrix} (p - c) & -c \\ c & (q + c) \end{bmatrix}^{-1} \begin{bmatrix} E\{x \mid \eta(x)\} \\ -E\{x \mid \eta(x)\} \end{bmatrix}$$

and

$$J^* = J(\alpha^*, \beta^*) = E\{(E(x \mid \eta(x)))^2\}((q-p)/[(p-c)(q+c)+c^2]) + c\sigma_x^2$$
$$= ((q-p)/[(p-c)(q+c)+c^2])\, \sigma_{\eta x}^2(\sigma_\eta^2)^{-1} + c\sigma_x^2.$$

Therefore, (i) if $\sigma_\eta^2 \to \infty$, that is, A and B only have a prior knowledge about x, we have that $J^* = c\sigma_x^2$; and (ii) if $\eta(x)$ and x are uncorrelated, then $J^* = c\sigma_x^2$. This is not surprising. It simply means that, if an information has nothing to do with the uncertainty involved in the problem, no one can gain anything from knowing this information. Also, (iii) if $q > p$ and $\sigma_{\eta x} \neq 0$, then J^* is an increasing function of $|\sigma_{\eta x}|$ and a decreasing function of σ_η^2. Furthermore, if we adopt the definition of better information proposed by Basar and Ho (Ref. 2), the informations are ordered by their variances, then (iii) says that A would like to have better information in spite of what B will know and $\eta(x)$ is valuable for A, if $\eta(x)$ and x are highly dependent.

The performance index in this case can be viewed as a simplest case of pursuer and evader game with only one move involved. Player A, an evader, wants to maximize the distance function $(x + a + b)^2$ subject to energy spent a^2, and Player B, *vice versa*. Then, c will be the cost attached to $(x + a + b)^2$, p and q will be the cost per energy for A and B, respectively. Hence, (iii) coincides with the intuition that knowing more about the state of the world is beneficial only if an evader has better ability, i.e., his cost of energy is cheaper.

3. Appendix: Statement and Solution of the Problem in the General Quadratic Gaussian Case

Consider that there are two teams A and B, where A consists of members $A_1, A_2, \ldots, A_{n_A}$ and B consists of members $B_1, B_2, \ldots, B_{n_B}$. Let $(\Omega, \mathscr{A})$ be a probability measurable space which represents the uncertainties of the external world and $\mathscr{P}$ be a known probability measure. x is a random vector from $(\Omega, \mathscr{A})$ to $(\mathbb{R}^n, \mathscr{F})$ and $\mathscr{P}_x$ is a Borel probability measure defined on $(\mathbb{R}^n, \mathscr{F})$. $y_i \in \mathbb{R}^{h_i}$ (or $z_j \in \mathbb{R}^{h_j}$) is the information received by A_i (or B_j) as measurements or observations for him to make decisions; $\eta_i(x)$ (or $\delta_j(x)$) is a known measurable function that maps $(\mathbb{R}^n, \mathscr{F})$ into $(\mathbb{R}^{h_i}, Y)$ (or $(\mathbb{R}^{k_j}, Z)$) such that $y_i = \eta_i(x)$ (or $z_j = \delta_j(x)$). $a_i \in \mathbb{R}^{l_i}$ (or $b_j \in \mathbb{R}^{m_j}$) is A_i's (or B_j's) control variable and $\alpha_i(y_i)$ (or $\beta_j(z_j)$) is his admissible control law that is a measurable function from $(\mathbb{R}^{h_i}, Y)$ (or $(\mathbb{R}^{k_j}, Z)$) into $(\mathbb{R}^{l_i}, \mathscr{F}_i^{(A)})$ [or $(\mathbb{R}^{m_j}, \mathscr{F}_j^{(B)})$] such that $\alpha_i(y_i) = a_i$ (or $\beta_j(z_j) = b_j$). Moreover, let $\mathscr{A}_{\eta_i}$ (or $\mathscr{B}_{\delta_j}$) be the collection of all such measurable function α_i's (or β_j's) of the available information $y_i = \eta_i(x)$ (or $z_j = \delta_j(x)$).

Then, we assume that the two teams are playing a zero-sum game with respect to a quadratic performance index

$$\begin{aligned} J(\alpha, \beta) &= E\{\mathscr{J}(x, a, b)\} \\ &= E\{\lambda(x) + 2\mu^t(x)\, a - 2s^t(x)\, b - a^tPa + b^tQb + 2a^tRb\}, \quad (1) \end{aligned}$$

where

(i) $\alpha = \begin{bmatrix} \alpha_1 \\ \vdots \\ \alpha_{n_A} \end{bmatrix}, \quad \beta = \begin{bmatrix} \beta_1 \\ \vdots \\ \beta_{n_B} \end{bmatrix}, \quad a = \begin{bmatrix} a_1 \\ \vdots \\ a_{n_A} \end{bmatrix}, \quad b = \begin{bmatrix} b_1 \\ \vdots \\ b_{n_B} \end{bmatrix},$

(ii) $\lambda(x)$, $\mu^t(x)$, and $s^t(x)$ are measurable functions over $(\mathbb{R}^n, \mathscr{F})$ with appropriate dimensions.

(iii) P, Q, and R are real matrices with appropriate dimensions, P and Q are positive definite.

Finally, we say that

$$(\alpha^*, \beta^*) \in \prod_{i=1}^{n_A} \mathscr{A}_{\eta_i} \times \prod_{j=1}^{n_B} \mathscr{B}_{\delta_j}$$

is an optimal decision pair iff

$$J(\alpha, \beta^*) \leqslant J(\alpha^*, \beta^*) \leqslant J(\alpha^*, \beta) \quad (2)$$

for any pair

$$(\alpha, \beta) \in \prod_{i=1}^{n_A} \mathscr{A}_{\eta_i} \times \prod_{j=1}^{n_B} \mathscr{B}_{\delta_j},$$

i.e., in the sense of saddle point.

To solve this problem, we first give the following definition.

Definition 3.1. For any decision pair $(\tilde{\alpha}, \tilde{\beta})$, let

$$\tilde{\alpha} = (\tilde{\alpha}_1, \tilde{\alpha}_2, \ldots, \tilde{\alpha}_{n_A}), \qquad \tilde{\beta} = (\tilde{\beta}_1, \tilde{\beta}_2, \ldots, \tilde{\beta}_{n_B}),$$

$$J_i(\alpha_i, \tilde{\alpha}, \tilde{\beta}) = J(\alpha_i, (\tilde{\alpha}_k \mid k \neq i), \tilde{\beta}), \qquad J_j(\tilde{\alpha}, \beta_j, \tilde{\beta}) = J(\tilde{\alpha}, \beta_j, (\tilde{\beta}_l \mid l \neq j))$$

for all $i = 1, \ldots, n_A$ and for all $j = 1, \ldots, n_B$. Then, we say that $(\tilde{\alpha}, \tilde{\beta})$ is person-by-person satisfactory if, for every i and every j,

$$J(\tilde{\alpha}, \tilde{\beta}) = \max_{\alpha_i \in \mathscr{A}_{\eta_i}} J_i(\alpha_i, \tilde{\alpha}, \tilde{\beta}) = \min_{\beta_j \in \mathscr{B}_{\delta_j}} J_j(\tilde{\alpha}, \beta_j, \tilde{\beta}).$$

The optimality in the sense of Definition 3.1 is certainly a necessary condition for the optimality relation in (2). However, for the quadratic performance index, these two types of optimality are equivalent. Thus, we have the following lemmas.

Lemma 3.1. For the performance index given as in (1), if (α^*, β^*) is person-by-person satisfactory, then it is a saddle-point optimal decision pair.

Proof. Suppose that team A (or team B) plays the game at the person-by-person satisfactory strategy α^* (or β^*). Then, the actions of team A (or team B) can be considered as part of the *state of the world*, and the problem facing team B (or team A) becomes a pure quadratic team decision problem. Hence, by Radner's result in the static quadratic team problem (Ref. 3), the person-by-person satisfactory strategy β^* (or α^*) for team B (or team A) is the optimal decision with respect to the performance index $J(\alpha^*, \beta)$ [or $J(\alpha, \beta^*)$]. That is ,

$$J(\alpha^*, \beta^*) \leqslant J(\alpha^*, \beta), \qquad \forall \beta \in \prod_{j=1}^{n_B} \mathscr{B}_{\delta_j} . \tag{3-1}$$

Similarly,

$$J(\alpha; \beta^*) \leqslant J(\alpha^*, \beta^*), \qquad \forall \alpha \in \prod_{i=1}^{n_A} \mathscr{A}_{n_i} . \tag{3-2}$$

Combining (3-1) and (3-2), we complete the proof.

Lemma 3.2. If the person-by-person satisfactory solution exists for the performance index described in (1), it is uniquely determined by the following simultaneous equations:

$$\alpha_i(y_i)\, p_{ii} + \sum_{k \neq i} p_{ik} E\{\alpha_k(y_k) | \, y_i\}$$
$$= E\{\mu_i(x) | \, y_i\} + \sum_{k=1}^{n_B} \gamma_{ik} E\{\beta_k(z_k) | \, y_i\}, \qquad \forall i = 1, \ldots, n_A , \tag{4-1}$$

$$\beta_j(z_j)\, q_{jj} + \sum_{l \neq j} q_{lj} E\{\beta_l(z_l) | \, z_j\}$$
$$= E\{s_j(x) | \, z_j\} - \sum_{l=1}^{n_A} \gamma_{lj} E\{\alpha_l(y_l) | \, z_j\}, \qquad \forall j = 1, \ldots, n_B , \tag{4-2}$$

where p_{ij} , q_{ij} , γ_{ij} , μ_i , and s_j are entries of matrices P, Q, R, μ, and s, respectively, and without loss of generality we have assumed that all a_i's and b_j's are scalars.

Proof. Let $\tilde{\alpha}$, $\tilde{\beta}$ be the person-by-person satisfactory solution. Then, by definition,

$$J_i(\tilde{\alpha}_i, \tilde{\alpha}, \tilde{\beta}) = \max_{\alpha_i} J_i(\alpha_i, \tilde{\alpha}, \tilde{\beta}), \qquad \forall i = 1,..., n_A,$$
$$J_j(\tilde{\alpha}, \tilde{\beta}_j, \tilde{\beta}) = \min_{\beta_i} J_j(\tilde{\alpha}, \beta_j, \tilde{\beta}), \qquad \forall j = 1,..., n_B.$$

Hence, for each player, the problem becomes a single-person decision problem. Furthermore, since the matrices P and Q are assumed to be positive definite, the person-by-person satisfactory solution $(\tilde{\alpha}, \tilde{\beta})$ is uniquely determined by the stationary conditions, that is,

$$\partial E\{\mathscr{J} \mid y_i\}/\partial a_i = 0 \quad \text{and} \quad \partial E\{\mathscr{J} \mid z_j\}/\partial b_j, \quad \forall i = 1,..., n_A \quad \text{and} \quad \forall j = 1,..., n_B, \tag{5}$$

where $a_i = \alpha_i(y_i)$, $b_j = \beta_j(z_j)$. Simple calculation of (5) yields the conditions in (4).

From Lemma 3.1 and Lemma 3.2, we can see that the existence of the optimal solution is totally determined by the existence of the solution of (4). In the following theorem, it is proved that, under certain conditions, the optimal solution exists and is affine. Without loss of generality,[2] we assume that $\eta(x)$ and $\delta(x)$ have zero means.

Theorem 3.1. If (i) $\mu(x)$ and $\eta(x)$ are jointly with mean $\begin{bmatrix} \bar{\mu} \\ \hline 0 \end{bmatrix}$ and covariance matrix

$$C_1 = \left[\begin{array}{c|c} \bar{D}_1 & \bar{D}_2 \\ \hline \bar{D}_2{}^t & \bar{D}_3 \end{array}\right] = \left[\begin{array}{c|c} [d_{ij}^{(1)}] & [d_{ij}^{(2)}] \\ \hline [d_{ij}^{(2)}]^t & [d_{ij}^{(3)}] \end{array}\right]$$
$$= \left[\begin{array}{c|c} [E\{(\mu_i(x) - \bar{\mu}_i)(\mu_j(x) - \bar{\mu}_j)\}] & [E\{(\mu_i(x) - \bar{\mu}_i)\,\eta_j{}^t(x)\}] \\ \hline [E\{(\mu_i(x) - \bar{\mu}_i)\,\eta_j{}^t(x)\}]^t & [E\{\eta_i(x)\,\eta_j{}^t(x)\}] \end{array}\right],$$

(ii) $s(x)$ and $\delta(x)$ are jointly normal with mean $\begin{bmatrix} \bar{s} \\ \hline 0 \end{bmatrix}$ and covariance matrix

$$\bar{C}_2 = \left[\begin{array}{c|c} \bar{D}_4 & \bar{D}_7 \\ \hline \bar{D}_7{}^t & \bar{D}_6 \end{array}\right] = \left[\begin{array}{c|c} (d_{ij}^{(4)}) & (d_{ij}^{(5)}) \\ \hline (d_{ij}^{(5)})^t & (d_{ij}^{(6)}) \end{array}\right]$$
$$= \left[\begin{array}{c|c} [E\{s_i(x) - \bar{s}_i)(s_j(x) - \bar{s}_j)\}] & [E\{(s_i(x) - \bar{s}_i)(\delta_j{}^t(x)\}] \\ \hline [E\{(s_i(x) - \bar{s}_i)\,\delta_j{}^t(x)\}]^t & [E\{\delta_i(x)\,\delta_j{}^t(x)\}] \end{array}\right],$$

[2] If $\eta(x)$ and $\delta(x)$ have nonzero means $\bar{\eta}$ and $\bar{\delta}$, then we can apply $\eta'(x) = \eta(x) - \bar{\eta}$ and $\delta'(x) = \delta(x) - \bar{\delta}$, which have zero means, and use η' and δ' as equivalent information.

(iii) $\eta(x)$ and $\delta(x)$ are jointly normal with mean $\begin{bmatrix} 0 \\ \hline 0 \end{bmatrix}$ and covariance matrix

$$\bar{C}_3 = \begin{bmatrix} \bar{D}_3 & \bar{D}_7 \\ \bar{D}_7{}^t & \bar{D}_6 \end{bmatrix} = \begin{bmatrix} [d_{ij}^{(3)}] & [d_{ij}^{(7)}] \\ [d_{ij}^{(7)}]^t & [d_{ij}^{(6)}] \end{bmatrix},$$

where $d_{ij}^{(7)} = E\{\eta_i(x)\, \delta_j^{(t)}(x)\}$, and (iv) $d_{ii}^{(3)}$ and $d_{ii}^{(6)}$ are positive definite, then the optimal decision pair (α^*, β^*) exists and is affine. That is,

$$\alpha_i{}^*(y_i) = c_i y_i + d_i\,, \qquad \text{for all} \quad i = 1, 2,..., n_A\,, \tag{6-1}$$

$$\beta_j{}^*(z_j) = e_j z_j + f_j\,, \qquad \text{for all} \quad j = 1, 2,..., n_B\,. \tag{6-2}$$

Furthermore, let

$$c = [c_1\,, c_2\,,..., c_{n_A}], \qquad d = [d_1\,,..., d_{n_A}]^t,$$

$$e = [e_1\,, e_2\,,..., e_{n_B}], \qquad f = [f_1\,, f_2\,,..., f_{n_B}]^t;$$

then, c_i's d_i's, e_j,'s and d_j's are uniquely determined by

$$\begin{bmatrix} c^t \\ \hline e^t \end{bmatrix} = \begin{bmatrix} (P \otimes \bar{D}_3)^t & -(R^t \otimes \bar{D}_7)^t \\ (R \otimes \bar{D}_7{}^t)^t & (Q \otimes \bar{D}_6)^t \end{bmatrix}^{-1} \begin{bmatrix} (\Delta\, \bar{D}_2)^t \\ (\Delta\, \bar{D}_5)^t \end{bmatrix}, \tag{7-1}$$

$$\begin{bmatrix} d \\ \hline f \end{bmatrix} = \begin{bmatrix} P & -R \\ R^t & Q \end{bmatrix}^{-1} \begin{bmatrix} \bar{\mu} \\ \bar{s} \end{bmatrix}, \tag{7-2}$$

where the symbol $\otimes$ represents Hadamard product under block-by-block multiplication, that is

$$P \otimes \bar{D}_3 = [\, p_{ij} E\{\eta_i(x)\, \eta_j{}^t(x)\}], \qquad \Delta\, \bar{D}_2 = [E\{(\mu_i(x) - \bar{\mu}_i)\, \eta_i{}^t(x)\}],$$
$$\Delta\, \bar{D}_5 = [E\{s_j(x) - \bar{s}_j)\, \delta_j{}^t\}].$$

Proof. From Lemma 3.1 and Lemma 3.2, we know that the optimal decision pair (α^*, β^*) is uniquely determined by (4). Hence, if (α^*, β^*) is to be linear as shown in (8), it should satisfy

$$(c_i y_i + d_i)\, p_{ii} + \sum_{k \neq i} p_{ik} E\{c_k y_k + d_k \mid y_i\}$$
$$= E\{\mu_i(x) \mid y_i\} + \sum_{k=1}^{n_B} \gamma_{ik} E\{e_k z_k + f_k \mid y_i\}, \qquad \forall\, i = 1,..., n_A\,, \tag{8-1}$$

$$(e_j z_j + f_j)\, q_{jj} + \sum_{l \neq j} q_{lj} E\{e_l z_l + f_l \mid z_j\}$$
$$= E\{s_j(x) \mid z_j\} - \sum_{l=1}^{n_A} \gamma_{lj} E\{c_l y_l + d_l \mid z_j\}, \qquad \forall\, j = 1,..., n_B\,. \tag{8-2}$$

Then, by assumptions of normality, we have

$$E\{y_j \mid y_i\} = d_{ij}^{(3)t}[d_{ii}^{(3)}]^{-1} y_i , \qquad E\{\mu_i \mid y_i\} = d_{ii}^{(2)}[d_{ii}^{(3)}]^{-1} y_i + \bar{\mu}_i ,$$

$$E\{z_j \mid y_i\} = d_{ij}^{(7)t}[d_{ii}^{(3)}]^{-1} y_i .$$

Using these identities, we can simplify (8-1) and obtain

$$\left(c_i p_{ii} + \sum_{k \neq i} p_{ik} c_k d_{ik}^{(3)t} (d_{ik}^{(3)})^{-1}\right) y_i + d_i p_{ii} + \sum_{k \neq i} p_{ik} d_k$$

$$= \left\{ d_{ii}^{(2)} (d_{ii}^{(3)})^{-1} \quad \sum_{k=1}^{n_B} \gamma_{ik} e_k d_{ik}^{(7)t} (d_{ii}^{(3)})^{-1} \right\} y_i + \sum_{k=1}^{n_B} \gamma_{ik} f_k + \bar{\mu}_i , \quad \forall i = 1,\ldots, n_A . \tag{9}$$

Since (9) should be true for all y_i's, we have

$$\sum_{k=1}^{n_A} p_{ik} c_k d_{ik}^{(3)t} = d_{ii}^{(2)} + \sum_{k=1}^{n_B} \gamma_{ik} e_k d_{ik}^{(7)t},$$

$$\sum_{k=1}^{n_A} p_{ik} d_k = \sum_{k=1}^{n_B} \gamma_{ik} f_k + \bar{\mu}_i , \qquad \forall\, i = 1,\ldots,n_A . \tag{10}$$

Using the Hadamard product under block-by-block multiplication, we can express (10) in vector form

$$C(P \otimes \bar{D}_3) = \Delta\, \bar{D}_2 + e[R^t \otimes \bar{D}_7], \qquad Pd = Rf + \bar{\mu}. \tag{11-1}$$

Similarly, from (8-2), we have

$$e(Q \otimes \bar{D}_6) = \Delta\, \bar{D}_5 - c[R \otimes \bar{D}_7{}^t], \qquad Qf = -R^t\, d + \bar{s}. \tag{11-2}$$

Rewriting (11) in vector form, we have

$$\begin{bmatrix} (P \otimes \bar{D}_3)^t & -(R^t \otimes \bar{D}_7)^t \\ (R \otimes \bar{D}_7{}^t)^t & (Q \otimes \bar{D}_6)^t \end{bmatrix} \begin{bmatrix} c^t \\ \overline{e^t} \end{bmatrix} = \begin{bmatrix} (\Delta\, \bar{D}_2)^t \\ (\Delta\, \bar{D}_5)^t \end{bmatrix}, \quad \begin{bmatrix} P & -R \\ R^t & Q \end{bmatrix} \begin{bmatrix} d \\ f \end{bmatrix} = \begin{bmatrix} \bar{\mu} \\ \bar{s} \end{bmatrix}. \tag{12}$$

Finally, to complete the proof we only need to prove that matrices

$$M = \begin{bmatrix} (P \otimes \bar{D}_3)^t & -(R^t \otimes \bar{D}_7)^t \\ (R \otimes \bar{D}_7{}^t)^t & (Q \otimes \bar{D}_6)^t \end{bmatrix}, \qquad N = \begin{bmatrix} P & -R \\ R^t & Q \end{bmatrix}$$

are nonsingular. Now, since the matrices P, Q, $d_{ii}^{(3)}$, and $d_{ii}^{(6)}$ are positive definite, it follows from the lemma in Radner's paper (Ref. 3, p. 870) that $P \otimes \bar{D}_3$ and $Q \otimes \bar{D}_6$ are positive definite. Furthermore, let $[l_1, l_2]$ be an arbitrary vector in an appropriate space. Then,

$$[l_1, l_2]M\begin{bmatrix} l_1^t \\ l_2^t \end{bmatrix} = l_1(P \otimes \bar{D}_3)^t l_1^2 + l_2(Q \otimes \bar{D}_6)^t l_2^t > 0.$$

Hence, M is nonsingular. Similarly, N is nonsingular.

From the discussion in Section 2, we see that the value of information need not always be positive. Hence, it is reasonable to consider the value of information in the context of both teams' information. Let (η^0, δ^0) denote the null information for which both players only have prior knowledge about x. We then formalize this idea and define the value of information $V_A(\eta(x)), \delta(x))$ (or $V_B(\eta(x), \delta(x)))$ for team A (or team B), when it has information $y = \eta(x)$ [or $z = \delta(x)$] and its opponent has $z = \delta(x)$ [or $y = \eta(x)$], to be

$$V_A(\eta(x), \delta(x)) = J(\alpha^*, \beta^*) - J(\alpha^0, \beta^0), \quad V_B(\eta(x), \delta(x)) = -V_A(\eta(x), \delta(x)), \tag{13}$$

where (α^*, β^*) and (α^0, β^0) are the optimal decision pairs corresponding to the information structure $(\eta(x), \delta(x))$ and (η^0, δ^0).

Obviously, V_A and V_B need not always be nonnegative. However, if one of the teams has its information fixed, then its opponent would like to know more. This is shown in the following lemma which is first proved by Witsenhausen (Ref. 1, p. 208).

Lemma 3.3. If $\mathscr{A}_{\eta'} \subseteq \mathscr{A}_\eta$ and $\mathscr{B}_{\delta'} = \mathscr{B}_\delta$, then

$$V_A(\eta'(x), \delta'(x)) \geqslant V_A(\eta(x), \delta(x)).$$

Proof. Let (α^*, β^*) and $(\alpha^{*\prime}, \beta^{*\prime})$ be the optimal decisions under $(\eta(x), \delta(x))$ and $(\eta'(x), \delta'(x))$, respectively. Then, by definition,

$$J(\alpha, \beta^*) \leqslant J(\alpha^*, \beta^*) \leqslant J(\alpha^*, \beta), \qquad \forall\, \alpha \in \mathscr{A}_\eta \text{ and } \beta \in \mathscr{B}_\delta, \tag{14}$$

$$J(\tilde{\alpha}, \beta^{*\prime}) \leqslant J(\alpha^{*\prime}, \beta^{*\prime}) \leqslant J(\alpha^{*\prime}, \tilde{\beta}), \qquad \forall\, \tilde{\alpha} \in \mathscr{A}_{\eta'} \text{ and } \tilde{\beta} \in \mathscr{B}_{\delta'}. \tag{15}$$

Now, since $\mathscr{B}_\delta = \mathscr{B}_{\delta'}$, we have that $\beta^{*\prime} \in \mathscr{B}_\delta$ and, by the second inequality in (14),

$$J(\alpha^*, \beta^*) \leqslant J(\alpha^*, \beta^{*\prime}).$$

Furthermore, since $\mathscr{A}_{\eta'} \subseteq \mathscr{A}_\eta$, we have that $\alpha^* \in \mathscr{A}_{\eta'}$ and, by the first inequality in (15),

$$J(\alpha^*, \beta^{*\prime}) \leqslant J(\alpha^{*\prime}, \beta^{*\prime}).$$

It follows that

$$J(\alpha^*, \beta^*) \leqslant J(\alpha^{*\prime}, \beta^{*\prime}).$$

Finally, since $J(\alpha^0, \beta^0)$ is underchanged in both cases, we have that

$$V_A(\eta'(x), \delta'(x)) \geqslant V_A(\eta(x), \delta(x)).$$

The immediate result of this lemma is that, when $y = \eta^*(x) = (x, x, \ldots, x)$ and $z = \delta^0$, V_A reaches its maximum $\hat{V}_A$ and V_B goes down to its minimum $\check{V}_B$. The reason is simple. By Lemma 3.3,

$$V_A(\eta^*(x), \delta^0(x)) \geqslant V_A(\eta(x), \delta^0(x)) \geqslant V_A(\eta(x), \delta(x))$$

for any possible information $(\eta(x), \delta(x))$. And, since

$$V_A(\eta^*(x), \delta^0(x)) = -V_B(\eta^*(x), \delta^0(x)),$$

then

$$V_B(\eta^*, \delta^0(x)) \leqslant V_B(\eta(x), \delta(x))$$

for any possible information $(\eta(x), \delta(x))$. Similarly, we can easily define $\hat{V}_B$ and $\check{V}_A$. Moreover, for the Gaussian quadratic performance index described as in (1), the form of $V_A(\eta(x), \delta(x))$ is given by a rather simple form. This is the next lemma.

Lemma 3.4. Given the performance index (1) and if the assumptions in the theorem hold,

$$V_A(\eta(x), \delta(x)) = E\left\{[\mu^t(x), -s^t(x)]\begin{bmatrix}\alpha^*\\ \beta^*\end{bmatrix}\right\} - E\{(\mu^t(x), -s^t(x)]\}\, E\left\{\begin{bmatrix}\alpha^*\\ \beta^*\end{bmatrix}\right\},$$

where (α^*, β^*) is the optimal decision pair under the information structure $(\eta(x), \delta(x))$.

Proof. By Theorem 3.1, (α^*, β^*) should satisfy (4). Since

$$E_y E\{\cdot \mid y\} = E\{\cdot\}, \qquad E_z E\{\cdot \mid z\} = E\{\cdot\},$$

multiplying both sides of (4) by $\alpha_i^*(y_i)$ and $\beta_j^*(z_j)$ and taking the total expectation for all $i = 1, \ldots, n_A$ and all $j = 1, \ldots, n_B$, we have

$$\sum_{k=1}^{n_A} p_{ik} E\{\alpha_i^*(y_i)\, \alpha_k^*(y_k)\}$$

$$= E\{\alpha_i^*(y_i)\, \mu_i(x)\} + \sum_{k=1}^{n_B} \gamma_{ik} E\{\alpha_i^*(y_i)\, \beta_k^*(z_k)\}, \quad \forall i = 1, \ldots n_A, \qquad (16\text{-}1)$$

$$\sum_{l=1}^{n_B} q_{lj} E\{\beta_j^*(z_j)\, \beta_l^*(z_l)\}$$

$$= E\{\beta_j^*(z_j)\, s_j(x)\} - \sum_{l=1}^{n_A} \gamma_{lj} E\{\beta_j^*(z_j)\, \alpha_l^*(y_l)\}, \quad \forall j = 1,..., n_B. \quad (16\text{-}2)$$

Adding up (16-1) for all i and (16-2) for all j, we have

$$E\left\{\sum_{i=1}^{n_A}\sum_{k=1}^{n_A} p_{ik}\alpha_i^*(y_i)\,\alpha_k^*(y_k)\right\}$$

$$= E\left\{\sum_{i=1}^{n_A} \alpha_i^*(y_i)\,\mu_i(x)\right\} + E\left\{\sum_{i=1}^{n_A}\sum_{k=1}^{n_B} \gamma_{ik}\alpha_i^*(y_i)\,\beta_k^*(z_k)\right\}, \quad (17\text{-}1)$$

$$E\left\{\sum_{j=1}^{n_B}\sum_{l=1}^{n_B} q_{lj}\beta_j^*(z_j)\,\beta_l^*(z_l)\right\}$$

$$= E\left\{\sum_{j=1}^{n_B} \beta_j(z_j)\, s_j(x)\right\} - \left\{E\sum_{j=1}^{n_B}\sum_{l=1}^{n_A} \gamma_{lj}\beta_j^*(z_j)\,\alpha_l^*(y_l)\right\}. \quad (17\text{-}2)$$

Writing (17) in matrix form and subtracting (17-1) from (17-2), we have

$$E\{-\alpha^{*t}(y)\, P\alpha^*(y) + \beta^{*t}(z)\, Q\beta^*(z)\}$$
$$= E\{-\mu^t(x)\,\alpha^*(y) + s^t(x)\,\beta^*(z) - 2\alpha^{*t}(y)\, R\beta^*(z)\}. \quad (18)$$

Using the identity (18), we obtain

$$J(\alpha^*, \beta^*) = E\{\lambda(x) + \mu^t(x)\,\alpha^*(y) - s^t(x)\,\beta^*(z)\}.$$

Therefore,

$$V(\eta(x), \delta(x)) = E\left\{[\mu^t(x), -s^t(x)]\begin{bmatrix}\alpha^*(y)\\ \beta^*(z)\end{bmatrix}\right\} - E\left\{[\mu^t(x), -s^t(x)]\begin{bmatrix}\alpha^0\\ \beta^0\end{bmatrix}\right\}. \quad (19)$$

Since $\begin{bmatrix}\alpha^0\\ \beta^0\end{bmatrix}$ is constant,

$$E\left\{\begin{bmatrix}\alpha^0\\ \beta^0\end{bmatrix}\right\} = \begin{bmatrix}\alpha^0\\ \beta^0\end{bmatrix}.$$

To complete the proof, we only need to prove that

$$E\left\{\begin{bmatrix}\alpha^*(y)\\ \beta^*(y)\end{bmatrix}\right\} = \begin{bmatrix}\alpha^0\\ \beta^0\end{bmatrix}.$$

To do this, taking the total expectation on both sides of (6), since y_i, z_j are zero mean, and the c_i's, d_j's, e_j's, f_j's are constant, we have

$$E\left\{\begin{bmatrix}\alpha^*\\ \beta^*\end{bmatrix}\right\} = \begin{bmatrix}P & -R\\ R^t & Q\end{bmatrix}^{-1}\begin{bmatrix}\bar{\mu}\\ \bar{s}\end{bmatrix} = E\left\{\begin{bmatrix}\alpha^0\\ \beta^0\end{bmatrix}\right\} = \begin{bmatrix}\alpha^0\\ \beta^0\end{bmatrix}. \tag{20}$$

Using (19) and (20), we complete the proof.

References

1. WITSENHAUSEN, H. S., *On the Relations Between the Values of a Game and Its Information Structure*, Information and Control, Vol. 19, No. 3, 1971.
2. HO, Y. C., and BASAR, T., *Note on Informational Properties of Games*, Journal of Economic Theory, Vol. 7, No. 4, 1974.
3. RADNER, R., *Team Decision Problem*, Annals of Mathematical Statistics, Vol. 33, No. 3, 1962.

XVIII

On the Minimax Principle and Zero-Sum Stochastic Differential Games[1]

Y. C. Ho

Abstract. The problem of prior and delayed commitment in zero-sum stochastic differential games is discussed. A new formulation and solution based on the delayed-commitment model is derived and its significant implications to stochastic games and control are considered.

1. Introduction

One of the fundamental tenets of game theory is the normalization principle of von Neumann which roughly says that, given an extensive game, one can always reduce it to an equivalent game in normal form involving only strategies and payoffs and where all dynamic and informational aspects of the original problem have been suppressed in the form of strategies by considering all the possible actions of all the players under all possible circumstances. As a conceptual simplification, this device is extremely useful. In fact, it is so useful that one can argue that it has disproportionately influenced the development of game theory in the past two decades with the result that very little work has been done on the extensive form of games. Recently, Aumann and Maschler (Ref. 1) reexamined the normalization principle and pointed out persuasively, via a simple counterexample, its inappropriateness under certain conditions. Their results have immediate and serious

[1] The research reported in this document was supported by the US Army Research Office, the US Air Force Office of Scientific Research, and the US Office of Naval Research under the Joint Services Electronics Program by Contracts Nos. N00014-67-A-0298-0006, 0005, and 0008, and by the National Science Foundation under Grant No. GK-31511.

consequences in stochastic control and differential game problems, since both are special cases of general extensive games. In this paper, we shall:

(i) present a counterexample in the same spirit as that of Ref. 1 but within the framework of a zero-sum stochastic two-person difference game; this example will point out the restricted circumstances under which earlier results on minimax strategies can be considered secure;

(ii) point out that (i) is actually a blessing in disguise and that, from our new viewpoint, we can actually solve the minimax problem for two-person zero-sum linear–quadratic Gaussian stochastic differential (difference) games much more effectively than before; a finite-dimensional minimax solution that is eminently computable will be presented;

(iii) show that the structure of the well-known optimal stochastic control law (Kalman–Bucy filter in cascade with a zero-memory linear map) for the LQG problem is in fact *optimal*[2] under circumstances which are neither Gaussian nor linear. This explains in part the incredible robustness of the LQG result in practical application and points the way to efficient solution of more general stochastic control problems.

2. Example

The notations used in this section are as follows: we write $\tilde{x}$ to denote the fact that we are considering it as a random variable, while plain x indicates a particular sample of $\tilde{x}$; then, $\bar{x}$ denotes the expected value of $\tilde{x}$; in particular, $\bar{x}'$ stands for the unconditional (prior) expectation of x, and $\bar{x}''$ stands for the conditional (posterior upon information obtained as the game evolved) expectation.

Consider the scalar two-stage dynamic systems

$$\tilde{x}_3 = \tilde{x}_2 + v = (\tilde{x}_1 + u) + v, \qquad \tilde{x}_1 \equiv \tilde{x} \sim N(0, \sigma), \tag{1}$$

where u and v are the controls of Players I and II, respectively. We have the performance criterion

$$\bar{J}' = \tfrac{1}{2}E\{(\tilde{x}_3)^2 + u^2 - 2v^2\} \tag{2}$$

which I attempts to minimize and II maximize. Player I is given the measurement

$$\tilde{z} = \tilde{x} + \tilde{w}, \qquad \tilde{w} \sim N(0, 1), \tag{3}$$

---

[2] In the sense to be explained more fully in Section 6.

$\tilde{w}$, $\tilde{x}$ are independent, while Player II receives no measurement. Both players know all the parameters and functional forms of (1)–(3). These are the common *prior information.*

The class of admissible strategies Γ for Player I is

$$\tilde{u} = \gamma(\tilde{z}), \qquad \gamma \in \Gamma = \text{class of all Borel measurable } \gamma : R \to R. \tag{4}$$

The class of admissible strategies for Player II is

$$v = c = \text{const}, \qquad c \in R. \tag{5}$$

Strictly speaking, of course, both γ and c depend on the common prior information, such as σ, *etc.* Such dependence, however will not be explicitly shown. The expectation in (2) is taken w.r.t. the Gaussian random variables $\tilde{x}$, $\tilde{w}$. Using (1) and (2), we can rewrite, equivalently,

$$\bar{J}' = \tfrac{1}{2}E\{2u^2 - v^2 + 2uv + 2v\tilde{x} + 2\tilde{x}u\}, \tag{6}$$

where the term $E[\tilde{x}^2]$ is a known constant σ and does not enter into the game. This simple zero-sum stochastic difference game can then be stated as follows: Find $\gamma^0 \in \Gamma$, $c^0 \in R$ such that (γ^0, c^0) constitutes a saddle-point for $\bar{J}'$ in (6). This is Problem (P–1).

It is not difficult to derive that (P–1) has a saddle-point in pure strategies with

$$\tilde{u}^0 = \gamma^0(\tilde{z}) = -\tfrac{1}{2}\bar{x}'' = -\tfrac{1}{2}[\sigma/(\sigma+1)]\tilde{z}, \tag{7}$$

$$v^0 = c^0 = 0. \tag{8}$$

For $v = 0$,

$$\min_{\gamma\in\Gamma} \bar{J}' = \min_{\gamma\in\Gamma} E_{\tilde{z}}E_{/\tilde{z}}[\tilde{J}] = E_{\tilde{z}} \min_{\gamma\in\Gamma} E_{/\tilde{z}}[\tilde{J}] = E_{\tilde{z}} \min_{u} E_{/\tilde{z}}[\tilde{J}] \tag{9}$$

implies

$$u^0 = -\tfrac{1}{2}E(\tilde{x}/z) = -\tfrac{1}{2}\bar{x}''. \tag{10}$$

Similarly, for

$$\tilde{u}^0 = \gamma^0(\tilde{z}) = -\tfrac{1}{2}[\sigma/(\sigma+1)]\tilde{z} \triangleq a\tilde{z},$$

$$\max_{c} \bar{J}' = \max_{v} \tfrac{1}{2}E\{2a^2\tilde{z}^2 - v^2 + (2a\tilde{z} + 2\tilde{x})\,v + 2a\tilde{z}\tilde{x}\}$$

for $\sigma = 1$,

$$= \max_{v} \tfrac{1}{2}\{4a^2 - v^2 + 2a\} \Rightarrow v^0 = c^0 = 0, \tag{11}$$

and

$$\bar{J}'(\gamma^0, c^0) = 2a^2 + a. \tag{12}$$

The saddle-point property of (γ^0, c^0) is thus established. Concomitant with this saddle-point property, it is often asserted or implied that, if

Player I chooses the strategy γ^0, then he is guaranteed a minimax expected payoff value of (12) above. This statement has to be interpreted with considerable care as the following discussion will show. Let us consider the situation facing Player I *after* he has received the information z but *before* anyone has acted. Instead of (6), his payoff is now evaluated by

$$\bar{J}'' = \tfrac{1}{2}E_{/z}\{2u^2 - v^2 + 2uv + 2v\tilde{x} + 2\tilde{x}u\}. \tag{13}$$

To be sure, *if* Player II uses $v^0 = c^0 = 0$, then the optimal act for Player I is still given by (10), that is,

$$u^0 = -\tfrac{1}{2}\bar{x}''.$$

However, *this action does not guarantee his security level* which is obtained by solving the following Problem (P–2):

$$\bar{J}''(u^*, v^*) = \min_{u\in R} \max_{v\in R} \bar{J}''.$$

Note that, in (P–2), z is no longer a random variable but a given number. To solve (P–2), we shall derive u^* and v^* as a saddle-point pair for $\bar{J}''$. For the purpose of solving the ZSTP game of (P–2), z can be regarded as part of the common prior information without violating the restriction of (5) on the class of admissible strategies for v. For fixed u, $\max_v \bar{J}''$ implies

$$v^* = u + \bar{x}''. \tag{14}$$

Substituting (12) into (11) and $\min_u \bar{J}''$ implies

$$\min_u \tfrac{1}{2}E_{/z}\{2u^2 - (u + \bar{x}'')^2 + 2u(u + \bar{x}'') + 2x(u + \bar{x}'') + 2\tilde{x}u\}, \tag{15}$$

which implies

$$u^* = -\tfrac{2}{3}\bar{x}'', \qquad v^* = u + \bar{x}'' = \tfrac{1}{3}\bar{x}'', \tag{16}$$

and

$$\bar{J}''(u^*, v^*) = -\tfrac{1}{6}[\bar{x}'']^2 = -\tfrac{1}{6}[\sigma^2/(\sigma + 1)^2]z^2. \tag{17}$$

Similarly, for fixed

$$v^* = \tfrac{1}{3}\bar{x}'' = \tfrac{1}{3}[\sigma/(\sigma + 1)]z,$$

we can directly verify that

$$u^* = -\tfrac{2}{3}\bar{x}''$$

is the optimal reply and yields the security level of (17). On the other hand, the strategy

$$\gamma^0 = -\tfrac{1}{2}\bar{x}''$$

against

$$v^* = u + \bar{x}''$$

produces a payoff

$$\bar{J}''(\gamma^0, v^*) = \tfrac{1}{8}(\bar{x}'')^2 > \bar{J}''(u^*, v^*) = -\tfrac{1}{6}(\bar{x}'')^2 \tag{18}$$

which is nonoptimal. The inequality in (18) is disconcerting. It says that, for all possible values of z, the strategy u^* is actually a safer strategy than γ^0. Unless Player I has reason to believe that Player II has irrevocably committed himself to $v^0 = c^0$ or that I can convince II that he has irrevocably committed himself to γ^0, there is no reason at all to play γ^0 when u^* is safer and available. The reason for this phenomenon, as pointed out by Harsanyi (Ref. 2) and Aumann and Maschler (Ref. 1), is the problem of prior and delayed (posterior) commitment. Put another way, after the information is received, we really have a nonzero-sum game facing the two players with (13) the payoff for Player I and (6) the payoff for Player II. The strategy pair (γ^0, c^0) is an equilibrium pair for I and II (in the Nash sense). However, it is well known that equilibrium strategies do not possess in general any minimax or guaranteed value properties in nonzero-sum games. The above example is simply one illustration of this fact. If the game takes place at a very fast time scale such that human reactions are not practical and mechanical decision is necessary, then the prior strategy pair (γ^0, c^0) represents a reasonable solution. On the other hand, in many socio-economic multistage games, the idea of a purely mechanistic decision procedure with no human intervention and irrevocable commitment to a strategy is rather untenable when confronted with the kind of evidence in (18). In such cases, the posterior strategy u^* seems much more preferable. Of course, one may counter with the argument that, since both the prior strategy and the posterior strategy for Player II *from his viewpoint* are the same ($v^0 = c^0 = 0$), we should expect him to play it; hence, Player I should play γ^0. This reasoning is defective on two accounts:

(i) Player I is dependent on II's intelligence (that is, II is clever enough to compute both the prior and posterior optimal strategies) for his payoff. But, what if II is dumb but lucky to play v^*?

(ii) Suppose that we endow II with the measurement

$$\tilde{y} = \tilde{x} + \tilde{\epsilon}, \qquad \tilde{\epsilon} \sim N(0, 1),\ \tilde{\epsilon},\ \tilde{w},\ \tilde{x} \text{ are independent.} \tag{19}$$

Then, in general, II will not have the same prior and posterior strategies. In fact, it can be shown that, *from the viewpoint of Player I*,

$$\tilde{u}^0 = \gamma^0(\tilde{z}) = -\{\sigma(1+2\sigma)/[2(\sigma+1)^2+\sigma^2]\}\tilde{z}, \tag{20-1}$$

$$\dot{\tilde{v}}^0 = \beta^0(\tilde{y}) = \{\sigma(\sigma+2)/[2(\sigma+1)^2+\sigma^2]\}\tilde{y}, \tag{20-2}$$

constitutes a saddle-point for $\bar{\tilde{J}}'$ and

$$u^* = -\tfrac{2}{3}\bar{x}'' = -[2\sigma/3(\sigma+1)]z, \tag{21-1}$$

$$v^* = u + E_{/y,z}[\tilde{x}] \equiv u + \bar{x}''' = [\sigma/(\sigma+1)]\,(\tfrac{1}{2}y - \tfrac{1}{6}z) \tag{21-2}$$

is a saddle-point pair for $\bar{J}''$. Note that

$$\gamma^0(z) \neq u^* \qquad \text{and} \qquad \bar{J}''(\gamma^0, v^*) > \bar{J}''(u^*, v^*).$$

Furthermore, from the viewpoint of Player II, he faces a payoff

$$\bar{J}''' = E_{/y}[\tilde{J}] \neq \bar{J}'' = E_{/z}[\tilde{J}]. \tag{22}$$

Since, Player I does not have knowledge of $\bar{J}'''$, there is no compelling reason to assume that II will play v^0 unless I believes in prior commitment. In fact, the *optimal* action from Player II's viewpoint may just turn out to be numerically equal to v^*. In other words, I need not assume that II is malicious in order to prepare for the worst.

3. Some Preliminaries to Stochastic Differential Games

At first glance, the result of Section 2 seems to spell disaster for practically all previous work on the stochastic (in particular, linear–quadratic–Gaussian) differential game problem. The *minimax* or saddle-point strategies that have been obtained are all of the *prior* variety. They are useful or reasonable only if we have firm belief that our opponent has made irrevocable prior commitments, before the game has begun. This severely limits their applicability, not to mention the fact that, in general, these strategies can only be realized with infinite-dimensional dynamic systems (Ref. 3) which are hardly practical. We would like to show in sections below that our new awareness is actually a blessing in disguise and that a secure *posterior* strategy can be derived for both players that is both simple and realizable by finite-dimensional linear systems.

Before we describe the problem formulation in the game situation, let us recall a few facts for the one-player linear–quadratic–Gaussian stochastic control problem which we shall require later.[3]

Consider the finite-dimensional linear stochastic dynamic system described by the Ito stochastic differential equation

$$dx = A(t)\,xdt + B(t)\,udt + C(t)\,dw(t), \qquad x(t_0) \sim N(\hat{x}_0, P_0), \tag{23}$$

$$dz = H(t)\,xdt + F(t)\,d\xi(t), \tag{24}$$

where A, B, C, H, F are known $n \times n$, $n \times m$, $n \times r$, $p \times n$, $p \times q$ matrices whose elements are continuous on $[t_0, t_f]$ and F is of full rank with $q > p$ for all t. $w(t)$ and $\xi(t)$ are independent standard Wiener processes. We also consider the payoff

$$\bar{J}' = E(\tilde{J}) = \tfrac{1}{2}E\left\{x(t_f)^T S_f x(t_f) + \int_{t_0}^{t_f} [u(t)^T Ru(t) + x(t)^T Mx(t)]\,dt\right\}, \tag{25}$$

where $S_f \geqslant 0$, $M(t) \geqslant 0$, $R(t) > 0$ are $n \times n$, $n \times n$, $m \times m$ symmetric matrices whose elements are continuous on $[t_0, t_f]$.

First, we have the following well-known result.

Result 3.1. $x(t)$ and $z(t)$ are measurable separable Gaussian random processes with values in R^n and R^p, respectively, and each having continuous sample paths with probability one (Ref. 4, pp. 135–136).

Next, we shall define the class of admissible control laws Γ(strategies). Let I denote $[t_0, t_f]$; $C[t_0, t]$ the space of continuous functions on $[t_0, t]$; Z_t the minimal σ-algebra generated by $z_t \in C[t_0, t]$, that is,

$$Z_t = \sigma\{Z(s), s \in [t_0, t]\}.$$

An admissible control law is a functional $\gamma : I \times C[t_0, t] \to R^m$ such that $\gamma(., z_t)$ is Lebesgue measurable for each $z_t \in C[t_0, t]$ and $\gamma(t, .)$ is Z_t-measurable for all $t \in I$. Essentially, this means that the control u at t can only depend on the past and present values of the measurement history z_t. With the above set-up, there follows the next two well-known results.

Result 3.2. *Kalman–Bucy Filtering* (Ref. 5). The conditional mean of $x(t)$ on Z_t, $\hat{x}(t) \triangleq E(x(t)/Z_t)$ is given by

$$d\hat{x} = (A(t)\,\hat{x} + B(t)u)\,dt + P(t)\,H^T(FF^T)^{-1}\,(dz - H(t)\,\hat{x}dt), \tag{26-1}$$

$$\hat{x}(t_0) = \hat{x}_0, \tag{26-2}$$

[3] Readers well versed in control theory or engineering can skip the following technical specifications and go directly to the next section.

where $P(t)$ satisfies the DE

$$\dot{P} = AP + PA^T + CC^T - PH^T(FF^T)^{-1} HP, \qquad P(t_0) = P_0 . \tag{27}$$

Corollary 3.1. (Ref. 5, pp. 70–72). If, in addition, (A, H) constitutes an observable pair, i.e.,

$$\int_t^{t_f} \Phi^T(t, t_f) H^T(FF^T)^{-1} H\Phi(t, t_f)\, dt > 0 \qquad \forall t < t_f , \tag{28}$$

where $\Phi(t, \tau)$ is the fundamental matrix associated with $A(t)$, then $P(t)$ exists and is bounded for all $t > t_0$.

Result 3.3. *Separation Principle.* The optimal control law $\gamma \in \Gamma$ which minimizes (25) (see Ref. 5, pp. 100–101) subject to (23)–(24) is given by

$$u(t) = \gamma(t, z_t) = -R^{-1} B^T S(t)\, \hat{x}(t), \tag{29}$$

where

$$\dot{S} = -A^T S - SA - M + SBR^{-1} B^T S, \qquad S(t_f) = S_f . \tag{30}$$

Corollary 3.2. (see Ref. 5, pp. 98–99). If, in addition, (A, B) constitutes a controlable pair, i.e.,

$$\int_{t_0}^{t} \Phi(t, t)\, BR^{-1} B^T\Phi^T(t, t)\, dt > 0 \qquad \forall t > t_0 , \tag{31}$$

then $S(t)$ exists and is bounded for all $t < t_f$.

Operationally, what these results say is that the optimal control law can be realized by linear combinations [Eq. (7)] of the state $\hat{x}(t)$ of a linear finite-dimensional dynamic systems [Eq. (4)] which has as its input $z(t)$. This is one of the most successful and widely-used results in control theory.

In the next section, we shall be using Results 3.2 and 3.3 extensively. In order to avoid cumbersome notation, we shall display these two results graphically to highlight their significance. This is done in Fig. 1. The optimal controlleı for the linear dynamics system (block ①) is another linear dynamic system of the same dimension (block ②) followed by a static linear map (block ③). Dotted lines indicate major parameter inputs to the controller which are precomputed via Eqs. (5) and (8). In the sequel, we shall only utilize Results 3.2 and 3.3 in the form of Fig. 1 and avoid spelling out the various detail parameter matrices associated with each block.

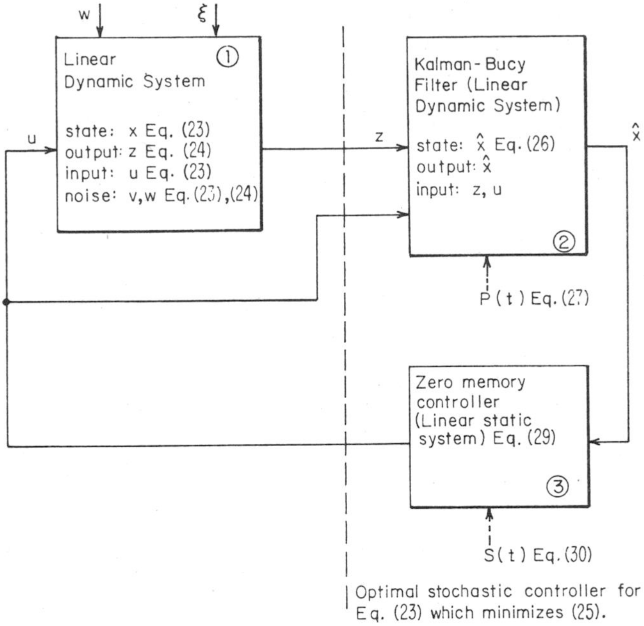

Fig 1 Graphical representation of Results 3.2 and 3.3.

4. New Formulation and Solution of the Linear–Quadratic–Gaussian Stochastic Differential Games

In the LQG games, instead of Eq. (23), we have

$$dx = (A(t)\,x + B(t)\,u + D(t)v)\,dt + C dw(t), \tag{32}$$

where D is an $n \times s$ matrix similarly defined for the control input v of Player II. Players I and II are endowed with measurements

$$dz = Hx\,dt + F d\xi(t), \tag{33-1}$$

$$dy = Gx\,dt + K d\epsilon(t), \tag{33-2}$$

respectively, where (33–2) is defined similarly to (24), with $\epsilon(t)$ an independent Wiener process, K is $k \times i$, $i \geqslant k$, and of full rank.

The payoff is similarly defined with

$$J' = \tfrac{1}{2} E \left\{ x^T(t_f)\, S_f x(t_f) + \int_{t_0}^{t_f} [u^T R u - v^T Q v + x^T M x]\, dt \right\}, \tag{34}$$

where $Q > 0$ for all $t \in I$ and the addition of the $-v^T Q v$ term is due to the fact that v is maximizing. The strategy class Γ_u for u is the same as

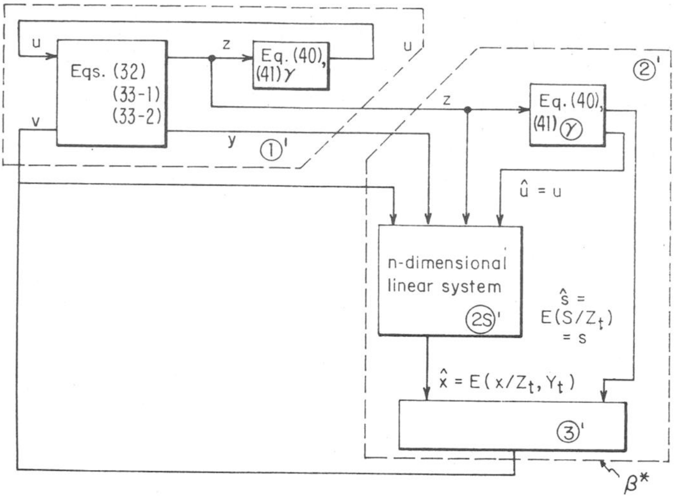

Fig. 2. Optimal controller $\beta_{\text{opt}} = \beta^*$.

before, and Γ_v is similarly defined for v, that is, $\beta(t, \cdot)$ is Y_t-measurable for all t and $\beta(\cdot, y_t)$ is Lebesgue measurable for each $y_t \in C[t_0, t]$. The minimax strategy pair (γ^0, β^0) has been formally obtained earlier in Ref. 3. They are infinite dimensional, in the sense that block ②$'$ in Fig. 2 for each player can only be realized by linear dynamic systems which are describable by partial, rather than ordinary, linear differential equations.

In terms of our discussions in Section 2, (γ^0, β^0) are strategies of the prior commitment type. After the game has started, at time t and from the viewpoint of Player I, the payoff now becomes

$$\bar{J}'' = \tfrac{1}{2}E_{/z_t}\{\tilde{J}\}. \tag{35}$$

While (γ^0, β^0) still retain their equilibrium property, they no longer are secure strategies. The question then arises as to what secure strategy can Player I adopt? Note that, in (4), for fixed γ, β, $\bar{J}''$ is parameterized by the observation history $z_t \in Z_t$. For the purpose of computing the security payoff of (4), z_t is merely part of the prior information. It is reasonable to base the computation on the knowledge of z_t, that is, we assume the admissible strategy class of β to include Z_t-measurable functions, in addition to being Y_t-measurable. This amounts to saying that, in calculating his control, we shall assume that Player II either through divine guidance or a perfect spy has access to Player I's information. Another way of putting it is to say that, if I *knows* that he is playing the game as specified by (4), then he should assume the same for II.

We submit that this is an eminently reasonable viewpoint to take *for the purpose of calculating Player I's security payoff*. To be sure, we may endow Player II with additional information pertaining to the problem, e.g., we may assume that II also knows $w(t)$ or $\xi(t)$. However, such assumptions are less than natural.

Summarizing then, we wish to find $\gamma^* \in \Gamma_u$, $\beta^* \in \Gamma_u \times \Gamma_v$, such that

$$\bar{J}''(\gamma^*, \beta^*) = \min_{\gamma \in \Gamma_u} \max_{\beta \in \Gamma_v \times \Gamma_v} [\bar{J}''].\tag{36}$$

Our overall approach to the solution of (36) is this. We shall arbitrarily fix γ^* and then use the result of Section 3 to solve

$$\bar{J}''(\gamma^*, \beta_{\text{opt}}) \geqslant \bar{J}''(\gamma^*, \beta) \qquad \forall \beta \in \Gamma_v \times \Gamma_u .\tag{37}$$

Let $\beta_{\text{opt}}(\gamma^*)$ be the optimal controller for Player II when Player I employs the fixed γ^*. Then, fix $\beta_{\text{opt}}(\gamma^*)$ and use the result of Section 3 again to solve

$$\bar{J}''(\gamma_{\text{opt}}, \beta_{\text{opt}}) \leqslant \bar{J}''(\gamma, \beta_{\text{opt}}) \qquad \forall \gamma \in \Gamma_u .\tag{38}$$

Let $\gamma_{\text{opt}}(\beta_{\text{opt}})$ be the solution. Consistency then requires us to solve the implicit equation

$$\gamma_{\text{opt}}(\beta_{\text{opt}}(\gamma^*)) = \gamma^*.\tag{39}$$

Thus, let $\gamma^*(\cdot, \cdot)$ be a particular strategy adopted by Player I. Let $\gamma^*(\cdot, \cdot)$ be realized by an n-dimensional linear dynamic system with state s, input z, and output u, that is,

$$ds = A^* s\,dt + B^* dz, \qquad s(t_0) = \hat{x}_0 ,\tag{40}$$

$$u = C^* s.\tag{41}$$

Then, Eqs. (32), (33), (40) appear as a combined linear dynamic system of dimension $2n$ with states (x, s) to Player II through the measurements (33–1) and (33–2). The payoff (36) for fixed γ^* becomes

$$\max_{\beta \in \Gamma_u \times \Gamma_v} \tfrac{1}{2} E\{(x^T S_f x)_{t_f} + \int_t^{t_f} (x^T M x + s^T C^{*T} R C^* s - v^T Q v)\, dt / Z_t , Y_t\}.\tag{42}$$

This is a standard LQG control problem to which Results 3.2 and 3.3 of Section 3 apply directly. The optimal controller $\beta_{\text{opt}}(t, z_t, y_t)$ is given as in Fig. 2.

The combined linear dynamic system is indicated by the block ①′ enclosed by the dotted line. This plays the same role as block ① in Fig. 1. The optimal controller, as in Fig. 1, consists of blocks ②′ and ③′.

The filtering part, block ②′, computes the estimate $\hat{x}$ and $\hat{s}$. It does this by reproducing $s(t)$ and $u(t)$ exactly, since both $s(t)$ and $u(t)$ are Z_t-measurable. Hence,

$$\hat{s}(t) \equiv E(s(t)/Z_t) = s(t), \qquad \hat{u}(t) \equiv E(u(t)/Z_t) = u(t).$$

The conditional mean

$$\hat{x}(t) \equiv E(x(t)/Z_t, Y_t)$$

is computed via an n-dimensional linear system via Result 3.2 (Kalman–Bucy filter ②S′). Block ③′ is a static linear map of $\hat{x}$ and s to v similar to ③ in Fig. 1, that is,

$$v(t) = S_1(t)\,\hat{x}(t) + S_2(t)\,\hat{s}(t). \tag{43}$$

Now, suppose that Player II fixed his strategy at $\beta_{\text{opt}}(\gamma^*) \equiv \beta^*$ as determined above; we shall show that Eq. (39) precisely defines the optimal strategy for γ_{opt}. Thus, γ^*, β^* constitutes a saddle-point pair to (26) and, consequently, solves the problem. To see this, let us consider the combined dynamic system (32)–(33) and blocks ②′ and ③′ as they appear to Player I. They constitute a $3n$-dimensional linear dynamic system with states $(x, \hat{x}, \hat{s})$: $2n$ from Eq. (1) and blocks ②S′ and ③′, and n from block ⓨ. Furthermore, using (43), we see that the payoff (35) becomes

$$\bar{J}' = \tfrac{1}{2}E\left\{(x^T S_f x)_{t_f} + \int_t^{t_f}\left(u^T R u + [x, \hat{x}, \hat{s}]^T\,[\Theta]\begin{bmatrix} x \\ \hat{x} \\ \hat{s}\end{bmatrix}\right) dt/Z_t\right\}, \tag{44}$$

where

$$\Theta = \begin{bmatrix} M & 0 & 0 \\ 0 & -S_1^T Q S_1 & -S_2^T Q S_1 \\ 0 & -S_1^T Q S_2 & -S_2^T Q S_2 \end{bmatrix}.$$

Thus, Player I sees for fixed β^* a standard LQG problem with $3n$ state variables to which Results 3.2 and 3.3 again apply.

We have

$$u(t) = K_1(t)\,E(x/Z_t) + K_2(t)\,E(\hat{x}/Z_t) + K_3(t)\,E(\hat{s}/Z_t). \tag{45}$$

However, since all outputs of block ⓨ are Z_t-measurable by construction, they are deterministic as far as Player I is concerned. In fact, by definition and the requirement of Eq. (39), they are also outputs of γ^* that we are in the process of determining. Thus, they need *not* be estimated or

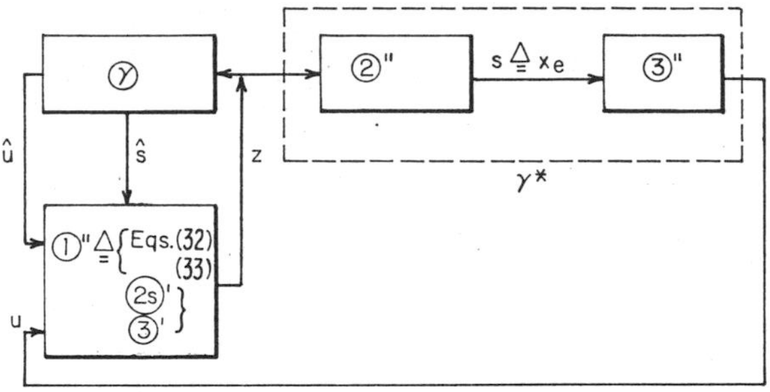

Fig. 3. Optimal controller γ^*.

computed. The states of ① ′ and ②S′, that is, x and $\hat{x}$, can be estimated via Result 3.2; that is, we have

$$x_e \triangleq E(x/Z_t), \qquad \hat{x}_e \triangleq E(\hat{x}(t)/Z_t) \triangleq E(E(x(t)/Z_t, Y_t)/Z_t) = E(x/Z_t) \triangleq x_e\,!,$$

which are computable via a block ②″ by regarding ①′, ②S′, and ③′ as a new block ①″. The states of ②″, an n-dimensional linear dynamic system, is x_e which, by construction and definition, is precisely the state $\hat{s}$ of the block ⓨ and is the conditional mean of *both* x and $\hat{x}$ given Z_t. Consequently, from Result 3.3, we conclude that the optimal control u can be produced using a linear combination of x_e only in a block ③″; that is, Eq. (11) becomes

$$u(t) = [K_1(t) + K_2(t) + K_3(t)]\, x_e(t).$$

This is shown in Fig. 3, which is simply a rearrangement of Fig. 2.

Finally, it is worthwhile to clarify the meaning of the strategy γ^* as compared to other strategies. Let (γ^0, β^0) be the minimax strategy pair determined according to Ref. 3 (the prior commitment model). At time $t = t_0$, if Player I has to make a commitment to a strategy for playing *the rest of the game*, γ^0 certainly represents a reasonable choice (similarly, for β^0), since[4]

$$\bar{J}'(\gamma^*, \beta^*) \geqslant \bar{J}'(\gamma^*, \beta^0) \geqslant \bar{J}'(\gamma^0, \beta^0). \tag{46}$$

On the other hand, as soon as the game has progressed for some time, we have at $t > t_0$

$$\max_{\beta \in \Gamma_u \times \Gamma_v} \bar{J}''(\gamma^0, \beta) \geqslant \bar{J}''(\gamma^0, \beta^*) \geqslant \bar{J}''(\gamma^*, \beta^*). \tag{47}$$

[4] Note that this is different from deciding what value to use for $u(t_0)$, $v(t_0)$. In fact, (γ^0, β^0) and (γ^*, β^*) will produce the same $u(t_0)$, since $Z_{t_0} = Y_{t_0}$.

From the vantage point of Player I at t, γ^0 becomes a rather unsafe strategy for the rest of the game compared to γ^*. To be sure, we still have

$$\bar{J}''(\gamma^*, \beta^*) \geqslant \bar{J}''(\gamma^0, \beta^0).$$

But there is no compelling reason to believe that Player II will definitely play β^0 as explained in Section 2. Conceptually, at $t > t_0$, we use (γ^*, β^*) for the purpose of determining $u(t)$ only. At $t' > t$, we have a different $\bar{J}''$ based on new information and a different minimax game to solve. A different (γ^*, β^*) will be used to determine $u(t')$. In general, this would require the solution of a TPZSG for each t. However, in the LQG game being discussed here, a great practical simplification occurs due to the fact that the parameters of γ^*, β^*, that is, S_1, S_2 in Eq. (9) and K_1, K_2, K_3 in Eq. (11) [see also Eqs. (57), (59), (65), (69), next section] are completely independent of z_t and y_t. Consequently, they can in fact be computed beforehand. In other words, the different (γ^*, β^*) pair Player I determines for each $t \geqslant t_0$ are in fact independent of the actual z_t. Note, however, that this does not mean that we advocate that I should commit himself to γ^* beforehand. Conceptually, he uses γ^* at t to compute $u(t)$ only. He then resolves for γ^* at each different t and uses the new (but identical) γ^* to compute the new $u(t)$. In practice, what this means is that he must have secrecy if he decides (i.e., commits himself) to adopt the posterior strategy γ^*. He should convince his opponent that his decisions are made as the need arises and that *all his options* are open at all times. If no secrecy is possible and he must announce his strategy beforehand then γ^0 should be his choice.[5]

Note that, under the fictitious[6] saddle-point condition, when (γ^*, β^*) are employed, the block ⓨ and γ^* are identical as well as the outputs s, $\hat{s}$ and $\hat{u}$. Of course, if we choose to use a different $\beta \neq \beta^*$ by (say) using $\gamma \neq \gamma^*$, in such a case $u \neq \hat{u}$ and $s \neq \hat{s}$, and $\hat{x}$, x_e can no longer be interpreted as conditional means. However,

$$\bar{J}''(\gamma^*, \beta^*) \geqslant \bar{J}''(\gamma^*, \beta)$$

in this case, by the derivation just given. Consequently, the minimax security level of (36) is achieved when we render β^* such that the γ block

[5] For example, if Player I suspects that II knows or guesses that he is playing γ^* and adopts a counterstrategy $\beta^{\#} = \arg\max \bar{J}'(\beta, \gamma^*)$, then he should really use γ^0 instead of γ^*, since the posterior strategy is not the *best* reply possible if one knows that his opponent has adopted a prior strategy. However, a discussion of the psychology of the pregame guessing and bluffing is outside the scope of this paper. The purpose here is to make us aware of the various alternatives.

[6] Fictitious in the sense that this game is solved only for the purpose of computing Player I's security payoff.

is identical to the ⓨ* block in Fig. 3. In other words, under the conditions stated, the worst that Player II can do to Player I is to use the strategy β^*, and the best counterstrategy is γ^* with $\bar{J}''(\gamma^*, \beta^*)$ the security level at time t. Of course, in real life, when Player II does not have available *both* the information $z(t)$ and $y(t)$, Player I can probably expect better returns than $\bar{J}''(\gamma^*, \beta^*)$.

5. Existence Questions and a Simple Example

So far, we have not addressed ourselves to the question of existence of the solution which was derived in the previous section, Since the solution is obtained by solving a pair of coupled stochastic control problems [Eqs. (37)–(39)], the existence question is directly dependent on the existence of solutions of a set of coupled Riccati equations associated with the control problems. The explicit form of these Riccati equations, while straightforward to write down, is rather cumbersome notationally in the general case. Nor is it possible to state simple and meaningful sufficient conditions to guarantee the existence of the solutions to these DE's. What we propose to do in this section is to carry out the derivation of the explicit solution for a very simple problem to show the various equations involved. The procedure is completely similar in the general case.

Let the scalar dynamic system and observations be[7]

$$\dot{x} = u + v, \qquad x(t_0) \sim N(\hat{x}_0, p_0), \tag{48}$$

$$dz = xdt + d\xi, \tag{49}$$

$$dy = xdt \mid d\epsilon, \tag{50}$$

and payoff

$$\tilde{J} = E\{\tfrac{1}{2}x^2(t_f) + \tfrac{1}{2}\int_{t_0}^{t_f} (u^2 - 2v^2)\, dt\}. \tag{51}$$

Let γ^* be given by

$$ds = as\, dt + bdz, \tag{52}$$

$$u = cs, \tag{53}$$

where a, b, c are parameters to be determined. From Player I's viewpoint of a secure strategy, II maximizes at $t \geqslant t_0$

$$E[\tilde{J}/Z_t, Y_t] = E\{\tfrac{1}{2}x^2(t_f) + \tfrac{1}{2}\int_t^{t_f} (c^2s^2 - 2v^2)\, dt/Z_t, Y_t\} \tag{54}$$

[7] ξ, ϵ are statistically independent standard Wiener processes with zero mean and variance $t - t_0$.

subject to (48), (52), (53). Using standard LQG results, we get, for all $t \geqslant t_0$,

$$d\hat{x} = (c\hat{s} + v)\,dt + p(dz + dy - 2\hat{x}dt), \qquad \hat{x}(t_0) = \hat{x}_0, \tag{55}$$

$$d\hat{s} = a\hat{s}dt + bdz, \qquad \hat{s}(t_0) = \hat{x}_0, \tag{56}$$

where

$$\dot{p} = -2p^2, \qquad p(t_0) = p_0, \tag{57}$$

and the control

$$v = \tfrac{1}{2}(S_{11}(t)\,\hat{x} + S_{12}(t)\hat{s}), \tag{58}$$

where

$$\dot{S}(t) \triangleq \begin{bmatrix} \dot{S}_{11} & \dot{S}_{12} \\ \dot{S}_{12} & \dot{S}_{22} \end{bmatrix} = -S\begin{bmatrix} 0 & c \\ b & a \end{bmatrix} - \begin{bmatrix} 0 & b \\ c & a \end{bmatrix} S - \begin{bmatrix} 0 & 0 \\ 0 & c^2 \end{bmatrix} - S\begin{bmatrix} \frac{1}{2} & 0 \\ 0 & 0 \end{bmatrix} S, \tag{59-1}$$

$$s(t_f) = \begin{bmatrix} 1 & 0 \\ 0 & 0 \end{bmatrix}. \tag{59-2}$$

Equations (55)–(59) define $\beta_{\text{opt}}(\gamma^*)$. Now, from the viewpoint of Player I, β_{opt} and (48) define a $3n$-dimensional linear dynamic system

$$dx = (\tfrac{1}{2}S_{11}\hat{x} + \tfrac{1}{2}S_{12}\hat{s} + u)\,dt, \qquad x(t_0) \sim N(\hat{x}_0, P_0), \tag{60}$$

$$d\hat{x} = (px + (\tfrac{1}{2}S_{11} - 2p)\,\hat{x} + (c + \tfrac{1}{2}S_{12})\hat{s})\,dt + pdz + pd\epsilon, \quad \hat{x}(t_0) = \hat{x}_2, \tag{61}$$

$$d\hat{s} = a\hat{s}dt + bdz, \qquad \hat{s}(t_0) = \hat{x}_0, \tag{62}$$

with a payoff at time $t \geqslant t_0$

$$E\{\tilde{J}/Z_t\} = E\left\{\tfrac{1}{2}x^2(t_f) + \int_t^{t_f}\left[u^2 - \tfrac{1}{2}[\hat{x}, \hat{s}]\begin{bmatrix} S_{11}^2 & S_{11}S_{12} \\ S_{11}S_{12} & S_{12}^2 \end{bmatrix}\begin{bmatrix} \hat{x} \\ \hat{s} \end{bmatrix}\right] dt/Z_t\right\} \tag{63}$$

to be minimized. Once again, using standard results, we first compute the conditional mean of $x(t)$ and $\hat{x}(t)$, as $x_e(t)$ and $\hat{x}_e(t)$. Note that, since s is Z_t-measurable, we have

$$dx_e = (\tfrac{1}{2}S_{11}\hat{x} + \tfrac{1}{2}S_{12}\,\hat{s} + u)\,dt + \Sigma_{11}(t)\,(dz - x_e dt), \qquad x_e(t_0) = \hat{x}_0, \tag{64-1}$$

$$d\hat{x}_e = (px_e + (\tfrac{1}{2}S_{11} - 2p)\,\hat{x}_e + \tfrac{1}{2}S_{12}\,\hat{s} + \hat{u})\,dt + pdz + \Sigma_{12}(t)\,(dz - x_e dt),$$
$$\hat{x}_e(t_0) = \hat{x}_0, \tag{64-2}$$

where

$$\dot{\Sigma}(t) \triangleq \begin{bmatrix} \dot{\Sigma}_{11} & \dot{\Sigma}_{12} \\ \dot{\Sigma}_{12} & \dot{\Sigma}_{22} \end{bmatrix} = \begin{bmatrix} 0 & \frac{1}{2}S_{11} \\ p & \frac{1}{2}S_{11} - 2p \end{bmatrix}\Sigma + \Sigma\begin{bmatrix} 0 & p \\ \frac{1}{2}S_{11} & \frac{1}{2}S_{11} - 2p \end{bmatrix}$$
$$- \begin{bmatrix} \Sigma_{11}^2 & \Sigma_{11}\Sigma_{12} \\ \Sigma_{11}\Sigma_{12} & \Sigma_{12}^2 \end{bmatrix} + \begin{bmatrix} 0 & 0 \\ 0 & p^2 \end{bmatrix}, \quad \Sigma(t_0) = \begin{bmatrix} p_0 & 0 \\ 0 & 0 \end{bmatrix}. \tag{65}$$

Now, setting by definition $\hat{s} = s = x_e$ and noting the easily checked identity

$$\Sigma_{11}(t) \equiv \Sigma_{12}(t) + P, \qquad \Sigma_{12}(t) \equiv \Sigma_{22}(t),$$

we can verify that

$$x_e(t) \equiv \hat{x}_e(t) \equiv E(x(t)/Z_t);$$

and we have, finally,

$$dx_e = (\tfrac{1}{2}(S_{11} + S_{12})\, x_e + u)\, dt + \Sigma_{11}(dz - x_e dt), \tag{66}$$

where

$$\dot{\Sigma}_{11} = \Sigma_{11} S_{11} - \Sigma_{11}^2 - p S_{11}, \qquad \Sigma_{11}(t_0) = p_0 . \tag{67}$$

Furthermore,

$$u = -(K_{11}(t)\, x_e + K_{12}(t)\, \hat{x}_e + K_{13}(t)\hat{s}) = -(K_{11} + K_{12} + K_{13})x_e, \tag{68}$$

where

$$\dot{K} = \begin{bmatrix} \dot{K}_{11} & \dot{K}_{12} & \dot{K}_{13} \\ \dot{K}_{12} & \dot{K}_{22} & \dot{K}_{23} \\ \dot{K}_{13} & \dot{K}_{23} & \dot{K}_{33} \end{bmatrix} = -KA - A^T K + \tfrac{1}{2} \begin{bmatrix} 0 & 0 & 0 \\ 0 & S_{11}^2 & S_{11}S_{12} \\ 0 & S_{11}S_{12} & S_{12}^2 \end{bmatrix}$$

$$+ K \begin{bmatrix} 1 & 0 & 0 \\ 0 & 0 & 0 \\ 0 & 0 & 0 \end{bmatrix} K, \qquad K(t_f) = \begin{bmatrix} 1 & 0 & 0 \\ 0 & 0 & 0 \\ 0 & 0 & 0 \end{bmatrix}, \tag{69}$$

and

$$A = \begin{bmatrix} 0 & \tfrac{1}{2}S_{11} & \tfrac{1}{2}S_{12} \\ 2p & \tfrac{1}{2}S_{11} - 2p & c + \tfrac{1}{2}S_{12} \\ b & 0 & a \end{bmatrix}.$$

The consistency requirement of Eq. (39) now specifies

$$c \equiv -(K_{11} + K_{12} + K_{13}), \tag{70}$$

$$b \equiv \Sigma_{11}, \tag{71}$$

$$a = \tfrac{1}{2}(S_{11} + S_{12}) - b + c. \tag{72}$$

If we substitute Eqs. (70)–(72) into Eqs. (57), (59), (67), (69), they form a set of coupled nonlinear differential equations of the Riccati type. Their solution completely specifies the secure strategy γ^* via Eqs. (52), (53), (70)–(72). Consequently, the existence of γ^* is equivalent to the existence of solutions of Eqs. (57), (59), (67), (69).

6. Practical Implications, Open Problems, and Conclusions

There are several implications of the results obtained that are worth further discussion.

First of all, it should be understood that the strategy we derived for u (that is, γ^*) is secure only with respect to a set of assumptions which we assert to be reasonable. Roughly speaking, *we allow our opponent to know everything that we may know*. This seems to be as pessimistic an assumption as one would like to use. It appears paranoid to assume that the other player can have access to knowledge concerning the choices of nature, i.e., values of $\xi(t)$, $\epsilon(t)$, $w(t)$, *etc.*, beyond the probabilistic knowledge that is already permitted in the statement of the original problem. Our assumption is also in line with other approaches to the control of uncertain systems (Refs. 6–7). They have taken the viewpoint that such problems may be regarded as a game against an opponent (nature), where the upper value of the game is sought. In other words, the opponent (nature or uncertainty) makes the moves knowing everything you have known and/or have done.

In this respect, the derived solution has an additional appealing feature. Consider the linear stochastic dynamic systems

$$dx = (Ax + Bu)\,dt + Cdw + vdt, \tag{73}$$

$$dz = Hxdt + d\xi, \tag{74}$$

where $v(t)$ represents terms which arise due to approximations and inaccuracies in modelling of the real (probably nonlinear and non-Gaussian) system. Now, if we consider a payoff

$$J = \tfrac{1}{2}E\left\{(x^T S_f x)_{t_f} + \int_{t_0}^{t_f} (u^T R u)\,dt\right\} \tag{75}$$

and a size-of-approximation constraint[8]

$$E\int_{t_0}^{t_f} v^T v\,dt \leqslant \lambda, \tag{76}$$

then the results in Section 4 state that a *good* control law from a minimax viewpoint is an n-dimensional linear dynamic system followed by a zero-memory linear map. This explains the almost unbelievable robustness of the *structure* of the well-known optimal law (Results 3.2 and 3.3) in widely diverse applications where linearity or Gaussianness has been

[8] The constraint (76) can be incorporated into (75) via a Lagrange multiplier q. We then have the problem

$$\min_u \max_v \left[J - qE\int_0^{t_f} v^T v dt\right],$$

which is treated in Section 4.

clearly violated. In other words, except for parameter values, the linear structure of γ^* remains appropriate (i.e., safe) in highly nonlinear and poorly defined situations. In fact, the above discussion implies that *"optimal" stochastic control of nonlinear system can now be attempted by finite-dimensional optimization on the parameters of* γ^*. *The engineering significance of this cannot be overstated.* Similarly, if we wish to consider a minimax estimator for nonlinear systems, then, from the posterior-strategic viewpoint, the structure of the Kalman–Bucy filter remains optimal. All that is needed is the adjustment or determination of parameter values. Of course, since no correction is applied in a pure estimation problem, we do not expect such a minimax estimator to be as robust as the minimax controller mentioned above. This seems in fact to have been the general experience in practice.

The recognition that, in the delayed commitment mode, all stochastic games in extensive form are nonzero sum raises interesting problems as well as possibilities. In this report we have only explored two solution concepts associated with NZS games, namely, Nash equilibrium and individual minimax solutions. There are many other solution concepts involving bargaining, coalitions, *etc.* For example, we can visualize that the two players may wish to enter into information exchange during the play of the game.

References

1. Aumann, R., and Maschler, P., *Some Thoughts on the Minimax Principle*, Management Science, Vol. 18, pp. 54–63, 1972.
2. Harsanyi, J., *Game with Incomplete Information Played by Bayesian Players—Part II*, Management Science, Vol. 14, pp. 320–334, 1968.
3. Willman, W., *Formal Solution of a Class of Stochastic Differential Games*, IEEE Transaction on Automatic Control, Vol. AC-14, pp. 504–509, 1969.
4. Wonham, W., *Random Differential Equations in Control Theory*, Probabilistic Methods in Applied Mathematics, Vol. 2, Edited by A. T. Barucha-Reid, Academic Press, New York, New York, 1970.
5. Bucy, R., and Joseph, P., *Filtering for Stochastic Processes with Applications to Guidance*, Interscience, New York, New York, 1968.
6. Witsenhausen, H. S., *A Minimax Control Problem for Sampled Linear Systems*, IEEE Transactions on Automatic Control, Vol. AC-13, pp. 5–21, 1968.
7. Bertsekas, D. P., and Rhodes, I. B., *On the Minimax Feedback Control of Uncertain Dynamic Systems*, Proceedings of 1971 IEEE Decision and Control Conference, Miami, Florida, pp. 451–455, 1971.

XIX

Role of Information in the Stochastic Zero-Sum Differential Game[1]

F. K. SUN AND Y. C. HO

Abstract. In this paper a linear quadratic Gaussian, zero-sum differential game is studied. "Maneuverability" is defined to measure players' strength. It is shown that a "more maneuverable" player would prefer a "more observable" information system. An example is given to show that a "more controllable" player might not prefer "more observable" measurements in the stochastic environment.

1. Introduction and Problem Formulation

In our previous work (Ref. 1) we have shown, through a static linear quadratic Gaussian model, that under a "broadcasting" information channel a "strong" player would prefer "better" information, while the opposite is true for a "weak" player, where (i) "broadcasting" means that both players have the same measurements, (ii) "strong" player refers to the one with the cheaper cost per energy spent, and (iii) "better" refers to the smaller variance of the measurement error. In this paper we consider a linear stochastic zero-sum differential game in which both players also obtain their information from a broadcasting

[1] The research reported in this paper was made possible through support extended to the Division of Engineering and Applied Physics, Harvard University, by the U.S. Office of Naval Research under the Joint Services Electronics Program by Contract N00014-75-c-0648 and by the National Science Foundation under Grant GK31511.

channel. Mathematically, this game may be characterized by the following equations:

System dynamics:

$$dx_t = (F_t x_t + G_{pt} u_t - G_{et} v_j)\, dt + R_{1t}\, dw_j . \tag{D}$$

Information structure:

$$dz_{pt} = dz_{et} = dz_t = H_t x_t\, dt + R_{2t}\, ds_t . \tag{I}$$

Here (i) $u_t \in \mathbb{R}^p$ and $v_t \in \mathbb{R}^e$ are respective controls of the minimizer and maximizer at time t. (ii) $x_t \in \mathbb{R}^n$ is the state vector at time t and x_0 is the initial state, which is normally distributed with $N(\bar{x}_0, M_0)$. (iii) w_t and s_t are independent normalized l- and m-vector-valued Wiener processes that are assumed to be statistically independent of x_0. (iv) z_{pt} and z_{et} are measurements taken by the minimizer and maximizer, respectively. Through the broadcasting channel, we have $z_{pt} = z_{et} = z_t \in \mathbb{R}^k$ at each time t. (v) The matrices G_{pt}, G_{et}, F_t, and H_t are nonrandom with the appropriate dimensions. (vi) $D_t = R_{2t}R_{2t}^t$ is positive definite (denoted by $D_t > 0$).

Objective function:

$$J = E\left\{x_{t_f}^t S_f{}^t S_f x_{t_f} + \int_0^{t_f} (u_\tau{}^t B_\tau u_\tau - v_\tau{}^t C_\tau v_\tau)\, d\tau\right\}, \tag{O}$$

where (i) $B_\tau > 0$, $C_\tau > 0$, and $S_f{}^t S_f \geqslant 0$ for all $\tau \in [0, t_f]$ are nonrandom and have the appropriate dimensions, (ii) the final time t_f is fixed and finite, and (iii) the expectation is taken over all random variables considered in this problem.

It should be noted that the above formulation covers the usual pursuer–evader game as a special case. To complete the statement of the problem, let I be the interval $[0, t_f]$, $C[0, t]$ be the space of continuous functions on $[0, t]$, and $\mathscr{Z}_t$ be the smallest σ-algebra generated by $Z_t = \{z_t\, ;\, z_t \in C[0, t]\}$. Then we say that an admissible control law for the minimizer is a functional $\gamma: I \times C[0, t] \to \mathbb{R}^p$ such that $\gamma_0 \cdot (z)$ is Lebesque-measurable for each $z \in C[0, t]$ and $\gamma_t(\cdot)$ is $\mathscr{Z}_t$-measurable for all $t \in I$ with $E\{\gamma_t{}^t \gamma_t\} < \infty$. Under the broadcasting channel the admissible laws, denoted as β's, for the maximizer are similarly defined. From now on, for notational simplicity, we write $\gamma_t(\mathscr{Z}_t) = \gamma_t$ and $\beta_t(\mathscr{Z}_t) = \beta_t$, and keep in mind that both are $\mathscr{Z}_t$-measurable. Furthermore, let $\hat{J}(\gamma_t, \beta_t)$ be the outcome of the game when the minimizer uses

γ_t and the maximizer uses β_t. Finally, we say that an admissible pair (γ_t^*, β_t^*) is a saddle point for this game iff

$$\hat{J}(\gamma_t, \beta_t^*) \leq \hat{J}(\gamma_t^*, \beta_t^*) \leq \hat{J}(\gamma_t^*, \beta_t), \tag{S}$$

where γ_t and β_t are any admissible control laws, and (γ_t^*, β_t^*) is called the optimal strategic pair. Thus our objectives are (i) to find such an optimal pair for the problem described by (D), (I), (O), and (S), which will be called the main problem, and (ii) to see the effect of different broadcasting information.

2. Optimal Solution for the Main Problem

The first objective with respect to the main problem has been studied intensively for the more general case in which both players have different information (Ref. 2). However, the solution becomes complicated and difficult to visualize. In our setup both players have the same information and so have the same σ-algebra.[2] This main problem may be easily solved by introducing the corresponding deterministic game (call it the D-game). More precisely, for the D-game we consider that $R_{1t} = 0$, $R_{2t} = 0$, and $H_t = I$ (identity matrix) in the main problem and the equations (D), (I), and (O) become

$$\dot{x}_t = F_t x_t + G_{pt} u_t - G_{et} v_t, \qquad x_0 = \bar{x}_0, \tag{D$'$}$$

$$z_t = x_t, \tag{I$'$}$$

$$J' = x_{t_f}^t S_f^t S_f x_{t_f} + \int_0^{t_f} (u_\tau^t B_\tau u_\tau - v_\tau^t G_\tau u_\tau)\, d\tau. \tag{O$'$}$$

The solution to this D-game is well known. In particular, by Proposition 1 in Rhodes and Luenberger's paper (Ref. 4), we have the necessary and sufficient condition for which the closed-loop control laws exist. Since these results are useful in the further development, we summarize them in the following statements.

Necessary and sufficient condition:

$$S_t = \left\{ I + S_f \int_t^{t_f} \Phi(t_f, \tau)[G_{p\tau} B_\tau^{-1} G_{p\tau} - G_{e\tau} C_\tau^{-1} G_{e\tau}]\, \Phi(t_f, \tau)\, d\tau S_f^t \right\} > 0 \tag{C1}$$

[2] Balakrishnan has obtained the saddle point for this game by using a function-space approach. However, the value of the game was not given in his result (Ref. 3).

for all $t \in [0, t_f]$, where $\Phi(t_f, \tau)$ is the transition matrix for the dynamic system (D′).

Optimal control laws:

$$\gamma_t^*(x_t) = -B_t^{-1}G_{pt}^t K_t x_t, \qquad \beta_t^*(x_t) = -C_t^{-1}G_{et}^t K_t x_t, \tag{1}$$

and K_t is the unique $n \times n$, symmetric, nonnegative-definite solution to the matrix Riccati equation

$$\dot{K}_t = -K_t F_t - F_t{}^t K_t + K_t[G_{pt}B_t^{-1}G_{pt}^t - G_{et}C_t^{-1}G_{et}^t]\, K_t \tag{K}$$

with the boundary condition $K_{t_f} = S_f{}^t S_f$.

Note that condition (C1) is used to guarantee that there exists a unique solution for (K). In order to obtain the solution for the main problem in addition to the result in the D-game, we need the following two lemmas.

Lemma 2.1. Let x_τ be defined as the main problem and K_τ be the unique solution to the equation (K); then

$$0 = E\left\{\int_0^{t_f} [x_\tau{}^t K_\tau x_\tau + 2x_\tau{}^t K_\tau(F_\tau x_\tau + G_{p\tau}u_\tau - G_{e\tau}v_\tau)]\, d\tau\right\}$$
$$+ E\left\{\int_0^{t_f} \mathrm{Tr}[K_\tau R_{1\tau} R_{1\tau}^t]\, d\tau\right\} - E\{x_\tau{}^t K_\tau x_\tau \,|_0^{t_f}\}. \tag{2}$$

Proof. This result can be immediately obtained by the Ito differential rule with respect to $dx_t{}^t K_t x_t$ and by the fact that

$$E\left\{\int_0^{t_f} x_\tau{}^t K_\tau R_{1\tau}\, dw_\tau\right\} = 0.$$

Next, by adding identity (2) to Equation (O), we have

$$J = E\left\{x_0{}^t K_0 x_0 + \int_0^{t_f} [\| u_\tau - B_\tau^{-1}G_{p\tau}^t K_\tau x_\tau \|_{B_\tau}^2 - \| v_\tau + C_\tau^{-1}G_{e\tau}^t K_\tau x_\tau \|_{C_\tau}^2]\, d\tau\right\}$$
$$+ \int_0^{t_f} [\mathrm{Tr}\, K_\tau R_{1\tau} R_{1\tau}^t]\, d\tau. \tag{3}$$

Considering that the minimizer uses γ_τ and the maximizer use β_τ, we can express $J(\gamma_\tau, \beta_\tau)$ as in the following lemma.

Lemma 2.2.

$$\hat{J}(\gamma_\tau, \beta_\tau) = E\left\{\int_0^{t_f} [\| \gamma_\tau + B_\tau^{-1} G_{p\tau} K_\tau \hat{x}_\tau \|_{B_\tau}^2 - \| \beta_\tau + C_\tau^{-1} G_{e\tau} K_\tau \hat{x}_\tau \|_{C_\tau}^2 \, d\tau\right\}$$
$$+ \int_0^{t_f} [\mathrm{Tr}(K_\tau R_{1\tau} R_{1\tau} + K_\tau (G_{p\tau} B_\tau^{-1} G_{p\tau}^t - G_{e\tau} C_\tau^{-1} G_{e\tau}^t) K_\tau P_\tau)] \, d\tau$$
$$+ \mathrm{Tr}\, K_0 M_0,$$

where (i) $\hat{x}_\tau = E\{x_\tau \mid \mathscr{Z}_\tau ; \gamma_\tau, \beta_\tau\}$,[3] and may be obtained from Kalman–Bucy filter,

$$d\hat{x}_\tau = (F_\tau \hat{x}_\tau + G_{p\tau} \gamma_\tau - G_{e\tau} \beta_\tau) \, d\tau + P_\tau H_\tau{}^t D_\tau^{-1} (dz_\tau - H_\tau \hat{x}_\tau \, d\tau) \tag{4}$$

with $\hat{x}_0 = \bar{x}_0$; and (ii) $P_\tau = E\{(x_\tau - \hat{x}_\tau)(x_\tau - \hat{x}_\tau)^t\}$ is the error covariance matrix and is the unique solution to the following matrix Riccati equation:

$$\dot{P}_\tau = F_\tau P_\tau + P_\tau F_\tau{}^t - P_\tau N(z_\tau) P_\tau + R_{1\tau} R_{1\tau}^t, \tag{5}$$

with $P_0 = M_0$, where $N(z_\tau) = H_\tau{}^t D_\tau^{-1} H_\tau$.

Proof. This result is obtained by substituting $x_\tau = \hat{x}_\tau + \tilde{x}_\tau$ into Eq. (3), where $\tilde{x}_\tau$ is the residual from the best linear estimation of x_τ, and using the fact that γ_τ, β_τ, and $\hat{x}_\tau$ are $\mathscr{Z}_\tau$-measurable and $\tilde{x}_\tau$ is uncorrelated with any $\mathscr{Z}_\tau$-measurable functions.

By the construction of Lemma 1 and 2, the solution to the main problem can be obtained immediately. We state it in the following theorem.

Theorem 2.1. If the optimal strategic pair to the D-game exists, then the optimal strategic pair to the main problem exists. Furthermore, the optimal strategic pair at each time t is

$$\gamma_t{}^* = -B_t^{-1} G_{pt}^t K_t \hat{x}_t, \qquad \beta_t{}^* = -C_t^{-1} G_{et}^t K_t \hat{x}_t, \tag{6}$$

where K_t satisfies Eq. (K) and $\hat{x}_t$ satisfies Eq. (4). Moreover,

$$J(\gamma_t{}^*, \beta_t{}^*) = \int_0^{t_f} \mathrm{Tr}\, K_\tau [R_{1\tau} R_{1\tau}^t + (G_{p\tau} B_\tau^{-1} G_{p\tau}^t - G_{e\tau} C_\tau^{-1} G_{e\tau}^t) K_\tau P_\tau] \, d\tau$$
$$+ \mathrm{Tr}\, K_0 M_0, \tag{7}$$

where P_τ is obtained from Eq. (5).

---

[3] $\hat{x}_\tau$ is the conditional mean only when γ_τ and β_τ are used. For actual implementation and interpretation, see p. 351 of Ref. 5.

Proof. The proof is completed by Lemmas 1 and 2 and the fact that γ_τ and β_τ are both $\mathscr{Z}_\tau$-measurable.

Note that from Eq. (7), P_τ, which is affected by the choice of information structure through $N(z_\tau)$, is separated from parameters of the system dynamics and the objective function. Hence, it enables us to compare the value of the game under two different information structures.

3. Comparison of Different Information Structures

Let

$$dz_t' = H_t' x_t \, dt + R_{2t}' \, ds_t' \tag{I'}$$

be another broadcasting information system, and $J'(\gamma_t^0, \beta_t^0)$ be the value of the game under the new information system, where s_t' is also a normalized Wiener process and has the same properties as s_t. Since we have only changed the information structure, to obtain $\hat{J}'(\gamma_\tau^0, \beta_\tau^0)$ we simply need to replace P_τ in $\hat{J}(\gamma_\tau^*, \beta_\tau^*)$ and Eq. (P) by P_τ', where P_τ' is the error covariance matrix under the new information. Thus, by taking the difference between $\hat{J}'(\gamma_\tau^0, \beta_\tau^0)$ and $\hat{J}(\gamma_\tau^*, \beta_\tau^*)$, we have

$$\hat{J}'(\gamma_\tau^0, \beta_\tau^0) - \hat{J}(\gamma_\tau^*, \beta_\tau^*) = \int_0^{t_f} \operatorname{Tr} L_\tau(P_\tau' - P_\tau) \, d\tau, \tag{8}$$

where $L_\tau = K_\tau(G_{p\tau}B_\tau^{-1}G_{p\tau}^t - G_{e\tau}C_\tau^{-1}G_{e\tau}^t)K_\tau$. Therefore, if $(G_{p\tau}B_\tau^{-1}G_{p\tau}^t - G_{e\tau}C_\tau^{-1}G_{e\tau}^t)$ is nonnegative definite for all $\tau \in [0, t_f]$, then L_τ is nonnegative definite. It follows that the sufficient condition for $[\hat{J}'(\gamma_\tau^0, \beta_\tau^0) - \hat{J}(\gamma_\tau^*, \beta_\tau^*)] \geq 0$, i.e., it pays the maximizer to use the new information, is $P_\tau' - P_\tau \geq 0$ for all $\tau \in [0, t_f]$. Furthermore, since $G_{p\tau}$ determines how much the minimizer could affect the state x_τ and B_τ is associated with the cost per energy spent, $G_{p\tau}B_\tau^{-1}G_{p\tau}^t = E_{p\tau}$ might be viewed as the minimizer's maneuverability. For more insight, let us consider the following proposition.

Proposition 3.1. Let P_τ and P_τ' be positive definite, and $P_0 = P_0' = M_0$. Assume that all the parameters in the main problem remain unchanged except that either (i) $N(z_\tau)$ is replaced by $N(z_\tau')$ and $N(z_\tau) \geq N(z_\tau')$, $\tau \in [0, t_f]$; or (ii) $R_{1\tau}$ is replaced by $R_{1\tau}'$ and $R_{1\tau}'R_{1\tau}'^t \geq R_{1\tau}R_{1\tau}^t$, $\tau \in [0, t_f]$; then $P_\tau' \geq P_\tau$, $\tau \in [0, t_f]$, where $N(z_\tau) = H_\tau^t D_\tau^{-1} H_\tau$ and $N(z_\tau') = H_\tau'^t D_\tau^{-1} H_\tau'$.

Proof. The result can be obtained immediately by taking the difference between the two Riccati equations associated with P_τ' and

P_τ and showing that the solution for the resultant equation is nonnegative definite.

Note that if H_τ, D_τ, H_τ', and D_τ' are scalar, then condition (i) in the above proposition is equivalent to the condition for better information proposed by Ho and Basar (Ref. 6). In fact, it can be shown that a linear information z_τ is more informative than a linear information z_τ' within the class of quadratic Gaussian problems, i.e., in a one-person decision problem using z_τ one cannot do worse than using z_τ' for any quadratic criteria with the assumption of normality on the state of the world, if and only if $N(z_\tau) \geq N(z_\tau')$. Thus the first part of the proposition, with the condition that $E_{p\tau} \geq E_{e\tau}$ for all $\tau \in [0, t_f]$, intuitively implies that in a zero-sum differential game the player with better maneuverability would like to have better information. The second part of the proposition implies that if the information remains unchanged, the weak player (the one with poorer maneuverability) would like the dynamic system to be as noisy as possible.

In the case where the system noise is equal to zero, i.e., $R_{1t} = 0$ for all $t \in [0, t_f]$, we can relate condition (i) in the above proposition with the condition of observability. We say that the system (D) with $R_{1t} = 0$ and (I) for all $t \in [0, t_f]$ is observable when

$$M(t, 0) = \int_0^t \Phi^t(\tau, 0)\, N(z_\tau)\, \Phi(\tau, 0)\, d\tau$$

is positive definite for all $t \in [0, t_f]$, where $\Phi(\tau, 0)$ is the transition matrix corresponding to F_t (Ref. 7). It is obvious that if z_t is better than z_t', i.e., $N(z_t) > N(z_t')$, then z_t should provide a more observable system in the sense that $\hat{M}(t, 0) = M(t, 0) - M'(t, 0)$ is positive definite. But the converse is not true in general. However, we have the following proposition.

Proposition 3.2. Let P_τ and P_τ' be positive definite and $R_{1t} = 0$, $\tau \in [0, t_f]$. If $E_{p\tau} \geq E_{e\tau}\ \forall \tau \in [0, t_f]$, then the minimizer would prefer the information that could provide a more observable system in the sense that $\hat{M}(\tau, 0) \geq 0\ \forall \tau \in [0, t_f]$.

Proof. It is obvious by the fact that $\hat{M}(\tau, 0) \geq 0$ implies $P_\tau \leq P_\tau'\ \forall \tau \in [0, t_f]$.

By the assumption of $B_\tau > 0$ and $C_\tau > 0\ \forall \tau \in [0, t_f]$, it is no loss of generality to assume $B_\tau = C_\tau = I$.[4] Then it follows that $E_{p\tau} =$

[4] B_τ and C_τ can therefore by absorbed into u and v by rescaling.

$G_{p\tau}G^t_{p\tau}$ and $E_{e\tau} = G_{e\tau}G^t_{e\tau}$. From Proposition 2, it is interesting to ask whether the condition of better maneuverability is equivalent to the condition that the relative controllability matrix be positive definite, which was proposed by Ho *et al.* (Ref. 8), i.e., $W_t = W_{pt} - W_{et} \geq 0$ $\forall \tau \in [0, t_f]$, where

$$W_{pt} = \int_t^{t_f} \Phi(t, \tau)\, E_{p\tau}\Phi^t(t, \tau)\, d\tau; \qquad W_{et} = \int_t^{t_f} \Phi(t, \tau)\, E_{e\tau}\Phi^t(t, \tau)\, d\tau.$$

Unfortunately, the answer is no, as can be seen from a specific example below. Before presenting it, however, let us examine the two conditions "more maneuverable" ($E_{p\tau} \geq E_{e\tau}$) and "more controllable" ($W_{p\tau} \geq W_{e\tau}$) intuitively. If a dynamic system is more maneuverable $\forall t \in [0, t_f]$, then it is certainly more controllable, but not vice versa. System 1 can be temporarily less maneuverable than 2 for some subinterval of $[0, t_f]$ and yet be more controllable over the entire interval. In deterministic cases, this says that so long as 1 is more controllable than 2, a temporary short-term disadvantage can always be made up over the long term. However, in the stochastic case, this is no longer so because one can no longer count on the future, which is now uncertain. One had better "make hay while the sun shines." It is for this reason that we cannot have the more aesthetically appealing assertion that a "more controllable" system prefers "more observable" measurements. Instead, we say a "more maneuverable" system prefers "more observable" measurements. The following example shows this.

4. Example

Consider a scalar time-variant case in which the dynamic equation is given by $\dot{x}_t = g_{pt}u_t - g_{et}v_t$, the performance index is given by $J = \dot{E}\{x_1^2 + \int_0^1 (u_t^2 - v_j^2)\, dt\}$, and the observation process is given by $dz_{pt} = dz_{et} = dz_t = h_t x_t\, dt + ds_t$. Let $g_t = g^2_{pt} - g^2_{et}$ and $\eta_t = h_t^2 > 0$ for all $0 \leq y \leq 1$. By the previous results we know that if

$$1 + \int_t g_s\, ds > 0 \tag{9}$$

for all $t \in [0, 1]$, then there exists a saddle point (γ^*, β^*) such that

$$J(\gamma^*, \beta^*) = \int_0^1 k_t^2 g_t\, p_t\, dt + k_0 m_0\,,$$

where

$$\dot{p}_t = -p_t^2\eta_t; \qquad p_0 = m_0 > 0; \tag{10}$$

$$\dot{k}_t = k_t^2 g_t; \qquad k_1 = 1. \tag{11}$$

Next, let $dz_t' = h_t' x_t\, dt + ds_t'$ be a more observable system, i.e.,

$$\int_0^t (\eta_\tau' - \eta_\tau)\, d\tau \geq 0, \qquad 0 \leq t \leq 1, \tag{12}$$

or, equivalently, $p_t \geq p_t'$ $\forall t \in [0, 1]$. Assume that the rest of the parameters remain the same. Let (γ', β') and $J(\gamma', \beta')$ be the saddle point and the value of the game under this new information system, respectively. Without loss of generality, let u be the one with better controllability in the sense that

$$\int_t^1 g_s\, ds > 0 \qquad \text{for all} \quad 0 \leq t \leq 1. \tag{13}$$

Thus, if u prefers a more observable system, then

$$J(\gamma^*, \beta^*) - J(\gamma', \beta') = \int_0^1 k_t^2 g_t (p_t - p_t')\, dt \geq 0. \tag{14}$$

To prove that this cannot be true in general, we need to choose g_{pt} and g_{et}, and hence g_t, such that (13) is satisfied. Meanwhile, choose η_t and η_t', and therefore p_t and p_t', such that (12) is satisfied, or, equivalently, $p_t - p_t' \geq 0$ for all $0 \leq t \leq 1$. Finally, prove under this condition that (14) is negative. Let g_{pt} and g_{et} be chosen such that

$$g_t = \begin{cases} -1 & \text{if } \frac{1}{2} - \epsilon \leq t \leq \frac{1}{2} + \epsilon, \\ 0 & \text{if } \frac{1}{2} + \epsilon < t \leq \frac{1}{2} + 2\epsilon, \\ 1 & \text{otherwise}; \end{cases}$$

$$\int_{(1/2)-\epsilon}^{(1/2)+\epsilon} g_s\, ds + \int_{(1/2)+2\epsilon}^{1} g_s\, ds > 0.$$

Next, let

$$\eta_t = \begin{cases} n_0 \\ K \\ 2K' \\ n_0 \end{cases} \quad \text{and} \quad \eta_t' = \begin{cases} n_0 & \text{if } t \in [0, \frac{1}{2} - \epsilon], \\ K' & \text{if } t \in [\frac{1}{2} - \epsilon, \frac{1}{2} + \epsilon], \\ 2K & \text{if } t \in [\frac{1}{2} + \epsilon, \frac{1}{2} + 2\epsilon], \\ n_0 & \text{if } t \in [\frac{1}{2} + 2\epsilon, 1], \end{cases}$$

where $K' > K > 0$ and $n_0 > 0$. By simple calculation, we have

$$\int_0^t (\eta_\tau' - \eta_\tau)\, d\tau = \hat{n}(t) \geq 0 \qquad \forall t \in [0, 1].$$

From Eqs. (10) and (11) we have

$$p_t = \left[m_0^{-1} + \int_0^t \eta_s \, ds\right]^{-1}, \qquad p_t' = \left[m_0^{-1} + \int_0^t \eta_s' \, ds\right[^{-1};$$

$$k_t = \left[1 + \int_t^1 g_s \, ds\right]^{-1}.$$

It is easy to check that $\Delta p_t = p_t - p_t'$ is nonnegative. Thus,

$$J(\gamma^*, \beta^*) - J(\gamma', \beta') = -\int_{(1/2)-\epsilon}^{(1/2)+\epsilon} k_t^2 D_1 \, dt \tag{15}$$

where $D_1 = [m_0^{-1} + K(t - \frac{1}{2} + \epsilon)]^{-1} - [m_0^{-1} + K'(t - \frac{1}{2} + \epsilon)]^{-1} > 0$. It follows that (15) is negative.

5. Conclusion

An intuitively reasonable guideline has been demonstrated through a stochastic differential game model. Roughly, it says that the stronger player can exploit new information better than his opponent. Furthermore, the noisier the environment is, the more myopic one's policy tends to be. This partially answers the question "When both players' information has been changed, who has the advantage?"

References

1. Ho, Y. C., and Sun, F. K., *Value of Information in Two Team Zero Sum Problems*, Journal of Optimization Theory and Applications, Vol. 14, No. 5, Nov. 1974.
2. Willman, W. W., *Differential Games with Measurement Uncertainties*, Harvard University, Ph.D. Thesis, 1969.
3. Balakrishnan, A. V., *Stochastic Differential Systems I: Filtering and Control, A Function Space Approach*, Springer-Verlag, Berlin, Heidelberg, New York, 1973.
4. Rhodes, I. B., and Luenberger, D. G., *Differential Games with Imperfect Information*, IEEE Trans. on Automatic Control, Vol. AC-14, No. 1, February 1969.
5. Ho, Y. C., *On the Minimax Principle and Zero-Sum Stochastic Differential Games*, Journal of Optimization Theory and Applications, Vol. 13, No. 3, March 1974.

6. BASAR, T., and HO, Y. C., *Information Properties of Nash Solutions of Two Stochastic Nonzero-Sum Games*, Journal of Economic Theory, Vol. 17, No. 4, April 1974.
7. BRYSON, A. E., and HO, Y. C., *Applied Optimal Control*, Ginn and Co., Waltham, Massachusetts, 1969.
8. HO, Y. C., BRYSON, A. E., and BARON, S., *Differential Games and Optimal Pursuit-Evasion Strategies*, IEEE Transactions on Automatic Control, Vol. AC-10, No. 4, October 1965.

XX

Note on the Linear Quadratic Gaussian, Zero-Sum Differential Game with Asymmetric Information Structures

F. K. SUN AND Y. C. HO

Abstract. In this chapter, we study a linear quadratic Gaussian, zero-sum differential game in which one player's information structure is part of another's. Using a result from Chapter XIX, the saddle-point strategic pair and the value of this game are derived.

1. Introduction

We consider a linear quadratic Gaussian, zero-sum differential game in which one player's information structure is part of another's. The saddle-point strategic pair for this game has been formally obtained by Ho (see Chapter XVIII). However, the value of the game was not given in his paper and the direct computation of this value from this optimal strategies seems difficult. Using the result for the case of broadcasting information structures (see Chapter XIX), we shall derive the saddle-point strategic pair and the value of this game here.

2. Results

For notational simplicity[1] we shall consider a game which might

[1] This formulation will be as general as those in Chapter XIX. If we redefine G_{pt} and G_{et} by a nonsingular transformation and absorb A_t and B_t into u_t and v_t (since A_t and B_t are positive definite by assumption) in Chapter XIX, then we shall obtain the same form as in this note without loss of generality.

be characterized by the following equations.

System dynamics:

$$dx_t = (G_{pt}u_t - G_{et}v_t)\,dt + R_{1t}\,dw_t\,. \tag{D}$$

Objective function:

$$J = E\left\{x_{t_f}^t K_f{}^t K_f x_{t_f} + \int_0^{t_f} (u_\tau{}^t u_\tau - v_\tau{}^t v_\tau)\,d\tau\right\}. \tag{O}$$

Information structures:

$$\begin{gathered} dz_t = H_t x_t\,dt + R_{2t}\,ds_t\,, \qquad dz_{1t} = H_{1t}x_t\,dt + R_{3t}\,ds_{1t}\,, \\ dz_{pt} = dz_t\,, \qquad dz_{et} = dz_{2t} = d\begin{bmatrix} z_t \\ z_{1t}\end{bmatrix}, \end{gathered} \tag{I}$$

where all the variables and coefficients are defined similarly to those in Chapter XIX; $z_{1t} \in \mathbb{R}^{k_1}$ at each time t, and s_{1t} is a normalized m_1-vector-valued Wiener process that is assumed to be statistically independent of x_0 and s_t; and $D_{1t} = R_{3t}R_{3t}^t > 0$. Let $\mathscr{Z}_t$ and $\mathscr{Z}_{2t}$ be the smallest σ-algebras generated by $Z_t = \{z_t\,;\, z_\tau \in C[0, t]\}$ and $Z_{2t} = \{z_{2t}\,;\, z_{2\tau} \in C[0, t]\}$, respectively. Then the admissible strategy γ_t for the minimizer P is defined in a way similar to that in Chapter XIX and the admissible strategy β_{2t} for the maximizer E is that of $\mathscr{Z}_{2t}$-measurable functions. Thus, the objective is to find such an admissible pair that it is a saddle point.

Obviously, we have $\mathscr{Z}_t \subset \mathscr{Z}_{2t}$ for all $t \in [0, t_f]$. If γ_t is $\mathscr{Z}_t$-measurable, then it is $\mathscr{Z}_{2t}$-measurable. From Lemmas 2.1 and 2.2 of Chapter XIX, in order to obtain the solution for this problem, we need the following lemmas.

Lemma 2.1. If γ_t and β_{2t} are $\mathscr{Z}_t$- and $\mathscr{Z}_{2t}$-measurable, respectively, then the objective function (O) can be rewritten as

$$\hat{J}(\gamma_t\,, \beta_{2t}) = E\left\{\int_0^{t_f} [\|\gamma_t + G_{pt}^t K_t \hat{x}_{2t}\|^2 - \|\beta_{2t} + G_{et}^t K_t \hat{x}_{2t}\|^2]\,dt\right\} + V, \tag{1}$$

where (i) K_t satisfies Eq. (K) with $F_t = 0$ in Chapter XIX; (ii) $\hat{x}_{2t} = E\{x_t \mid \mathscr{Z}_{2t}\,;\, \gamma_t\,, \beta_{2t}\}$ and may be obtained from Kalman–Bucy filter as follows:

$$d\hat{x}_{2t} = (G_{pt}\gamma_t - G_{et}\beta_{2t})\,dt + P_{2t}H_{2t}^t D_{2t}^{-1}(dz_{2t} - H_{2t}\hat{x}_{2t}\,dt), \tag{2}$$

with

$$H_{2t} = \begin{bmatrix} H_t \\ H_{1t}\end{bmatrix} \quad \text{and} \quad D_{2t} = \begin{bmatrix} D_t & 0 \\ 0 & D_{1t}\end{bmatrix} > 0;$$

(iii) $P_{2t} = E\{(x_t - \hat{x}_{2t})(x_t - \hat{x}_{2t})^t\} = E\{\tilde{x}_{2t}\tilde{x}_{2t}^t\}$, which is the solution to the following Riccati equation:

$$\begin{aligned} \dot{P}_{2t} &= -P_{2t}N(z_{2t})\,P_{2t} + R_{1t}R_{1t}^t\,, \qquad P_{20} = M_0\,, \\ N(z_{2t}) &= H_{2t}^t D_{2t}^{-1} H_{2t} = (N(z_t) + N(z_{1t})); \end{aligned} \tag{3}$$

and (iv)

$$V = \int_0^{t_f} [\mathrm{Tr}(K_t R_{1t} R_{1t}^1 + K_t(E_{pt} - E_{et})\,K_t P_{2t}]\,dt + \mathrm{Tr}\,K_0 M_0;$$

E_{pt} and E_{et} are defined as in Chapter XIX.

Proof. This result is the direct consequence of Lemmas 2.1 and 2.2 in Chapter XIX.

Note that V is the value of the game in which both players obtain the information z_{2t} from a broadcasting channel. Now, if we let $\tilde{z}_{2t} = z_{2t} - \hat{z}_{2t}$ and $d\hat{z}_{2t} = H_{2t}\hat{x}_{2t}\,dt$, then it is well known that $\tilde{z}_{2t}$ has increments which are independent and have zero mean values and the incremental covariance $D_{2t}\,dt$, and z_{2t} can be rewritten in terms of $\hat{x}_{2t}$ and $\tilde{z}_{2t}$ (see Chapter XIX). That is,

$$dz_t = d[I, 0]\,z_{2t} = [I, 0]\{H_{2t}\hat{x}_{2t}\,dt + d\tilde{z}_{2t}\} = H_t\hat{x}_{2t}\,dt + [I, 0]\,d\tilde{z}_{2t}\,. \tag{4}$$

Furthermore, let $\hat{x}_{2t}^0 = E\{x_t \mid \mathscr{Z}_{2t}\,;\,\gamma_t\,,\,\beta_{2t}^0\}$, $\beta_{2t}^0 = -G_{et}^t(K_t\hat{y}_t + C_t(\hat{x}_{2t}^0 - \hat{y}_t))$, and $\hat{y}_t = E\{\hat{x}_{2t}^0 \mid \mathscr{Z}_t\,;\,\gamma_t\,,\,\beta_{2t}^0\}$, where C_t is a nonrandom matrix with appropriate dimension.[2] Then we have the following lemma.

Lemma 2.2. $\hat{x}_{2t}^0$ and $\hat{y}_t$ are solutions to the following stochastic differential equations for any $\mathscr{Z}_t$-measurable function γ_t:

$$\begin{aligned} d\hat{x}_{2t}^0 &= E_{et}C_t\hat{x}_{2t}^0\,dt + E_{2t}(K_t - C_t)\,\hat{y}_t\,dt + G_{pt}\gamma_t \\ &\quad + P_{2t}H_{2t}^t D_{2t}^{-1}(dz_{2t} - H_{2t}\hat{x}_{2t}^0\,dt), \end{aligned} \tag{5}$$

$$d\hat{y}_t = E_{et}K_t\hat{y}_t + G_{pt}\gamma_t + (P_{yt} + P_{2t})\,H_t{}^t D_t^{-1}(dz_t - H_t\hat{y}_t\,dt), \tag{6}$$

where $\hat{y} = \bar{x}_0$, and $P_{yt} = E\{\hat{x}_{2t}^0 - \hat{y}_t)(\hat{x}_{2t}^0 - \hat{y}_t)^t\}$, which is the solution to the following Riccati equation:

$$\dot{P}_{yt} = E_{et}C_tP_{yt} + P_{yt}C_t{}^tE_{et} - (P_{yt} + P_{2t})\,N(z_t)(P_{yt} + P_{2t}) + P_{2t}N(z_{2t})\,P_{2t}\,. \tag{7}$$

---

[2] Since $\hat{y}_t$ is $\mathscr{Z}_t$-measurable, intuitively C_t can be taken as a weighting parameter for E from the fact that he knows more than P does by the information z_{1t}. Below, C_t will be clearly determined.

Proof. This result will be obtained immediately by applying the previous lemma, and looking at Eq. (5) as the state equation and Eq. (4) with $\hat{x}_{2t}$ replaced by $\hat{x}^0_{2t}$ and $\tilde{z}_{2t}$ replaced by $\tilde{z}^0_{2t}$ as the measurement equation, where $\tilde{z}^0_{2t} = z_{2t} - z^0_{2t}$ and $dz^0_{2t} = H_{2t}\hat{x}^0_{2t}\, dt$.

Intuitively, since player P cannot know $\hat{x}_{2t}$ precisely, he can only estimate player E's estimate by using his information z_t and assuming that E will use β^0_{2t}. Furthermore, we have the following lemma.

Lemma 2.3. If γ_t and β_{2t} are $\mathscr{Z}_t$- and $\mathscr{Z}_{2t}$-measurable, then

$$\begin{aligned}\hat{J}(\gamma_t, \beta_{2t}) &= \hat{J}(\gamma_t, \delta_t)\\ &= V + E\left\{\int_0^{t_f} \| \gamma_t + G_{pt}K_t\hat{y}_t \|^2 + 2(\gamma_t + G^t_{pt}K_t\hat{y}_t)^t G^t_{pt}K_t\tilde{x}_{2yt}\right.\\ &\quad \left. + \| G^t_{pt}K_t\tilde{x}_{2yt} \|^2 - \| \delta_t + G^t_{et}K_t\tilde{x}_{2yt} \|^2\, dt\right\}; \qquad (8)\end{aligned}$$

$\tilde{x}_{2yt}$ is the solution to the following stochastic differential equation:

$$d\tilde{x}_{2yt} = F_t{}'\tilde{x}_{2yt}\, dt - G_{et}\delta_t\, dt + Q_t\, d\tilde{z}_{2t}, \qquad \tilde{x}_{2yt} = 0, \qquad (9)$$

where (i) $F_t{}' = -(P_{yt} + P_{2t})\, N(z_t)$ and $Q_t = -[P_{2t}H^t_{2t}D^{-1}_{2t} - (P_{yt} + P_{2t})\, H_tD^{-1}_t[I, 0]]$, (ii) $\delta_t = \beta_t + G^t_{et}K_t\hat{y}_t$, and (iii) $\tilde{x}_{2yt} = \hat{x}_{2t} - \hat{y}_t$.

Proof. The result will be obtained by straightforward computation.

Intuitively, due to the fact that B does not have y_{1t}, $\tilde{x}_{2yt}$ is the difference of estimates of x_t obtained by P and E when P assumes that E will use β^0_{2t}. Consequently, the problem becomes that of finding a saddle point (γ_t, δ_t) for the system described by state equations (6) and (9) and performance index (8) under the situation that E has complete state information and P has only incomplete state information. Namely, E knows $\tilde{x}_{2yt}$ and $\hat{y}_t$, and P only knows $\hat{y}_t$. From the above lemmas, we have the following proposition.

Proposition 2.1. If (i) main (M1) problem has the saddle point as shown in Theorem 2.1 in Chapter XIX and (ii) the Riccati equation (7) and

$$\dot{S}_t = -F_t{}'S_t - S_tF'^t_t - K_t(E_{pt} - E_{et})K_t - (K_t + S_t)\, E_{et}(K_t + S_t), \qquad S_{t_f} = 0, \qquad (10)$$

have solutions, then the saddle point $(\gamma_t{}^0, \beta_t{}^0)$ for the present problem exists, and is given as

$$\gamma_t{}^0 = -G_{pt}K_t\hat{y}_t, \qquad \beta^0_{2t} = -G^t_{et}(K_t\hat{y}_t + C_t(\hat{x}^0_{2t} - \hat{y}_t)),$$

where (i) $\hat{y}_t$ satisfies Eq. (6), (ii) K_t is the solution to Eq. (K) in Chapter XIX with $F_t = 0$, (iii) $\hat{x}_{2t}^0 = E\{x_t \mid \mathscr{Z}_{2t}; \gamma_t^0, \beta_{2t}^0\}$, and is the solution to Eq. (5), and (iv) $C_t = S_t + K_t$, and S_t is the solution to Eq. (10). Moreover, the value of the game is

$$\hat{J}(\gamma_t^0, \beta_{2t}^0) = \int_0^{t_f} \mathrm{Tr}[K_t R_{1t} R_{1t}^t + K_t(E_{pt} - E_{et}) K_t P_{2t} + S_t Q_t D_{2t} Q_t^t] + T_\gamma K_0 M_0 .$$

Proof. Applying the same technique as in Chapter XIX and using the fact that $\hat{x}_{2t}^0 - \hat{y}_t$ is an innovation process which is independent of all $\mathscr{Z}_t$-measurable functions, we can therefore complete the proof.

However, it should be noticed that since the Riccati equations (7), (10), and (K) (Chapter XIX) are all nonlinearly coupled together, the necessary and sufficient condition for the existence of saddle point for this problem might not be easily obtained.

XXI

An Extension of Radner's Theorem[1]

W. H. Hartman

Abstract. Radner's theorem states that the optimal solution for a static linear quadratic Gaussian (LQG) team is linear. In this paper, we find the optimal solution for a static LQG team in which each player knows which observations he has, but in which the observation set that a player receives (how many, which measuring device, *etc.*) is random before the team acts. Via the concept of nesting, the result extends to the dynamic case and includes teams in which the order of play of the team members as well as their sets of data are random. We also include random changes in the cost function which depend on the randomness of the observation system but are independent of the stochastic process that the team is observing and controlling.

1. Introduction

We wish to pose and solve a more general class of team problems than has been treated to date.[2] We will motivate our mathematical formulation by the following example.

Consider a military reconnaissance operation in mountainous and/or

[1] The research reported in this paper was supported by the US Office of Naval Research under the Joint Services Electronics Program, Contract No. N00014-67-A-0298-0006, and by the National Science Foundation, Grant No. GK-31511. The author would like to thank Professor Y. C. Ho of Harvard University for suggesting this problem and for many helpful conversations during the course of this work.

[2] Note that the classical information pattern specialization of the following has been treated by different methods by Bellman and Kalaba (Ref. 1) and Eaton (Ref. 2).

jungle terrain. To minimize game theoretic aspects, we assume that the territory is neutral, unknown or poorly known to both sides, and that the main enemy influence is via random air surveillance and attack. The object of the team is to locate certain features, such as a cave suitable for a supply depot or a path through the terrain suitable for armored equipment. In order to search efficiently, the team will be broken into several small groups. To avoid calling down enemy fire upon themselves, radio communication between these groups will be severely limited, so that they will have to do their decision making in a decentralized fashion.

The team will have prior knowledge of the local features that they are trying to identify via old surveys or satellite photographs, which will presumably be on too large a scale to eliminate uncertainty concerning these features. The team input, their limited resources of personnel and transportation, will be used to reduce this uncertainty. However, there is an important second source of randomness in the problem: the quantity of information that the team produces and the time at which it becomes available will both be random. The time that it takes a group of men to reach a certain objective will have delays due to unforeseeable events such as swampy ground, rock slides, or stormy weather. Time will also be spent taking cover from enemy planes. Failure to do so may necessitate sending in another group of men and/or increasing the risk of discovery for the whole team. The point which we wish to make is that the two types of randomness that we have described are independent. Delays due to weather, for example, have nothing to do with the military value of the local geography.

Another important point is that the order in which decisions will be made cannot be known in advance. Two groups of men may be searching for the same type of object and we do not know whether and in what order both will succeed or what decisions they will make concerning committing men and material to guard or develop their finds. If one group of men has been given two objectives, we do not know which they will be able to accomplish first.

Finally, the second sort of randomness may affect the team's payoff, as when too many delays force the exploration to be abandoned or too little or too much enemy activity in the area, as reflected in team casualties and delays to avoid enemy surveillance, change the military value of the team's activities. Note that the information structure, as well as the payoff, must respond to these random events. The useless risk of sending a message to a group that has been wiped out should be avoided when possible, and the chain of information flow should not be broken by such accidents.

The model below assumes a fixed period of operation for the team, but this could be interpreted as a sure upper limit on a random stopping time that might, for example, reflect uncertainty in the overall time schedule that is not strongly related to the team problem we wish to isolate.

In the next section, we abstract the above considerations into a mathematical problem based on Marschak's concept of the team (Ref. 3) assuming linear observations, quadratic cost, and Gaussian distribution of the process being observed and controlled. Given the independence of this process and the randomness in the information-gathering system, we would expect the optimal control to be linear in the observation values received, as in the deterministic observation case first treated by Radner (Ref. 4). This is shown for the static team in the third section. The proof is parallel to the proof of Radner (Ref. 4) for the theorem mentioned in the abstract. The fourth section considers the dynamic case using the concept of nesting[3] of Ho and Chu (Ref. 5) and Chu (Ref. 6). The final section briefly considers some implementation difficulties.

2. Problem Formulation

We consider a static team of n decision makers, the ith decision maker, abbreviated DMi, controlling the scalar variable u_i. The state of the world is described by two random vectors: $\xi = (\xi_1, \dots, \xi_m)^t$, which is Gaussian with mean zero and covariance matrix $V > 0$; and $\delta = (\delta_1, \dots, \delta_{n+l})^t$, which is a discrete random vector independent of ξ. The cost for a given set of actions u and state of the world ξ and δ is

$$c(\xi, \delta, u) = \tfrac{1}{2} u^t Q(\delta)\, u + u^t\{S(\delta)\, \xi + r(\delta)\}, \tag{1}$$

with $Q(\delta) = Q(\delta)^t > 0$ for all possible values of δ.

The first n components of δ are the random variables determining which data set the decision makers see. In particular, δ_i can take on values $1, 2, \dots, k_i$ and DMi sees δ_i and $A^i(\delta_i)\xi$, with $A^i(1), \dots, A^i(k_i)$ all known to DMi. To avoid degeneracy, we assume that $A^i(\delta_i)$ has full rank, hence that its number of rows may vary, and that $\text{prob}(\delta_i = j) > 0$ for all $i, j \in \{1, 2, \dots, k_i\}$.

In the next section we will prove the following theorem.

[3] In a nested information structure, if DM1's action affects DM2's data, then DM2 knows all the data that determined DM1's action. The symbol DM stands for decision maker.

Theorem 2.1. The optimal strategies $u_i = \gamma_i(\delta_i, A^i(\delta_i)\xi)$ are of the form

$$u_i = \sigma^i(\delta_i)^t A^i(\delta_i) \xi + \tau_i(\delta_i). \tag{2}$$

We might call Eq. (2) a quasi-affine strategy, since, for a given value of δ_i, it is an affine function of the real-valued random variables that DMi observes. Since the cost is convex in the control vector u and the set of admissible strategies is convex, the following stationarity condition [Eq. (3)] is sufficient as well as necessary for optimality:

$$(\forall i) \quad 0 = E\left[\sum_{j=1}^{n} Q_{ij}(\delta)\, \gamma_j(\delta_j, A^j(\delta_j)\xi) + \sum_{j=1}^{m} S_{ij}(\delta)\, \xi_j + r_i(\delta) \,\middle|\, \delta_i, A^i(\delta_i)\xi\right]. \tag{3}$$

For the class of strategies described in Theorem 2.1, the above becomes two linear equations; and, paralleling Radner, we will show that these are solvable by showing that the appropriate matrices are positive definite.

3. Proof of Theorem 2.1

We will need the following three lemmas.

Lemma 3.1. $E[f(\delta) \mid \delta_i, A^i(\delta_i)\xi] = E[f(\delta) \mid \delta_i]$.

Proof. Let g be any function such that $g(A^i(\delta_i)\xi)$ is integrable. Then,

$$\begin{aligned} E[f(\delta)\, g(A^i(\delta_i)\xi) \mid \delta_i] &= E[E[f(\delta)\, g(A^i(\delta_i)\xi) \mid \delta_i, \xi] \mid \delta_i] \\ &= E[g(A^i(\delta_i)\xi)\, E[f(\delta) \mid \delta_i, \xi] \mid \delta_i] \\ &= E[g(A^i(\delta_i)\xi)\, E[f(\delta) \mid \delta_i] \mid \delta_i] \\ &= E[f(\delta) \mid \delta_i]\, E[g(A^i(\delta_i)\xi) \mid \delta_i], \end{aligned}$$

where we have used the independence of δ and ξ to obtain the third equality. The appropriate extension of Loève's Theorem A of Ref. 7, Section 25.3, gives the desired result.

Similar use of the properties of conditioning with respect to nested σ-fields yields the following lemma.

Lemma 3.2.

$$E[f(\delta)\, A^j(\delta_j)\xi \mid \delta_i, A^i(\delta_i)\xi] = E[f(\delta)\, A^j(\delta_j) \mid \delta_i]\, E[\xi \mid \delta_i, A^i(\delta_i)\xi].$$

Finally, by appealing to the fundamental definition of a conditional expectation we have, recalling $E[\xi\xi^t] = V$, the third lemma.

Lemma 3.3.

$$E[\xi \mid \delta_i, A^i(\delta_i)\xi] = VA^i(\delta_i)^t \{A^i(\delta_i) VA^i(\delta_i)^t\}^{-1} A^i(\delta_i)\xi.$$

The stationarity condition can now be combined with the proposed solution form to give

$$0 = \sum_{j=1}^{n} E[Q_{ij}(\delta) \sigma^j(\delta_j)^t A^j(\delta_j) | \delta_i] E[\xi \mid \delta_i, A^i(\delta_i)\xi]$$
$$+ \sum_{j=1}^{m} E[S_{ij}(\delta) | \delta_i] E[\xi_j \mid \delta_i, A^i(\delta_i)\xi]$$
$$+ \sum_{j=1}^{n} E[Q_{ij}(\delta) \tau_j(\delta_j) | \delta_i] + E[r_i(\delta) | \delta_i], \qquad i = 1, 2,..., n. \tag{4}$$

From Lemma 3.3, we see that, for each i, we have an expression of the form

$$(\alpha^i)^t A^i(\delta_i) \xi + \beta_i = 0,$$

which, since $A^i(\delta_i)$ and $V = E[\xi\xi^t]$ are of full rank, can hold iff

$$\alpha^i = 0, \qquad \beta_i = 0.$$

Hence, we have two sets of equations, one involving only the $\sigma^i(\cdot)$, the other involving only the $\tau_i(\cdot)$.

Since δ is a discrete random variable, we can write the set of equations for the $\tau_i(\cdot)$ as

$$0 = \sum_{j=1}^{n} \sum_{\delta=d\in\Delta} Q_{ij}(d) \tau_j(d_j) \operatorname{prob}(\delta = d \mid \delta_i = \bar{\imath})$$
$$+ \sum_{\delta=d\in\Delta} r_i(d) \operatorname{prob}(\delta = d \mid \delta_i = \bar{\imath}), \qquad i = 1, 2,..., n, \bar{\imath} = 1, 2,..., k_i. \tag{5}$$

The range set Δ of the random vector δ is taken to be the cross-product of the range sets for the components of δ. Using, for all i, $\bar{\imath}$, $\operatorname{prob}(\delta_i = \bar{\imath}) > 0$, we can multiply Eq. (5) by this quantity to obtain an equivalent set of equations

$$0 = \sum_{j=1}^{n} \sum_{\delta=d\in\Delta} Q_{ij}(d) \tau_j(d_j) \operatorname{prob}(\delta = d, \delta_i = \bar{\imath})$$
$$+ \sum_{\delta=d\in\Delta} r_i(d) \operatorname{prob}(\delta = d, \delta_i = \bar{\imath}), \qquad i = 1, 2,..., n, \bar{\imath} = 1, 2,..., k_i. \tag{6}$$

We would like to regard the $\tau_i(\bar{\imath})$ as a single vector of unknowns, say

$$\tau = (\tau_1(1), \tau_1(2),..., \tau_1(k_1), \tau_2(1),..., \tau_n(k_n))^t,$$

to be found by solving an equation

$$0 = \tau^t H + \alpha^t. \tag{7}$$

We do this by noting that

$$\tau_j(d_j)\ \mathrm{prob}(\delta = d, \delta_i = \bar{\imath}) = \sum_{\bar{\jmath}=1}^{k_j} \tau_j(\bar{\jmath})\ \mathrm{prob}(\delta = d, \delta_i = \bar{\imath}, \delta_j = \bar{\jmath}),$$

which yields

$$0 = \sum_{j=1}^{n} \sum_{\bar{\jmath}=1}^{k_j} \tau_j(\bar{\jmath}) \sum_{\delta=d\in\Delta} Q_{ij}(d)\ \mathrm{prob}(\delta = d, \delta_i = \bar{\imath}, \delta_j = \bar{\jmath})$$

$$+ \sum_{\delta=d\in\Delta} r_i(d)\ \mathrm{prob}(\delta = d, \delta_i = \bar{\imath}), \qquad i = 1, 2,..., n, \bar{\imath} = 1, 2,..., k_i. \tag{8}$$

We can now make the identification

$$H_{j\bar{\jmath},i\bar{\imath}} = \sum_{\delta=d\in\Delta} Q_{ij}(d)\ \mathrm{prob}(\delta = d, \delta_i = \bar{\imath}, \delta_j = \bar{\jmath}),$$

$$\alpha_{i\bar{\imath}} = \sum_{\delta=d\in\Delta} r_i(d)\ \mathrm{prob}(\delta = d, \delta_i = \bar{\imath}),$$

where the double index $i\bar{\imath}$ refers to the $\bar{\imath}$th possible observation set of DMi. Note that H is obviously symmetric.

$$\tau^t H \tau = \sum_{i=1}^{n} \sum_{\bar{\imath}=1}^{k_i} \sum_{j=1}^{n} \sum_{\bar{\jmath}=1}^{k_j} \tau_j(\bar{\jmath})$$

$$* \left\{ \sum_{\delta=d\in\Delta} Q_{ij}(d)\ \mathrm{prob}(\delta = d, \delta_i = \bar{\imath}, \delta_j = \bar{\jmath}) \right\} \tau_i(\bar{\imath})$$

$$= \sum_{\delta=d\in\Delta} \mathrm{prob}(\delta = d) \sum_{j=1}^{n} \sum_{i=1}^{n} \tau_j(d)\, Q_{ij}(d)\, \tau_i(d),$$

where, in the last expression, we regard

$$\tau(d) = (\tau_1(d),..., \tau_n(d))^t$$

as an n-vector-valued function of the random vector δ. Since $Q(d) > 0$ and $\mathrm{prob}(\delta = d) \geqslant 0$, we have that $H \geqslant 0$ and $\tau^t H \tau = 0$ iff $\mathrm{prob}(\delta = d) > 0$ implies that $\tau(d) = 0$. But, for any i, $\bar{\imath}$, $\mathrm{prob}(\delta_i = \bar{\imath}) > 0$, so that we can find a vector d with $d_i = \bar{\imath}$ and $\mathrm{prob}(\delta = d) > 0$. Thus, $\tau^t H \tau = 0$ iff $\tau_i(\bar{\imath}) = 0$ $i = 1, 2,..., n$, $\bar{\imath} = 1, 2,..., k_i$, giving $H > 0$, and Eq. (8) solvable.

Except for a proliferation of subscripts, we can similarly show that the equations involving the $\sigma^i(\bar{\imath})$ can be put in the form $0 = \sigma^t K + \beta^t$, with $K = K^t > 0$.

Note that the above would also hold if we gave each DMi an additional observation $a^i(\delta)$ as well as δ_i and $A^i(\delta_i)\xi$. This might correspond to acquiring information about the time or probability of arrival of some future observation or about the effect of δ on the quantities $Q(\delta)$, $S(\delta)$, and $r(\delta)$ in the cost function.

4. Dynamic Teams

In a dynamic version of the static problem of Section 2, DMi would receive the observations δ_i, $a_i(\delta)$, $A^i(\delta_i)\xi + B^i(\delta)u$. Note that DM$i$ might not know more about the matrix $B^i(\delta)$ than its number of rows. That is, DMi knows, at the time when he receives his observations, how many he has and what measuring system they come from; but he might not know who has acted before him [i.e., what information sets appear in the controls appearing in $B^i(\delta)u$]. We assume, of course, that the vector δ and the set of matrices $B^i(\delta)$ are causal: for any particular value of δ, if DMi is a precedent of DMj, then DMj is not a precedent of DMi (see Ref. 5 for a definition of precedent). A discussion of causality in control and game theoretic formulations is given in Witsenhausen (Ref. 8).

The above information structure is equivalent to a static structure when $B^i(\delta)$ depends only on δ_i and δ_i, $a^i(\delta)$, $A^i(\delta_i)\xi + B^i(\delta_i)u$ contain enough information to allow DMi to compute all values u_j appearing in $B^i(\delta_i)u$ and also to compute all the observation values used by DMj to determine u_j. The only technical difficulty in constructing a nested information structure is ensuring causality.

Note that to achieve nesting, given random order of play, may require an increase, relative to the case of deterministic order of play, in the communication cost of the team, and/or a decrease in decentralization. In the deterministic order case, we can ensure nesting if each DMi sends one message, in general a vector quantity, to each DMj of whom DMi is a direct predecessor (u_i appears in DMj's observation values, written DM$i <_{\partial p}$ DMj) but such that there is no DMk with DM$i <_{\partial p}$ DM$k <_{\partial p}$ DMj. In the random order case, a DM does not know of whom he will be the direct predecessor, hence to whom to send his data. If communication costs are low, he could simply send the data to all DMs whose action might follow his. An alternative would be to establish a number of data banks; when a DM's turn to play comes he would simply tap the appropriate bank to find out whom his predecessors were, what they knew, and hence what they did. Even if this cuts communication costs there would be additional equipment costs as well as a possibly undesirable increase in centralization.

5. Implementation

The computational requirements for implementing the optimal solution are qualitatively the same as for the certain observation set case. Instead of computing an affine function of an observation vector, we now must choose one among a finite set of affine functions which we then compute. The logic to do the choosing should be negligible compared to that which computes the affine function; the difficulty lies in possibly having too many such functions to store. If DMi might receive any subset of a set of ten observations, he would need to store 1,024 vectors of constants.

In practice, things might not be nearly this bad: if, for example, the ten observations came from three different DMs, each sending all or none of their data, there would be eight vectors to store. Another possibility might be that each of DMi's observation is of, e.g., ξ_{13}, say $\xi_{13} + \xi_{13+j}$, $j = 1, 2,..., 10$, with $\{\xi_j\}_{j=13}^{23}$ independent and $\{\xi_j\}_{j=14}^{23}$ having the same variance. Then, DMi's control might depend only on how many observations he receives, giving 11 vectors. If the number of vectors to be stored is still too large, the above two examples might be used to generate suboptimal controls based on similar groupings of observations.

References

1. Bellman, R., and Kalaba, R., *A Note on Interrupted Stochastic Control Processes*, Information and Control, Vol. 4, No. 4, 1961.
2. Eaton, J. H., *Discrete-Time Interrupted Stochastic Control Processes*, Journal of Mathematical Analysis and Applications, Vol. 5, No. 2, 1962.
3. Marschak, J., *Elements for a Theory of Teams*, Management Science, Vol. 1, No. 2, 1955.
4. Radner, R., *Team Decision Problems*, Annals of Mathematical Statistics, Vol. 33, No. 3, 1962.
5. Ho, Y. C., and Chu, K. C., *Team Decision Theory and Information Structures in Optimal Control Problems, Part 1*, IEEE Transactions on Automatic Control, Vol. AC-17, No. 1, 1972.

6. Chu, K. C., *Team Decision Theory and Information Structures in Optimal Control Problems, Part 2*, IEEE Transactions on Automatic Control, Vol. AC-17, No. 1, 1972.
7. Loeve, M., *Probability Theory*, Van Nostrand Reinhold Company, New York, New York, 1963.
8. Witsenhausen, H., *On Information Structures, Feedback, and Causality*, SIAM Journal on Control, Vol. 9, No. 2, 1971.

XXII

Linear Differential Games with Delayed and Noisy Information

K. MORI AND E. SHIMEMURA

Abstract. This paper deals with a linear–quadratic–Gaussian zero-sum game in which one player has delayed and noisy information and the other has perfect information. Assuming that the player with perfect information can deduce his opponent's state estimate, the optimal closed-loop control laws are derived. Then, it is shown that the separation theorem is satisfied for the player with imperfect information and his optimal state estimate is given by a delay-differential equation.

1. Introduction

In control and game theories, the information structure plays an important role in determining the optimal decisions. Recently, the interplay of information and decision has been generally discussed in Refs. 1–2. Analytical discussions have been made for linear–quadratic–Gaussian zero-sum games (Refs. 3–6). In Refs. 3–4, it has been shown that the separation theorem is satisfied for the game in which one player's measurement is additively corrupted by white Gaussian noise and the other player has perfect information or no measurement. On the other hand, the game in which both players have noisy measurements (Refs. 5–6) cannot be easily solved but it possesses a certainty-coincidence property.

In dynamic games of large-complex systems, an important type of nonclassical information pattern is the delayed information pattern, which does not appear in static games. In physical situations, the information is usually obtained with time delay because of transmitting and processing the data and computing the control. In Ref. 7, a min-max solution to the game with delayed measurement is derived under the condition that one player with delayed information selects a control first and informs his opponent with perfect information of his control.

The game considered here is a linear–quadratic–Gaussian zero-sum game in which Player I has delayed measurement additively corrupted by white Gaussian noise and Player II has perfect information. Assuming that Player II can derive the error of Player I's state estimate, it will be shown that the optimal closed-loop control laws are obtained and the separation theorem for Player I's optimal closed-loop control law is satisfied.

2. Statement of the Problem

We consider a special but interesting problem such that Player I has delayed and noisy measurement and Player II has perfect information. The system is described by the equations

$$\dot{x}(t) = A(t)x(t) + B(t)u(t) + C(t)v(t), \qquad t_0 \leqslant t \leqslant t_f, \tag{1}$$

$$x(t_0) = x_0, \tag{2}$$

and the observation data available at time t are given by

$$y_I(t) = 0, \qquad t_0 \leqslant t \leqslant t_0 + \theta, \tag{3}$$

$$\dot{y}_I(t) = H_I(t-\theta)x(t-\theta) + W_I(t-\theta)\dot{w}_I(t-\theta), \qquad t_0 + \theta \leqslant t \leqslant t_f, \tag{4}$$

$$y_{II}(t) = x(t), \qquad t_0 \leqslant t \leqslant t_f. \tag{5}$$

Here, the n-vector $x(t)$ is the system state; $u(t) \in R^q$ and $v(t) \in R^r$ are the controls of Player I and Player II; $y_I(t) \in R^m$ and $y_{II}(t) \in R^n$ are the observations of Player I and Player II; $w_I(t) \in R^p$ is a standard Wiener process with

$$E[(w_I(t) - w_I(s))(w_I(t) - w_I(s))'] = |t - s| I$$

and (4) is interpreted as the meaning of

$$y_I(t) = \int_{t_0+\theta}^{t} H_I(\tau-\theta)x(\tau-\theta)\, d\tau + \int_{t_0+\theta}^{t} W_I(\tau-\theta)\, dw_I(\tau-\theta); \tag{6}$$

θ is the information delay; $A(t)$, $B(t)$, and $C(t)$ are $n \times n$, $n \times q$, and $n \times r$ piecewise-continuous matrices; $H_I(t)$ and $W_I(t)$ are $m \times n$ and

$m \times p$ differentiable matrices and $W_I(t)\, W_I'(t)$ is positive definite. It is assumed that Player I knows the initial state to be Gaussian with mean and covariance given by

$$E[x(t_0)] = \bar{x}_0\,, \qquad \operatorname{cov}[x(t_0), x(t_0)] = M_0\,;$$

it is also assumed that Player II knows these statistics of the initial state and that $x(t_0)$ is independent of the increment $w_I(t) - w_I(t_0)$, $t_0 < t \leqslant t_f - \theta$. All random processes are defined on a measurable space $(\Omega, \mathscr{B})$. The cost, which Player I wishes to minimize and Player II wishes to maximize, is given by

$$J_{t_0}(u, v) = x'(t_f)P'\,Px(t_f) + \int_{t_0}^{t_f} [u'(\tau)\,Q_I(\tau)\,u(\tau) - v'(\tau)\,Q_{II}(\tau)\,v(\tau)]\,d\tau, \tag{7}$$

where $Q_I(t)$ and $Q_{II}(t)$ are symmetric and positive-definite piecewise-continuous matrices; P is an $s \times n$ matrix.

Let $C_m[t_0\,, t_f]$ and $C_n[t_0\,, t_f]$ denote the classes of continuous functions defined on $[t_0\,, t_f]$ with values in R^m and R^n, respectively; and let the past accumulative data be given by

$$(\pi_t y_I)(s) = \begin{cases} y_I(s), & t_0 \leqslant s \leqslant t, \\ y_I(t), & t \leqslant s \leqslant t_f\,, \end{cases} \tag{8}$$

$$(\pi_t y_{II})(s) = \begin{cases} y_{II}(s), & t_0 \leqslant s \leqslant t, \\ y_{II}(t), & t \leqslant s \leqslant t_f\,. \end{cases} \tag{9}$$

Since $y_I(t) \in C_m[t_0\,, t_f]$ and $y_{II}(t) \in C_n[t_0\,, t_f]$, it follows that $\pi_t y_I \in C_m[t_0\,, t_f]$ and $\pi_t y_{II} \in C_n[t_0\,, t_f]$. We denote by $\mathscr{Y}_{It} \subset \mathscr{B}$ the minimal σ-algebra induced by the observation $\{\pi_t y_I\}$.

It is assumed that Player II can deduce Player I's state estimate $\hat{x}(t \mid t)$, that is,

$$\hat{x}(t \mid t) = E(x(t) \mid \mathscr{Y}_{It}). \tag{10}$$

Let $\mathscr{Y}_{IIt} \subset \mathscr{B}$ be the minimal σ-algebra induced by $\{\pi_t y_{II}\,, \hat{x}(t \mid t)\}$. Note that Player I has no observation in the interval $[t_0\,, t_0 + \theta)$ and then

$$\mathscr{Y}_{It} = \mathscr{Y}_{It_0}\,, \qquad t_0 \leqslant t \leqslant t_0 + \theta. \tag{11}$$

The admissible controls are defined as the closed-loop controls

$$u(t) = \varphi(t, \pi_t y_I), \tag{12}$$

$$v(t) = \psi(t, \pi_t y_{II}\,, \hat{x}(t \mid t)), \tag{13}$$

where the mappings

$$\varphi(\cdot, \cdot) : [t_0, t_f] \times C_m[t_0, t_f] \to R^q,$$

$$\psi(\cdot, \cdot, \cdot) : [t_0, t_f] \times C_n[t_0, t_f] \times R^n \to R^r$$

satisfy the Lipschitz conditions

$$| \varphi(t, \xi_1) - \varphi(t, \xi_2) | \leqslant K_1 \| \xi_1 - \xi_2 \|, \quad \xi_1, \xi_2 \in C_m[t_0, t_f], \quad t_0 \leqslant t \leqslant t_f, \tag{14}$$

$$| \psi(t, \zeta_1) - \psi(t, \zeta_2)| \leqslant K_2 \| \zeta_1 - \zeta_2 \|, \quad \zeta_1, \zeta_2 \in C_n[t_0, t_f] \times R^n, \quad t_0 \leqslant t \leqslant t_f. \tag{15}$$

Under these conditions, there exists a unique solution $(x(t), y_I(t))$ to (1)–(4) (see Ref. 8).

Our problem is to find the admissible controls (u^*, v^*) such that, for any admissible controls u and v,

$$E\{J_{t_0}(u^*, v^*)| \mathscr{Y}_{It_0}\} \leqslant E\{J_{t_0}(u, v^*)| \mathscr{Y}_{It_0}\}, \tag{16}$$

$$E\{J_{t_0}(u^*, v)| \mathscr{Y}_{IIt_0}\} \leqslant E\{J_{t_0}(u^*, v^*)| \mathscr{Y}_{IIt_0}\}. \tag{17}$$

These closed-loop controls u^* and v^* are said to be optimal. From (16) and (17), we get the saddle-point condition

$$E\{J_{t_0}(u^*, v)\} \leqslant E\{J_{t_0}(u^*, v^*)\} \leqslant E\{J_{t_0}(u, v^*)\} \tag{18}$$

for any admissible controls u and v.

3. Estimate of the State and Optimality Criterion

The zero-sum games have the property of equivalence and interchangeability of all the solutions satisfying the saddle-point condition (Ref. 3). This suggests the derivation of the optimal controls as follows. First, guess the admissible controls (u^*, v^*) and solve the optimal control problems such that

$$E\{J_{t_0}(u^{**}, v^*)| \mathscr{Y}_{It_0}\} \leqslant E\{J_{t_0}(u, v^*)| \mathscr{Y}_{It_0}\}, \tag{19}$$

$$E\{J_{t_0}(u^*, v)| \mathscr{Y}_{IIt_0}\} \leqslant E\{J_{t_0}(u^*, v^{**})| \mathscr{Y}_{IIt_0}\} \tag{20}$$

for any admissible controls (u, v). Second, find the condition such that the following equalities are satisfied simultaneously:

$$u^{**} = u^*, \tag{21}$$

$$v^{**} = v^*. \tag{22}$$

Then, u^* and v^* are optimal.

Let Φ and Ψ be the classes of functions

$$\Phi = \{\hat{\varphi} : [t_0, t_f] \times R^n \to R^q\}, \tag{23}$$

$$\Psi = \{\hat{\psi} : [t_0, t_f] \times R^n \times R^n \to R^r\}, \tag{24}$$

such that

$$| \hat{\varphi}(t, \lambda_1) - \hat{\varphi}(t, \lambda_2)| \leqslant K_3 | \lambda_1 - \lambda_2 |, \qquad \lambda_1, \lambda_2 \in R^n, \qquad t_0 \leqslant t \leqslant t_f, \tag{25}$$

$$| \hat{\psi}(t, \mu_1) - \hat{\psi}(t, \mu_2)| < K_4 | \mu_1 - \mu_2 |, \qquad \mu_1, \mu_2 \in R^{2n}, \qquad t_0 \leqslant t \leqslant t_f. \tag{26}$$

Now, we assume that the optimal controls are given in Φ and Ψ by

$$u^*(t) = W(t)\hat{x}(t \mid t), \tag{27}$$

$$v^*(t) = S(t)x(t) + T(t)\tilde{x}(t \mid t), \qquad t_0 \leqslant t \leqslant t_f, \tag{28}$$

where $\tilde{x}(t \mid t) = x(t) - \hat{x}(t \mid t)$.

When Player II's control is assumed to be given by (28), Player I's optimal state estimate $\hat{x}(t \mid t)$ is easily derived by a delay-differential equation

$$d\hat{x}(t \mid t)/dt = \{A(t) + C(t)\, S(t)\}\, \hat{x}(t \mid t) + B(t)\, \varphi(t, \pi_t y_I)$$
$$+ \kappa(t)\, \Phi(t, t - \theta)\, G(t)\{\dot{y}_I(t) - H_I(t - \theta)\, \hat{x}(t - \theta \mid t)\}, \tag{29}$$

$$\hat{x}(t_0 \mid t_0) = \bar{x}_0, \qquad t_0 \leqslant t \leqslant t_f, \tag{30}$$

$$\hat{x}(t - \theta \mid t) = \Phi(t - \theta, t)\, \hat{x}(t \mid t) - \int_{t-\theta}^{t} \Phi(t - \theta, \tau)\, \{B(\tau)\, \varphi(\tau, \pi_\tau y_I)$$
$$- C(\tau)\, T(\tau)\, \hat{x}(\tau \mid \tau)\}\, d\tau, \qquad t_0 + \theta \leqslant t \leqslant t_f, \tag{31}$$

where

$$\hat{x}(t - \theta \mid t) = E\{x(t - \theta) \mid \mathscr{Y}_{It}\},$$

$$G(t) = M(t - \theta \mid t)\, H_I'(t - \theta)(W_I(t - \theta)\, W_I'(t - \theta))^{-1},$$

and

$$\kappa(t) = \begin{cases} 0, & t_0 \leqslant t < t_0 + \theta, \\ 1, & t_0 + \theta \leqslant t \leqslant t_f; \end{cases} \tag{32}$$

$\Phi(t, \tau)$ is the transition matrix given by

$$\partial\Phi(t, \tau)/\partial t = [A(t) + C(t)\, \{S(t) + T(t)\}]\, \Phi(t, \tau), \tag{33}$$

$$\Phi(\tau, \tau) = I; \tag{34}$$

and the conditional covariance $M(t - \theta \mid t)$ on $[t_0 + \theta, t_f]$ is defined by

$$M(t - \theta \mid t) = E\{(x(t - \theta) - \hat{x}(t - \theta \mid t))(x(t - \theta) - \hat{x}(t - \theta \mid t))' \mid \mathscr{Y}_{It}\} \quad (35)$$

and it is the solution to the equation

$$\begin{aligned} dM(t - \theta \mid t)/dt \\ &= [A(t - \theta) + C(t - \theta)\{S(t - \theta) + T(t - \theta)\}]\, M(t - \theta \mid t) \\ &\quad + M(t - \theta \mid t)[A'(t - \theta) + \{S'(t - \theta) + T'(t - \theta)\}\, C'(t - \theta)] \\ &\quad - G(t)\, W_I(t - \theta)\, W_I'(t - \theta)\, G'(t), \end{aligned} \quad (36)$$

$$M(t_0 \mid t_0 + \theta) = M_0, \qquad t_0 + \theta \leqslant t \leqslant t_f. \quad (37)$$

Since $G(t)$ is differentiable, it is easily shown that the solution $\hat{x}(t \mid t)$ to (29)–(31) satisfies the Lipschitz condition in $\pi_t y_I$. This implies that the assumed controls (27) and (28) are admissible. In (29), the process $\dot{\nu}(t - \theta)$ given by

$$\dot{\nu}(t - \theta) = \dot{y}_I(t) - H_I(t - \theta)\, \hat{x}(t - \theta \mid t), \qquad t_0 + \theta \leqslant t \leqslant t_f, \quad (38)$$

is called an innovation process, that is, $\dot{\nu}(t)$ is a white Gaussian process

$$E\{\dot{\nu}(t)\} = 0, \quad \operatorname{cov}\{\dot{\nu}(t), \dot{\nu}(\tau)\} = W_I(t)\, W_I'(t)\, \delta(t - \tau), \quad t_0 + \theta \leqslant t \leqslant t_f. \quad (39)$$

From (36) and (37), the covariance matrix $M(t - \theta \mid t)$ is independent of Player I's data $\pi_t y_I$ and his control $u(t)$. These facts imply that $\mathscr{Y}_{It}$ and $\hat{x}(t \mid t)$ are equivalent statistics (Ref. 9). Thus, Player I's information is nested in Player II's information, that is,

$$\mathscr{Y}_{It} \subset \mathscr{Y}_{IIt}, \qquad t_0 \leqslant t \leqslant t_f, \quad (40)$$

so that, for any $\mathscr{Y}_{It}$-measurable function $h(t)$,

$$E\{h(t) \mid \mathscr{Y}_{IIt}\} = h(t); \quad (41)$$

and, from (16) and (17), we have the saddle-point condition

$$E\{J_t(u^*, v) \mid \mathscr{Y}_{It}\} \leqslant E\{J_t(u^*, v^*) \mid \mathscr{Y}_{It}\} \leqslant E\{J_t(u, v^*) \mid \mathscr{Y}_{It}\} \quad (42)$$

for any $t \in [t_0, t_f]$ and any admissible controls (u, v). Then, the closure problem (Ref. 3) is not raised, and Player I can derive $E\{x(t) \mid \mathscr{Y}_{It}\}$ based on only an estimator (29) which depends on the data $y_I(t)$ and the history of the state estimate $\hat{x}(\tau \mid \tau)$, $\tau \in (t - \theta, t)$.

Player I's state estimate $\hat{x}(t \mid t)$ given by (29)–(31) is derived on the assumption that Player II uses the optimal control (28). Suppose that Player II does not use (28); $\hat{x}(t \mid t)$ obtained by (29)–(31) is different from (10). Then, the error of $\hat{x}(t \mid t)$ given by (29)–(31) is derived by a couple of equations, namely,

$$\begin{aligned} d\tilde{x}(t \mid t)/dt &= A(t)\, \tilde{x}(t \mid t) + C(t)\, [\psi(t, \pi_t y_{II}, \hat{x}(t \mid t)) - S(t)\, \{x(t) - \tilde{x}(t \mid t)\}] \\ &\quad - \kappa(t)\Phi(t, t-\theta)G(t)\{H_I(t-\theta)\tilde{x}(t-\theta \mid t) + W_I(t-\theta)\dot{w}_I(t-\theta)\}, \end{aligned} \tag{43}$$

$$\tilde{x}(t_0 \mid t_0) = x_0 - \bar{x}_0, \qquad t_0 \leqslant t \leqslant t_f, \tag{44}$$

$$\begin{aligned} d\tilde{x}(t-\theta \mid t)/dt &= A(t-\theta)\tilde{x}(t-\theta \mid t) + C(t-\theta)[\psi(t-\theta, \pi_{t-\theta} y_{II}, \hat{x}(t-\theta \mid t-\theta)) \\ &\quad - \{S(t-\theta) + T(t-\theta)\}\{x(t-\theta) - \tilde{x}(t-\theta \mid t)\} + T(t-\theta)\{x(t-\theta) \\ &\quad - \tilde{x}(t-\theta \mid t-\theta)\}] - G(t)\{H_I(t-\theta)\tilde{x}(t-\theta \mid t) + W_I(t-\theta)\dot{w}_I(t-\theta)\}, \end{aligned} \tag{45}$$

$$\tilde{x}(t_0 \mid t_0 + \theta) = x_0 - \bar{x}_0, \qquad t_0 + \theta \leqslant t \leqslant t_f, \tag{46}$$

where

$$\tilde{x}(t-\theta \mid t) = x(t-\theta) - \hat{x}(t-\theta \mid t).$$

Consider two optimal control problems such that Player I and II select their controls satisfying (19) and (20), respectively, where their controls are restricted to be in the classes Φ and Ψ, that is,

$$u(t) = \hat{\varphi}(t, \hat{x}(t \mid t)), \tag{47}$$

$$v(t) = \hat{\psi}(t, x(t), \tilde{x}(t \mid t)). \tag{48}$$

Here, each player chooses his optimal control under the assumption that his opponent is using the optimal control. Hence, the estimate $\hat{x}(t \mid t)$ which the players use in their closed-loop controls (47) and (48) is the solution to (29)–(31). Let the functionals $V_I(t, \hat{x}(t \mid t))$ and $V_{II}(t, x(t), \tilde{x}(t \mid t))$ be defined by

$$\begin{aligned} V_I(t, \hat{x}(t \mid t)) &= \min_{\hat{\varphi} \in \Phi} E[x'(t_f)\, P'Px(t_f) \\ &\quad + \int_t^{t_f} \hat{\varphi}'(\tau, \hat{x}(\tau \mid \tau))\, Q_I(\tau)\, \hat{\varphi}(\tau, \hat{x}(\tau \mid \tau)) - \{S(\tau)\, x(\tau) + T(\tau)\, \tilde{x}(\tau \mid \tau)\}' \\ &\quad \cdot Q_{II}(\tau)\, \{S(\tau)x(\tau) + T(\tau)\tilde{x}(\tau \mid \tau)\}\, d\tau \mid \mathscr{Y}_{It}], \qquad t_0 \leqslant t \leqslant t_f, \end{aligned} \tag{49}$$

$$\begin{aligned} V_{II}(t, x(t), \tilde{x}(t \mid t)) &= \max_{\hat{\psi} \in \Psi} E[x'(t_f)\, P'Px(t_f) \\ &\quad + \int_t^{t_f} \{\hat{x}'(\tau \mid \tau)\, W'(\tau)\, Q_I(\tau)\, W(\tau)\, \hat{x}(\tau \mid \tau) - \hat{\psi}'(\tau, x(\tau), \tilde{x}(\tau \mid \tau)) \\ &\quad \cdot Q_{II}(\tau)\, \hat{\psi}(\tau, x(\tau), \tilde{x}(\tau \mid \tau))\}\, d\tau \mid \mathscr{Y}_{IIt}], \qquad t_0 \leqslant t \leqslant t_f. \end{aligned} \tag{50}$$

By the principle of optimality and Ito–Dynkin's formula (Ref. 10), the functional equations are obtained by

$$\min_{\phi\in\Phi} [\hat{\varphi}'(t, \hat{x}(t \mid t)) Q_I(t) \hat{\varphi}(t, \hat{x}(t \mid t)) - \hat{L}(t, \hat{x}(t \mid t)) + V_{It}(t, \hat{x}(t \mid t)) + \mathscr{L}_{\hat{\varphi}} V_I(t, \hat{x}(t \mid t))] = 0, \qquad t_0 \leqslant t \leqslant t_f, \tag{51}$$

$$\max_{\hat{\psi}\in\Psi} [(x(t) - \tilde{x}(t \mid t))' W'(t) Q_I(t) W(t) (x(t) - \tilde{x}(t \mid t)) - \hat{\psi}'(t, x(t), \tilde{x}(t \mid t)) Q_{II}(t) \hat{\psi}(t, x(t), \tilde{x}(t \mid t)) + V_{IIt}(t, x(t), \tilde{x}(t \mid t)) + \mathscr{L}_{\hat{\psi}} V_{II}(t, x(t), \tilde{x}(t \mid t))] = 0, \qquad t_0 \leqslant t \leqslant t_f. \tag{52}$$

Here, V_{it}, $i = I, II$, denotes $\partial V_i/\partial t$; also,

$$\hat{L}(t, \hat{x}(t \mid t)) = \hat{x}'(t \mid t) S'(t) Q_{II}(t) S(t) \hat{x}(t \mid t) + \operatorname{tr}[M(t \mid t)\{S(t) + T(t)\}' Q_{II}(t) \{S(t) + T(t)\}], \tag{53}$$

$$\mathscr{L}_{\hat{\varphi}} V_I(t, \hat{x}(t \mid t)) = \{A(t) \hat{x}(t \mid t) + B(t) \hat{\varphi}(t, \hat{x}(t \mid t)) + C(t) S(t) \hat{x}(t \mid t)\}' \cdot [\partial V_I(t, \hat{x}(t \mid t))/\partial \hat{x}(t \mid t)] + \tfrac{1}{2} \kappa(t) \operatorname{tr}\{[\partial^2 V_I(t, \hat{x}(t \mid t))/\partial \hat{x}^2(t \mid t)]\, \Phi(t, t - \theta) \cdot G(t) W_I(t - \theta) W_I'(t - \theta) G'(t) \Phi'(t, t - \theta)\}, \tag{54}$$

$$\mathscr{L}_{\hat{\psi}} V_{II}(t, x(t), \tilde{x}(t \mid t)) = \{A(t) x(t) + B(t) W(t)(x(t) - \tilde{x}(t \mid t)) + C(t) \hat{\psi}(t, x(t), \tilde{x}(t \mid t))\}' [\partial V_{II}(t, x(t), \tilde{x}(t \mid t))/\partial x(t)] + \{A(t) \tilde{x}(t \mid t) + C(t) \hat{\psi}(t, x(t), \tilde{x}(t \mid t)) - C(t) S(t) (x(t) - \tilde{x}(t \mid t)) - \kappa(t) \Phi(t, t - \theta) G(t) H_I(t - \theta) \tilde{x}(t - \theta \mid t)\}' \cdot [\partial V_{II}(t, x(t), \tilde{x}(t \mid t))/\partial \tilde{x}(t \mid t)] + \tfrac{1}{2} \kappa(t) \operatorname{tr}\{[\partial^2 V_{II}(t, x(t), \tilde{x}(t \mid t))/\partial \tilde{x}^2(t \mid t)] \cdot \Phi(t, t - \theta) G(t) W_I(t - \theta) W_I'(t - \theta) G'(t) \Phi'(t, t - \theta)\}. \tag{55}$$

For these optimal control problems, we get the following sufficient conditions for optimality.

Lemma 3.1. Let $V_I(t, \hat{x}(t \mid t))$ and $V_{II}(t, x(t), \tilde{x}(t \mid t))$ be functions such that $V_I : [t_0, t_f] \times R^n \to R^1$ and $V_{II} : [t_0, t_f] \times R^n \times R^n \to R^1$; V_I, V_{It}, $V_{I\hat{x}}$, $V_{I\hat{x}\hat{x}}$ and V_{II}, V_{IIt}, V_{IIx}, $V_{II\tilde{x}}$, $V_{II\tilde{x}\tilde{x}}$ are continuous; and, for some K_5 and K_6,

$$| V_I | + | V_{It} | + | \hat{x} | \, | V_{I\hat{x}} | + | V_{I\hat{x}\hat{x}} | \leqslant K_5(1 + | \hat{x} |^2), \tag{56}$$

$$| V_{II} | + | V_{IIt} | + (| x | + | \tilde{x} |)(| V_{IIx} | + | V_{II\tilde{x}} |) + | V_{II\tilde{x}\tilde{x}} | \leqslant K_6(1 + (| x | + | \tilde{x} |)^2). \tag{57}$$

Suppose there exist closed-loop control laws $\hat{\varphi}^{**} \in \Phi$ and $\hat{\psi}^{**} \in \Psi$ such that, for any $(t, \hat{x}, x, \tilde{x}) \in [t_0, t_f] \times R^n \times R^n \times R^n$ and any admissible controls φ and ψ,

$$\begin{aligned} 0 &= V_{It}(t, \hat{x}(t \mid t)) + \mathscr{L}_{\hat{\varphi}^{**}} V_I(t, \hat{x}(t \mid t)) \\ &\quad + \hat{\varphi}^{**\prime}(t, \hat{x}(t \mid t))\, Q_I(t)\, \hat{\varphi}^{**}(t, \hat{x}(t \mid t)) - \hat{L}(t, \hat{x}(t \mid t)) \\ &\leqslant V_{It}(t, \hat{x}(t \mid t)) + \mathscr{L}_{\varphi} V_I(t, \hat{x}(t \mid t)) \\ &\quad + \varphi'(t, \pi_t y_I)\, Q_I(t)\, \varphi(t, \pi_t y_I) - \hat{L}(t, \hat{x}(t \mid t)), \end{aligned} \tag{58}$$

$$V_I(t_f, \hat{x}(t_f \mid t_f)) = \hat{x}'(t_f \mid t_f) P' P \hat{x}(t_f \mid t_f) + \operatorname{tr}\{M(t_f \mid t_f) P' P\}, \tag{59}$$

$$\begin{aligned} 0 &= V_{IIt}(t, x(t), \tilde{x}(t \mid t)) + \mathscr{L}_{\hat{\psi}^{**}} V_{II}(t, x(t), \tilde{x}(t \mid t)) \\ &\quad + (x(t) - \tilde{x}(t \mid t))'\, W'(t) Q_I(t)\, W(t)\, (x(t) - \tilde{x}(t \mid t)) \\ &\quad - \hat{\psi}^{**\prime}(t, x(t), \tilde{x}(t \mid t))\, Q_{II}(t)\, \hat{\psi}^{**}(t, x(t), \tilde{x}(t \mid t)) \\ &\geqslant V_{IIt}(t, x(t), \tilde{x}(t \mid t)) + \mathscr{L}_{\psi} V_{II}(t, x(t), \tilde{x}(t \mid t)) \\ &\quad + (x(t) - \tilde{x}(t \mid t))'\, W'(t) Q_I(t)\, W(t)\, (x(t) - \tilde{x}(t \mid t)) \\ &\quad - \psi'(t, \pi_t y_{II}, \hat{x}(t \mid t))\, Q_{II}(t)\, \psi(t, \pi_t y_{II}, \hat{x}(t \mid t)), \end{aligned} \tag{60}$$

$$V_{II}(t_f, x(t_f), \tilde{x}(t_f \mid t_f)) = x'(t_f) P' P x(t_f). \tag{61}$$

Then, the controls $\hat{\varphi}^{**}$ and $\hat{\psi}^{**}$ are optimal for any admissible controls.

This lemma is proved by the method analogous to Ref. 11. Suppose that the conditions (21) and (22) are satisfied simultaneously; it follows from this lemma that Player I's and Player II's optimal closed-loop control laws can be constructed by the state estimate $\hat{x}(t \mid t)$ and by the state $x(t)$ and the estimation error $\tilde{x}(t \mid t)$, respectively.

4. Game with Delayed and Noisy Information

Lemma 3.1 shows that the optimal closed-loop control laws are derived from (51) and (52). Then, we get

$$u^{**}(t) = -\tfrac{1}{2} Q_I^{-1}(t) B'(t)\, [\partial V_I(t, \hat{x}(t \mid t)) / \partial \hat{x}(t \mid t)], \tag{62}$$

$$\begin{aligned} v^{**}(t) &= \tfrac{1}{2} Q_{II}^{-1}(t) C'(t)\, \{\partial V_{II}(t, x(t), \tilde{x}(t \mid t)) / \partial x(t) \\ &\qquad + \partial V_{II}(t, x(t), \tilde{x}(t \mid t)) / \partial \tilde{x}(t \mid t)\}. \end{aligned} \tag{63}$$

Substituting (62), (63) into (51), (52), Bellman's equations are obtained. Suppose that the solution to this Bellman's equation for Player II is given by

$$\begin{aligned} V_{II}(t, x(t), \tilde{x}(t \mid t)) &= x'(t)\, R(t)\, x(t) + x'(t)\, L(t)\, \tilde{x}(t \mid t) \\ &\qquad + \tilde{x}'(t \mid t)\, N(t)\, \tilde{x}(t \mid t) + r(t). \end{aligned} \tag{64}$$

Since $\mathscr{Y}_{It} \subset \mathscr{Y}_{IIt}$, we get

$$V_I(t, \hat{x}(t \mid t)) = E\{V_{II}(t, x(t, x(t), \tilde{x}(t \mid t)) \mid \mathscr{Y}_{It}\}$$
$$= \hat{x}'(t \mid t)\, R(t)\, \hat{x}(t \mid t) + \mathrm{tr}[M(t \mid t)\{R(t) + L(t) + N(t)\}] + r(t). \tag{65}$$

Then, we get the following theorem (Appendix A).

Theorem 4.1. The optimal controls which satisfy (16) and (17) subject to (1)–(5) and the assumption that Player II can deduce the value of $\hat{x}(t \mid t)$ at time $t \in [t_0, t_f]$, are given by

$$u^*(t) = -Q_I^{-1}(t)\, B'(t)\, R(t)\, \hat{x}(t \mid t), \tag{66}$$

$$v^*(t) = Q_{II}^{-1}(t)\, C'(t)\{R(t)\, x(t) + N(t)\, \tilde{x}(t \mid t)\}, \qquad t_0 \leqslant t \leqslant t_f; \tag{67}$$

and the optimal cost from t_0 to t_f is given by

$$J_{t_0}(u^*, v^*) = x'(t_0)\, R(t_0)\, x(t_0) + \tilde{x}'(t_0 \mid t_0)\, N(t_0)\, \tilde{x}(t_0 \mid t_0) + r(t_0), \tag{68}$$

where $\hat{x}(t \mid t)$, $\tilde{x}(t \mid t)$, and $M(t - \theta \mid t)$ are, respectively, the solutions to (29)–(31), (43)–(46), and (36)–(37) into which (66) and (67) are substituted; the symmetric matrices $R(t)$ and $N(t)$ are the solutions of

$$dR(t)/dt + A'(t)\, R(t) + R(t)\, A(t) - R(t)\, B(t)\, Q_I^{-1}(t)\, B'(t)\, R(t)$$
$$+ R(t)\, C(t)\, Q_{II}^{-1}(t)\, C'(t)\, R(t) = 0, \tag{69}$$

$$R(t_f) = P'P, \tag{70}$$

$$dN(t)/dt + A'(t)\, N(t) + N(t)\, A(t) + N(t)\, C(t)\, Q_{II}^{-1}(t)\, C'(t)\, R(t)$$
$$+ R(t)\, C(t)\, Q_{II}^{-1}(t)\, C'(t)\, N(t) + N(t)\, C(t)\, Q_{II}^{-1}(t)\, C'(t)\, N(t)$$
$$+ R(t)\, B(t)\, Q_I^{-1}(t)\, B'(t)\, R(t)$$
$$- \kappa(t)\, N(t)\, \Phi(t, t - \theta)\, G(t)\, H_I(t - \theta)\, \Phi(t - \theta, t)$$
$$- \kappa(t)\, \Phi'(t - \theta, t)\, H_I'(t - \theta)\, G'(t)\, \Phi'(t, t - \theta)\, N(t) = 0, \tag{71}$$

$$N(t_f) = 0; \tag{72}$$

the scalar $r(t)$ satisfies the relations

$$dr(t)/dt + \kappa(t)\, \mathrm{tr}[N(t)\, \Phi(t, t - \theta)\, G(t)\, W_I(t - \theta)\, W_I'(t - \theta)$$
$$\cdot\, G'(t)\, \Phi'(t, t - \theta)] = 0, \tag{73}$$

$$r(t_f) = 0. \tag{74}$$

It is easily shown that, from (43)–(46), (66), and (67), we get

$$\tilde{x}(t \mid t) = \Phi(t, t-\theta)\, \tilde{x}(t-\theta \mid t), \tag{75}$$

$$M(t \mid t) = \Phi(t, t-\theta)\, M(t-\theta \mid t)\, \Phi'(t, t-\theta), \qquad t_0 + \theta \leqslant t \leqslant t_f. \tag{76}$$

Player I's optimal control law (66) shows the separation theorem to be satisfied, in the sense that the state in the optimal control law for the deterministic game (Ref. 12) is replaced by the optimal estimate $\hat{x}(t \mid t)$. In view of (68), the optimal cost from t_0 to t_f consists of three terms. The first term is the optimal cost from t_0 to t_f corresponding to the deterministic game with the initial state x_0 (Ref. 12). The second term depends on the initial error $\tilde{x}(t_0 \mid t_0)$ in Player I's state estimate. The third term is due to the noise in Player I's measurement from $t_0 + \theta$ to t_f. In the interval $[t_0, t_0 + \theta)$, Player I has no observation, so that $\hat{x}(t \mid t) = \hat{x}(t \mid t_0)$ and his control coincides with the open-loop control. Suppose that $\theta > t_f - t_0$, this game is reduced to the one in which one player has perfect information and the other has no observation. In the case where the information delay is reduced to zero, the above results are easily shown to coincide with the results obtained from the game of imperfect information without delay (Refs. 3–4).

Theorem 4.1 is derived on the assumption that Player II can deduce Player I's state estimate $\hat{x}(t \mid t)$. Some of the conditions needed to satisfy this assumption for the game of imperfect information without delay have been discussed variously (Refs. 3–4). These conditions are applicable to this game with delayed information. In the time-invariant case, we get a less restrictive condition as follows.

Lemma 4.1. Consider the time-invariant system (1)–(5), (7). Suppose that Player I uses his optimal control on $[t_0, t_f]$. Player II can deduce $\hat{x}(t \mid t)$ on $[t_0, t_f]$ if it holds that

$$\operatorname{rank}[F_0'(t), F_1'(t), \ldots, F_i'(t)] = n, \qquad t_0 \leqslant t < t_0 + \theta, \tag{77}$$

$$\begin{aligned}
&\mathscr{R}(R(t)\, BQ_I^{-1}B' + N(t)\, CQ_{II}^{-1}C') + [\{R(t)\, A + \kappa(t)\, \Phi'(t-\theta, t)\, H_I'G'(t) \\
&\quad \cdot \Phi'(t, t-\theta)\, R(t)\}\, BQ_I^{-1}B' + \{N(t)\, A + N(t)\, CQ_{II}^{-1}C'R(t) \\
&\quad + R(t)\, BQ_I^{-1}B'\, N(t) + N(t)\, CQ_{II}^{-1}C'\, N(t) + R(t)\, BQ_I^{-1}B'\, R(t) \\
&\quad - \kappa(t)\, N(t)\, \Phi(t, t-\theta)\, G(t)\, H_I\, \Phi(t-\theta, t)\}\, CQ_{II}^{-1}C'] \\
&\quad \cdot \mathscr{N}\{W_I'G'(t)\, \Phi'(t, t-\theta)\, (R(t)\, BQ_I^{-1}B' + N(t)\, CQ_{II}^{-1}C')\} = R^n, \\
&\qquad\qquad\qquad\qquad t_0 + \theta \leqslant t \leqslant t_f,
\end{aligned} \tag{78}$$

where i is some finite integer;

$$F_{i+1}(t) = (d/dt) F_i(t) + F_i(t) \{A - BQ_I^{-1}B' R(t) + CQ_{II}^{-1}C' R(t)\},$$
$$F_0(t) = BQ_I^{-1}B' R(t) + CQ_{II}^{-1}C' N(t);$$

$\mathscr{R}(Y)$ and $\mathscr{N}(Y)$ are the range and the null space of Y, respectively.

This lemma is proved in Appendix B. Player I has no observation in the time interval $[t_0, t_0 + \theta)$, so that $\hat{x}(t \mid t)$ on $[t_0, t_0 + \theta)$ obeys the ordinary differential equation (29). On the other hand, $\hat{x}(t \mid t)$ on $[t_0 + \theta, t_f]$ is the solution to the stochastic differential equation (29) and (31). Hence, the differing types of conditions [(77) and (78)] obtain for $t \in [t_0, t_0 + \theta)$ and $[t_0 + \theta, t_f]$.

5. Conclusions

In this paper, we have formulated a linear-quadratic–Gaussian zero-sum game in which Player I has delayed and noisy information and Player II has perfect information. Here, it is assumed that Player II can derive Player I's state estimate, which means that Player I's information is nested in Player II's information. Then, it is shown that the solution to this problem requires an estimator given by a delay-differential equation and Player I's optimal closed-loop control law satisfies the separation theorem.

6. Appendix A: Proof of Theorem 4.1

Substituting (64)–(65) into (58)–(61), we can get the optimal controls

$$u^*(t) = -Q_I^{-1}(t) B'(t) R(t) \hat{x}(t \mid t), \tag{79}$$

$$v^*(t) = Q_{II}^{-1}(t) C'(t) \{(R(t) + \tfrac{1}{2} L'(t)) x(t) + (N(t) + \tfrac{1}{2} L(t)) \tilde{x}(t \mid t)\}, \tag{80}$$

where

$$\begin{aligned} dR(t)/dt &+ A'(t) R(t) + R(t) A(t) - R(t) B(t) Q_I^{-1}(t) B'(t) R(t) \\ &+ R(t) C(t) Q_{II}^{-1}(t) C'(t) R(t) - \tfrac{1}{4} L(t) C(t) Q_{II}^{-1}(t) C'(t) L'(t) = 0, \end{aligned} \tag{81}$$

$$R(t_f) = P'P, \tag{82}$$

$$\begin{aligned} dL(t)/dt &+ A'(t) L(t) + L(t) A(t) - R(t) B(t) Q_I^{-1}(t) B'(t) L(t) \\ &+ R(t) C(t) Q_{II}^{-1}(t) C'(t) L(t) + L(t) C(t) Q_{II}^{-1}(t) C'(t) R(t) \\ &+ \tfrac{1}{2} L(t) C(t) Q_{II}^{-1}(t) C'(t) L(t) + \tfrac{1}{2} L(t) C(t) Q_{II}^{-1}(t) C'(t) L'(t) \\ &- \kappa(t) L(t) \Phi(t, t - \theta) G(t) H_I(t - \theta) \Phi(t - \theta, t) = 0, \end{aligned} \tag{83}$$

$$L(t_f) = 0, \tag{84}$$

$$
\begin{aligned}
&dN(t)/dt + A'(t)\,N(t) + N(t)\,A(t) + N(t)\,C(t)\,Q_{II}^{-1}(t)\,C'(t)\,R(t) \\
&\quad + R(t)\,C(t)\,Q_{II}^{-1}(t)\,C'(t)\,N(t) + N(t)\,C(t)\,Q_{II}^{-1}(t)C'(t)\,N(t) \\
&\quad + R(t)\,B(t)\,Q_I^{-1}(t)\,B'(t)\,R(t) - \kappa(t)\,N(t)\,\Phi(t, t-\theta)\,G(t) \\
&\quad\ \cdot H_I(t-\theta)\,\Phi(t-\theta, t) - \kappa(t)\,\Phi'(t-\theta, t)\,H_I'(t-\theta)\,G'(t) \\
&\quad\ \cdot \Phi'(t, t-\theta)\,N(t) + \tfrac{1}{2}\,(L(t) + L'(t))\,C(t)\,Q_{II}^{-1}(t)\,C'(t)\,N(t) \\
&\quad + \tfrac{1}{2}\,N(t)\,C(t)\,Q_{II}^{-1}(t)\,C'(t)\,(L(t) + L'(t)) \\
&\quad + \tfrac{1}{2}\,R(t)\,B(t)\,Q_I^{-1}(t)\,B'(t)\,L(t) + \tfrac{1}{2}\,L'(t)\,B(t)\,Q_I^{-1}(t)\,B'(t)\,R(t) \\
&\quad + \tfrac{1}{4}\,L'(t)\,C(t)\,Q_{II}^{-1}(t)\,C'(t)\,L(t) = 0, \qquad (85)
\end{aligned}
$$

$$N(t_f) = 0, \qquad (86)$$

$$
\begin{aligned}
&dr(t)/dt + \kappa(t)\,\mathrm{tr}[N(t)\,\Phi(t, t-\theta)\,G(t)\,W_I(t-\theta)\,W_I'(t-\theta) \\
&\quad \cdot G'(t)\,\Phi'(t, t-\theta)] = 0, \qquad (87)
\end{aligned}
$$

$$r(t_f) = 0. \qquad (88)$$

From (83)–(84), we get

$$L(t) = 0, \qquad t_0 \leqslant t \leqslant t_f\,. \qquad (89)$$

This completes the proof.

7. Appendix B: Proof of Lemma 4.1

Since Player II knows $x(t)$ and $dx(t)/dt$, he can utilize the observation q_0 defined by

$$
\begin{aligned}
q_0(t) &= -dx(t)/dt + \{A + CQ_{II}^{-1}C'(R(t) + N(t))\}\,x(t) \\
&= \{BQ_I^{-1}B'R(t) + CQ_{II}^{-1}C'\,N(t)\}\,\hat{x}(t \mid t) \equiv F_0(t)\,\hat{x}(t \mid t). \qquad (90)
\end{aligned}
$$

Thus, we can recognize the projection of $\hat{x}(t \mid t)$ on the subspace

$$\mathcal{N}(BQ_I^{-1}B'\,R(t) + CQ_{II}^{-1}C'\,N(t))^{\perp} = \mathcal{R}(R(t)\,BQ_I^{-1}B' + N(t)\,CQ_{II}^{-1}C'). \qquad (91)$$

More knowledge of $\hat{x}(t \mid t)$ can be obtained by taking the derivative of (90) and substituting (29), (69), (71), and (76), as follows:

$$
\begin{aligned}
&[-BQ_I^{-1}B'\,A'R(t) - CQ_{II}^{-1}C'\{A'N(t) + R(t)\,CQ_{II}^{-1}C'\,N(t) + N(t)\,BQ_I^{-1}B'\,R(t) \\
&\quad + N(t)\,CQ_{II}^{-1}C'\,N(t) + R(t)\,BQ_I^{-1}B'\,R(t) \\
&\quad - \kappa(t)\,\Phi'(t-\theta, t)\,H_I'G'(t)\,\Phi'(t, t-\theta)\,N(t)\} \\
&\quad - \kappa(t)\,BQ_I^{-1}B'\,R(t)\,\Phi(t, t-\theta)\,G(t)\,H_I\,\Phi(t-\theta, t)]\,\hat{x}(t \mid t) \\
&\quad = \dot{q}_0(t) - \kappa(t)\,\{BQ_I^{-1}B'\,R(t) + CQ_{II}^{-1}C'\,N(t)\}\,\Phi(t, t-\theta)\,G(t) \\
&\quad\ \cdot \{H_I\,\Phi(t-\theta, t)\,x(t) + W_I\dot{w}_I(t-\theta)\} \equiv F_1(t)\,\hat{x}(t \mid t). \qquad (92)
\end{aligned}
$$

For each t on $[t_0 + \theta, t_f]$, let us define the projection $P(t)$ on the subspace

$$\mathscr{R}\{(BQ_I^{-1}B' R(t) + CQ_{II}^{-1}C' N(t)) \Phi(t, t-\theta) G(t)W_I\}^{\perp}$$
$$= \mathscr{N}\{W_I' G'(t) \Phi'(t, t-\theta) (R(t) BQ_I^{-1}B' + N(t) CQ_{II}^{-1}C')\}. \quad (93)$$

Here, we define the process $q_2(t)$ as follows:

$$q_2(t) = \int_{t_0}^{t} P(s) F_1(s) \hat{x}(s \mid s) \, ds$$
$$= \int_{t_0}^{t} P(s) [\dot{q}_0(s) - \kappa(s) \{BQ_I^{-1}B' R(s) + CQ_{II}^{-1}C' N(s)\}$$
$$\cdot \Phi(s, s-\theta) G(s) H_I \Phi(s-\theta, s) x(s)] \, ds. \quad (94)$$

Since $q_2(t)$ is equal to the middle term of (94), it must be differentiable, and so Player II can get the observation $q_1(t)$ by differentiating $q_2(t)$. Thus, he can recognize the projection of $\hat{x}(t \mid t)$ on the subspace

$$\mathscr{N}(P(t) F_1(t))^{\perp} = F_1'(t) \mathscr{R}(P'(t)). \quad (95)$$

From the definition of $P(t)$, the right-hand side of (95) is reduced to

$$F_1'(t) \mathscr{N}\{W_I' G'(t) \Phi'(t, t-\theta) (R(t) BQ_I^{-1}B' + N(t) CQ_{II}^{-1}C')\}. \quad (96)$$

Then, from (91) and (96), we can get the condition (78) for $t \in [t_0 + \theta, t_f]$. In the interval $[t_0, t_0 + \theta)$, $\kappa(t) = 0$, and so $P(t) = I$. Then, iterating the above procedure, we see that the condition (77) is obtained.

References

1. Witsenhausen, H. S., *On Information Structures, Feedback, and Causality*, SIAM Journal on Control, Vol. 9, No. 2, 1971.
2. Witsenhausen, H. S., *Separation of Estimation and Control for Discrete Time System*, Proceedings of the IEEE, Vol. 59, No. 11, 1971.
3. Behn, R. D., and Ho, Y. C., *On a Class of Linear Stochastic Differential Games*, IEEE Transactions on Automatic Control, Vol. AC-13, No. 3, 1968.
4. Rhodes, I. B., and Luenberger, D. G., *Differential Games with Imperfect State Information*, IEEE Transactions on Automatic Control, Vol. AC-14, No. 1, 1969.
5. Rhodes, I. B., and Luenberger, D. G., *Stochastic Differential Games with Constrained State Estimators*, IEEE Transactions on Automatic Control, Vol. AC-14, No. 5, 1969.

XXIII

Mixed Strategy Solutions for Quadratic Games

D. J. WILSON

Abstract. Mixed strategy solutions are given for two-person, zero-sum games with payoff functions consisting of quadratic, bilinear, and linear terms, and strategy spaces consisting of closed balls in a Hilbert space. The results are applied to linear–quadratic differential games with no information, and with quadratic integral constraints on the control functions.

1. Introduction

Conditions satisfied by open-loop saddle-points of two-person, zero-sum, linear-quadratic, differential games have been obtained by Schmitendorf, Rekasius, and the author (Refs. 1–3). Lukes and Russell (Ref. 4) have used a general theory of n-person games with quadratic payoff functions defined on Hilbert spaces to obtain open-loop equilibrium point solutions of n-person, linear–quadratic, differential games. However, all these results have been obtained for *unbounded* control functions. If solutions are to exist under these circumstances, it is necessary that the term in a player's payoff function which is quadratic in his control is also semidefinite (i.e., positive for a minimizing player, or negative for a maximizing player).

In this paper, we consider two-person, zero-sum games with integral constraints on the control functions. Under these conditions, mixed strategy solutions always exist, and a procedure for obtaining these solutions can be given.

The terminology and notation we shall use are introduced below.

Hilbert Space and Linear Operators. The *inner product* of two elements u, v of a Hilbert Space H will be denoted throughout by (u, v), and the *norm* $(w, w)^{1/2}$ of $w \in H$ will be denoted by $\| w \|$. The *orthogonal complement* of a subset C of H is denoted by $C^{\perp}$. A bounded, self-adjoint (linear) operator $R : H \to H$ is called *positive* if $(u, Ru) \geqslant 0$ for all $u \in H$ and is called *positive definite* if $(u, Ru) > 0$ for all nonzero $u \in H$. An operator S is *negative* (*definite*) if $-S$ is positive (definite). The *adjoint* of a continuous linear mapping M of a Hilbert space H_1 into a Hilbert space H_2 is denoted by M^*. M^* is defined by the condition

$$(M^*u, v) = (u, Mv)$$

for all $u \in H_2$, $v \in H_1$. The *spectrum* of an operator $R : H \to H$ is denoted by $\sigma(R)$. A number λ is called an *approximate eigenvalue* of the operator R if, for every $\epsilon > 0$, there exists a nonzero $x \in H$ such that

$$\|(R - \lambda)x\| < \epsilon \| x \|.$$

In this case, any nonzero element of the line $\{\mu x; \mu \text{ scalar}\}$ is called an *ϵ-eigenvector* of R corresponding to the approximate eigenvalue λ. We shall represent a scalar multiple λI of the identity operator I simply by the scalar λ.

The following well-known results (see Refs. 5, 6) will be found useful, and we quote them here without proof.

Let R be a bounded, self-adjoint operator on a Hilbert space. Then, the following statements hold.

(a) $\sigma(R)$ is a compact set of real numbers (Ref. 5, pp. 903, 907);

(b) $\inf \sigma(R) \geqslant 0$ iff R is positive (Ref. 5, p. 907). As a consequence, $R - \inf(\sigma(R))$ is always positive.

(c) $\lambda \in \sigma(R)$ iff λ is an approximate eigenvalue of R (Ref. 6, pp. 37, 40).

(d) If R is positive, then (u, Ru) is a convex function of u. This follows from the fact that $(u, Ru)^{1/2}$ is a seminorm.

(e) A closed, convex subset of a Hilbert space is weakly closed (Ref. 5, p. 422).

(f) The closed unit ball of a Hilbert space is weakly compact (Ref. 5, p. 425).

Quadratic Game. Henceforth, H_1, H_2 will denote two *real* Hilbert spaces, and the following symbols shall retain a fixed meaning which we now assign to them: U, V are the closed unit balls of H_1,

H_2, respectively; R, S are bounded, self-adjoint operators on H_1, H_2; a, b are elements of H_1, H_2; M is a continuous, linear mapping from H_2 into H_1; also,

$$\lambda^* = \sup \sigma(R), \qquad \mu_* = \inf \sigma(S).$$

We now denote by G the zero-sum, two-person game with *strategy sets* U, V and *payoff* J (to the first player) defined on $U \times V$ by

$$J(u, v) = (u, Ru) + (v, Sv) + (u, Mv) + (a, u) + (b, v).$$

Mixed Strategies. A *mixed strategy* for Player 1 (resp., for Player 2) in G is a Borel probability measure on U (resp., on V). If μ, ν are mixed strategies for Players 1 and 2, respectively, we shall write

$$J(\mu, \nu) = \int_{U\times V} J(u, v)\, d\mu(u)\, d\nu(v).$$

The mixed strategies which appear explicitly will all be *atomic* mixed strategies, which assign a positive probability to each of a finite set of (pure) strategies. We will denote the probability measure which is concentrated at the single atom x by $\delta(x)$. Thus, the mixed strategy which selects pure strategies x_k, $k = 1, 2,\ldots, n$, with probabilities α_k, respectively, with

$$\alpha_k > 0, \qquad \sum_{k=1}^{n} \alpha_k = 1,$$

will be denoted by

$$\sum_{k=1}^{n} \alpha_k\, \delta(x_k).$$

We have then

$$J\left(\sum_{k=1}^{n} \alpha_k\, \delta(u_k), \sum_{j=1}^{m} \beta_j\, \delta(v_j)\right) = \sum_{k=1}^{n} \sum_{j=1}^{m} \alpha_k \beta_j\, J(u_k, v_j).$$

Since

$$J(\delta(u), \delta(v)) = J(u, v) \qquad \text{for all} \quad u, v,$$

we shall identify the mixed strategies $\delta(u)$, $\delta(v)$ with the (pure) strategies u, v respectively.

If

$$\sup_\mu \inf_\nu J(\mu, \nu) = \inf_\nu \sup_\mu J(\mu, \nu) = V^*,$$

the suprema and infima being taken over all mixed strategies of Players 1 and 2, respectively, then V^* will be called the *value* of G. An ϵ-saddle-*point* of G is a pair $\mu(\epsilon)$, $\nu(\epsilon)$ of mixed strategies such that

$$J(\mu(\epsilon), \nu) + \epsilon/2 \geqslant J(\mu, \nu(\epsilon)) - \epsilon/2$$

for all mixed strategies μ, ν. A necessary and sufficient condition for G to have a value is that there exist an ϵ-saddle-point of G for every positive ϵ. A *saddle-point* of G is simply a 0-saddle-point.

2. Pure Strategy Saddle-Points

When the operators R, S are negative and positive, respectively, then the function J is concave–convex and, as a consequence, G has a pure strategy saddle-point. In the general case, the operators $R - \lambda^*$ and $S - \mu_*$ are negative and positive, respectively, so that the modified payoff J^*, defined by

$$J^*(u, v) = J(u, v) - \lambda^*(u, u) - \mu_*(v, v),$$

has a saddle-point. The saddle-point of this modified payoff will be used in the next section, to construct *mixed strategy* ϵ-saddle-points of G for arbitrarily small positive ϵ.

The following result is due to Auslender (Ref. 7).

Lemma 2.1. A continuous, convex, real-valued function f on a Hilbert space is weakly lower semicontinuous. Similarly, a continuous, concave function is weakly upper semicontinuous.

Proof. Let α be any real number. Then, since f is continuous and convex, the set $\{x; f(x) \leqslant \alpha\}$ is closed and convex and, consequently, it is weakly closed. Since α was an arbitrary real number, this implies that f is lower semicontinuous.

Theorem 2.1. If R is negative and S positive, then G has a saddle-point consisting of pure strategies. That is, there exist $u^* \in U$, $v^* \in V$ such that

$$J(u, v^*) \leqslant J(u^*, v^*) \leqslant J(u^*, v) \qquad \text{for all} \quad u \in U, v \in V.$$

Proof. Since R is negative, the function $:u \to (u, Ru)$ is concave and, therefore, weakly upper semicontinuous by Lemma 2.1 (since R

is a continuous operator). If v is a fixed element of V, then the weak upper semicontinuity of (u, Ru) implies that $J(., v)$ is weakly upper semicontinuous [since the linear terms of $J(., v)$ are *weakly continuous*]. Similarly, $J(u, .)$ is weakly lower semicontinuous for each fixed $u \in U$. Since U, V are weakly compact, it follows from Fan's minimax theorem (Ref. 8) that J has a saddle-point.

Because of the special form of J, it is possible to write down a necessary and sufficient condition for a strategy pair to be a saddle-point. We shall require the following lemma.

Lemma 2.2. Let L be a bounded, self-adjoint operator on a Hilbert space H, and put $\rho_* = \inf \sigma(L)$. Let N denote the closed, unit ball of H and d a fixed element of H. Define a real-valued function f on N by

$$f(x) = (x, Lx) + (d, x).$$

(a) If $\epsilon \geqslant 0$, $\rho \leqslant \min(0, \rho_*)$, $x_0 \in N$, and the conditions

$$\| 2(L - \rho)x_0 + d \| \leqslant \epsilon/2, \tag{1}$$

$$\rho(1 - \| x_0 \|^2) = 0 \tag{2}$$

are satisfied, then

$$f(x_0) \leqslant f(x) + \epsilon \qquad \text{for all} \quad x \in N.$$

(b) For an element x_0 of N to minimize f over N, it is necessary and sufficient that there exist a real number $\rho \leqslant \min(0, \rho_*)$ such that

$$2(L - \rho)x_0 + d = 0, \qquad \rho(1 - \| x_0 \|^2) - 0. \tag{3}$$

Proof. (a) If $x \in N$, then

$$\begin{aligned} f(x) &= (x, Lx) + (d, x) \\ &= (x_0, Lx_0) + (d, x_0) - (x_0 - x, 2(L - \rho)x_0 + d) \\ &\quad + (x_0 - x, (L - \rho)(x_0 - x)) - \rho(\| x_0 \|^2 - \| x \|^2) \\ &= f(x_0) - (x_0 - x, 2(L - \rho)x_0 + d) \\ &\quad + (x_0 - x, (L - \rho)(x_0 - x)) - \rho(1 - \| x \|^2), \end{aligned}$$

since

$$\rho \| x_0 \|^2 = \rho$$

from Eq. (2). Now, from (1), we see that

$$(x_0 - x, 2(L - \rho)x_0 + d) \leqslant \| x_0 - x \| \epsilon/2 \leqslant \epsilon,$$

since both x_0 and x lie in N. Also, since $\rho \leqslant \rho_*$, then $L - \rho$ is positive and

$$(x_0 - x, (L - \rho)(x_0 - x)) \geqslant 0.$$

Lastly, since $\rho \leqslant 0$ and $x \in N$, then

$$\rho(1 - \| x \|^2) \leqslant 0.$$

Substituting these inequalities in the expression for $f(x)$ above, we obtain

$$f(x) \geqslant f(x_0) - \epsilon,$$

which proves the first result, since x was arbitrary.

(b) The sufficiency of (3) follows immediately from (a) by putting $\epsilon = 0$.

Now, suppose that x_0 minimizes f over N. Since f is differentiable, it follows from the usual necessary conditions (Ref. 9, p. 178) that there exists a nonpositive, real number ρ such that

$$2(L - \rho)x_0 + d = 0.$$

Furthermore, if $\| x_0 \| < 1$, then x_0 lies in the interior of N and must satisfy the equation

$$2Lx_0 + d = 0.$$

Thus, the number ρ satisfies Eq. (2). It now only remains to show that $\rho \leqslant \rho_*$.

Let $\epsilon > 0$, and let x_1 be an ϵ-eigenvector of L corresponding to the approximate eigenvalue ρ_*. Then,

$$(x_1, (L - \rho_* - \epsilon)x_1) < 0.$$

By choosing the appropriate sign and a sufficiently small positive η in the expression

$$x_2 = \pm x_1 - \eta x_0,$$

we can always ensure that

$$(x_2, x_0) < 0, \qquad (x_2, (L - \rho_* - \epsilon)x_2) < 0.$$

Then, by appropriate choice of positive t, we can find $x_3 = tx_2$ such that

$$\| x_0 + x_3 \| = 1, \qquad (x_3, (L - \rho_* - \epsilon)x_3) < 0.$$

We then have

$$(x_0, Lx_0) + (d, x_0) \leqslant (x_0 + x_3, L(x_0 + x_3)) + (d, x_0 + x_3),$$

that is,[1]

$$\begin{aligned} 0 &\leqslant (2Lx_0 + d, x_3) + (x_3, Lx_3) \\ &= 2\rho(x_0, x_3) + (x_3, (L - \rho_*)x_3) + \rho_*(x_3, x_3) \\ &\leqslant \rho((x_0, x_0) - 1) + 2\rho(x_0, x_3) + (\rho_* + \epsilon)(x_3, x_3) \\ &= \rho(\| x_0 + x_3 \|^2 - 1) + (\rho_* - \rho + \epsilon)(x_3, x_3) \\ &= (\rho_* - \rho + \epsilon)(x_3, x_3). \end{aligned}$$

Thus, $\rho \leqslant \rho_* + \epsilon$; and, since ϵ was an arbitrary positive number, then $\rho \leqslant \rho_*$. This completes the proof of (b).

We now immediately obtain necessary and sufficient conditions for (u_0, v_0) to be a (pure strategy) saddle-point of G, by applying Lemma 2.2 to the functions $J(., v_0)$ and $J(u_0, .)$.

Theorem 2.2. For the pair $(u_0, v_0) \in U \times V$ to be a saddle-point of G, it is necessary and sufficient that there exist real numbers

$$\lambda \geqslant \max(\lambda^*, 0), \qquad \mu \leqslant \min(\mu_*, 0)$$

such that

$$2(R - \lambda)u_0 + Mv_0 + a = 0, \tag{4-1}$$

$$M^*u_0 + 2(S - \mu)v_0 + b = 0, \tag{4-2}$$

$$\lambda(1 - \| u_0 \|) = \mu(1 - \| v_0 \|) = 0. \tag{4-3}$$

3. Mixed Strategy Solutions

We now use the results of the previous section to construct a (mixed strategy) ϵ-saddle-point of the general game G for every positive ϵ (no matter how small). In general, G will have a *saddle-point* only when λ^* and μ_* are *eigenvalues* of the operators R and S respectively.

Lemma 3.1. There exists a strategy pair $(u_0, v_0) \in U \times V$ and real numbers

$$\lambda \geqslant \max(\lambda^*, 0), \qquad \mu \leqslant \min(\mu_*, 0)$$

[1] The result on the second line is obtained from (3) and that on the third line is obtained using (2).

satisfying the equations

$$2(R - \lambda)\, u_0 + Mv_0 + a = 0, \tag{5-1}$$

$$M^* u_0 + 2(S - \mu)\, v_0 + b = 0, \tag{5-2}$$

$$\{\lambda - \max(\lambda^*, 0)\} \cdot (1 - \| u_0 \|) = 0, \tag{5-3}$$

$$\{\min(\mu_*, 0) - \mu\} \cdot (1 - \| v_0 \|) = 0. \tag{5-4}$$

Proof. Since $R - \max(\lambda^*, 0)$ is negative and $S - \min(\mu_*, 0)$ is positive, the function J^* defined by

$$J^*(u, v) = J(u, v) - \max(\lambda^*, 0)\,(u, u) - \min(\mu_*, 0)\,(v, v)$$

has a saddle-point (u_0, v_0), by Theorem 2.1. By Theorem 2.2, this saddle-point must satisfy Eqs. (5) for some $\lambda \geqslant \max(\lambda^*, 0)$ and $\mu \leqslant \min(\mu_*, 0)$.

We now proceed to construct the ϵ-optimal mixed strategies of the players of G. Let u_0, v_0 satisfy (5) for some real numbers $\lambda \geqslant \max(\lambda^*, 0)$ and $\mu \leqslant \min(\mu_*, 0)$.

If $\lambda > \lambda^*$, put $u_1 = u_2 = u_0$.

If $\lambda = \lambda^*$, let u_ϵ be an $(\epsilon/4)$-eigenvector of R corresponding to the approximate eigenvalue λ^*, and let $\beta = \beta_1, \beta_2$ be the roots of the equation

$$\| u_0 + \beta u_\epsilon \|^2 = 1.$$

Note that β_1, β_2 are either both equal to zero, or are of opposite sign. Then, put

$$u_1 = u_0 + \beta_1 u_\epsilon, \qquad u_2 = u_0 + \beta_2 u_\epsilon.$$

Now, whatever the values of u_1, u_2, it is possible to find $\nu \epsilon [0, 1]$ such that

$$\nu u_1 + (1 - \nu) u_2 = u_0.$$

We then put

$$\sigma^* = \nu \delta(u_1) + (1 - \nu)\, \delta(u_2).$$

Similarly, if $\mu < \mu_*$, put $v_1 = v_2 = v_0$. If $\mu = \mu_*$, put

$$v_1 = v_0 + \alpha_1 v_\epsilon, \qquad v_2 = v_0 + \alpha_2 v_\epsilon,$$

where v_ϵ is an $(\epsilon/4)$-eigenvector of S corresponding to the approximate eigenvalue μ_*, and $\alpha = \alpha_1, \alpha_2$ are the roots of the equation

$$\| v_0 + \alpha v_\epsilon \|^2 = 1.$$

Again, we choose $\zeta \in [0, 1]$ such that

$$\zeta v_1 + (1 - \zeta) v_2 = v_0$$

and define

$$\tau^* = \zeta \delta(v_1) + (1 - \zeta)\, \delta(v_2).$$

Note that u_i, v_i satisfy the conditions

$$\| 2(R - \lambda) u_i + M v_0 + a \| < \epsilon/4, \qquad \lambda \cdot (1 - \| u_i \|) = 0, \qquad i = 1, 2, \tag{6}$$

$$\| M^* u_0 + 2(S - \mu) v_i + b \| < \epsilon/4, \qquad \mu \cdot (1 - \| v_i \|) = 0, \qquad i = 1, 2. \tag{7}$$

We are now in a position to show that (σ^*, τ^*) is an ϵ-saddle-point of G.

Theorem 3.1. For every positive ϵ, G has an ϵ-saddle-point.

Proof. Let σ^*, τ^* be defined as above. From Eq. (6) and Lemma 2.2(a), we conclude that

$$J(\delta(u_i), \tau^*) \geqslant \sup_{u \in U} J(\delta(u), \tau^*) - \epsilon/2, \qquad i = 1, 2.$$

Thus,

$$J(\sigma^*, \tau^*) = \nu J(\delta(u_1), \tau^*) + (1 - \nu)\, J(\delta(u_2), \tau^*) \geqslant J(\delta(u), \tau^*) - \epsilon/2$$

for all $u \in U$ and, hence, also

$$J(\sigma^*, \tau^*) \geqslant J(\sigma, \tau^*) - \epsilon/2$$

for all mixed strategies σ of Player 1.

Similarly, from Eq. (7) and Lemma 2.2(a), we can show that

$$J(\sigma^*, \tau^*) \leqslant J(\sigma^*, \tau) + \epsilon/2$$

for all mixed strategies τ of Player 2. Combining the above inequalities, we see that (σ^*, τ^*) is indeed an ϵ-saddle-point of G, as was to be proved.

4. Applications

Consider a differential game whose state $x \in R^n$ is governed by the differential equation(s)

$$\dot{x}(t) = A(t)\, x(t) + B(t)\, u(t) + C(t)\, v(t), \tag{8-1}$$

$$x(0) = x_0 . \tag{8-2}$$

The control variables $u(t) \in R^p$, $v(t) \in R^q$ are chosen by two different players who have no knowledge of the current or past values of the state variable, except for its initial value x_0. The control functions are restricted, therefore, to be functions of time only; that is, open-loop controls. The control functions are also required to satisfy the quadratic *energy* constraints

$$\int_0^T u(t)'\, u(t)\, dt \leqslant 1, \qquad \int_0^T v(t)'\, v(t)\, dt \leqslant 1,$$

where T is a fixed positive time. The controller of u wishes to maximize, and the controller of v to minimize a cost function

$$J(u, v) = x'(T)\, Qx(T) + \int_0^T \{u'(t)\, R(t)\, u(t) + v'(t)\, S(t)\, v(t)\}\, dt.$$

The matrix functions A, B, C, R, S are continuous and of orders $n \times n$, $n \times p$, $n \times q$, $p \times p$, $q \times q$, respectively. The matrices $R(t)$, $S(t)$ are symmetric for every t, and Q is a symmetric $n \times n$ matrix.

Let M be the *fundamental matrix* of the differential equation (8), which satisfies

$$\dot{M}(t) = -M(t)\, A(t), \qquad M(0) = I.$$

Then, at time t, the state variable is

$$x(t) = M^{-1}(t)\, x_0 + M^{-1}(t) \int_0^t M(s)\, \{B(s)\, u(s) + C(s)\, v(s)\}\, ds.$$

The players are required to choose their control functions from the unit balls of the spaces L_p^2, L_q^2 of square-integrable p-vector and q-vector functions, respectively. If we denote the inner products of both of these spaces by (., .), and put

$$\Phi(t) = B'(t)\, M'(t)\, [M^{-1}(T)]',$$

$$\Psi(t) = C'(t)\, M'(t)\, [M^{-1}(T)]',$$

then the payoff J may be written as

$$J(u, v) = (u, Ru) + (v, Sv) + (u, Mv) + (\alpha, u) + (\beta, v) + \gamma,$$

where

$$(Ru)(t) = R(t)\, u(t) + \int_0^T \Phi(t)\, Q\Phi'(s)\, u(s)\, ds,$$

$$(Sv)(t) = S(t)\, v(t) + \int_0^T \Psi(t)\, Q\Psi'(s)\, v(s)\, ds,$$

$$(Mv)(t) = 2 \int_0^T \Phi(t)\, Q\Psi'(s)\, v(s)\, ds,$$

$$\alpha(t) = 2\Phi(t)\, QM^{-1}(T) x_0\,,$$

$$\beta(t) = 2\Psi(t)\, QM^{-1}(T) x_0\,,$$

and

$$\gamma = x_0'[M^{-1}(T)]' QM^{-1}(T)x_0$$

is a constant. Thus, this differential game is a quadratic game of the class defined in Section 1 and must possess, therefore, a value and ϵ-optimal strategies (possibly mixed) for every positive ϵ. A possible procedure for solving the game is as follows.

(a) Find $\lambda^* = \sup \sigma(R)$ and the corresponding (approximate) eigenvectors. This is equivalent to solving the one-sided optimal control problem

$$\text{maximize } J = x'(T)Qx(T) + \int_0^T u'(t)\, R(t)\, u(t)\, dt,$$

$$\begin{aligned} \text{subject to}\quad & \dot{x}(t) = A(t)\, x(t) + B(t)\, u(t), \\ & x(0) = 0, \\ & \int_0^T u'(t)\, u(t)\, dt \leqslant 1. \end{aligned}$$

(b) Find $\mu_* = \inf \sigma(S)$ and the corresponding (approximate) eigenvectors. Again, this is equivalent to the solution of an optimal control problem.

(c) Try to solve the linear integral equations

$$2(R - \lambda^*)u + Mv + \alpha = 0, \tag{9-1}$$

$$M^*u + 2(S - \mu_*)\, v + \beta = 0 \tag{9-2}$$

for functions u, v satisfying

$$\int_0^T u'(t)\, u(t)\, dt \leqslant 1, \qquad \int_0^T v'(t)\, v(t)\, dt \leqslant 1.$$

These integral equations are separable, and, *once* λ^*, μ^* *are known*, their solution, if possible, is in principle quite simple.

(d) If the integral equations (9) cannot be solved for appropriate u and v, then we either have to replace λ^* in (9) by $\lambda > \lambda^*$ and use the condition

$$\int_0^T u'(t)\, u(t)\, dt = 1$$

to determine this extra variable λ, or we have to replace μ_* by $\mu < \mu_*$ and use the condition

$$\int_0^T v'(t)\, v(t)\, dt = 1;$$

or, if none of these methods work, we are faced with the solution of the double-eigenvalue problem

$$2(R - \lambda)\, u + Mv + \alpha = 0,$$
$$M^* u + 2(S - \mu)\, v + \beta = 0,$$

where

$$\lambda > \lambda^*, \qquad \mu < \mu_*,$$
$$\int_0^T u'(t)\, u(t)\, dt = \int_0^T v'(t)\, v(t)\, dt = 1.$$

No theory or general method of solution of double-eigenvalue problems such as the above is known to the author. However, in this case, their solution may be facilitated by the fact that the game has a saddle-point in *pure strategies*.

Once Steps (a)–(b) and either (c) or (d) have been carried out, the (approximately) optimal strategies may be constructed as described in Section 3. As an illustration, we solve the following simple example.

Example. Consider the one-dimensional differential game with state equation

$$\dot{x}(t) = ax(t) + u(t) + bv(t), \qquad x(0) = x_0, \tag{10}$$

and payoff function

$$J = cx^2(T) + \int_0^T [u^2(t) - dv^2(t)]\, dt, \tag{11}$$

where a, b, c, d are nonzero constants, $c > 0$, $a > 0$, and the control functions u, v are required to satisfy

$$\int_0^T u(t)^2\, dt \leqslant 1, \qquad \int_0^T v(t)^2\, dt \leqslant 1.$$

At each instant t, the controls $u(t)$, $v(t)$ are chosen by the two players with no knowledge of the present or past values of the state variable, except for its initial value x_0. The controller of u wishes to maximize J, and the controller of v wishes to minimize it. Using the same notation as for the general case above, we have

$$(Ru)(t) = u(t) + c \exp[a(2T - t)] \int_0^T \exp(-as)\, u(s)\, ds,$$

$$(Sv)(t) = -dv(t) + cb^2 \exp[a(2T - t)] \int_0^T \exp(-as)\, v(s)\, ds,$$

$$(Mv)(t) = 2cb \exp[a(2T - t)] \int_0^T \exp(-as)\, v(s)\, ds,$$

$$\alpha(t) = 2cx_0 \exp[a(2T - t)] = \beta(t)/b,$$

$$\gamma = cx_0^{\,2} \exp(2aT).$$

The spectrum of the operator R consists of two eigenvalues, 1 and $1 + c[\exp(2aT) - 1]/2a$. The eigenspace of the eigenvalue 1 consists of all functions orthogonal to $\exp(-as)$, while the eigenvalue $1 + c[\exp(2aT) - 1]/2a$ has the single eigenfunction $\exp(-as)$. Thus,

$$\sup \sigma(R) = 1 + c[\exp(2aT) - 1]/2a,$$

the larger of the two eigenvalues. Similarly, the operator S has eigenvalues $-d$ and $-d + cb^2[\exp(2aT) - 1]/2a$. The smaller of these eigenvalues is $-d = \inf \sigma(S)$, which has an eigenspace consisting of all functions orthogonal to $\exp(-as)$. Equations (9) may now be written as

$$u(t)\, [1 - \exp(2aT)]/2a + \exp[a(2T - t)] \times \left\{\int_0^T \exp(-as)\, u(s)\, ds + b \int_0^T \exp(-as)\, v(s)\, ds + x_0\right\} = 0,$$

$$\int_0^T \exp(-as)\, u(s)\, ds + b \int_0^T \exp(-as)\, v(s)\, ds + x_0 = 0,$$

These equations immediately give $u(t) \equiv 0$ and

$$\int_0^T \exp(-as)\, v(s)\, ds = -x_0/b.$$

This second relation can be satisfied by a feasible control function v iff

$$\int_0^T \exp(-2as)\, ds \geqslant (x_0/b)^2,$$

that is,

$$k^2 = [1 - \exp(-2aT)]/2a\, (x_0/b)^2 \geqslant 1$$

Suppose that this last inequality *is* satisfied. Then, by applying the procedure of Section 3, we find that the first player's optimal strategy is to choose each of the control functions

$$(2a)^{1/2}\exp(-as)/[1-\exp(-2aT)]^{1/2}, \quad -(2a)^{1/2}\exp(-as)/[1-\exp(-2aT)]^{1/2}$$

with probability $\frac{1}{2}$. The second player has an optimal *pure* strategy

$$-x_0(2a)\exp(-as)/b\,[1-\exp(-2aT)]+\sqrt{[1-(1/k)^2]}\,v(s),$$

where $v(s)$ is any normalized function which is orthogonal to $\exp(-as)$. For example,

$$v(s)=(2a)^{3/2}\,(s-T/2)\exp(as)/\{[(Ta-1)^2+1]\exp(2aT)-(Ta+1)^2-1\}^{1/2}$$

will do.

If

$$(x_0/b)^2>[1-\exp(-2aT)]/2a,$$

then Eqs. (9) have no solution, and it is easy to show that the only solution to the game is obtained by replacing λ^* by $\lambda>\lambda^*$ and μ_* by $\mu<\mu_*$ in Eqs. (9). In this case,

$$\int_0^T u(t)^2\,dt=\int_0^T v(t)^2\,dt=1,$$

and the integrand in the payoff (11) is simply $1-d$. The optimal controls u_0, v_0 of Players 1 and 2, respectively, are then given by

$$u_0(t)=(2a)^{1/2}\operatorname{sign}x_0\exp(-as)/[1-\exp(-2aT)]^{1/2},$$

$$v_0(t)=-\operatorname{sign}(x_0/b)\,(2a)^{1/2}\exp(-as)/[1-\exp(-2aT)]^{1/2}.$$

5. Conclusions

An extension of most of the results given here to n-person games with quadratic payoff functions is not too difficult. Necessary and sufficient conditions for (Nash) equilibrium points of such games, similar to the conditions of Theorem 2.2, can be given. The construction of mixed strategy equilibria can also be carried out along the lines of the procedure given in Section 3 *if the appropriate necessary conditions can be solved.* However, current fixed point theorems (Ref. 10) seem insufficient to guarantee the existence of a pure strategy equilibrium point in the case when each player's payoff function is concave. Consequently, the author is unable to assert the existence of a mixed strategy equilibrium of the general n-person game.

References

1. Schmitendorf, W. E., *Existence of Optimal Open-Loop Strategies for a Class of Differential Games*, Journal of Optimization Theory and Applications, Vol. 5, No. 5, 1970.
2. Rekasius, Z. V., *On Open-Loop and Closed-Loop Solutions of Linear Differential Games*, Paper presented at the First International Conference on the Theory and Applications of Differential Games, Amherst, Massachusetts, 1969.
3. Wilson, D. J., *Differential Games with no Information*, University of Adelaide, Department of Mathematics, Ph.D. Thesis, 1972.
4. Lukes, D. L., and Russel, D. L., *A Global Theory for Linear–Quadratic Differential Games*, Journal of Mathematical Analysis and Application, Vol. 33, No. 1, 1971.
5. Dunford, N., and Schwartz, J. T., *Linear Operators*, John Wiley and Sons (Interscience Publishers), New York, New York, 1963.
6. Halmos, P. R., *A Hilbert Space Problem Book*, Van Nostrand Company, Reinhold, New York, New York, 1967.
7. Auslender, A., *Recherche des Points de Selle d'une Fonction*, Paper presented at the Third International Colloquium on Optimization, Nice, France, 1969.
8. Fan, K., *Minimax Theorems*, Proceedings of the National Academy of Sciences of the USA, Vol. 39, No. 1, 1953.
9. Luenberger, D. G., *Optimization by Vector Space Methods*, John Wiley and Sons, New York, New York, 1969.
10. Bohnenblust, M. F., and Karlin, S., *On a Theorem of Ville*, Contributions to the Theory of Games, Vol. 1, Edited by H. W. Kuhn and A. W. Tucker, Princeton University Press, Princeton, New Jersey, 1950.

XXIV

Zero-Sum Games with Incompletely Defined Payoff[1]

A. Sprzeuzkouski

Abstract. It is shown that a saddle-point solution exists in a two-person, zero-sum game whose payoff is given by a matrix which is not completely defined. On the other hand, we show that such games do not always have a value, so that a saddle-point solution is not necessarily an optimal solution.

1. Introduction

In this paper, we shall study two-person, zero-sum games in which the payoff is not defined for every possible outcome of the game. This situation arises, in particular, in some dynamical games in which one of the aims of both players is to reach a given set at the end of the game or for which the state of the game is subject to some constraints. First, we shall give definitions concerning optimality in this type of game. Then, limiting ourselves to matrix games in which some elements of the payoff matrix are not defined, we shall show, by introducing mixed strategies, that such games always possess a saddle point. This problem has been studied from a different point of view in Refs. 1 and 2.

Let X and Y be two sets, $\mathscr{D}$ a subset of $X \times Y$, and $f\colon \mathscr{D} \to \mathbb{R}$ a real-valued function on $\mathscr{D}$. For every $x \in X$ and $y \in Y$, we shall set

$$\mathscr{X}(y) \triangleq \{x \in X\colon \quad (x, y) \in \mathscr{D}\}, \qquad \mathscr{Y}(x) \triangleq \{y \in Y\colon \quad (x, y) \in \mathscr{D}\},$$

[1] This work was supported by the Centre d'Etudes Atomiques, Saclay, France.

and we shall assume that

$$\mathscr{X}(y) \neq \varnothing, \qquad \mathscr{Y}(x) \neq \varnothing.$$

The word *game* will mean a quadruple

$$\Gamma = \langle X, Y, \mathscr{D}, f \rangle.$$

X and Y represent the strategy sets of Players 1 and 2, respectively, and f is the payoff function of Player 1. Player 1 tries to maximize f, while Player 2 tries to minimize it, so that $-f$ is the payoff function of Player 2.

We shall set

$$V^+(\Gamma) = \sup_{x \in X} \inf_{y \in \mathscr{Y}(x)} f(x, y), \qquad V^-(\Gamma) = \inf_{y \in Y} \sup_{x \in \mathscr{X}(y)} f(x, y).$$

Definition 1.1. The pair $(x_*, y_*) \in \mathscr{D}$ will be a saddle point of Γ if it satisfies the inequalities

$$f(x, y_*) \leqslant f(x_*, y_*) \leqslant f(x_*, y)$$

for every $x \in \mathscr{X}(y_*)$ and every $y \in \mathscr{Y}(x_*)$.

Definition 1.2. The game Γ will have a value $V_\Gamma{}^*$ if

$$V_\Gamma{}^* = V^+(\Gamma) = V^-(\Gamma).$$

Definition 1.3. We say that a game Γ possesses an optimal pair (x^*, y^*) if

(i) Γ has a value $V_\Gamma{}^*$.

(ii) There exists $(x^*, y^*) \in \mathscr{D}$ such that

$$\inf_{y \in \mathscr{Y}(x^*)} f(x^*, y) = \sup_{x \in \mathscr{X}(y^*)} f(x, y^*) = f(x^*, y^*) = V_\Gamma{}^*.$$

Theorem 1.1. Every optimal pair is a saddle point.
This can be readily seen from Definition 1.3.

Theorem 1.2. If Γ is such that the following condition (C):

$$V^+(\Gamma) \leqslant V^-(\Gamma)$$

holds, then every saddle point of Γ is also an optimal pair.

Proof. If (x_*, y_*) is a saddle point, then, by Definition 1.1, we have

$$f(x, y_*) \leqslant f(x_*, y_*) \leqslant f(x_*, y)$$

for every $x \in \mathscr{X}(y_*)$ and every $y \in \mathscr{Y}(x_*)$. Thus,

$$\sup_{x \in \mathscr{X}(y_*)} f(x, y_*) \leqslant f(x_*, y_*) \leqslant \inf_{y \in \mathscr{Y}(x_*)} f(x_*, y),$$

which leads to

$$\begin{aligned} \inf_{y \in Y} \sup_{x \in \mathscr{X}(y_*)} f(x, y) &\leqslant \sup_{x \in \mathscr{X}(y_*)} f(x, y_*) \leqslant f(x_*, y_*) \\ &\leqslant \inf_{y \in \mathscr{Y}(x_*)} f(x_*, y) \leqslant \sup_{x \in X} \inf_{y \in \mathscr{Y}(x_*)} f(x_*, y), \end{aligned}$$

so that, if condition (C) is satisfied, we have

$$V^+(\Gamma) = V^-(\Gamma) = V_\Gamma{}^*$$

and

$$\sup_{x \in \mathscr{X}(y_*)} f(x, y_*) = \inf_{y \in \mathscr{Y}(x_*)} f(x_*, y) = f(x_*, y_*) = V_\Gamma{}^*.$$

Therefore, from Definitions 1.2 and 1.3, (x_*, y_*) is an optimal pair.

We note that, contrary to ordinary two-person, zero-sum games, condition (C) is not always true, as will be shown in an example in the next section.

2. Matrix Games

Let A be a rectangular $n \times m$ array and D a subset of $I \times J$, where

$$I \triangleq \{1, 2, ..., n\}, \qquad J = \{1, 2, ..., m\}.$$

We assume that an element a_{ij} of A is defined only if $(i, j) \in D$. For every $i \in I$ and $j \in J$, we shall set

$$T(i) \triangleq \{j \in J\colon \ (i, j) \in D\}, \qquad S(j) \triangleq \{i \in I\colon \ (i, j) \in D\}.$$

In addition, we assume that, for every $i \in I$ and $j \in J$, the sets $T(i)$ and $S(j)$ are nonempty (one can always take out those rows and columns of A for which this assumption is not true and define new sets I and J).

We then define the game $\Gamma_A = \langle I, J, D, A \rangle$, where I and J represent the strategy sets of Players 1 and 2, respectively, and a_{ij} (resp., $-a_{ij}$) is the payoff to the first (resp., second) player when the pair $(i, j) \in D$ is played.

Applying Definitions 1.1–1.3 to the game Γ_A, we shall say that the pair $(i_*, j_*) \in D$ is a saddle point of Γ_A if it satisfies

$$a_{ij_*} \leqslant a_{i_*j_*} \leqslant a_{i_*j}$$

for every $i \in S(j_*)$ and every $j \in T(i_*)$. Moreover, we shall say that the pair $(i^*, j^*) \in D$ is an optimal pair of Γ_A if

(i) Γ_A has a value, i.e., there exists $v_A{}^*$ such that

$$v_A{}^* = V^+(\Gamma_A) = V^-(\Gamma_A),$$

(ii) $\max_{i \in S(j^*)} a_{ij^*} = \min_{j \in T(i^*)} a_{i^*j} = a_{i^*j^*} = v_A{}^*$.

We now give an example of a matrix game Γ_A for which

$$V^+(\Gamma_A) > V^-(\Gamma_A).$$

Let the game $\Gamma_A = \langle I, J, D, A \rangle$ be such that

$$I = J = \{1, 2\}, \qquad D = \{(1, 1), (2, 2)\}, \qquad a_{11} = 3, \qquad a_{22} = 5.$$

Player 2

Player 1		1	2
	1	3	
	2		5

We can see that

$$\max_{i \in I} \min_{j \in T(i)} a_{ij} = 5, \qquad \min_{j \in J} \max_{i \in S(j)} a_{ij} = 3,$$

so that we have

$$V^+(\Gamma_A) > V^-(\Gamma_A).$$

We can also see that the pairs (1, 1) and (2, 2) are both saddle points of the game Γ_A, which shows that, for the games that we study here, a saddle point is not necessarily an optimal pair and that two different saddle points do not always give the same payoff.

3. Mixed Strategies

Let P be the set of all $p = (p_1, p_2, \ldots, p_n)$ where $p_i \geqslant 0$ and $\sum_{i \in I} p_1 = 1$; and let Q be the set of all $q = (q_1, q_2, \ldots, q_m)$ where $q_j \geqslant 0$

and $\sum_{j \in J} q_j = 1$. We shall say that P and Q are the sets of mixed strategies for Players 1 and 2, respectively.

Let A^0 be the $n \times m$ matrix, the elements of which are given by

$$a_{ij}^0 = a_{ij} \quad \text{if} \quad (i, j) \in D,$$
$$a_{ij}^0 = 0 \quad \text{if} \quad (i, j) \notin D,$$

and let B^0 be the $n \times m$ matrix whose elements are such that

$$b_{ij}^0 = 1 \quad \text{if} \quad (i, j) \in D,$$
$$b_{ij}^0 = 0 \quad \text{if} \quad (i, j) \notin D.$$

Given a row vector $x = (x_1, x_2, ..., x_n)$ and a matrix M, $\tilde{x}$ will denote the column vector transpose of x, and $\tilde{M}$ the matrix transpose of M. We have M_j as the jth column of M and $|x|$ as the norm of x, i.e., $|x| = |x_1| + |x_2| + \cdots + |x_n|$.

The elements of P (resp., Q), in which all the components are zero except for the ith component (resp., jth component), which is equal to one, will be denoted by p^i (resp., q^j) and will be called pure strategies of Player 1 (resp., Player 2).

Let

$$R \triangleq \{(p, q) \in P \times Q: \quad pB^0\tilde{q} \neq 0\}.$$

To each $(p, q) \in R$ we attribute a payoff

$$F_A(p, q) = \Big[\sum_{(i,j) \in D} a_{ij} p_i q_j\Big] \Big/ \Big[\sum_{(i,j) \in D} p_i q_j\Big] = pA^0\tilde{q}/pB^0\tilde{q}.$$

Then, we can interpret the mixed strategies and the payoff defined above as follows: Player 1 (resp., Player 2) plays his pure strategy p^i (resp., q^j) with probability p_i (resp., q_j). If the outcome (i, j) of a trial belongs to D, then this yields a payoff a_{ij} to Player 1; and if (i, j) does not belong to D, then the players disregard this trial.[2] Thus, $F_A(p, q)$ represents Player 1's expected payoff, given that D occurs.

From the initial game Γ_A, we can thus define a new game

$$\hat{\Gamma}_A = \langle P, Q, R, F_A \rangle$$

such that

$$F_A(p^i, q^j) = a_{ij}, \qquad (i, j) \in D.$$

[2] In Refs. 1 and 2, the payoff is $L_A(p, q) = pA^0\tilde{q}$, which means that if the outcome (i, j) of a trial does not belong to D, then the corresponding payoff is assumed to be zero.

We shall set

$$\mathscr{S}(q) \triangleq \{p \in P: \ (p, q) \in R\}, \qquad \mathscr{T}(p) \triangleq \{q \in Q: \ (p, q) \in R\}.$$

According to Definitions 1.1–1.3, a pair $(p_*, q_*) \in R$ will be a saddle point for $\hat{\Gamma}_A$ if

$$F_A(p, q_*) \leqslant F_A(p_*, q_*) \leqslant F_A(p_*, q)$$

for every $p \in \mathscr{S}(q_*)$ and every $q \in \mathscr{T}(p_*)$; and a pair $(p^*, q^*) \in R$ will be an optimal pair for $\hat{\Gamma}_A$ if

(i) $\hat{\Gamma}_A$ has a value, that is, there exists $V_A{}^*$ such that

$$V_A{}^* = V^+(\hat{\Gamma}_A) = V^-(\hat{\Gamma}_A).$$

(ii) $\sup_{p \in \mathscr{S}(q^*)} F_A(p, q^*) = \inf_{q \in \mathscr{T}(p^*)} F_A(p^*, q^*) = F_A(p^*, q^*) = V_A{}^*$.

Let $\bar{P}$ and $\bar{Q}$ be the pure strategy sets of Players 1 and 2, respectively. $\bar{P}$ and $\bar{Q}$ are finite subsets of P and Q. For $(p, q) \in P \times Q$ and $(p^i, q^j) \in \bar{P} \times \bar{Q}$, we shall set

$$S(q) \triangleq \{i \in I: \ q\tilde{B}_i{}^0 \neq 0\}, \qquad T(p) \triangleq \{j \in J: \ pB_j{}^0 \neq 0\},$$

$$F_A(p, j) \triangleq F_A(p, q^j), \qquad F_A(i, q) \triangleq F_A(p^i, q),$$

$$\bar{\mathscr{S}}(q) \triangleq \mathscr{S}(q) \cap \bar{P}, \qquad \bar{\mathscr{T}}(p) \triangleq \mathscr{T}(p) \cap \bar{Q}.$$

We have

$$\bar{\mathscr{S}}(q) \subset \mathscr{S}(q) \quad \text{and} \quad \bar{\mathscr{T}}(p) \subset \mathscr{T}(p).$$

Theorem 3.1. Every saddle point of Γ_A is also a saddle point of $\hat{\Gamma}_A$.

Proof. If $(i_*, j_*) \in D$ is a saddle point of Γ_A, then we have

$$a_{ij_*} \leqslant a_{i_*j_*} \leqslant a_{i_*j} \tag{1}$$

for every $i \in S(j_*)$ and every $j \in T(i_*)$; and if $q = (q_1, q_2, \ldots, q_n) \in Q$, from (1) we obtain

$$\Big[\sum_{j \in T(i_*)} q_j a_{i_*j}\Big] \Big/ \Big[\sum_{j \in T(i_*)} q_j\Big] \geqslant a_{i_*j_*}$$

for every $q \in Q$ such that

$$\sum_{j \in T(i_*)} q_j > 0.$$

That is, we have

$$q\tilde{A}^0_{i_*}/q\tilde{B}^0_{i_*} \geqslant a_{i_*j_*} \quad \text{for every} \quad q \in Q \quad \text{such that} \quad q\tilde{B}^0_{i_*} > 0,$$

or

$$F_A(i_*, q) \geqslant F_A(i_*, j_*) \quad \text{for every} \quad q \in \mathscr{T}(i_*).$$

Similarly, one could show that

$$F_A(p, j_*) \leqslant F_A(i_*, j_*) \quad \text{for every} \quad p \in \mathscr{S}(j_*).$$

Therefore, the pair (p^{i_*}, q^{j_*}) of pure strategies is a saddle point of $\hat{\Gamma}_A$.

Conversely, one could see that, if (p^{i_*}, q^{j_*}) is a saddle point of pure strategies for $\hat{\Gamma}_A$, then the pair (i_*, j_*) is a saddle point of Γ_A.

Theorem 3.2. If the games Γ_A and $\hat{\Gamma}_A$ have values, then these values are equal and every optimal pair for Γ_A is also an optimal pair for $\hat{\Gamma}_A$.

Proof. The first assertion results from the fact that $V^+(\Gamma_A) \leqslant V^+(\hat{\Gamma}_A)$ and $V^-(\Gamma_A) \geqslant V^-(\hat{\Gamma}_A)$. Now, let (i^*, j^*) be an optimal pair for Γ_A; then, from Theorem 1.1, (i^*, j^*) is also a saddle point for Γ_A, so that, from Theorem 3.1 (i^*, j^*) is a saddle point for $\hat{\Gamma}_A$; and, using Theorem 1.2, we conclude that (p^{i^*}, q^{j^*}) is an optimal pair for $\hat{\Gamma}_A$.

Theorem 3.3. For every $p \in P$, we have

$$\inf_{q \in \mathscr{T}(p)} F_A(p, q) = \min_{j \in T(p)} F_A(p, j).$$

Proof. First, we note that, since $\bar{\mathscr{T}}(p) \subset \mathscr{T}(p)$, we have

$$\inf_{q \in \mathscr{T}(p)} F_A(p, q) \leqslant \min_{q \in \bar{\mathscr{T}}(p)} F_A(p, q). \tag{2}$$

Setting

$$\lambda = \min_{q \in \bar{\mathscr{T}}(p)} F_A(p, q) = \min_{j \in T(p)} F_A(p, j),$$

we obtain

$$F_A(p, j) \geqslant \lambda \quad \text{for every} \quad j \in T(p),$$

that is,

$$pA_j^0/pB_j^0 \geqslant \lambda \quad \text{for every} \quad j \in T(p),$$

or

$$pA_j^0 \geqslant \lambda pB_j^0 \quad \text{for every} \quad j \in T(p).$$

Moreover, one can easily verify that

$$pB_j^0 = 0 \Rightarrow pA_j^0 = 0.$$

Thus, we have

$$pA_j^0 \geqslant \lambda p B_j^0 \qquad \text{for every} \quad j \in J$$

and

$$pA^0\tilde{q} \geqslant \lambda p B^0 \tilde{q} \qquad \text{for every} \quad q \in \mathscr{T}(p),$$

or

$$F_A(p, q) \geqslant \lambda \qquad \text{for every} \quad q \in \mathscr{T}(p),$$

which gives us

$$\inf_{q \in \mathscr{T}(p)} F_A(p, q) \geqslant \min_{j \in T(p)} F_A(p, j).$$

This last inequality, together with Ineq. (2), leads to

$$\inf_{q \in \mathscr{T}(p)} F_A(p, q) = \min_{j \in T(p)} F_A(p, j).$$

Similarly, it can be shown that

$$\sup_{p \in \mathscr{S}(q)} F_A(p, q) = \max_{i \in S(q)} F_A(i, q) \qquad \text{for every} \quad q \in Q.$$

Corollary 3.3. If the game $\hat{\Gamma}_A$ has a value $V_A{}^*$ and an optimal pair (p^{i^*}, q^{j^*}) of pure strategies, then the game Γ_A has a value equal to $V_A{}^*$ and (i^*, j^*) is an optimal pair for the game Γ_A.

Proof. First, we notice that $(i^*, j^*) \in D$ since $(p^{i^*}, q^{j^*}) \in R$. From Definition 1.3 and Theorem 3.3, we have

$$\min_{j \in T(p^{i^*})} F_A(p^{i^*}, j) = \max_{i \in S(q^{j^*})} F_A(i, q^{j^*}) = V_A{}^*$$

or

$$\min_{j \in T(i^*)} a_{i^*j} = \max_{i \in S(j^*)} a_{ij^*} = V_A{}^*,$$

from which we deduce that (i^*, j^*) is a saddle point for Γ_A and

$$V^+(\Gamma_A) = \max_{i \in I} \min_{j \in T(i)} a_{ij} \geqslant V_A{}^* \geqslant \min_{j \in J} \max_{i \in S(j)} a_{ij} = V^-(\Gamma_A).$$

Then, since $V^-(\Gamma_A) \geqslant V^-(\hat{\Gamma}_A) = V_A{}^* = V^+(\hat{\Gamma}_A) \geqslant V^+(\Gamma_A)$, we have

$$V^+(\Gamma_A) = V_A{}^* = V^-(\Gamma_A).$$

Thus, the game Γ_A has a value equal to V_A^+, and, using Theorem 1.2, we conclude that (i^*, j^*) is an optimal pair for Γ_A.

Theorem 3.4. The following result holds:

$$\sup_{p\in P} \min_{j\in T(p)} F_A(p,j) = \max_{p\in P} \min_{j\in T(p)} F_A(p,j).$$

Proof. Since P is a compact subset of $\mathbb{R}^n$, we need only to show that the function

$$G_A(p) = \min_{j\in T(p)} F_A(p,j)$$

is upper semicontinuous. This can be done in the following manner. Let $\{p^{(n)}\}$ be a converging sequence in P, and let p be its limit. Assume that

$$G_A(p^{(n)}) \geqslant \mu, \qquad n = 1, 2, \ldots,$$

so that

$$p^{(n)}A_j^0/p^{(n)}B_j^0 \geqslant \mu \qquad \text{for every} \quad j \in T(p^{(n)}) \quad \text{and every} \quad n = 1, 2, \ldots;$$

and, since

$$p^{(n)}B_j^0 = 0 \Rightarrow p^{(n)}A_j^0 = 0,$$

we have

$$p^{(n)}A_j^0 \geqslant \mu p^{(n)}B_j^0 \qquad \text{for every} \quad j \in J \quad \text{and} \qquad n = 1, 2, \ldots .$$

Now, let n go to infinity. Then, we obtain

$$pA_j^0 \geqslant \mu p B_j^0 \qquad \text{for every} \quad j \in J,$$

which implies that

$$pA_j^0/pB_j^0 \geqslant \mu \qquad \text{for every} \quad j \in T(p),$$

or

$$F_A(p,j) \geqslant \mu \qquad \text{for every} \quad j \in T(p),$$

and we obtain

$$G_A(p) = \min_{j\in T(p)} F_A(p,j) \geqslant \mu.$$

Therefore, the function G_A is upper semicontinuous.

Similarly, it can be seen that the function

$$H_A(q) = \max_{i\in S(q)} F_A(i,q)$$

is lower semicontinuous and thus we also have

$$\inf_{q\in Q} \max_{i\in S(q)} F_A(i, q) = \min_{q\in Q} \max_{i\in S(q)} F_A(i, q).$$

If we set

$$V_A^+ = V^+(\hat{\Gamma}_A), \qquad V_A^- = V^-(\hat{\Gamma}_A),$$

then from Theorems 3.4 and 3.5 we obtain

$$V_A^+ = \max_{p\in P} \min_{j\in T(p)} F_A(p, j), \qquad V_A^- = \min_{q\in Q} \max_{i\in S(q)} F_A(i, q).$$

4. Existence of a Saddle Point

We shall now prove the existence of a saddle point in the game $\hat{\Gamma}_A$. To this end, we shall consider the following problems.

Problem 4.1. $w^* = \max w, \qquad p(A^0 - wB^0) \geqslant 0, \qquad p \in P.$

Problem 4.2. $z^* = \min z, \qquad (A^0 - zB^0)\tilde{q} \leqslant 0, \qquad q \in Q.$

Theorem 4.1. If w^* is the solution of Problem 4.1, then we have

$$w^* = V_A^+.$$

Proof. Let w^* be the solution of Problem 4.1, and let $p_* \in P$ be such that

$$p_*(A^0 - w^*B^0) \geqslant 0.$$

Then, we have

$$w^* \leqslant F_A(p_*, j) \qquad \text{for every} \quad j \in T(p_*),$$

so that

$$w^* \leqslant V_A^+.$$

Suppose that

$$w^* < V_A^+;$$

then, there exists $\bar{p} \in P$ such that

$$F_A(\bar{p}, j) > w^*, \qquad j \in T(\bar{p}). \tag{3}$$

Now, let $\bar{w}$ be the solution of the following problem:

$$\bar{w} = \max w, \qquad \bar{p}(A^0 - wB^0) \geqslant 0.$$

We have

$$\bar{w} \leqslant w^*. \tag{4}$$

Moreover, there exists $j_0 \in T(\bar{p})$ such that

$$\bar{p}(A_{j_0}^0 - \bar{w}B_{j_0}^0) = 0.$$

Thus,

$$\bar{w} = \bar{p}A_{j_0}^0/\bar{p}B_{j_0}^0 = F_A(\bar{p}, j_0);$$

and, from (3), we obtain

$$\bar{w} > w^*,$$

which contradicts (4). Therefore, we have

$$w^* = V_A{}^+.$$

Similarly, it can be shown that if z^* is the solution of Problem 4.2, then

$$z^* = V_A{}^-.$$

If we set

$$\mathscr{L}^* \triangleq \{p \in P: p(A^0 - w^*B^0) \geqslant 0\}, \quad K(p) \triangleq \{k \in J: p(A_k{}^0 - w^*B_k{}^0) = 0\},$$

we have the following lemma.

Lemma 4.1. If $p_* \in \mathscr{L}^*$, then $K(p_*) \cap T(p_*) \neq \varnothing$.

Proof. Assume that $K(p_*) \cap T(p_*) = \varnothing$, with $p_* \in \mathscr{L}^*$; then, we have

$$p_*(A_k{}^0 - w^*B_k{}^0) > 0 \qquad \text{for every} \quad k \in T(p_*).$$

This gives

$$F_A(p_*, k) = p_*A_k{}^0/p_*B_k{}^0 > w^* \qquad \text{for every} \quad k \in T(p_*).$$

Hence,

$$V_A{}^* > w^*,$$

which contradicts Theorem 4.1.

Definition 4.1. Let $p_* \in \mathscr{L}^*$; we shall say that p_* satisfies Property ($\mathscr{P}$) if

$$\lambda \in \mathbb{R}, \quad x \in \mathbb{R}^n, \quad x \geqslant 0, \quad -\lambda p_*B_k{}^0 + x(A_k{}^0 - w^*B_k{}^0) \geqslant 0, \quad \forall k \in K(p_*),$$

implies that $\lambda \geqslant 0$.

Theorem 4.2. Let $p_* \in \mathscr{L}^*$; then, one of the following is true: (i) if $T(p_*) = J$, then p_* satisfies $(\mathscr{P})$; (ii) if $T(p_*) \neq J$ and p_* does not satisfy $(\mathscr{P})$, then there exists $p_{**} \in \mathscr{L}^*$ such that

$$T(p_{**}) \supset T(p_*) \qquad \text{with} \quad T(p_{**}) \neq T(p_*).$$

Proof. Suppose that p_* belongs to $\mathscr{L}^*$ and does not satisfy $(\mathscr{P})$; then, there exists $\bar{x} \in \mathbb{R}^n$, $\bar{x} \geqslant 0$, and $\lambda > 0$ such that

$$-\lambda p_* B_k^0 + \bar{x}(A_k^0 - w^* B_k^0) \geqslant 0 \qquad \text{for every} \quad k \in K(p_*).$$

Thus, we have

$$\bar{x}(A_k^0 - w^* B_k^0) \geqslant 0, \qquad k \in K(p_*), \tag{5}$$

$$\bar{x}(A_k^0 - w^* B_k^0) > 0, \qquad k \in K(p_*) \cap T(p_*). \tag{6}$$

Since $p_* \in \mathscr{L}^*$, then, from Lemma 4.1,

$$K(p_*) \cap T(p_*) \neq \varnothing.$$

Thus, from (6), we have $|\,\bar{x}\,| \neq 0$. Setting $\bar{p} = \bar{x}/|\,\bar{x}\,|$, we have $\bar{p} \in P$; and, from (5) and (6), we obtain

$$\bar{p}(A_k^0 - w^* B_k^0) \geqslant 0, \qquad k \in K(p_*), \tag{7}$$

$$\bar{p}(A_k^0 - w^* B_k^0) > 0, \qquad k \in K(p_*) \cap T(p_*). \tag{8}$$

Set

$$\bar{K}(p_*) \triangleq \{k \in J: \quad k \notin K(p_*)\}.$$

Then,

(a) if $\bar{K}(p_*) = \varnothing$ [i.e., $K(p_*) = J$], we shall take $p_{**} = \bar{p}$, (9)

(b) if $\bar{K}(p_*) \neq \varnothing$, we shall take $p_{**} = \mu p_* + (1 - \mu)\,\bar{p}$, (10)

where

$$\mu = 0 \qquad \text{if} \quad \nu' > 0.$$

$$\nu'/(\nu' - \nu) < \mu < 1 \qquad \text{if} \quad \nu' \leqslant 0,$$

with

$$\nu = \min_{k \in \bar{K}(p_*)} p_*(A_k^0 - w^* B_k^0) > 0, \qquad \nu' = \min_{k \in \bar{K}(p_*)} \bar{p}(A_k^0 - w^* B_k^0).$$

P is a convex set and p_*, $\bar{p} \in P$, so that p_{**} in (10) belongs to P. Then, using (7) and (8), we can verify that p_{**}, as defined in (9) or (10), satisfies the following inequalities:

$$\begin{aligned} p_{**}(A_k^0 - w^* B_k^0) &\geqslant 0, \qquad k \in J, \\ p_{**}(A_k^0 - w^* B_k^0) &> 0, \qquad k \in T(p_*). \end{aligned} \tag{11}$$

From these inequalities, we see that

$$T(p_{**}) \supset T(p_*) \qquad \text{and} \qquad p_{**} \in \mathscr{L}^*.$$

Therefore, from Lemma 4.1, we have

$$K(p_{**}) \cap T(p_{**}) \neq \varnothing.$$

This, together with (11), shows that there exists $k \in T(p_{**})$ such that $k \notin T(p_*)$. Thus, the second part of our theorem is proven.

Now, if $T(p_*) = J$, then (11) shows that

$$K(p_{**}) = \varnothing,$$

which contradicts Lemma 4.1, since $p_{**} \in \mathscr{L}^*$. We thus have the first assertion of our theorem.

Theorem 4.3. There exists $p_* \in \mathscr{L}^*$ such that $(\mathscr{P})$ holds true.

Proof. Let $p_*' \in \mathscr{L}^*$; then, from Theorem 4.2, we have: (i) if $T(p_*') = J$, then p_*' satisfies Property $(\mathscr{P})$ and we can take $p_* = p_*'$; (ii) if $T(p_*') \neq J$ and does not satisfy $(\mathscr{P})$, then there exists $p_*'' \in \mathscr{L}^*$ such that

$$T(p_*'') \supset T(p_*') \qquad \text{and} \qquad T(p_*'') \neq T(p_*').$$

It is easy to see that, by repeating this process a finite number of times, we will obtain a p_* such that $(\mathscr{P})$ holds.

Now, in order to prove the existence of a saddle point for a game $\hat{\Gamma}_A$, we shall state the following well-known lemma.

Lemma 4.2. (Farkas, Ref. 3). Let M be a matrix and r a row vector; if, for every row vector x such that $M\tilde{x} \geqslant 0$, we have $r\tilde{x} \geqslant 0$, there exists a nonnegative row vector u such that $r = uM$.

Theorem 4.4. $\hat{\Gamma}_A$ has a saddle point (p_*, q_*) such that

$$F_A(p_*, q_*) = V_A^+.$$

Proof. Theorem 4.3 tells us that there exists $p_* \in \mathscr{L}^*$ such that $\{\lambda \in \mathbb{R},\ x \in \mathbb{R}^n,\ x \geqslant 0,\ -\lambda p_* B_k{}^0 + x(A_k{}^0 - w^* B_k{}^0) \geqslant 0,\ k \in K(p_*)\} \Rightarrow \lambda \leqslant 0.$ Therefore, from Lemma 4.2, there exists

$$y = (y_1, y_2, \ldots, y_{|K(p_*)|}), \qquad y_k \geqslant 0,$$

such that

$$\sum_{k \in K(p_*)} y_k(p_* B_k{}^0) = 1, \tag{12}$$

$$\sum_{k \in K(p_*)} y_k(a_{ik}^0 - w^* b_{ik}^0) \geqslant 0, \qquad i \in I. \tag{13}$$

From (12), we have

$$v = \sum_{k \in K(p_*)} y_k > 0.$$

Then, we can set

$$\begin{aligned} q_{*j} &= y_j/v && \text{if } \ j \in K(p_*), \\ q_{*j} &= 0 && \text{if } \ j \in J, \ \ j \notin K(p_*), \end{aligned}$$

and we have

$$q_* = (q_{*1}, q_{*2}, \ldots, q_{*m}) \in Q.$$

Thus, (12) gives

$$p_* B^0 \tilde{q}_* > 0.$$

Hence, $(p_*, q_*) \in R$. Moreover, (13) becomes

$$(A^0 - w^* B^0)\, \tilde{q}_* \leqslant 0. \tag{14}$$

On the other hand, since $p_* \in \mathscr{L}^*$, we have

$$p_*(A^0 - w^* B^0) \geqslant 0. \tag{15}$$

From (14) and (15), we obtain

$$pA^0\tilde{q}_*/pB^0\tilde{q}_* \leqslant w^* \leqslant p_* A^0 \tilde{q}/p_* B^0 \tilde{q}$$

for every $p \in \mathscr{S}(q_*)$ and every $q \in \mathscr{T}(p_*)$; and, since $(p_*, q_*) \in R$, then we have

$$F_A(p, q_*) \leqslant w^* = F_A(p_*, q_*) \leqslant F_A(p_*, q)$$

for every $p \in \mathscr{S}(q_*)$ and every $q \in \mathscr{T}(p_*)$. Therefore, (p_*, q_*) is a saddle point such that

$$F_A(p_*, q_*) = w^* = V_A^+.$$

Similarly, we can show that there exists a saddle point $(\bar{p}_*, \bar{q}_*)$ such that

$$F_A(\bar{p}_*, \bar{q}_*) = V_A^-.$$

Therefore, if the following condition (C_A) holds:

$$V_A^+ \leqslant V_A^-,$$

then, from Theorem 1.2, the game $\hat{\Gamma}_A$ will have an optimal pair. But condition (C_A) is not always satisfied, as we see in the following example.

Let a game Γ_A be described by the following array:

		Player 2 1	2	3
Player 1	1	5		
	2		13	2
	3		3	12

If $\bar{p} = (0, \frac{1}{2}, \frac{1}{2})$, then $T(\bar{p}) = \{2, 3\}$ and

$$F_A(\bar{p}, 2) = 8, \qquad F_A(\bar{p}, 3) = 7.$$

Hence,

$$\min_{j \in T(\bar{p})} F_A(\bar{p}, j) = 7 \qquad \text{and} \qquad V_A^+ \geqslant 7.$$

If $\bar{q} = (1, 0, 0)$, then

$$S(\bar{q}) = \{1\} \qquad \text{and} \qquad F_A(1, \bar{q}) = 5.$$

Thus,

$$\max_{i \in S(q)} F_A(i, \bar{q}) = 5 \qquad \text{and} \qquad V_A^- \leqslant 5.$$

Therefore,

$$V_A^+ > V_A^-.$$

On the other hand, we can easily verify that the pair (1, 1) is an optimal pair for Γ_A.

This last remark shows once more that many of the well-known properties of ordinary, two-person, zero-sum games do not hold true when the payoff is not defined for every possible outcome.

In order to make the theory of matrix games with incompletely defined payoff more complete, one needs a condition for the existence of an optimal pair [i.e, for condition (C_A) to hold]. This will be the object of a forthcoming study.

References

1. Blaquière, A., *Quantitative Games: Problem Statement and Examples, New Geometric Aspects*, Advanced Study Institute on the Theory and Application of Differential Games, Control Theory Centre, University of Warwick, Coventry, England, 1974.
2. Blaquière, A., *An Example of Multistage Stochastic Game*, paper presented at the Symposium on Optimization Problems in Engineering and Economics, Naples, Italy, 1974.
3. Berge, C., *Espaces Topologiques, Fonctions Multivoques*, Dunod, Paris, France, 1966.

XXV

On a Problem of Stochastic Differential Games

P. T. Liu

Abstract. The process of bargaining between management and union during a strike is modelled by a nonlinear stochastic differential game. It is assumed that the two sides bargain in the mood of a cooperative game. A pair of Pareto-optimal strategies is obtained.

1. Introduction

In Ref. 1, the process of bargaining between management and union during a strike was modeled by a nonzero-sum differential game. One of the two cases considered in that paper is when the two sides bargain in the "mood" of a cooperative game. Some results were obtained for Pareto-optimal strategies when the model was specialized to a linear scalar differential game.

In the present paper, we consider a more general model for the same process, i.e., a nonlinear, stochastic differential game. The nonlinearity that we are going to introduce in the model is such that if the initial offer and demand from the two sides are too far apart, their difference would decrease fairly rapidly. If the difference is sufficiently small, the nonlinear terms become relatively unimportant. The behavior of the two sides, as far as seeking optimal strategies is concerned, would be close to the behavior with a linear model, i.e., the Pareto-optimal strategies obtained in Ref. 1.

We also assume that the bargaining process is subject to interruptions due to some random factor beyond the control of either side. To take this into consideration in our model, we introduce a Poisson process with two states, 0 and 1. When the Poisson process is in the state 1, the bargaining process is going on, and when it is in the state 0, the process is temporarily "broken up." The whole process of negotiation is therefore represented by two Markov processes which stop when their difference takes on a certain fixed value.

2. Mathematical Formulation

Let $\{N(t), t \geqslant 0\}$ be a Poisson process with two states 0 and 1, and $N(0) = 1$, with probability 1. The transition probability of $N(t)$ is given by

$$P\{N(t + \Delta t) = 0 \mid N(t) = 1\} = \mu \, \Delta t + o(\Delta t) \tag{1}$$

and

$$P\{N(t + \Delta t) = 1 \mid N(t) = 0\} = \lambda \, \Delta t + o(\Delta t), \tag{2}$$

where μ and λ are positive constants.

Consider a differential game described by two scalar stochastic differential equations

$$\dot{x} = [u(y - x) + \epsilon_1(y - x)^2] \, N(t), \tag{3}$$

$$\dot{y} = -[v(y - x) + \epsilon_2(y - x)^2] \, N(t), \tag{4}$$

where $\epsilon_i > 0$, $i = 1, 2$, $0 \leqslant u \leqslant a$, and $0 \leqslant v \leqslant b$. For the purpose of minimizing the payoffs to be given later, it is convenient to combine (3) and (4) and have

$$\dot{z} = -[(u + v)z + \epsilon z^2] \, N(t), \tag{5}$$

where $z = y - x$ and $\epsilon = \epsilon_1 + \epsilon_2$. We assume that $z(0) = y(0) - x(0) > 0$ and the game is terminated at $t = T$, the first instant when $z(t) = y(t) - x(t) = m$, where $0 < m < z(0)$.

An admissible strategy pair (u, v) is a function of z and N, i.e., $u = u(z, N)$ and $v = v(z, N)$, $z \in [m, \infty[$, $N = 0, 1$, such that, for given (deterministic) $x(0)$, $y(0)$, and $z(0)$, there exists at least one solution $x(t)$, $y(t)$, and $z(t)$ to (3), (4), and (5), respectively, satisfying

(i) $z(T) = m$ for some (random) T;

(ii) $0 \leqslant u(z(t), N(t)) \leqslant a$ and $0 \leqslant v(z(t), N(t)) \leqslant b$, for all $t \in [0, T]$;

with probability 1. Of course, when $N(t) = 0$ in an interval the choices for $u(z, 0)$ and $v(z, 0)$ are irrelevant, since $x(t)$, $y(t)$, and $z(t)$ are equal to constants in that interval. The two players therefore choose their strategies $u(z, 1)$ and $v(z, 1)$ so as to minimize the following payoffs, respectively:

$$J_1 = E\{k_1 T + x(T) \mid N(0) = 1\}, \tag{6}$$

$$J_2 = E\{k_2 T - y(T) \mid N(0) = 1\}, \tag{7}$$

where the k_i $(i = 1, 2)$ are positive constants. Equations (6) and (7) can also be written in the integral form as

$$J_1 = E\left\{\int_0^T [k_1 + (u(z(t), N(t))\, z(t) + \epsilon_1 z^2(t))\, N(t)]\, dt \mid N(0) = 1\right\} - x(0), \tag{8}$$

$$J_2 = E\left\{\int_0^T [k_2 + (v(z(t), N(t))\, z(t) + \epsilon_2 z^2(t))\, N(t)]\, dt \mid N(0) = 1\right\} - y(0). \tag{9}$$

Equations (3), (4), (8), and (9) describe a nonzero-sum stochastic differential game. We assume that the game is cooperative and a Pareto-optimal strategy pair $(u^*(z, 1), v^*(z, 1))$ is to be sought, i.e.,

$$J(u^*(z, 1), v^*(z, 1)) \leqslant J(u(z, 1), v(z, 1)) \tag{10}$$

for all admissible pairs (u, v), where $J = \gamma_1 J_1 + \gamma_2 J_2$ and $\gamma_i > 0$, $i = 1, 2$.

We further assume that $\gamma_1 > \gamma_2$. Minimizing $J = \gamma_1 J_1 + \gamma_2 J_2$ is then equivalent to minimizing[1]

$$C = E\left\{\int_0^T [k + (\gamma u(z(t), N(t)) + \gamma \epsilon_1 z^2(t))\, N(t)]\, dt \mid N(0) = 1\right\}, \tag{11}$$

where $\gamma = \gamma_1 - \gamma_2 > 0$ and $k = \gamma_1 k_1 + \gamma_2 k_2$. Thus, instead of dealing with (3), (4), (8), and (9), we seek to minimize (11) subject to (5).

3. Pareto-Optimal Strategies

Let $z(t)$ be the solution of (5) corresponding to an admissible pair $(u(z, N), v(z, N))$. Clearly $\{z(t), t \geqslant 0\}$ is a Markov process, which is

[1] In our original paper, which appeared in Journal of Optimization Theory and Applications, January 1976, a factor of γ was inadvertently dropped from (11) and subsequent equations. The corrections in this paper lead to slightly different results in the next section. In particular, the optimal strategies cannot have more than one switching.

continuous; with probability 1. Let $V = V(z, N)$ be continuously differentiable in z for $N = 0, 1$. Define an operator D on V by

$$DV(z(t), N(t)) = \lim_{h\to 0} (1/h)\, E\{V(z(t+h), N(t+h)) - V(z(t), N(t)) \mid z(t), N(t)\}.$$

Then, according to Dynkin's formula (see Ref. 2), we have, for any Markov time[2] τ

$$E\{V(z(\tau), N(\tau)) \mid z(0), N(0)\} - V(z(0), N(0)) = E\left\{\int_0^\tau DV(z(t), N(t))\, dt \mid z(0), N(0)\right\}. \tag{12}$$

Also, taking (1) and (2) into account, we can show

$$DV(z(t), 0) = \lambda(V(z(t), 1) - V(z(t), 0)) \tag{13}$$

and

$$DV(z(t), 1) = -(\partial V/\partial z)(z(t), 1)\{[u(z(t), 1) + v(z(t), 1)]\, z(t) + \epsilon z^2(t)\} + \mu(V(z(t), 0) - V(z(t), 1)). \tag{14}$$

With the aid of (12)–(14), the following sufficiency theorem for an optimal strategy pair can be easily established.

Theorem. If there exist a function $V = V(z, N)$, continuously differentiable in z for $N = 1$, and $V(m, N) = 0$ for $N = 0, 1$, and an admissible strategy pair $(u^*(z, N), v^*(z, N))$ such that

$$\lambda(V(z, 1) - V(z, 0)) + k = 0 \tag{15}$$

and

$$\begin{aligned}-(\partial V/\partial z)(z, 1)[(\alpha + \beta)z + \epsilon z^2] + \mu[V(z, 0) - V(z, 1)] + k + \gamma\alpha z + \gamma\epsilon_1 z^2 \\ \geqslant -(\partial V/\partial z)(z, 1)[(u^*(z, 1) + v^*(z, 1))z + \epsilon z^2] + \mu[V(z, 0) \\ - V(z, 1)] + k + \gamma u^*(z, 1)z + \gamma\epsilon_1 z^2 = 0\end{aligned} \tag{16}$$

for all $0 \leqslant \alpha \leqslant a$ and $0 \leqslant \beta \leqslant b$, then $(u^*(z, N), v^*(z, N))$ is an optimal strategy pair and

$$V(z(0), 1) = \min_{(u,v)} C(u, v). \tag{17}$$

[2] Any first exit time of a Markov process is a Markov time.

The proof of this theorem can be found in Ref. 3, p. 118, and is omitted here.

When taken together, (15) and (16), call for the minimization of

$$-(\partial V/\partial z)(z, 1)[(\alpha + \beta)z + \epsilon z^2] + K + \gamma\alpha z + \gamma\epsilon_1 z^2 \tag{18}$$

with respect to (α, β), where $K = k(1 + (\mu/\lambda))$, and the minimizing pair (u^*, v^*) must be such that

$$-(\partial V/\partial z)[z(u^*(z, 1) + v^*(z, 1)) + \epsilon z^2] + K + \gamma u^*(z, 1)z + \epsilon_1 z^2 = 0. \tag{19}$$

Let z_1 be the (unique) positive root of the quadratic equation $\gamma\epsilon_2 z^2 + \gamma bz - K = 0$ and suppose

$$V(z, 1) = V_1(z), \qquad m < z < z_1,$$

$$V(z, 1) = V_2(z), \qquad z_1 < z,$$

where $V_1(z)$ and $V_2(z)$ are the solutions of

$$dV_1/dz = (K + \gamma az + \epsilon_1 z^2)/[(a + b)z + \epsilon z^2], \qquad m < z < z_1, \tag{20}$$

with $V_1(m) = 0$ and

$$dV_2/dz = (K + \epsilon_1 z^2)/(bz + \epsilon z^2), \qquad z > z_1, \tag{21}$$

with $V_2(z_1) = V_1(z_1)$. The solutions to (20) and (21) can be found by simple techniques of integration. Since they do not affect $(u^*(z, 1), v^*(z, 1))$, we do not give them here. The function $V(z, 1)$ thus defined is obviously continuously differentiable in z. We then have, from the minimization of (18),

$$u^*(z, 1) = \begin{cases} a, & m < z < z_1, \\ 0, & z_1 < z, \end{cases} \tag{22}$$

and

$$v^*(z, 1) = b, \qquad m < z, \tag{23}$$

since

$$(\partial V/\partial z)(z, 1) > \gamma, \qquad m < z < z_1,$$

$$(\partial V/\partial z)(z, 1) < \gamma, \qquad z_1 < z,$$

and

$$(\partial V/\partial z)(z, 1) > 0, \qquad m < z.$$

Furthermore, with (u^*, v^*) given by (22) and (23), (19) is satisfied. Hence, (22) and (23) give the optimal strategies in this case, when $N(t) = 1$.

If $z_1 < m$, there is no switching for $u^*(z, 1)$, i.e., $u^*(z, 1) = 0$ for all $z > m$ and $V(z, 1) = V_2(z)$ for all $z > m$, while V_2 satisfies (21) with $V_2(m) = 0$.

The other case, $\gamma_1 < \gamma_2$, can be handled in exactly the same way and similar results can be obtained, i.e., $u^*(z, 1) = a$ for all $z > m$ and there is either one or no switching for $v^*(z, 1)$.

4. Discussion

As in Ref. 1, x and y can represent, respectively, the offer from management and the demand from the union. We have found, as in Ref. 1, that, if $\gamma_1 > \gamma_2$, the union should be continuously making concessions by lowering the demand at the maximum rate possible whenever the negotiation is going on, whereas the optimal strategy for management could switch between 0 and a. The following remarks are noteworthy.

1. If m is very small, the optimal strategy for management, when z is sufficiently close to m, is $u^*(z, 1) = a$ in both cases discussed in the previous section. With the linear model in Ref. 1 such was also the case. Thus, the optimal strategies in the linear and the nonlinear models coincide for sufficiently small z.
2. The optimal switching times depend on, among other things, the ratio μ/λ and as long as $\mu/\lambda \ll 1$, the prospect of a possible "breakup" of the bargaining process affects the optimal strategies only slightly.

The choices for γ_1 and γ_2 are, of course, crucial to the determination of optimal strategies. This is related to the bargaining power of both sides, i.e., the ability of each side to threaten his opponent. For details, see Ref. 4.

References

1. Leitmann, G., and Liu, P. T., *A Differential Game Model of Labor-Management Negotiation During a Strike*, Journal of Optimization Theory and Applications, Vol. 13, No. 4, April 1974.
2. Dynkin, E. B., *Markov Processes*, Vol. 1, Springer-Verlag, Berlin, 1965.
3. Kushner, H., *Stochastic Stability and Control*, Academic Press, New York, 1967.
4. Liu, P. T., *Optimal Threat Strategies in Differential Games*, Journal of Mathematical Analysis and Applications, Vol. 43, No. 1, 1973.

XXVI

The Game of Two Identical Cars[1]

A. W. Merz

Abstract. This paper describes a third-order pursuit–evasion game in which both players have the same speed and minimum turn radius. The game of kind is first solved for the *barrier* or envelope of capturable states. When capture is possible, the game of degree is then solved for the optimal controls of the two players as functions of the relative position. The solution is found to include a universal surface for the pursuer and a dispersal surface for the evader.

1. Introduction

The two-car differential game problem was originally defined and examined by Isaacs in Ref. 1. In this pursuit–evasion game, the pursuer P and the evader E both have positive minimum-turn radii and constant speeds, and motion is restricted to a plane. The state vector of the game has three components, which are chosen as the Cartesian coordinates x, y of E's position relative to P and the angle θ between the two velocities. The *game of kind* terminates when the range-rate is zero, and in this game, P seeks to minimize the final range while E strives to maximize it. In the *game of degree*, termination occurs when E's separation from P becomes less than a specified capture radius, and it is the termination time that P seeks to minimize and E to maximize. The general two-car problem has three independent parameters: the speed ratio and the two ratios of capture radius to minimum-turn radius.

The present study is a specialization to the case of two identical cars; i.e., both P and E have unit velocity and unit maximum turn rate, so that only one parameter remains, which is the ratio β of capture

[1] The author is grateful to Professor J. V. Breakwell, Stanford University, for constructive criticism of the manuscript. Financial support for the research was provided at Stanford University under Air Force Contract No. F33615-70-C-1637.

radius to the common minimum-turn radius. It is assumed that the roles of P and E do not change during the game, so that, when capture can occur, a necessary condition to be satisfied by the saddle-point controls of the players is the Hamiltonian (which can be derived as in Ref. 1)

$$\min_{\sigma_1} \max_{\sigma_2}[V_x \dot{x} + V_y \dot{y} + V_\theta \dot{\theta} + 1] = 0. \tag{1}$$

Here, the controls are the normalized turn rates of P and E, σ_1 and σ_2, where $| \sigma_i | \leqslant 1$, which are to be found as functions of the state vector, $[x, y, \theta]$. The adjoint vector $[V_x, V_y, V_\theta]$ is known in terms of the terminal conditions of the game, so the overall game can be viewed generically as a two-point boundary-value problem.

1.1. Preliminary Remarks. Isaacs' treatment of this problem is restricted to the case of a faster pursuer, and is further limited to a study of the optimal strategies and trajectories on the *barrier*. This is a surface in $xy\theta$-space representing solutions to the *game of kind*. This form of the game is appropriate when P and E respectively seek to minimize and maximize the final value of the range (i.e., the miss-distance), which occurs when the range-rate is zero. By making the miss-distance a parameter, the solution to the game of kind can be presented as in the collision-avoidance problem (Ref. 2), for which either or both of the vehicles maneuver so as to maximize the miss-distance. When a particular final value of range is defined as the *capture radius*, the corresponding barrier separates the local state space into capture and escape regions. That is, the optimal capture time is discontinuous across the barrier, and on this surface the optimal paths contact the capture circle tangentially, without penetrating, as shown in Fig. 9.2.3 of Ref. 1. For arbitrary parameter values, the barrier may be open or closed; and, when it is closed, capture or termination of the game is possible under optimal play only for those initial states bounded between the capture circle and the barrier.

In the specialized case analyzed here, capture does not necessarily occur, because the players' speeds and maximum turn rates are equal. If the initial velocities are parallel (for example, $\theta = 0$), the equations of relative motion show that E can maintain the initial radial separation forever, by simply duplicating P's strategy. The barrier is therefore closed, and the *game of kind* is concerned with the determination of this surface.

The calculation of the barrier begins with the determination of a terminal relation among the state components, which expresses the tangential relative velocity condition. Values of the normalized adjoints are also known in terms of the state at this time. The controls of both players at termination are thus known. Retrogressive integration of the state equations is then possible in terms of the terminal states. Sketching a typical pair of trajectories in realistic space, however, shows that, when E is sufficiently far from P on this path, E could avoid P entirely by adopting a different policy, despite the fact that neither player's switch function has changed sign at this retrograde time. This initially puzzling feature of the solutions was found to have a precedent in the second-order homicidal chauffeur game. In this game, for certain parameter values, the right and left barriers intersect ahead of the capture circle at an evader's dispersal point, and the barrier loci beyond this point are discarded (Ref. 1, p. 235). In the present third-order problem, the two barrier surfaces are found to intersect along an evader's dispersal *line*, and the retrograde solutions beyond this line are no longer optimal.

When capture is possible; i.e., when E cannot maintain the range greater than the capture radius, termination must occur with E entering the capture circle between the two barrier surfaces. Solutions to the *game of degree* are then found by treating the *time* to capture as the performance criterion to be minimized and maximized by P and E. This is done by first determining strategies for P and E as functions of the state at the time of capture. This results in four possible sets of terminal controls, each of which is optimal in a specific area of the terminal region of the state space. As with the game of kind, the corresponding sets of terminal conditions can be integrated retrogressively, away from the capture circle, so as fill the capture volume with optimal trajectories. This procedure is complicated by the presence of a dispersal surface for E and a universal surface for P, but these surfaces intersect the capture circle along the lines which subdivide the terminal area, as mentioned above. Consequently, their influence on the shapes of the various strategy regions can be predicted qualitatively as these surfaces are generated away from the capture region. The actual calculation of these surfaces is by simultaneous numerical solution of a set of transcendental equations, which are the general solutions, in retrogressive time, to the differential equations of the game.

These surfaces essentially provide the solution to the game, by giving those states from which capture is possible, and by specifying optimal strategies for both players as functions of the state when capture is possible. The results are finally shown as $\theta =$ const sections of the three-dimensional state space, because this seems to be the simplest and clearest way of presenting these *strategy volumes*.

1.2. Equations of Motion and Terminal Conditions. With the notations shown in Fig. 1, the equations of relative motion of the game are found to be

$$\dot{x} = -\sigma_1 y + \sin\theta, \qquad \dot{y} = -1 + \sigma_1 x + \cos\theta, \qquad \dot{\theta} = -\sigma_1 + \sigma_2 . \quad (2)$$

Terminal conditions of the states are expressible as the vector

$$\mathbf{x}_0 = [\beta \sin\phi_0 , \beta\cos\phi_0 , \theta_0], \quad (3)$$

where the polar coordinate ϕ_0 must be such that the radial velocity at termination is nonpositive. That is,

$$\dot{r}_0 = \cos(\theta_0 - \phi_0) - \cos\phi_0 \leqslant 0. \quad (4)$$

The terminal condition of the game of kind is given by the equality in (4), so that a *safe-contact* trajectory which touches the capture circle tangentially must satisfy either

$$\theta_0 = 2\phi_0 \quad (5)$$

or

$$\theta_0 = 0. \quad (6)$$

These terminal conditions denote the *boundary of the usable part* (or BUP), as shown in Fig. 2, and the game of degree terminates on the portion of the capture circle between these lines, where $\dot{r}_0 < 0$. Note that, because the speeds of the players are equal, the BUP does not intersect the xy-plane at $\phi_0 = \pi/2$, as is the case when P is faster (Ref. 1, Fig. 9.2.2).

The *main equation* or Hamiltonian for the game of kind is written as

$$\min_{\sigma_1} \max_{\sigma_2} [\nu_x \dot{x} + \nu_y \dot{y} + \nu_\theta \dot{\theta}] = 0, \quad (7)$$

where the adjoint vector in this case can be initially normalized to unit

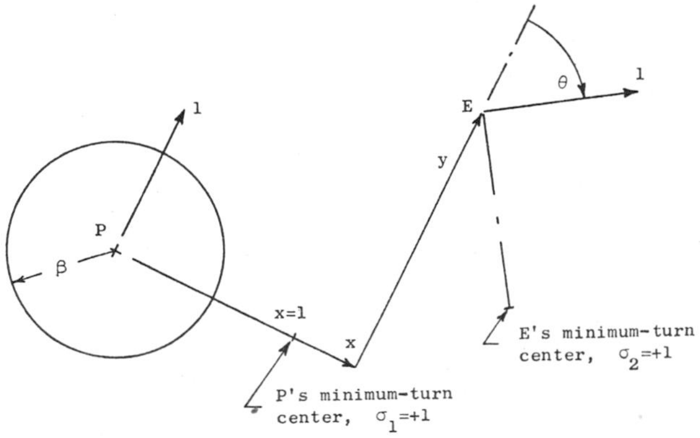

Fig. 1. Notations and coordinates.

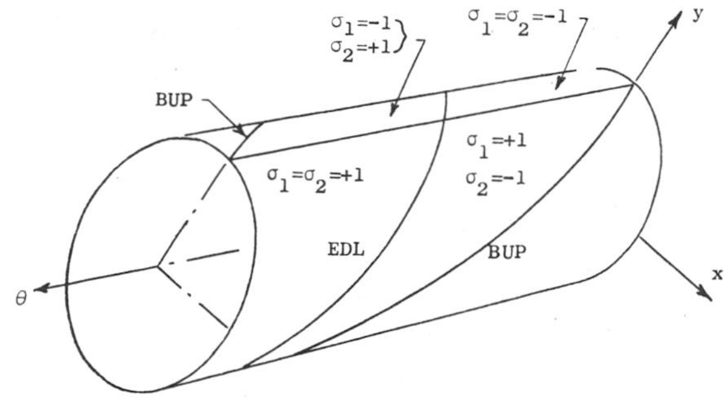

Fig. 2. Terminal strategies on capture circle.

magnitude as in Refs. 1 and 2. Substituting (2) into (7) and performing the indicated operations yields

$$\sigma_1 = \operatorname{sign} S \triangleq \operatorname{sign}(\nu_x y - \nu_y x + \nu_\theta), \qquad \sigma_2 = \operatorname{sign} \nu_\theta . \tag{8}$$

The adjoints in turn satisfy the equations $\nabla\dot{\boldsymbol{\nu}} = -\partial H/\partial \mathbf{x}$ or

$$\dot{\nu}_x = -\sigma_1 \nu_y , \qquad \dot{\nu}_y = \sigma_1 \nu_x , \qquad \dot{\nu}_\theta = -\nu_x \cos\theta + \nu_y \sin\theta, \tag{9}$$

where the values at termination of the game of kind are written vectorially as

$$\nabla \boldsymbol{\nu}_0 = [\sin\phi_0 , \cos\phi_0 , 0]. \tag{10}$$

The terminal states in the game of degree are as given by (3), where Ineq. (4) holds. Strategies are given in terms of the adjoints V_x, V_y, V_θ, which are the gradients given in (1). The adjoints satisfy the equations

$$\dot{V}_x = -\sigma_1 V_y , \qquad \dot{V}_y = \sigma_1 V_x , \qquad \dot{V}_\theta = -V_x \cos\theta + V_y \sin\theta, \tag{11}$$

and the terminal values are easily found in terms of the radial relative velocity; i.e., at termination, the main equation can be expressed in terms of polar coordinates, so that

$$V_{r_0} = -1/\dot{r}_0 , \tag{12}$$

where $\dot{r}_0$ is given by (4). Hence, denoting the retrograde derivative by a superscript circle, we have

$$V_{x_0} = \sin(\phi_0)/\mathring{r}_0 , \qquad V_{y_0} = \cos(\phi_0)/\mathring{r}_0 , \qquad V_{\theta_0} = 0, \tag{13}$$

where $\mathring{r}_0$ is positive when $\phi_0 < 2\theta_0$.

1.3. Terminal Strategies. When the barrier ends with $\theta_0 = 0$, two cases must be distinguished. For $\phi_0 \neq 0$, the terminal state vector is

$$\mathbf{x}_0 = [\beta \sin\phi_0 , \beta \cos\phi_0 , 0], \tag{14}$$

which, with (8) and (10), shows that both switch functions are zero at this time. The retrograde derivatives, however, give

$$dS/d(-t) = \mathring{S}_0 = \nu_{x_0} = \sin \phi_0 , \qquad d\nu_\theta/d(-t) = \mathring{\nu}_{\theta_0} = \sin \phi_0 , \tag{15}$$

so that, just before tangency, $\sigma_1 = \sigma_2 = \text{sign } x_0$. That is, P and E are turning in the same direction.

The second possibility when $\theta_0 = 0$ is that $\phi_0 = 0$, and here (15) shows that the first derivatives are also zero. The second retrograde derivatives, however, show that the strategies just before termination are $\sigma_1 = \text{sign } \sigma_1$ and $\sigma_2 = \text{sign } \sigma_2$. Consequently, three possible strategies for each player ($\sigma_i = -1, 0, +1$) satisfy the necessary conditions, and further tests of these nine possible strategy pairs will be made using the retrograde solutions of the next section.

For barrier termination when $\theta_0 = 2\phi_0 \neq 0$, the terminal state is

$$\mathbf{x}_0 = [\beta \sin \phi_0 , \beta \cos \phi_0 , 2\phi_0], \tag{16}$$

while the terminal adjoint is as given in (10). Again, both switch functions are zero at termination, and the retrograde derivatives yield

$$\mathring{S} = \sin \phi_0 , \qquad \mathring{\nu}_{\theta_0} = \sin(\phi_0 - \theta_0) = -\sin \phi_0 . \tag{17}$$

These results imply that, if E is not directly in front of or behind P, $\sigma_1 = \text{sign } \mathring{S}_0 = \text{sign } x_0$; that is, P is turning toward E just before tangency, as might be expected. Furthermore, $\sigma_2 = -\sigma_1$, so that E is turning in the opposite direction, or toward the outward normal to the capture circle. This is also physically appealing; and it may be noted that, in the homicidal chauffeur game, E's velocity is parallel to the terminal radius vector; analogies between these two games are often useful.

For the game of degree, the switch functions are identical in form to (8); i.e., the strategies are determined by

$$\sigma_1 = \text{sign}(V_x y - V_y x + V_\theta), \qquad \sigma_2 = \text{sign } V_\theta , \tag{18}$$

and (3) and (13) show that both switch functions are zero at termination. The retrograde derivatives, however, are

$$\mathring{S}_0 = V_{x_0} = \sin(\phi_0)/\mathring{r}_0 , \qquad \mathring{V}_{\theta_0} = -\sin(\theta_0 - \phi_0)/\mathring{r}_0 . \tag{19}$$

For $x_0 > 0$, these imply that P's strategy is $\sigma_1 = +1$ (P is turning right), while E's switch function is $\sigma_2 = -\text{sign} \sin(\theta_0 - \phi_0)$. That is, $\sigma_2 = -1$ if $2\phi_0 < \theta_0 < \phi_0 + \pi$, and $\sigma_2 = +1$ if $\phi_0 + \pi < \theta_0 < 2\pi$. The line $\theta_0 = \phi_0 + \pi$ on the capture circle (see Fig. 2) thus represents a *dispersal line* for E (or EDL), and either extreme strategy is optimal here.

On the other hand, if $x_0 = \phi_0 = 0$, the second retrograde derivative gives P's strategy as

$$\sigma_1 = \text{sign}\ \overset{\circ\circ}{S}_0 = \text{sign}\ \sigma_1\,, \tag{20}$$

which implies that $\sigma_1 = 0$ or ± 1. For this same condition, E's strategy is unique, $\sigma_2 = -\text{sign}\ \sin\theta_0$, while the multiple strategies for P are a consequence of the *universal surface* which intersects the front of the capture circle along the line $x = 0$, $y = \beta$. P's strategy changes for termination on either side of this line, as shown in Fig. 2.

1.4. General Solutions. We have shown that optimal controls are at extreme values, unless a switch function is identically zero. For reference purposes, the solutions corresponding to these controls are given here. When $\sigma_2 = -\sigma_1 = \pm 1$, so that E is turning in a direction opposite to P's, the retrograde solutions to (2) are found in terms of the terminal state as

$$\begin{aligned} x &= x_0 \cos\tau + \sigma_1[1 - \cos\tau + y_0 \sin\tau + \cos\theta - \cos(\theta_0 + \sigma_1\tau)], \\ y &= y_0 \cos\tau + \sin\tau - \sigma_1[x_0 \sin\tau + \sin\theta - \sin(\theta_0 + \sigma_1\tau)], \\ \theta &= \theta_0 + 2\sigma_1\tau. \end{aligned} \tag{21}$$

Similarly, when $\sigma_2 = \sigma_1 = \pm 1$, the retrograde solutions are

$$\begin{aligned} x &= x_0 \cos\tau + \sigma_1[1 - \cos\tau + y_0 \sin\tau + \cos(\theta_0 + \sigma_1\tau) - \cos\theta_0], \\ y &= y_0 \cos\tau + \sin\tau - \sigma_1[x_0 \sin\tau + \sin(\theta_0 + \sigma_1\tau) - \sin\theta_0], \\ \theta &= \theta_0\,. \end{aligned} \tag{22}$$

We will find that singular arcs for P are sometimes optimal, such that $\sigma_1 = 0$ and $\sigma_2 = \pm 1$. For this case, the paths are given by

$$\begin{aligned} x &= x_0 - \sigma_2(\cos\theta - \cos\theta_0), \\ y &= y_0 + \tau + \sigma_2(\sin\theta - \sin\theta_0), \\ \theta &= \theta_0 - \sigma_2\tau. \end{aligned} \tag{23}$$

While it is not hard to write solutions to the adjoint equations (11), it is found that discontinuities in the adjoints occur in the retrograde paths before the switch functions change sign. This is due to the dispersal surface for E (Ref. 1, Chapter 6), which intersects the capture circle at the

EDL shown in Fig. 2. The implication is that strategies (and resulting solutions) can be found without explicit knowledge of the adjoint vector components.

2. Game of Kind

We have shown that the barrier encloses the capture circle, and we have mentioned that an evader's dispersal line intercepts the retrograde solution for the barrier. Consequently, as mentioned in Section 1.3, the general retrograde solutions can be used, with the various terminal conditions (14) and (16), to determine possible trajectory surfaces and the associated switching lines for the two players.

2.1. Termination with $\theta_0 = 0$. We have determined in (15) that, unless $\phi_0 = 0$, $\sigma_1 = \sigma_2 = \text{sign } x_0$. Using the general solutions (22) for $x_0 = \beta \sin \phi_0 > 0$ yields the retrograde motion as

$$x = \beta \sin(\phi_0 + \tau), \qquad y = \beta \cos(\phi_0 + \tau), \tag{24}$$

which shows that the radial separation remains constant in this *safe-contact* motion (Fig. 3a). The real-space interpretation is more easily understood (Fig. 3b). It is readily shown that, when E is initially behind P, it is E who chooses the direction of travel on the capture circle. That is, the point $\mathbf{x} = [0, -\beta, 0]$ is an evader's dispersal point, since, if $|\sigma_2| \neq 1$ and $\sigma_1 = \pm 1$, the equations of motion will show that $\dot{r} < 0$. Therefore, E must choose $\sigma_2 = \pm 1$ to avoid penetrating the capture circle, and P's strategy must be $\sigma_1 = \sigma_2$ in order to keep $\dot{r} = 0$; that is, if $\sigma_1 \neq \sigma_2$, then $\dot{r} > 0$, and P would then lose contact with E.

In the special terminal case $\theta = \phi_0 = 0$, as mentioned in Section 1.3, nine possible sets of strategies satisfy the necessary conditions. The retrograde solutions of Section 1.4 can be used to show that $r(\tau) \geqslant \beta$ for six of these strategy pairs (namely, $\sigma_2 = \pm 1$ and $\sigma_1 = 0$ or ± 1). The case $\sigma_1 = \sigma_2 = 0$ obviously leaves $r(\tau) = \beta$, and the remaining

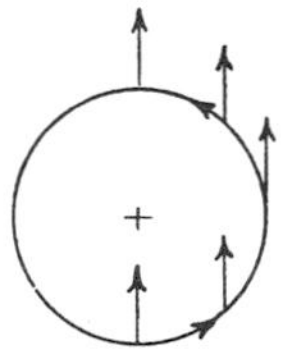

Fig. 3a. Safe contact with $\sigma_1 = \sigma_2 = +1$ (relative space).

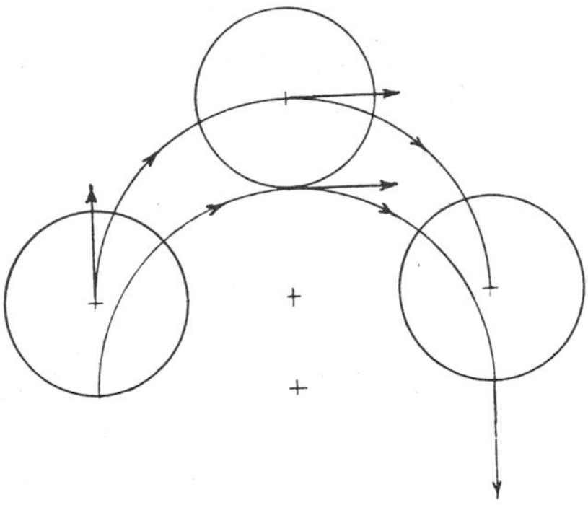

Fig. 3b. Safe contact with $\sigma_1 = \sigma_2 = +1$ (real space).

two cases are related to the strategies $\sigma_2 = 0$ and $\sigma_1 = \pm 1$. The solutions corresponding to these strategies show that $\mathring{r}_0 = \mathring{\mathring{r}}_0 = 0$ and that the third retrograde derivative is *negative*, which implies that these strategies cannot be optimal for the terminal condition being examined.

When the barrier terminates at $\theta_0 = \phi_0 = 0$, with $\sigma_1 = 0$ and $\sigma_2 = \pm 1$, the paths in relative and real space will resemble those shown in Figs. 4a and 4b for $\sigma_2 = +1$. Here, P is following a *singular arc*, and E's trajectory in relative space is termed a universal line (or UL). As mentioned in Section 1.1, the retrograde barrier paths terminate at an evader's dispersal point. This means that, at a certain point on the retrograde path shown in Figs. 4a and 4b, E can instead choose to turn *left*, causing P to turn *right* and resulting in safe contact, as will be discussed in the next section.

The final possibility for termination of the barrier at $\mathbf{x} = [0, \beta, 0]$ is given by the strategies $\sigma_1 = -\sigma_2 = \pm 1$. Here, P and E are turning in opposite directions before termination, and again the retrograde

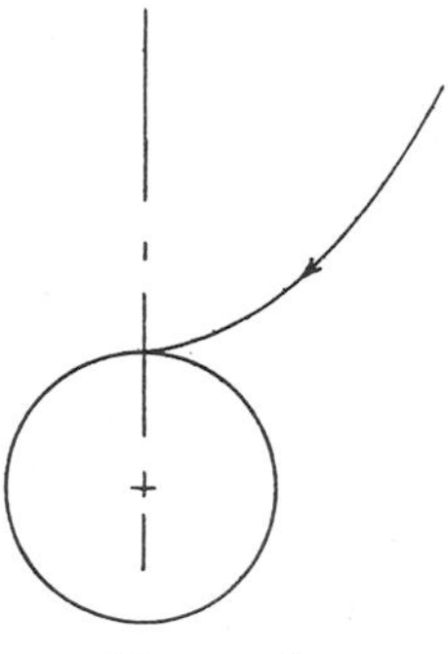

Fig. 4a. Safe contact with $\sigma_1 = 0$, $\sigma_2 = +1$ (relative space).

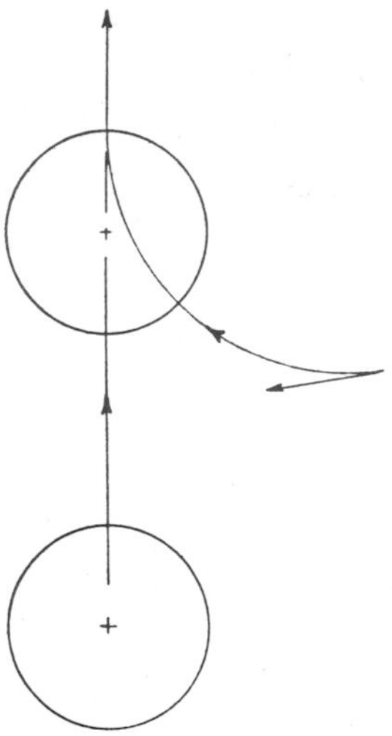

Fig. 4b. Safe contact with $\sigma_1 = 0$, $\sigma_2 = +1$ (real space).

solutions given by (21) are curtailed at an evader's dispersal line. The determination of this EDL will be discussed in Section 2.3.

2.2. Termination with $\theta_0 = 2\phi_0$. When the barrier intersects the capture circle at a nonzero angle ϕ_0, we have seen in (5) that the angular state must be $\theta_0 = 2\phi_0$. The strategies at termination were found in (17) to be $\sigma_1 = -\sigma_2 = \text{sign } x_0$; so, if $x_0 > 0$, the retrograde solutions are given by (21) as

$$\begin{aligned} x &= 1 - \cos\tau + \beta\sin(\phi_0 + \tau) - \cos(\theta_0 + \tau) + \cos\theta, \\ y &= \sin\tau + \beta\cos(\phi_0 + \tau) + \sin(\theta_0 + \tau) - \sin\theta, \\ \theta &= \theta_0 + 2\tau. \end{aligned} \tag{25}$$

As in the other barrier trajectories discussed, a dispersal point marks the end of this trajectory, as is suggested by Fig. 5. This is an illustration in real space of the dispersal point phenomenon, drawn for $\beta = 0.5$. In this typical case, E contacts the capture circle at the angular coordinate ϕ_0 by choosing $\sigma_2 = -1$, or at the coordinate $\phi_0 = 0$ by choosing $\sigma_2 = +1$. P's strategies will be $\sigma_1 = +1$ for the first choice, while the second choice entails a two-part strategy for P ($\sigma_1 = -1, 0$), the terminal portion of which includes the UL discussed in Section 2.1.

This example illustrates that barrier termination points at the EDP require simultaneous consideration of the various combinations of

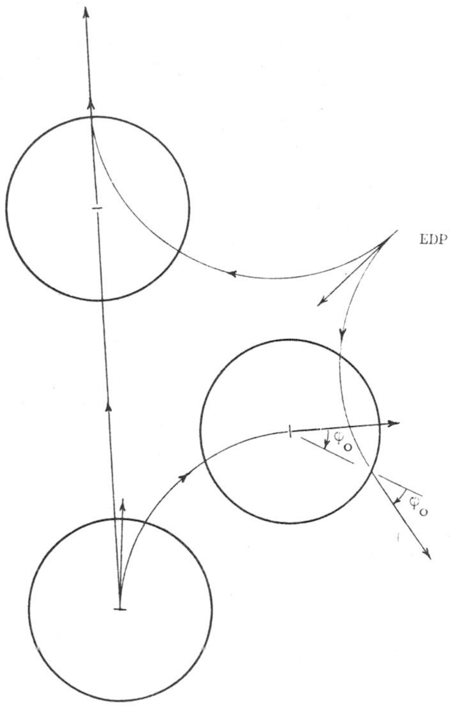

Fig. 5. Evader's dispersal point, $\theta = -3\pi/4$.

safe-contact strategies determined in Section 1.3. That is, the retrograde paths given by (25) are optimal only until an EDP is reached, and the adjoints and both switch functions are discontinuous at this point. Notice that the time-to-go from the EDP is different for the two paths in Fig. 5, but that this characteristic applies only to dispersal points on the *barrier*.

2.3. Barrier. By combining the notions of the two previous sections, the barrier (or envelope of capturable states) can be constructed. We first consider the determination of a point on the barrier for which $x = 0$ and $\theta = \pi$. Here, the players are initially moving toward each other, and obviously E must choose either extreme strategy, $\sigma_2 = \pm 1$. P's initial strategy is $\sigma_1 = -\sigma_2$, and this strategy holds until the time-to-go is $\tau = \tau_1$, when P's velocity is tangent to E's minimum-turn circle. Subsequent relative motion is along the UL, and P maintains $\sigma_1 = 0$ until tangential contact at $x_0 = \phi_0 = 0$ (Fig. 6a). The symmetry here implies that the optimal times-to-go are equal for either choice of σ_2.

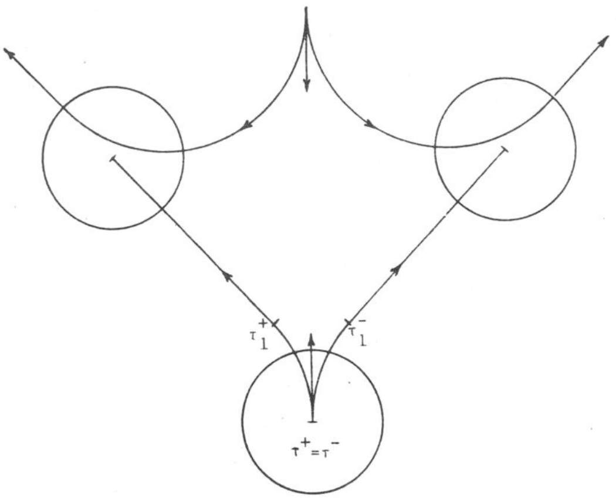

Fig. 6a. Barrier strategies, $\theta = \pi$, $x = 0$.

That is, at this point on the EDL, $\tau^+ = \tau^-$, where the superscripts correspond to E's choice of turn rate, $\sigma_2 = \pm 1$. Also shown in the figure are the points at which $\sigma_1 = 0$; these are labelled $\tau_1{}^+$ and $\tau_1{}^-$. The superscript notation is used to distinguish two optimal strategies and paths which begin at a dispersal point. A process of elimination shows that, for this initial condition, barrier termination must occur at $\phi_0 = 0$, to be in accord with the various possibilities discussed in Section 2.1.

For a *nearby* point on the barrier, where E's initial velocity is oriented by the same angle $\theta = \pi$, the terminal state $\mathbf{x}_0 = [0, \beta, 0]$ will be the same, as shown in Fig. 6b. This *nearby* point is shown for $x > 0$, and E's strategy here is unique, since the EDL at $\theta = \pi$ is located at $x = 0$.

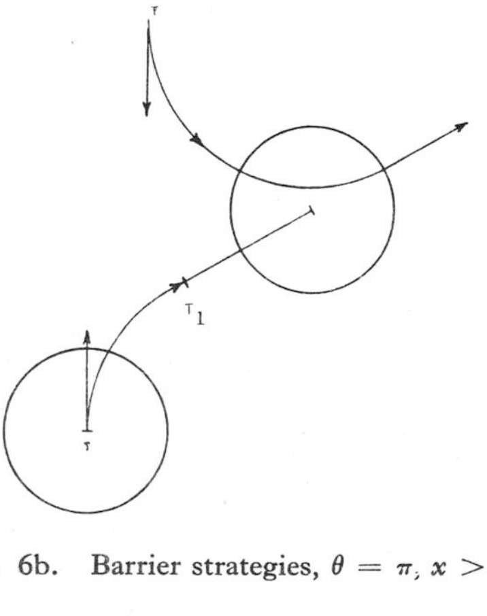

Fig. 6b. Barrier strategies, $\theta = \pi$, $x > 0$.

The small changes in τ_1 and τ due to this small change in x can be noted by comparing Fig. 6b to the right-hand paths of Fig. 6a.

As P's switch time τ_1 is reduced, the locus of barrier initial conditions at $\theta = \pi$ will change; and, when this time-to-go is $\tau_1 = 0$, P does not use $\sigma_1 = 0$. E's initial condition for this circumstance is $\mathbf{x} = [\beta, 2, \pi]$, as can be seen by sketching the real-space paths of P and E for the strategies $\sigma_1 = +1$ and $\sigma_2 = -1$. The subsequent barrier locus at $\theta = \pi$ is given by a vertical line in relative space, which is tangent to the right edge of the capture circle. For initial conditions on this vertical line, E contacts the capture circle with the terminal sta*e given by (16) of Section 1.3, that is, $\phi_0 = \theta_0/2 > 0$.

Thus, a section of the barrier at $\theta = \pi$ has the Gothic arch shape shown in Fig. 7. For $\theta = \pi$ and for other values of θ not too far from π, P's path will include a UL ($\sigma_1 = 0$) for either of E's strategies from the EDL. The calculation for such two-part strategies begins by fixing the initial value of θ and solving two simultaneous equations for τ^+ and τ^-.

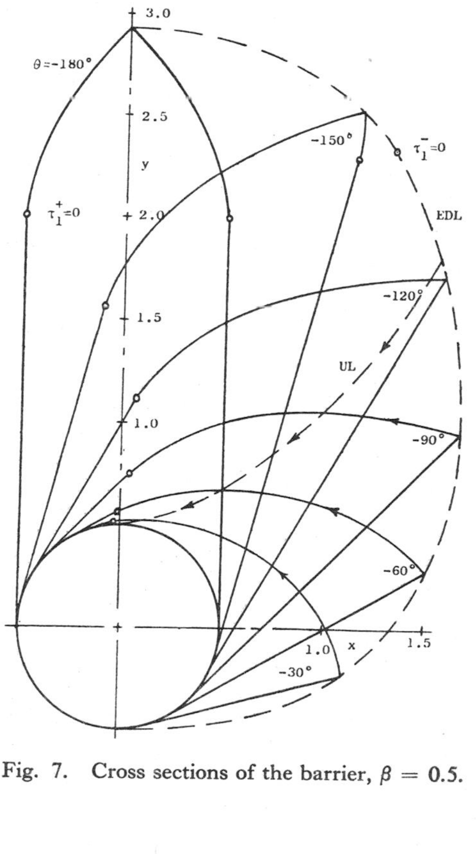

Fig. 7. Cross sections of the barrier, $\beta = 0.5$.

Thus, using (21) and (23) successively for the left-hand path, we have

$$\theta^+(\tau_1^+, \tau^+) = \tau_1^+ - 2\tau^+ = \theta, \tag{26}$$

where τ_1^+ is the time-to-go when P switches from $\sigma_1 = -1$ to $\sigma_1 = 0$. The right-hand path, for which E chooses $\sigma_2 = -1$ at the EDL, also ends at $\mathbf{x} = [0, \beta, 0]$, but different switch and termination times are implied; i.e., (22) and (23) give

$$\theta^-(\tau_1^-, \tau^-) = -\tau_1^- + 2\tau^- = \theta. \tag{27}$$

These two equations are solved for τ_1^+ and τ_1^-, and the coordinates of the dispersal point are given by equating $x^+ = x^-$ and $y^+ = y^-$. Thus, using (21)–(23), we have

$$\begin{aligned} x^+(\tau_1^+, \tau^+) &= -1 - (\beta + \tau_1^+)\sin(\tau^+ - \tau_1^+) + 2\cos(\tau^+ - \tau_1^+) - \cos(2\tau^+ - \tau_1^+), \\ x^-(\tau_1^-, \tau^-) &= 1 + (\beta + \tau_1^-)\sin(\tau^- - \tau_1^-) - 2\cos(\tau^- - \tau_1^-) + \cos(2\tau^- - \tau_1^-). \end{aligned} \tag{28}$$

Substituting from (26) and (27) for τ_1^+ and τ_1^- then gives

$$\begin{aligned} x^+(\tau^+, \theta) &= -1 - \cos\theta + (\beta + \theta + 2\tau^+)\sin(\theta + \tau^+) + 2\cos(\theta + \tau^+), \\ x^-(\tau^-, \theta) &= 1 + \cos\theta + (\beta - \theta + 2\tau^-)\sin(\theta - \tau^-) - 2\cos(\theta - \tau^-). \end{aligned} \tag{29}$$

Similar equations are found for $y^+(\tau^+)$ and $y^-(\tau^-)$; and, when the coordinates are equated, a numerical solution is possible for τ^+ and τ^-, and the dispersal point $x(\theta)$ and $y(\theta)$.

When $\tau_1^- \leqslant 0$, the implication is that, for this value of θ, E does not arrive at the UL, and $\sigma_1 \neq 0$. In place of (27), we then have, using (25),

$$\theta^-(\phi_0, \tau^-) = 2\phi_0 + 2\tau^- = \theta. \tag{30}$$

For $\beta = 0.5$, this change in the terminal condition turns out to be required when θ is in the interval $-123° < \theta \leqslant 0°$, approximately, and here both P and E have constant strategies from the EDL to termination.

The result of the barrier-computation procedure which has been described is shown in Fig. 7, and the following details are noteworthy: (i) the sections of the capture region are shown for $-\pi \leqslant \theta \leqslant 0$, and symmetry can give the contours for $0 \leqslant \theta \leqslant \pi$; (ii) when E is on the UL, shown as a dashed line intersecting several of the θ-contours, P's control is $\sigma_1 = 0$, while E takes $\sigma_2 = +1$ until termination at $\phi_0 = \theta_0 = 0$, as in Fig. 4; (iii) when E is initially on the curved

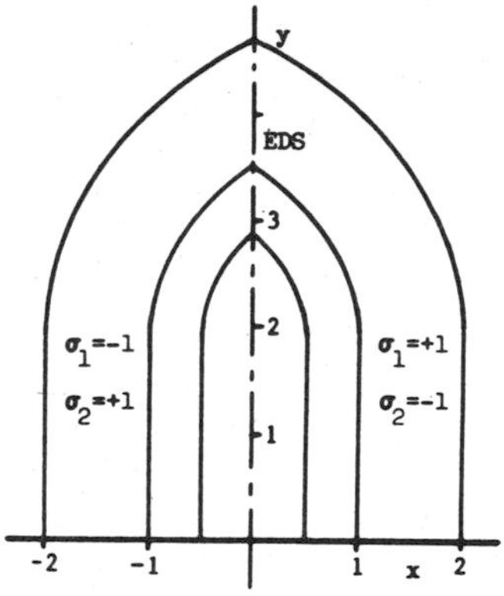

Fig. 8a. Optimal strategies in the game of kind, $\theta = 180°$.

portion of a $\theta =$ const section, the resulting two-stage trajectories include the UL, and terminate at $\phi_0 = \theta_0 = 0$; (iv) when E is initially on the straight portion of a section, P's control is $\sigma_1 = \pm 1$ until termination at $\phi_0 = \theta_0/2 \neq 0$.

This figure also implies the barrier strategies of both P and E. That is, if E is on the barrier to the left (counter-clockwise) of the EDL, then $\sigma_2 = +1$; and, if E is to the right of the EDL, $\sigma_2 = -1$. P's strategy is $\sigma_1 = +1$ when E is below the UL, and $\sigma_1 = -1$ if E is above this line. The UL trajectories can be preceded by paths for which $\sigma_1 = \pm 1$. However, since $\sigma_2 = +1$, tributary paths are shown in Fig. 7 only for $\sigma_1 = +1$, such that θ is constant. The tributary paths which join the UL from above are not shown, since for these paths θ is changing with time.

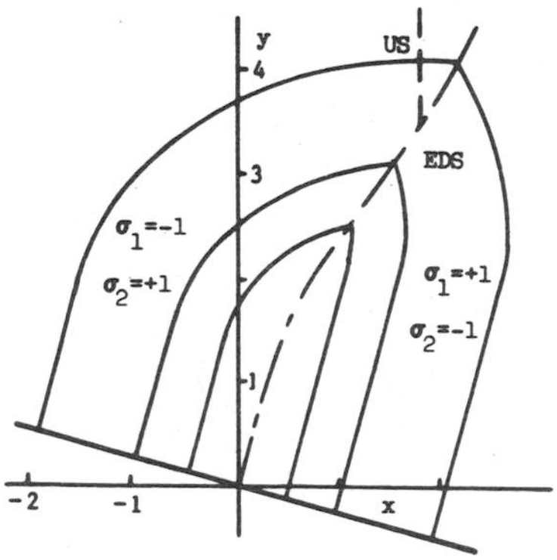

Fig. 8b. Optimal strategies in the game of kind, $\theta = -150°$.

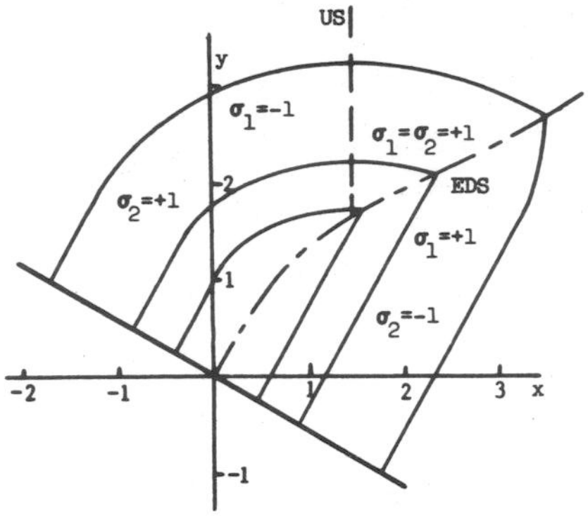

Fig. 8c. Optimal strategies in the game of kind, $\theta = -120°$.

The miss-distance performance criterion, β, can be varied through a range of values, and the optimal maneuvers for P and E displayed as functions of the relative heading. This has been done in Fig. 8, for $\beta = .5$, 1, and 2. These diagrams indicate, for example, that for the head-on initial geometry ($x = 0$, $\theta = 180°$), E can guarantee that the final range is greater than 1 only if the initial range is greater than $y = 3.6$. These graphical results can be compared with the similar charts given in Ref. 2 for the cooperative and non-cooperative collision-avoidance problems. The principal qualitative difference between the min max and max max problems is that P's universal surface is replaced by a dispersal surface in the collision avoidance problem, and that optimal collision avoidance requires the identical vehicles to turn hard right or hard left until minimum range is reached.

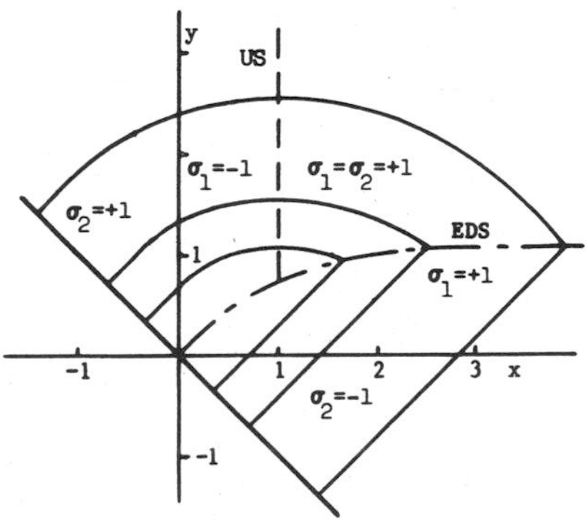

Fig. 8d. Optimal strategies in the game of kind, $\theta = -90°$.

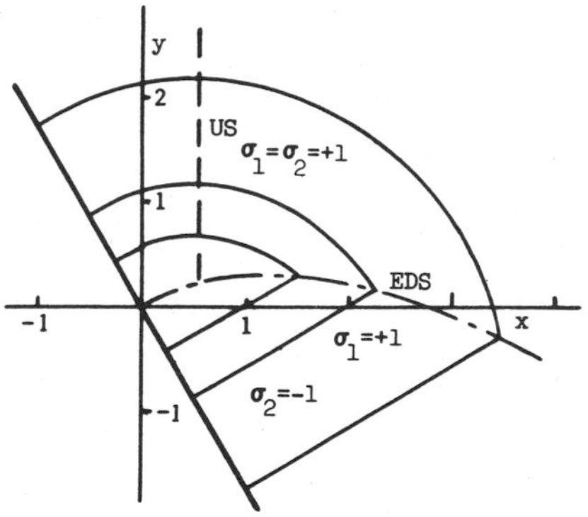

Fig. 8e. Optimal strategies in the game of kind, $\theta = -60°$.

3. Game of Degree

When **x** is inside the barrier, capture of E is possible, and the optimal strategies of P and E are such as to minimize and maximize the capture time, respectively. Retrograde solutions will be used to locate a dispersal surface for E and a universal surface for P, such that the capture region is subdivided into various subregions. The result is that the players' controls are known as functions of $\mathbf{x} = [x, y, \theta]$.

3.1. Evader's Dispersal Surface. As in the game of kind, it is intuitively clear that, when E is initially on the y axis and headed toward P, E must choose $\sigma_2 = \pm 1$. That is, the y axis coincides with the evader's dispersal surface (or EDS) when $\theta = \pi$. P's minimizing

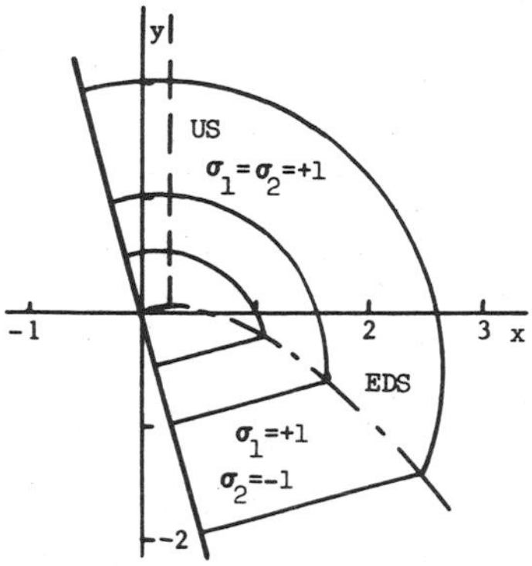

Fig. 8f. Optimal strategies in the game of kind, $\theta = -30°$.

strategy is to turn in E's direction ($\sigma_1 = -\sigma_2$), then switching to $\sigma_1 = 0$ such that capture occurs with E at the most forward position of the capture circle, but with $\theta_0 \neq 0$. These strategies are optimal, according to the implications of (19) and (20). That is, if P did *not* switch to $\sigma_1 = 0$, capture would occur with $\sigma_1 = -\text{sign}\, x_0$, which contradicts the terminal condition of (19).

For other nearby values of θ, the EDS is located by using the equations which are implied by the equal times-to-go (τ) through use of two different strategies. That is, $\tau^+ = \tau^- = \tau$, and

$$x^+(\tau_1^+, \theta_0^+, \tau) = x^-(\tau_1^-, \theta_0^-, \tau), \qquad y^+(\tau_1^+, \theta_0^+, \tau) = y^-(\tau_1^-, \theta_0^-, \tau), \quad (31)$$

where τ_1^+ and τ_1^- are the times-to-go when P's strategy becomes $\sigma_1 = 0$. Using the solutions to the twn angular equations corresponding to these two strategy pairs, we have

$$\theta^+ = \theta_0^+ - 2\tau + \tau_1^+, \qquad \theta^- = \theta_0^- + 2\tau - \tau_1^-. \quad (32)$$

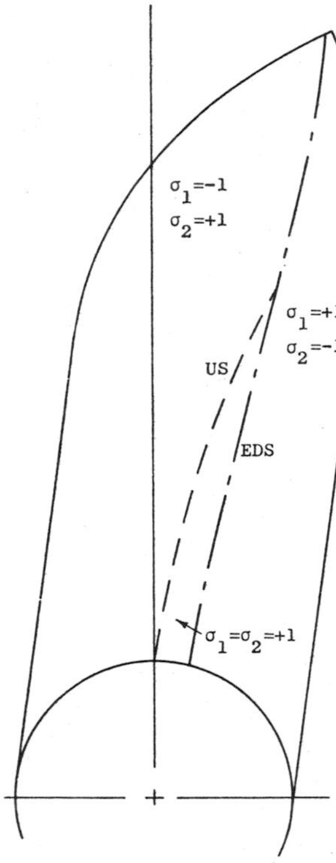

Fig. 9a. Optimal strategies in the capture region, $\beta = 0.5$, $\theta = -165°$.

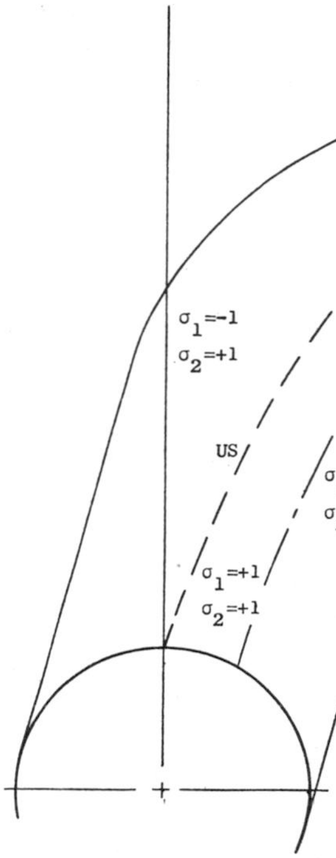

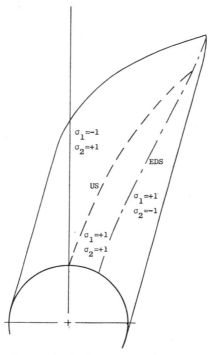

Fig. 9b. Optimal strategies in the capture region, $\beta = 0.5$, $\theta = -150°$.

Thus, θ_0^+ and θ_0^- can be eliminated from (31), so that, with $\theta^+ = \theta^- = \theta$,

$$\begin{aligned} x^- - x^+ = {} & 2[1 + (1 - \cos\tau)\cos\theta] + (\beta + \tau_1^-)\sin(\tau - \tau_1^-) \\ & + (\beta + \tau_1^+)\sin(\tau - \tau_1^+) - \cos(\tau - \tau_1^-) - \cos(\tau - \tau_1^+) = 0, \\ y^- - y^+ = {} & -2(1 - \cos\tau)\sin\theta + (\beta + \tau_1^-)\cos(\tau - \tau_1^-) \\ & - (\beta + \tau_1^+)\cos(\tau - \tau_1^+) + \sin(\tau - \tau_1^-) - \sin(\tau - \tau_1^+) = 0. \end{aligned}$$

Numerical solution of these equations for τ_1^+ and τ_1^- as functions of (θ, τ) then produces a dispersal point locus (x, y), as long as both switch times turn out to fall in the interval $0 \leqslant \tau_1^+, \tau_1^- \leqslant \tau$. Otherwise, P may initially turn right regardless of E's strategy from the EDS, then switching to $\sigma_1 = 0$ or not, depending on whether or not $\tau_1 > 0$. The numerical calculations in these various cases are essentially the same, however. That is, they all use the equalities $x^+ = x^-$, $y^+ = y^-$, and $\theta^+ = \theta^- = \theta$ as required on the EDS, and it is only necessary to use the proper sets of solutions with $\tau^+ = \tau^-$ to eliminate the intermediate unknowns, which are θ_0^+, θ_0^- and either $\tau_1^\pm$ or $\phi_0^\pm$.

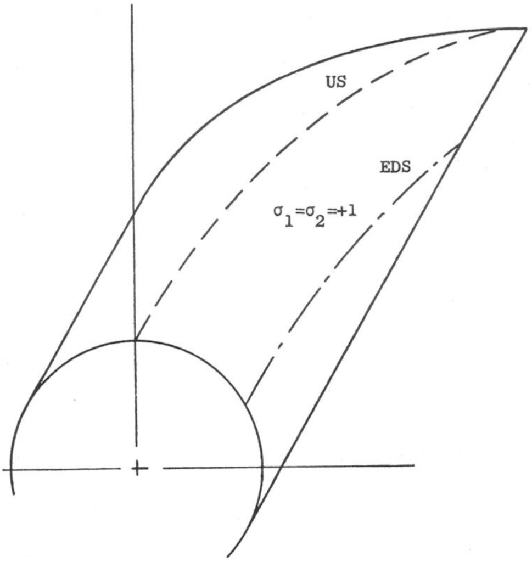

Fig. 9c. Optimal strategies in the capture region, $\beta = 0.5$, $\theta = -120°$.

3.2. Pursuer's Universal Surface. A universal surface (or US) for P exists at those points for which P's optimal strategy is $\sigma_1 = 0$. On this surface, the equations of relative motion have the solutions given in (23). But we have shown that the terminal conditions when $\sigma_1 = 0$ must be $x_0 = 0$, $y_0 = \beta$, with θ_0 arbitrary. Restricting attention to $\sigma_2 = 1$ (since for $-\pi < \theta_0 \leqslant 0$, this US originates to the left of the EDS, as in Fig. 2), we see that the retrograde solutions are

$$x = \cos\theta_0 - \cos(\theta_0 - \tau), \quad y = \beta + \tau - \sin\theta_0 + \sin(\theta_0 - \tau), \quad \theta = \theta_0 - \tau, \tag{33}$$

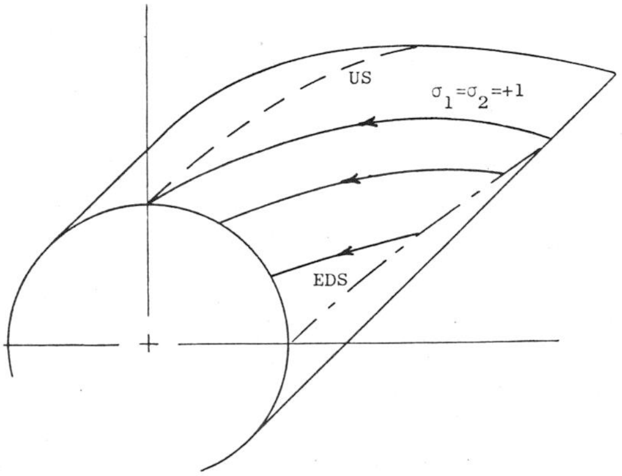

Fig. 9d. Optimal strategies in the capture region, $\beta = 0.5$, $\theta = -90°$.

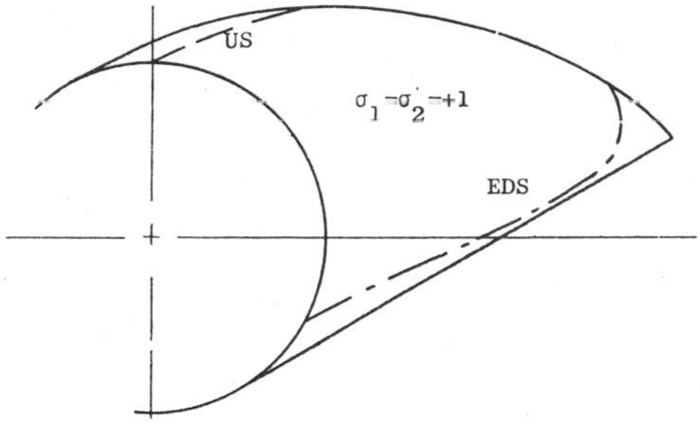

Fig. 9e. Optimal strategies in the capture region, $\beta = 0.5$, $\theta = -60°$.

and these paths are optimal from termination at θ_0 back to the EDS or the barrier.

In order to display the intersection of the US with planes of constant θ, as was done for the EDS, θ_0 is eliminated from Eqs. (33-1) and (33-2). The resulting line $x(\theta, \tau)$, $y(\theta, \tau)$ can then be drawn through each section of the capture region between the capture circle and the EDS or the barrier. When E is above the US, P turns hard left, and conversely if E is below the US. When E arrives at the US, $\sigma_1 = 0$, and the resulting path (a singular arc in the game of degree) remains on the US until capture at $x_0 = 0$.

3.3. Strategies of the Game of Degree. Results of the previous calculations are presented as $\theta =$ const sections of the state space. As shown in Figs. 9a–9f for $\beta = 0.5$, the EDS and US are smooth surfaces which vary continuously with θ. Across the EDS, σ_2 switches, and across the US, σ_1 switches.

Certain interesting features of the surface may be noted. Thus, for θ_0 near π, the US extends from the point $[0, \beta, \theta_0]$ only to the EDS, as

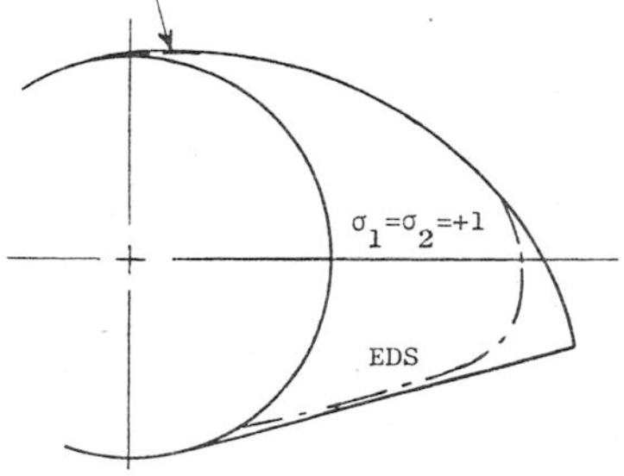

Fig. 9f. Optimal strategies in the capture region, $\beta = 0.5$, $\theta = -30°$.

shown in Figs. 9a and 9b. At $\theta = \pi$, the US has length zero, and the EDS coincides with the y axis, as noted earlier. The EDS extends from the EDL shown on the capture cylinder in Fig. 2 to the vicinity of the upper corner of the capture volume. This corner is the EDL in the game of kind. For θ_0 between 0° and $-140°$, approximately, the US and EDS no longer intersect, as shown in Fig. 9c. Trajectories, of course, can be shown in these sectional figures only when $\sigma_1 = \sigma_2 = 1$, such that $\dot{\theta} = 0$; representative paths for these strategies are shown in Fig. 9d. The sharp curve in the EDS for θ near $-60°$ is an unexpected result which can be verified by plotting real-space trajectories of the players, when E's initial condition is on this line.

4. Conclusions

The game of two identical cars represents a first step in the generalization of the second-order homicidal chauffeur game. The present study shows that optimal strategies for P and E in both the game of kind and the game of degree can be determined as functions of the third-order state. It is found that the higher order of the problem does not require the introduction of any new types of *exceptional* surfaces. Thus, when the capture region is finite, the only exceptional surfaces appear to be a universal surface for P and a dispersal surface for E. This is exactly the result found for the homicidal chauffeur game, when the capture region is finite and the speed ratio is between $\sqrt{(1 - \beta^2)}$ and 1 (Ref. 3).

Many of the exceptional lines which occurred in the homicidal chauffeur game were in fact possible only because of the pedestrian's unlimited maneuverability. It is conjectured that the general solution to the *smoother* game of two cars will involve fewer exceptional surfaces, even when P and E have different speeds and turn radii.

References

1. Isaacs, R., *Differential Games*, John Wiley and Sons, New York, 1965.
2. Merz, A., *Optimal Evasive Maneuvers in Maritime Collision Avoidance*, NAVIGATION, J. Inst. of Navigation, Vol. 20, No. 2, Summer, 1973, p. 144.
3. Merz, A., *The Homicidal Chauffeur—A Differential Game*, Stanford University, Department of Aeronautics and Astronautics, Ph.D. Thesis, 1971.

XXVII

A Differential Game with Two Pursuers and One Evader

P. HAGEDORN AND J. V. BREAKWELL

Abstract. This paper is concerned with a coplanar pursuit–evasion problem in which a faster evader E with constant speed $w > 1$ must pass between two pursuers P_1 and P_2 having unit speed, the payoff being the distance of closest approach to either one of the pursuers. This problem is of the two-players zero-sum type. The control variables are the directions of the velocities of P_1, P_2, and E. The path equations are integrated and a closed-form solution is obtained in terms of elliptic functions of the first and second kinds. A closed-loop solution is given graphically in several diagrams for different values of w.

1. Introduction

In this paper a pursuit problem is examined which is closely related to Isaacs' "deadline game" (Ref. 1). We have two pursuers P_1 and P_2 and one evader E moving in the plane, the evader having speed $w > 1$ and both pursuers having unit speed. The evader seeks to pass between the two pursuers crossing the line P_1P_2 from left to right and maximizing the minimum distance of approach to P_1 and P_2. As we shall see, this minimum distance is only reached to the right of the line P_1P_2. This means that after E passes the line P_1P_2 the distance of approach to his closest pursuer will continue to decrease. The control variables are the directions of the three velocity vectors, or the angles ψ, ϕ_1, and ϕ_2 in Fig. 1. We suppose that initially E is to the left of the line P_1P_2. It is clear that E will not be able to "escape," i.e., to pass between P_1 and P_2 if initially E is too far to the left of P_1P_2.

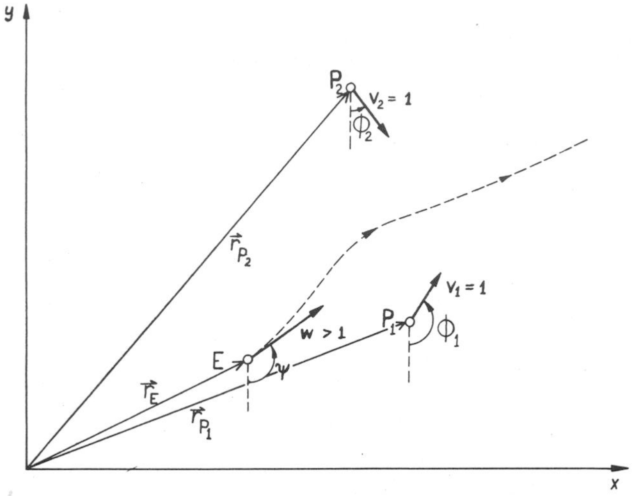

Fig. 1. The pursuit game in the fixed reference system.

The technique used for solving this problem will be the one of constructing barriers for a problem associated with ours. This associated problem corresponds to a "game of kind." It is the one of finding the initial positions of E, P_1, and P_2 so that E will succeed in passing between P_1 and P_2 maintaining a distance of approach $\geqslant l$, for some positive $l > 0$. The boundary of this region of escape is called a "barrier." Only on the barriers is there a well-defined and unique optimal strategy for the pursuers and the evader and on the barriers we can obtain the trajectories associated with optimal play in our original problem.

2. Construction of the Barriers

From Fig. 1 we obtain the kinematic equations

$$\begin{aligned} \dot{x}_E &= w \sin \psi, & \dot{y}_E &= -w \cos \psi, \\ \dot{x}_{P_1} &= \sin \phi_1 , & \dot{y}_{P_1} &= -\cos \phi_1 , \\ \dot{x}_{P_2} &= \sin \phi_2 , & \dot{y}_{P_2} &= -\cos \phi_2 . \end{aligned} \tag{1}$$

If the unit vectors in the direction of the velocities of P_1, P_2, and E are denoted by $\boldsymbol{\beta}_1$, $\boldsymbol{\beta}_2$, and $\boldsymbol{\beta}_E$, respectively, the main equation is obtained as

$$\operatorname*{Max}_{\boldsymbol{\beta}_E} \operatorname*{Min}_{\boldsymbol{\beta}_1, \boldsymbol{\beta}_2} \{ w \mathbf{J}_{\mathbf{r}_E} \boldsymbol{\beta}_E + \mathbf{J}_{\mathbf{r}_1} \boldsymbol{\beta}_1 + \mathbf{J}_{\mathbf{r}_2} \boldsymbol{\beta}_2 \} = 0, \tag{2}$$

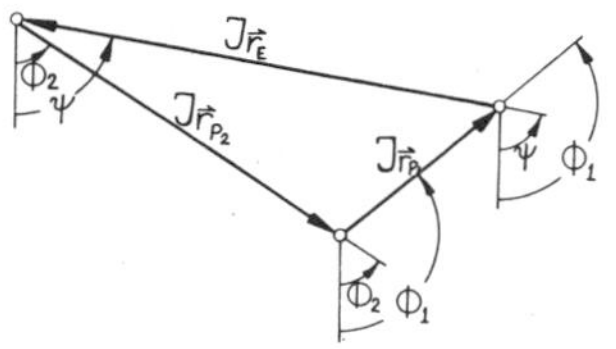

Fig. 2. Relation between the optimal controls.

which gives

$$w \cdot |\mathbf{J}_{\mathbf{r}_E}| - |\mathbf{J}_{\mathbf{r}_1}| - |\mathbf{J}_{\mathbf{r}_2}| = 0, \tag{3}$$

with $\boldsymbol{\beta}_E$ parallel to $\mathbf{J}_{\mathbf{r}_E}$, $\boldsymbol{\beta}_1$ parallel to $-\mathbf{J}_{\mathbf{r}_1}$, and $\boldsymbol{\beta}_2$ parallel to $-\mathbf{J}_{\mathbf{r}_2}$. Since $\mathbf{J}$ is a function just of the two vector differences $\mathbf{r}_1 = \mathbf{r}_E - \mathbf{r}_{P_1}$ and $\mathbf{r}_2 = \mathbf{r}_E - \mathbf{r}_{P_2}$, as can be seen easily, it follows that

$$-\mathbf{J}_{\mathbf{r}_E} = \mathbf{J}_{\mathbf{r}_1} + \mathbf{J}_{\mathbf{r}_2}. \tag{4}$$

Applying now the sine law to the triangle formed by $\mathbf{J}_{\mathbf{r}_E}$, $\mathbf{J}_{\mathbf{r}_1}$, and $\mathbf{J}_{\mathbf{r}_2}$ (Fig. 2), we obtain from the main equation

$$w \sin(\phi_1 - \phi_2) = \sin(\phi_1 - \psi) + \sin(\psi - \phi_2) \tag{5}$$

or

$$w \cos[(\phi_1 - \phi_2)/2] = \cos[\psi - \tfrac{1}{2}(\phi_1 + \phi_2)]. \tag{6}$$

This is a simple relation between the control variables, which will be used later. The adjoint equations reduce to $\dot{\mathbf{J}}_{\mathbf{r}_1} = \dot{\mathbf{J}}_{\mathbf{r}_2} = \mathbf{0}$, so that ψ, ϕ_1, and ϕ_2 are constant as long as the constraints $|P_1E| \geqslant l$ and $|P_2E| \geqslant l$ are inactive. The pursuers as well as the evader will therefore travel on straight lines as long as $|P_1E| > l$ and $|P_2E| > l$, where l is the distance of minimum approach. After meeting one of the constraints, for example $|P_1E| = l$, there may be a phase during which the paths of P_1 and E are no longer straight but curved and where a constant distance l is maintained.

During the final phase of the game in which $|P_1E| = l = \text{const}$ we must have in addition

$$w \cos(\theta - \psi) = \cos(\theta - \phi_1), \tag{7}$$

as can be seen from Fig. 3a. We try to solve (6) and (7) by equating both sides of these equations:

$$\psi - \theta = \tfrac{1}{2}(\phi_2 - \phi_1), \tag{8}$$

$$\psi - \tfrac{1}{2}(\phi_1 + \phi_2) = \theta - \phi_1 = \cos^{-1}[w \cos \tfrac{1}{2}(\phi_1 - \phi_2)], \tag{9}$$

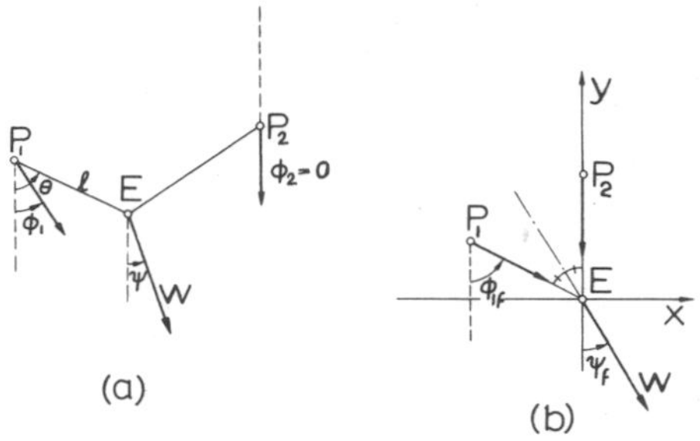

Fig. 3. The end conditions.

from which we obtain

$$\begin{aligned} \theta &= \phi_1 + \cos^{-1}[w \cos \tfrac{1}{2}(\phi_1 - \phi_2)], \\ \psi &= \tfrac{1}{2}(\phi_1 + \phi_2) + \cos^{-1}[w \cos \tfrac{1}{2}(\phi_1 - \phi_2)]. \end{aligned} \tag{10}$$

Now the last position in which P_1 still can maintain $\dot{l} = 0$ occurs when P_1 runs directly toward E, so that $\theta_f = \phi_{1_f}$, the subscript f indicating the final value. When this last situation is reached, the distance $| P_2E |$ must also be equal to l, because otherwise E could have done better by maintaining a larger distance to P_1 and decreasing his distance to P_2. Therefore, the final configuration is the one in Fig. 3b and for the final values of the angles ϕ_1 and ψ we obtain

$$\phi_{1_f} = 2 \cos^{-1}(1/w), \qquad \psi_f = \cos^{-1}(1/w), \tag{11}$$

if we set ϕ_2 equal to zero. Obviously this does not restrict the generality of the present discussion because we are always free to choose the direction of the y axis in Fig. 1 in the proper way; the final result will be given in a relative coordinate system in which the rotation of the line P_1P_2 is eliminated.

We are now able to integrate the equations of the constrained path, taking ϕ_1 as independent parameter. We note that

$$l\dot{\theta} = \sin(\theta - \phi_1) - w \sin(\theta - \psi), \tag{12}$$

which, taking into account the relations (9), can be written as

$$l\dot{\theta} = \sqrt{(1 - w^2c^2)} - ws, \tag{13}$$

where $c = \cos(\phi_1/2)$ and $s = \sin(\phi_1/2)$. On the other hand, from (10) we obtain

$$\dot{\theta} = \dot{\phi}_1 - [ws/2\sqrt{(1 - w^2c^2)}]\,\dot{\phi}_1, \tag{14}$$

and comparing (12) and (14) we finally get

$$dt = l \frac{\sqrt{(1 - w^2c^2)} + \frac{1}{2}ws}{\sqrt{(1 - w^2c^2)}[\sqrt{(1 - w^2c^2)} - ws]} d\phi_1$$

$$= l \frac{[\sqrt{(1 - w^2c^2)} + \frac{1}{2}ws][\sqrt{(1 - w^2c^2)} + ws]}{\sqrt{(1 - w^2c^2)}(1 - w^2)} d\phi_1 \tag{15}$$

This expression can be integrated, and we have

$$t = l\Big[-\frac{2}{w}F + \frac{3w}{w^2 - 1}E + \frac{3w}{w^2 - 1}c\Big] + \text{const}, \tag{16}$$

where

$$F = F(\sin^{-1}(wc), 1/w), \qquad E = E(\sin^{-1}(wc), 1/w),$$

the elliptic integrals of the second and first kind:

$$F(x, k) = \int_0^x \frac{d\alpha}{\sqrt{(1 - k^2 \sin^2 \alpha)}}, \qquad E(x, k) = \int_0^x \sqrt{(1 - k^2 \sin^2 \alpha)}\, d\alpha.$$

Using (9) and (15) with $\phi_2 = 0$, we can rewrite the kinematic equations (1) with ϕ_1 as the independent variable:

$$\dot{x}_E = w \sin \psi = w \sin \tfrac{1}{2}\phi_1 + \cos^{-1}(wc) = w^2sc + wc\sqrt{(1 - w^2c^2)},$$

$$\frac{dx_E}{d\phi_1} = \frac{\dot{x}_E}{\dot{\phi}_1} = l[w^2sc + wc\sqrt{(1 - w^2c^2)}] \frac{\sqrt{(1 - w^2c^2)} + \frac{1}{2}ws}{\sqrt{(1 - w^2c^2)}[\sqrt{(1 - w^2c^2)} - ws]}, \tag{17}$$

and similarly for $\dot{y}_E$, $\dot{x}_{P_1}$, $\dot{y}_{P_1}$, $\dot{y}_{P_2}$, and $\dot{x}_{P_2}$. Using the final conditions (Fig. 3b)

$$x_E(\phi_{1_f}) = y_E(\phi_{1_f}) = 0,$$

$$x_{P_1}(\phi_{1_f}) = -l \sin \phi_{1_f} = -2l[\sqrt{(w^2 - 1)}]/w^2,$$

$$y_{P_1}(\phi_{1_f}) = l \cos \phi_{1_f} = l[(2/w^2) - 1], \tag{18}$$

$$x_{P_2}(\phi_{1_f}) = 0,$$

$$y_{P_2}(\phi_{1_f}) = l,$$

we obtain after a somewhat cumbersome but routine integration

$$x_E(\phi_1) = \frac{l}{w^2 - 1}[\sqrt{(1 - w^2c^2)}(2w^2c^2 - w^2 - 1) - 2w^3s^3 + 2w(w^2 - 1)s]_{\phi_{1_f}}^{\phi_1}, \tag{19}$$

$$y_E(\phi_1) = \frac{l}{w^2 - 1}\left[2w^2 sc\sqrt{(1 - w^2c^2)} - 2w^3c^3 + w(2 + w^2)c + wE\right]_{\phi_{1_f}}^{\phi_1}, \tag{20}$$

$$x_{P_1}(\phi_1) = -\frac{2l}{w^2}\sqrt{(w^2 - 1)} - \frac{2l}{w^2 - 1}\left[s^2\sqrt{(1 - w^2c^2)} + ws^3\right]_{\phi_{1_f}}^{\phi_1}, \tag{21}$$

$$y_{P_1}(\phi_1) = l\left(\frac{2}{w^2} - 1\right) + \frac{l}{w^2 - 1}\left[3wc + 2sc\sqrt{(1 - w^2c^2)} - 2wc^3 + wE\right]_{\phi_{1_f}}^{\phi_1}, \tag{22}$$

$$x_{P_2}(\phi_1) = 0, \tag{23}$$

$$y_{P_2}(\phi_1) = l - l\left[+\frac{3w}{w^2 - 1}c - \frac{2}{w}F + \frac{3w}{w^2 - 1}E\right]_{\phi_{1_f}}^{\phi_1}, \tag{24}$$

where E and F are again the elliptic functions mentioned above.

These are the path equations on the constraint $|EP_1| = l$. In Fig. 4 we see how E and P_1 travel on spirals maintaining their relative distance constant and equal to l. Beginning from the final situation, i.e., $\phi_1 = \phi_{1_f}$, and constructing these spirals backward, we discard them as E crosses for the second time the straight line joining P_1 and P_2 because during the game we want E to cross this line only once and from left to right. We therefore discard the trajectory of E at E', the corresponding positions of the pursuers being P_1' and P_2'.

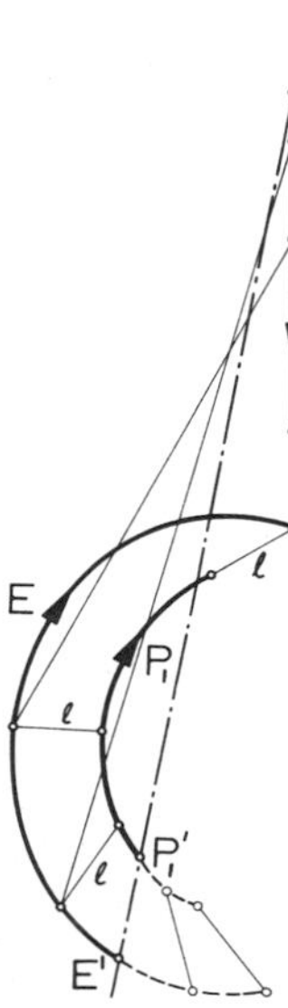

Fig. 4. Typical trajectories (final stage of the game; $w = 2$).

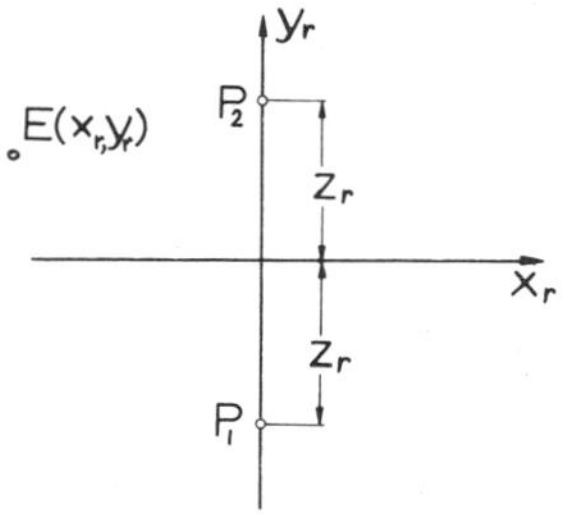

Fig. 5. The "relative" state variables x_r, y_r, and z_r.

It is now easy to construct the barrier $B(l)$. We simply take the tangents to the trajectories in the phase space:

$$x_E(\phi_1, \tau) = x_E(\phi_1) - \dot{x}_E(\phi_1)\tau, \tag{25}$$

and similarly for y_E, x_{P_1}, y_{P_1}, x_{P_2}, and y_{P_2}. In this way we obtain

$$x_E(\phi, \tau) = \frac{l}{w^2 - 1}[\sqrt{(1 - w^2c^2)}(2w^2c^2 - w^2 - 1) - 2w^3s^3 + 2w(w^2 - 1)s]_{\phi_{1_f}}^{\phi_1} - [w^2sc - wc\sqrt{(1 - w^2c^2)}]\tau, \tag{26}$$

$$y_E(\phi_1, \tau) = \frac{l}{w^2 - 1}[wE + 2w^2sc\sqrt{(1 - w^2c^2)} - 2w^3c^3 + wc(2 + w^2)]_{\phi_{1_f}}^{\phi_1} - 2sc\tau + [w^2c^2 - ws\sqrt{(1 - w^2c^2)}]\tau, \tag{27}$$

$$x_{P_1}(\phi_1, \tau) = -2l\frac{\sqrt{(w^2 - 1)}}{w^2} - \frac{2l}{w^2 - 1}[s^2\sqrt{(1 - w^2c^2)} + ws^3]_{\phi_{1_f}}^{\phi_1} - 2sc\tau, \tag{28}$$

$$y_{P_1}(\phi_1, \tau) = l\left(\frac{2}{w^2} - 1\right) + \frac{l}{w^2 - 1}[wE + 3wc + 2sc\sqrt{(1 - w^2c^2)} - 2wc^3]_{\phi_{1_f}}^{\phi_1} - (2c^2 - 1)\tau, \tag{29}$$

$$x_{P_2}(\phi_1, \tau) = 0, \tag{30}$$

$$t_{P_2}(\phi_1, \tau) = l - l\left[-\frac{2}{w}F + \frac{3w}{w^2 - 1}E + \frac{3w}{w^2 - 1}c\right]_{\phi_{1_f}}^{\phi_1} + \tau. \tag{31}$$

Equations (26)–(31) describe the two-dimensional barrier for a given l in terms of the parameters $\phi_1 \leqslant \phi_{1_f}$ and $\tau \geqslant 0$.

In order to see that the barrier actually divides the state space into two separate regions, it is convenient to go to the three-dimensional "relative" state space, where a state is characterized by the three state variables x_r, y_r, and z_r (see Fig. 5). In these coordinates the distances $|EP_1|$ and $|EP_2|$ are given, respectively, by

$$|EP_1| = \sqrt{[x_r^2 + (z_r + y_r)^2]}, \qquad |EP_2| = \sqrt{[x_r^2 + (z_r - y_r)^2]}; \tag{32}$$

the loci $|EP_1| = l$ and $|EP_2| = l$ therefore correspond to cylinders with axis $z_r = y_r$ and $z_r = -y_r$. The intersection of each of these cylindrical surfaces with the planes $x_r y_r$ and $x_r z_r$ gives a circle with radius l. The intersection with a plane perpendicular to the cylinder axis gives an ellipse.

In Fig. 6 the two constraint surfaces are shown. The final phase of the game, i.e., the constrained motion, corresponds to the spiral on one of the cylindrical surfaces. Since the picture is symmetric in y_r and $-y_r$, only the spiral for $y_r > 0$ was drawn. The tangents to this curve generate the barrier which intersects the plane $x_r z_r$ on the straight line ML. This line is also the intersection between the two symmetric parts of the barrier. The coordinates of the terminal point L are $x_r = l \sin(\phi_{1f}/2) = l/\sqrt{(1 - 1/w^2)}$, $y_r = 0$, $z_r = l \cos(\phi_{1_f}/2) = l/w$, as is easy to see from Fig. 3b. It can now be seen that we really have a barrier $B(l)$ which divides the state space into two regions. From initial points above $B(l)$ in Fig. 6 the evader will succeed in passing between the two pursuers maintaining a distance of approach larger than l. From points below $B(l)$ he will not be able to pass between the pursuers with minimum distance of approach $\geqslant l$.

Given an initial point on the barrier, the state will evolve along the corresponding tangent line until it reaches the cylindrical constraint surface. Then the state evolves along the curved line on the cylindrical surface until the terminal point L is reached. Since we can construct a barrier for each positive value of l, the problem initially posed can be considered as solved. In Fig. 7 typical trajectories of optimal play are shown in the "absolute" or fixed state space.

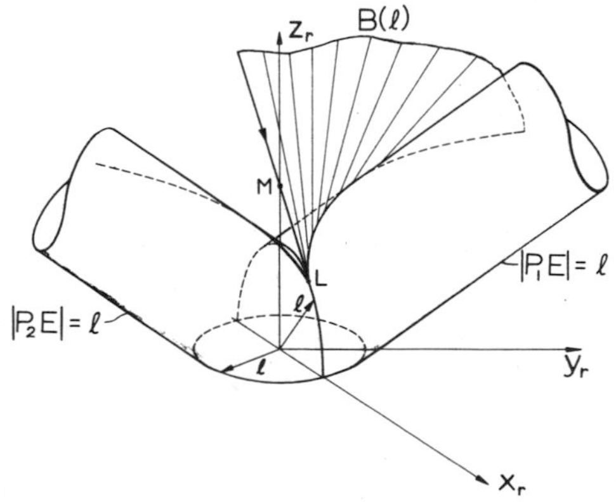

Fig. 6. Constraint surfaces and the barrier $B(l)$ in relative coordinates.

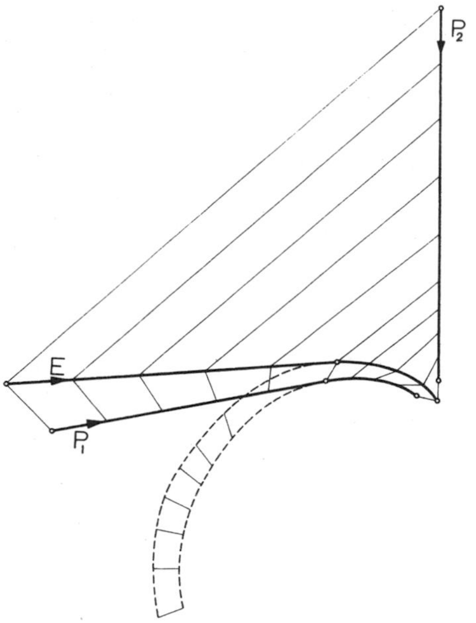

Fig. 7. Typical trajectories of optimal play ($w = 1.2$).

3. Analysis of the Results

It would be advantageous to represent the results graphically in such a way that the players can obtain the optimal value of the control variables at each state from a diagram. For this purpose we will use the relative coordinates defined in Fig. 5, normalizing them further by setting $X = x_r/z_r$, $Y = y_r/z_r$. In these "normalized" coordinates the position of the two pursuers P_1 and P_2 is fixed: They are located at the points (0, -1) and (0, $+1$), respectively. We have to note, however, that straight lines in the absolute reference system will now no longer be straight lines, but curves.

The trajectories of optimal play can easily be obtained in the coordinates X, Y from Eqs. (26)–(31) by setting

$$
\begin{aligned}
X &= \left(x_E - \frac{x_{P_1} + x_{P_2}}{2}\right)\frac{y_{P_2} - y_{P_1}}{2z_r} - \left(y_E - \frac{y_{P_1} + y_{P_2}}{2}\right)\frac{x_{P_2} - x_{P_1}}{2z_r}, \\
Y &= \left(x_E - \frac{x_{P_1} + x_{P_2}}{2}\right)\frac{x_{P_2} - x_{P_1}}{2z_r} - \left(y_E - \frac{y_{P_1} + y_{P_2}}{2}\right)\frac{y_{P_2} - y_{P_1}}{2z_r},
\end{aligned}
\tag{33}
$$

with

$$z_r = \tfrac{1}{2}\sqrt{[(x_{P_2} - x_{P_1})^2 + (y_{P_2} - y_{P_1})^2]}.$$

The doubly hatched region in Fig. 8b corresponds to the domain in which optimal play in the above sense is uniquely determined. Points on the line ABC correspond to initial conditions such that with optimal play the evader will meet both pursuers at $l = 0$. If the initial conditions are such that the point (X, Y) is in the singly hatched region, then the evader will not succeed in passing (at a finite distance) between the pursuers if they play optimally. If the initial conditions correspond to a point in the nonhatched region, then the evader can escape in such a way that his distance to the closest pursuer always increases. He can do that even by passing between them, if he is initially in the nonhatched part of the region $X < 0$. The corresponding controls are not uniquely defined.

On the other hand, if the initial conditions are in the doubly hatched region, then the evader will travel along a trajectory from left to right in such a way that the distance to the closest pursuer decreases steadily until the minimum l is reached. Then comes a phase during which the distance to his closest pursuer remains constant and equal to l, whereas the distance to the other pursuer decreases. When finally the distance to each of the pursuers is equal to l, the point L is reached.

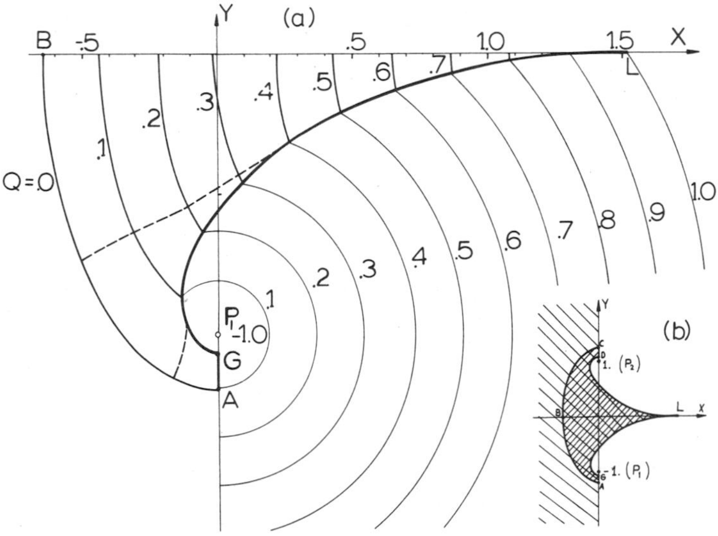

Fig. 8. The domain of optimal play in "normalized" coordinates ($w = 1.2$; $l_M = 1.809$).

Here the point then leaves the domain of optimal play, and the distance to his pursuers increases again. Note that for points with $X > 0$ in the doubly hatched region the minimum distance of approach has not yet been attained, even though the evader has already passed between the two pursuers.

Also shown in Fig. 8a are two optimal trajectories (dashed lines). Since the diagram is symmetric about the X axis, only the lower part is given. The final phase of the game, which corresponds to the motion on the cylindrical surface of Fig. 6, is given by the motion along the curve GL in Fig. 8a (the point G of Fig. 8 corresponds to E' in Fig. 4). The whole curve GL corresponds to points that have the same distance to the pursuer P_1 and along this curve the angle ϕ_1 decreases in the direction from G to L.

The maximum distance of approach which cannot increase immediately with $z_r = 0$, is equal to $\bar{l}_M = w/\sqrt{(w^2 - 1)}$. We now define the coefficient

$$Q = l/z_r \bar{l}_M . \tag{34}$$

If the players know the value of Q at the point corresponding to the initial conditions, they know the optimum distance of minimum approach given by $l = zQ\bar{l}_M$. Obviously the curve $Q = 0$ corresponds to the line ABC in Fig. 8b and $Q = 1$ to the point L. The curves $Q =$ const are constructed by setting $l = \bar{l} = Q\bar{l}_M$ in Eqs. (26)–(31). The condition $z_r = 1$ then gives a quadratic equation in τ, from which $\tau(\phi_1, Q)$ is obtained. Now τ can be used to calculate x_E, y_E, x_{P_1}, y_{P_1}, x_{P_2}, and y_{P_2} as a function of ϕ_1 and Q and finally the curves $Q =$ const are obtained in the coordinates X, Y through (33).

Some curves $Q =$ const and some optimal trajectories are drawn for different values of w in Figs. 8–11.

The curves of constant Q can also be extended to the nonhatched region. We know that from points in the nonhatched region E can "escape" in such a way that its distance to the closest pursuer is a non-decreasing function of time. The original game has no unique solution for positions of E in this region. The minimum distance of approach therefore corresponds to the initial conditions and the curves $Q =$ const are circles with center at the points $X = 0$, $Y = \pm 1$.

In Fig. 10 it can be seen how the diagram tends toward the limiting case $w \to \infty$. In this case the constrained trajectory in the X–Y plane corresponds to a semicircle with unit radius and center at P_i ($i = 1, 2$). The curves $Q =$ const are semicircles with center at the origin in the region of optimal play, which tends toward the whole left half-plane, and they are arcs of circles with center at P_i ($i = 1, 2$) in the right half-plane.

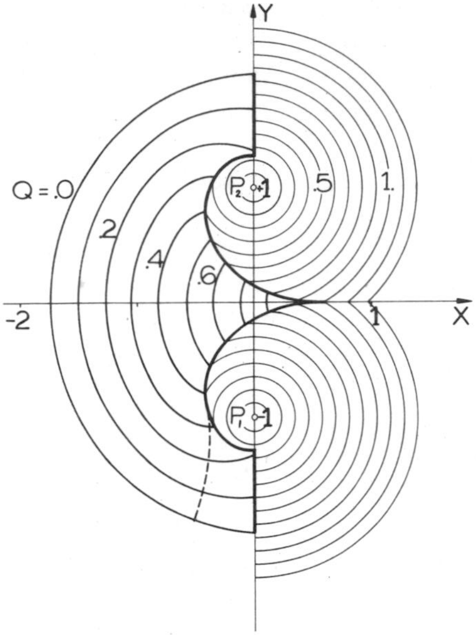

Fig. 9. The domain of optimal play in "normalized" coordinates ($w = 2.0$; $\bar{l}_M = 1.155$).

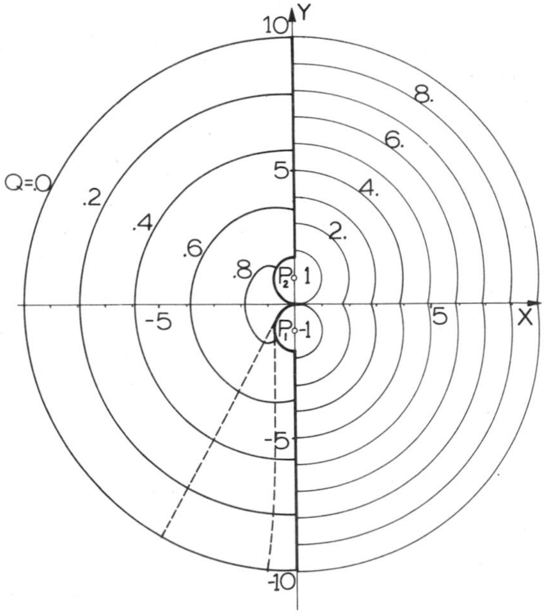

Fig. 10. The domain of optimal play in "normalized" coordinates ($w = 10.0$; $\bar{l}_M = 1.005$).

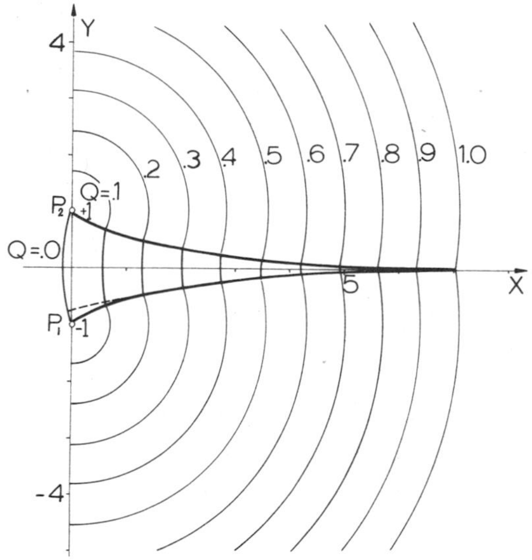

Fig. 11. The domain of optimal play in "normalized" coordinates ($w = 1.01$; $\bar{l}_M = 7.124$).

In Fig. 11 it is shown how the diagram tends toward the limiting case $w \to 1^+$. In this case the region of optimal play tends toward the half-strip $X > 0$, $|Y| \leqslant L$.

The evader's optimal strategy is readily apparent in Figs. 8–11. He should direct his absolute velocity vector along the gradient of Q, i.e, perpendicular to the lines Q = const.[1] Due to the fact that the X–Y coordinate system is moving in a complicated way, the relative path of E in this reference system will, however, *not* be perpendicular to the lines Q = const.

The optimal controls of the pursuers cannot be seen directly from these diagrams, so that these figures are of no immediate use to the pursuers. We can, however, draw analogous diagrams in a reference system in which the evader E is always at the origin and one of the pursuers, e.g., P_2, is at the point (0, 1). For different positions of P_1 we

[1] It is easy to see that this is true: let $\mathbf{r}_E$, $\mathbf{r}_{P_1}$, and $\mathbf{r}_{P_2}$ be the position vectors in the original, fixed reference system. We know that in optimal play the velocity vector of E should be parallel to $\mathbf{J}_{\mathbf{r}_E}$. Now we introduce the "normalized" position vector

$$\mathbf{r}_E{}^* = [2\mathbf{r}_E - (\mathbf{r}_{P_1} - \mathbf{r}_{P_2})]/|\,\mathbf{r}_{P_1} - \mathbf{r}_{P_2}\,|$$

with components X, Y. Clearly, Q is a function of $\mathbf{r}_E{}^*$ only and it can easily be shown that $\mathbf{Q}^*_{\mathbf{r}_E}$ is parallel to $\mathbf{J}_{\mathbf{r}_E}$. Therefore E's velocity must be parallel to $Q^*_{\mathbf{r}_E}$.

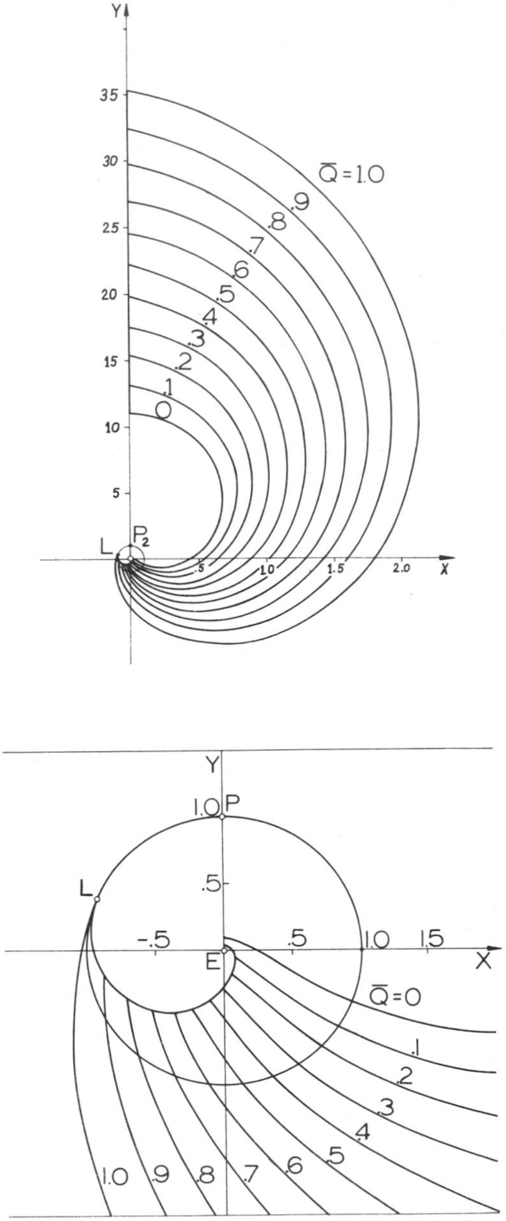

Fig. 12. The domain of optimal play in coordinates normalized with respect to one pursuer ($w = 1.20$).

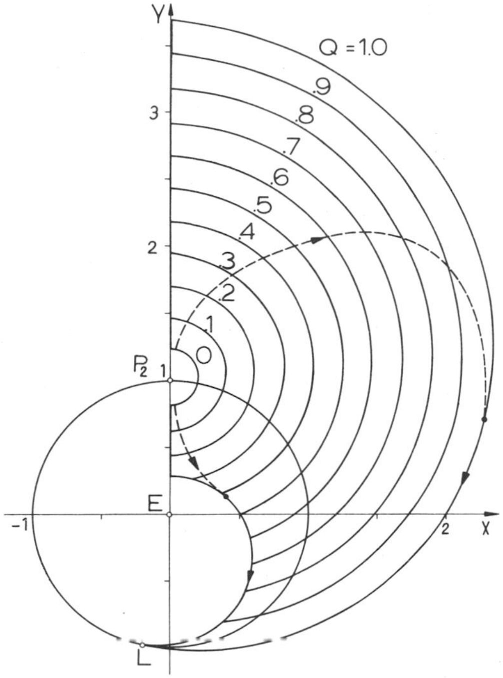

Fig. 13. The domain of optimal play in coordinates normalized with respect to one pursuer ($w = 10.00$).

now compute the "value" l. Defining the dimensionless quantity $\bar{Q}$ as $\bar{Q} = l/|\,\overline{EP}_2\,|$, we obtain curves of $\bar{Q} = \text{const}$ (see Figs. 12 and 13). The optimal strategy of P_1 is now readily apparent: P_1 should at each point direct his absolute velocity vector perpendicular to the lines $\bar{Q} = \text{const}$, in the direction of decreasing $\bar{Q}$.

In both diagrams the point L corresponds to the state in which the minimum distance of approach is reached. In Fig. 13 a sample trajectory is also represented twice: once normalized with respect to the nearer pursuer and once with respect to the pursuer who is farther away.

With the diagrams so constructed we now have a closed-loop solution for the optimal controls of both the evader and the pursuers.

Reference

1. Isaacs, R., *Differential Games*, John Wiley and Sons, New York, 1967, p. 260.

Index